ENGINEERING ECONOMIC PRINCIPLES

McGraw-Hill Series in Industrial Engineering and Management Science

Consulting Editors

Kenneth E. Case, *Department of Industrial Engineering and Management, Oklahoma State University*

Philip M. Wolfe, *Department of Industrial and Management Systems Engineering, Arizona State University*

Barnes: *Statistical Analysis for Engineers and Scientists: A Computer-Based Approach*
Bedworth, Henderson, and Wolfe: *Computer-Integrated Design and Manufacturing*
Black: *The Design of the Factory with a Future*
Blank: *Engineering Economy*
Blank: *Statistical Procedures for Engineering, Management, and Science*
Bridger: *Introduction to Ergonomics*
Denton: *Safety Management: Improving Performance*
Grant and Leavenworth: *Statistical Quality Control*
Hicks: *Industrial Engineering and Management: A New Perspective*
Hillier and Lieberman: *Introduction to Mathematical Programming*
Hillier and Lieberman: *Introduction to Operations Research*
Huchingson: *New Horizons for Human Factors in Design*
Juran and Gryna: *Quality Planning and Analysis: From Product Development through Use*
Khoshnevis: *Discrete Systems Simulation*
Kolarik: *Creating Quality: Concepts, Systems, Strategies, and Tools*
Law and Kelton: *Simulation Modeling and Analysis*
Marshall and Oliver: *Decision Making and Forecasting*
Moen, Nolan, and Provost: *Improving Quality through Planned Experimentation*
Nash and Sofer: *Linear and Nonlinear Programming*
Nelson: *Stochastic Modeling: Analysis and Simulation*
Niebel, Draper, and Wysk: *Modern Manufacturing Process Engineering*
Pegden: *Introduction to Simulation Using SIMAN*
Riggs, Bedworth, and Randhawa: *Engineering Economics*
Steiner: *Engineering Economic Principles*
Taguchi, Elsayed, and Hsiang: *Quality Engineering in Production Systems*
Wu and Coppins: *Linear Programming and Extensions*

ENGINEERING ECONOMIC PRINCIPLES

Second Edition

Henry Malcolm Steiner

The George Washington University

The McGraw-Hill Companies, Inc.

New York St. Louis San Francisco Auckland Bogotà Caracas Lisbon
London Madrid Mexico City Milan Montreal New Delhi
San Juan Singapore Sydney Tokyo Toronto

McGraw-Hill
A Division of The ***McGraw-Hill*** *Companies*

The editors were Eric M. Munson and Scott Amerman;
the production supervisor was Denise L. Puryear.
R. R. Donnelley & Sons Company was printer and binder.

ENGINEERING ECONOMIC PRINCIPLES

1 2 3 4 5 6 7 8 9 0 DOC DOC 9 0 9 8 7 6 5

ISBN 0-07-061256-0

Library of Congress Catalog Card Number: 95-81665

ABOUT THE AUTHOR

HENRY MALCOLM STEINER received a B.A. in engineering with a specialization in aeronautical engineering from Stanford University, an M.S. in construction engineering from Stanford, and a Ph.D. in engineering-economic planning, also from Stanford. After 15 years' experience as a practicing engineer, he taught at the University of the Americas in Mexico, at the Escuela Superiór de Administración de Negócios (ESAN) in Peru, where he gave courses in classical economics and economic development under the auspices of the Graduate School of Business at Stanford. In the United States, he taught in the management department of the School of Business at the University of Texas at Austin, before joining the department of engineering administration at The George Washington University in Washington, D.C. He has served as a visiting professor at Stanford, both in the Graduate School of Business and the Engineering School. Most recently he served in the department of industrial engineering and operations research at the University of California, Berkeley, as a visiting professor and scholar. Dr. Steiner was Fullbright professor at the University of Guadalajara, Facultad de Economia.

Professor Steiner is the author of four previous books, the most recent being *Public and Private Investments.* He has written many case studies, articles, and papers. The founder and first president of the international chapter of the Transportation Research Forum, he has served as consultant to the World Bank, U.S. Army Corps of Engineers, Bureau of Indian Affairs, and many private companies. His most recent assignment was as a consultant to the World Bank from 1991 to 1994 on a major highway loan to Mexico.

He is a member of Tau Beta Pi, Sigma Xi, the Transportation Research Forum, and the American Society of Civil Engineers.

To my beloved wife

CONTENTS

PREFACE TO THE FIRST EDITION

Why has another book on engineering economy come into existence? The usual reasons are present: to include the most recent advances in the subject, to remove extraneous material better treated in other disciplines, and to introduce new areas of study. But there is an additional reason: The advent of the personal computer provides a new tool which widens the scope of what can be done with the concepts of engineering economy.

Adaptation to the personal computer is not enough to justify a new textbook, even though, as a tool, it is far more useful than the electronic calculator. It makes problems like the determination of the internal rate of return of a cash flow simple to solve. With only an unprogrammed calculator and tables at hand, such problems take more time. For example, study of sensitivity analysis, one of the most useful notions in helping an analyst to make decisions, is much facilitated by the use of a personal computer. Calculations which take many hours can be reduced to seconds. Chapter 18 discusses sensitivity analysis. Appendix A, in conjunction with the computer diskette provided with this book, shows an important example from that chapter and the knowledge that can be gained from it when a computer is available. However, it should be mentioned that this book can be used without a computer, personal or otherwise. The usual compound interest tables are available in App. C. A diskette is provided, but it may be ignored.

This textbook includes new subjects and new expositions of old ones:

The area of loans is new to engineering economy books, at least as a subject in itself. Variable loan rates, differing loan periods, balloon payments, interest-only loans, and other variations have made the subject worthy of treatment. How to choose among loan alternatives is fully covered in this book.

The effect of the viewpoint of the analyst, and its implications for engineering economic analysis, is rarely treated in textbooks on the subject. In the cases where it is discussed, the very important implications are not explored. Analysis of projects in the public sector is particularly affected by the concepts of welfare economics, which are closely associated with the concept of viewpoint. The proper consideration of viewpoint can cause vital changes in project decisions, changing a yes to a no, or vice versa, and make all the difference between a naive and a sophisticated analysis.

Capital budgeting is an area that is relatively new to engineering economy and one that has not been properly appreciated. It also happens to be one that requires the use of mathematical programming, and thus the use of the computer, for the solution of problems. Of particular importance is the effect of the capital budget on the opportunity cost of capital, an area not completely covered by existing engineering economy textbooks, to the best of this writer's knowledge.

The title of this book, *Engineering Economic Principles,* describes exactly what it is about. However, it may be useful to delineate precisely where the subject matter of this text lies in the much larger area of economics. Economics, whose subject is the optimal use of resources in society, is divided into two parts, microeconomics and macroeconomics. Microeconomics is the study of the producing unit, the firm, whether public or private, in its various aspects and activities, among them, demand for its product, supply of resources to produce the product, and capital investment to provide the land, buildings, and machinery for production. Macroeconomics is the study of the economics of a society as a whole; we will not be concerned with it here except tangentially in the chapters on inflation and sector analysis and viewpoint.

Engineering economy is a special area of microeconomics, specifically the part concerned with capital investment. In other, more or less similar guises, it is known as economic analysis by the economists, capital budgeting—something of a misnomer—by people in business, management accounting by the accountants, and other titles not connected with any particular group or profession, such as life-cycle costing, value engineering, and discounted cash flow analysis. In its origins, engineering economy was concerned with the relationship between engineering and economics. Asking not how to design a dam or how to build it, engineering economy considered the question, Should we build this dam at all in this spot at this time, and how much will it pay off if we do? Such a question could be applied to many engineering situations: alternative railroad locations, choices among building heights, whether to buy or lease one make of machine or another, and so forth. No matter what the situation, engineering economy was involved with the capital investment question, but only in the microeconomic area.

Later on, it became obvious that a great many problems of engineering economy were also encountered in ordinary life. Should I keep my present car or buy a new one? is a question faced by countless car owners in the modern world. The methods of engineering economy can be used to answer it. Given the projected inflation rate over the next few years, should I invest in this piece of real estate or in that? is another question often asked these days. Engineering economy can answer it. Thus the reader will see in this book many examples and problems that reflect the economic questions we all confront in our daily lives and that cannot be classified under the practice of engineering. So, engineering economy has use not only in engineering but also in our personal economic decisions.

Now that economics and engineering economy have been dealt with, it is time to ask the meaning of *principles*. It should be emphasized that this text is limited, as its title indicates, to basics. The material explained could easily be amplified, in every chapter, to cover the controversies surrounding the subject. The book would then overreach its intended scope. For this reason, the author had to restrain himself with the question, posed time after time, Is this idea basic to the subject? Many times, the answer was no.

The principles of this subject are those aspects of it that must be well understood in order to solve those problems most likely to be encountered in the practice of engineering. Evident as this approach to the subject may seem, the fact is that considerable room exists for personal opinion in this matter. Not surprisingly, a list of basics for one engineer may differ markedly from a list compiled by another. The degree of importance attached to each entry in a list of essentials is also a matter of opinion. For example, I have included a chapter on viewpoint. Some recognized experts in engineering economy may not only disagree on including viewpoint, but also argue that it should not be present at all in a book on fundamentals. Others, indeed, may not even be aware that viewpoint can be a part of the subject. It appears that some sort of consensus on what is basic to the subject is itself a basic.

In an attempt to decide the question, I have carefully read the prefaces to many texts in engineering economy and tried to assemble a consensus. Then I have allowed my own opinion to enter. The table of contents is the result.

A possible selection of topics for a one-semester course of 3 hours per week is the following:

Chapter	
1	Introduction
2	Investment Choice
3	Equivalence
4	Interest and Financial Mathematics
5	Present Worth
6	Annual Worth
7	Benefit/Cost Ratio
8	Internal Rate of Return
9	Multiple Alternatives
10	Effect of Income Tax on Economic Analysis
11	Inflation
12	Risk
13	Loans
15	Opportunity and Financial Costs of Capital
16	Replacement
17	Sector Analysis and Viewpoint

This selection leaves out only the chapters on sensitivity analysis and capital budgeting. Both are considered to be advanced topics by most writers on the subject.

A less intensive course might exclude the chapters on loans and replacement analysis from the above list. It is possible, too, to spend less time on each topic.

The book contains 143 solved examples. Many of these refer to situations encountered in developing countries. In these countries, costs, revenues, rates of return, inflation rates, tax laws, and so forth are often different from those found in the United States. More and more students from such countries are attending engineering schools in the United States. It is hoped that such examples will help them understand the subject in the context of their homelands. For the American student, in this age of multinational corporations, familiarity with conditions overseas may be of considerable assistance in an engineering career.

At the ends of the chapters are 261 problems, some with answers supplied. An *Instructor's Manual* is available from the publisher. It provides solutions to all problems.

Appendix A contains instructions on how to use the diskette, available with this textbook, in a personal computer to assist in solving problems. The diskette may be used with an IBM AT, XT, PS/2, Macintosh, or compatible machine, and operates in conjunction with Lotus 1-2-3®, Quattro®, or Excel®. The capital budgeting chapter uses Lindo® as well. A step-by-step instruction in the use of the personal computer in engineering economic analysis is provided. It employs selected examples from the text, solves them, and allows space for the reader's problem solutions.

My thanks go to the many university instructors on the subject who reviewed the chapters in this book as they were written. Their comments were invaluable. I wish to thank especially Thanos Tsimberdonis for his help in creating the software of App. A. My editor, Aileen Landeros, deserves my gratitude for her dedicated work on this book. James Fay, Mohammed Mustafa, Charles G. Vandervoort, Carlos A. Velarde, and Thawat Watanatada gave valuable criticism.

My many students at the Escuela Superiór de Administración de Negocios (ESAN), Stanford University, the University of Texas at Austin, the George Washington University, and the University of California at Berkeley also helped me. Each class I gave was, in a real sense, an examination in the subject, and the students were the examiners, commentators, and advisers. I thank them.

My oldest debt I owe to Prof. Eugene L. Grant, who taught me the subject during my undergraduate days at Stanford and who elaborated on it during my later master's and doctoral studies there. Professor Clarkson H. Oglesby, also of Stanford, deserves equal gratitude for his interest and encouragement.

Inevitably I have forgotten some who helped. Those I also thank.

Henry Malcolm Steiner

PREFACE TO THE SECOND EDITION

A number of important changes and additions have been made to the text for this, the second edition. All changes to the income tax laws up until January 1995 have been included. They will be found in a completely revised Chap. 10. The MACRS depreciation methods, now required, have been made up to date and incorporated into the chapter from their former position as an appendix. The sum-of-the-years'-digits method, used before 1981, is covered even though it is no longer allowed for new property. It is still in use, however, for property placed in service in its time period. Two pre-1981 methods—declining-balance and straight-line—are explained because they are still required by the Internal Revenue Service for certain classes of property. Chapter 16, on replacement has been rewritten using only the equivalent uniform annual-cost approach to this difficult area of the subject. New examples have been included. The chapter on loans still remains one of the most interesting new areas of engineering economy, not adequately covered in any other textbook that I am aware of.

The number of problems appearing at the end of each chapter has been greatly increased, to 598 in all. As a consequence, the *Instructor's Manual* has been expanded to include solutions to all problems. This manual is available from the publisher.

Answers to selected problems have also been provided in the body of the text. Some instructors may wish for more answers to allow the student more opportunity for home study; others for less, so that more problems will be available for homework and test assignment. This decision is left to the individual instructor. Those who wish more answers may, of course, give them to the students from the *Instructor's Manual*.

A new feature of the textbook, introduced in this edition and not to be found in competing textbooks, is end-of-chapter problems based on chapter examples. That is, each example in the text—and there are more than 146—is accompanied by at least one problem based upon it, but with some data altered. Assigning such problems for homework means that the student will be required, at the very least, to read the relevant example, which in turn illustrates an important conceptual point explained in the material directly preceding it. This is a powerful teaching tool. In addition, the instructor can unmistakeably direct the students' attention to the concepts the instructor considers most important.

Moreover, this direct association between example and problem makes it possible to self-study the entire text with a reasonable expectation that the points of interest have been understood.

The problems based on examples require that the examples be tested for *sensitivity* to certain variables. Thus, although not formally introduced as such until Chap. 18, sensitivity analysis appears throughout the book. However, some instructors may wish to have sensitivity analysis formally presented much earlier in the text than it is now. There is no reason why an instructor cannot require part or all of Chap. 18 at any point in the course. To assign all of it, including problems, students will have to have completed at least through Chap. 12 on risk. It is, of course, perfectly possible to assign the first portion of Chap. 18, up to multivariable analysis, at any point after Chap. 8 on the internal rate of return.

The opportunity cost of capital, also known as the minimum attractive rate of return, discount rate, and other names, is discussed immediately after the capital budgeting chapter because a direct link exists between the two. Except where mandated by higher authority, the MARR depends on the capital budget.

Also, a great deal more emphasis has been given to understanding concepts by including questions on them in the problems at the end of the chapter.

Compound interest tables in App. C have been revised. They now include additional tables for ¼, ½, ¾, 1¼ and 1¾ percent. Values for n in these tables have been expanded to include 36, 48, 72, 84, 96, 120, 240, 360, and 480 to help with problems based on months. The A/G and P/G factors are now shown as columns in the main tables and not as separate tables. Values of factors for n equal to zero and infinity are now shown.

I wish to thank Eric Munson, my editor at McGraw-Hill, for his unfaltering assistance, Dr. John Wenzelberger for his excellent suggestions for improvements in the text, and my students for many helpful comments and corrections. Ms. Kristsana Nithikethkul gave unstinting assistance in computer difficulties. My thanks also go to Greer Nelson for all her help. I also wish to express my appreciation to those who reviewed and commented on this edition; they were a great help: Daniel Babcock, University of Missouri, Rolla; Richard H. Bernhard, North Carolina State University; Eleanor R. Fapohunda, State University of New York at Farmingdale; Thomas Lagerstrom, University of Nebraska; Robert Lundquist, The Ohio State University; Mansooreh Mollaghasemi, University of Central Florida; William C. Moor, Arizona State University; and James A. Rice, University of Illinois, Chicago.

Henry Malcolm Steiner

ENGINEERING ECONOMIC PRINCIPLES

CHAPTER 1

INTRODUCTION: INVESTMENTS EXPLAINED

What is engineering economy? Until recently the answer was easy—it is the application of certain principles of economics to the problem of investments, principally engineering-related investments. Today, this simple definition needs to be expanded considerably, as the subject itself has expanded over the years since the last century when Wellington wrote of railway location in the United States. To economics must be added mathematics—and particularly the applied version known as *operations research.* When more complicated problems are encountered—planning the entire transport system of a developing country, for example—sociology, anthropology, urban planning, and other disciplines must be combined with economics and mathematics in order to arrive at satisfactory solutions. Although such a comprehensive approach will not be treated in an elementary text such as this one, the reader should be aware of the necessity for such multidisciplinary approaches in order to avoid an overly narrow view of the subject. A good rule is that a broad enough view should be taken, by including disciplines other than engineering and economics, that an optimal solution of the problem can be determined. To clarify this remark, let us imagine a situation in which a crosstown freeway is proposed in a major city. The engineering and economic studies completed, the city government is called upon to decide whether the project will be built. A member of the city government asks a question concerning the effect of the proposed freeway on the form of the city: Will the project cause the city to spread out even farther, thus increasing transportation distances between working places and residences? Will the freeway erect barriers to movement across it, create poverty pockets, or bisect existing neighborhoods?

Evidently some aspects of the effects of the project have been disregarded. An urban planner is needed to combine views with those of the engineers and economists so that an optimal solution may even have the opportunity of being discovered.

In this book, we concentrate on the economic aspect of projects, leaving aside the engineering design problems and those of all other disciplines, but never forgetting that they exist.

Investment means using resources to create an addition to the present facilities. In the example of the preceding paragraph, a certain number of miles of urban freeway are to be added to the existing mileage. If we decide to divide the economy of a nation into two parts, public and private, the freeway investment is a *public sector* project because the necessary funds will come from public monies. The purchase of 10 new Euclid end-dump trucks by Nevada Construction Company is an example of a *private sector* investment because the addition to the supply of dump trucks for this company is financed from private funds. Such a conception of investment is not the one entertained by many people, that investment is the purchase of real estate, or stocks, or bonds in order to receive a return of so much per year. The latter is, of course, also investment, but in a narrower definition.

Investment from an economist's standpoint is the diversion of resources from consumption to uses that will improve the efficiency of the production process. Consumption is postponed so that more consumption will be possible in the future. The traditional example of Robinson Crusoe and his fishhooks makes the matter clear. Crusoe consumes fish, which he catches with a spear. It takes all his waking hours to feed himself by this primitive method. If he forgoes spear fishing for one day, thus going to bed hungry, he can make enough hooks to improve his production of edible fish by a large amount and thus have more time for rest and the improvement of his living conditions. The time and effort of one day in the production of fishhooks are invested. Consumption of production that would have resulted from the time and effort is forgone, but the production process is improved. This is investment, in simple terms.

Notice that money is in no way involved in Robinson Crusoe's world, yet investment is possible. The general term *resources* is a more exact description of Crusoe's time and effort. The fishhooks cost, not money, but time taken away from fishing and, more particularly, the amount of fish not caught. The *opportunity cost* of the fishhooks, that is, the investment cost, is one day's catch. Opportunity cost, therefore, is the benefit given up by Crusoe in order to make his investment in fishhooks. We will use this concept frequently in our study of engineering economy. It may occur to some readers that investment and the capital goods resulting from it are measurable in terms of the amount of human effort required to produce them. This is Karl Marx's *labor theory of value*. Unfortunately for its continuing validity, an objection is apparent when one inquires into the case of a capital good immediately available for use as soon as produced, in distinction from one that must be stored for a time before it can be used. Lumber, for example, must be dried out in yards and wine must be aged in casks or bottles, before they can be used. Take one example, a good such as fishhooks that requires 1,000 units of labor to produce. Another example, wine, also takes 1,000 units of labor to produce. The wine must be aged; the fishhooks need not be. Should both be costed in proportion to 1,000 units of labor? Of course, the answer is no, the wine should cost more. Why? Because the 1,000 units of labor invested in the wine were not available for use until a certain amount of time had passed. The 1,000 units of labor became capital whose use must be paid for. In general, investment requires a forgoing of consumption; it is measurable in terms of the consumption forgone; and it is not always directly measurable by the amount of labor involved in making the investment good.

Evidently the benefits of an investment must exceed the costs if the investment is to be approved. If Crusoe estimates that fishhooks will be less efficient than a fish spear in catching fish, he will certainly not invest in them. Time is inevitably involved in this judgment. Imagine that fishhooks are more efficient than fish spears in fish catching. But how does Crusoe know this? Observations must be made over time. He will compare, for example, the amount of fish he will catch in a day or a week or a year by hooks with the amount of fish he will catch in the same period by spear. The value of capital goods is measured by not only their first cost but also how long they will last. If Crusoe's hooks must be made of wood and will last for only one fish each, they will certainly be less attractive than those made of metal that will last a year.

What Crusoe really does, then, is compare the benefits and costs over time of continuing the present method of fishing with the benefits and costs of a new method. In other words, he compares alternatives over a period of time. This is typical of all investment decisions: At least two alternative courses of action are involved, and they are compared for the same period.

MUTUALLY EXCLUSIVE INVESTMENTS

Investments in engineering works are made in order to solve a problem. A river must be crossed, for example. A bridge must therefore be built. Imagine that the alternative designs consist of one in steel and one in reinforced concrete. These are *mutually exclusive investments* because only one of the alternatives proposed will be built. It would make no sense to build both alternatives. Even clearer is the example of a multistory building. Suppose that a selection must be made among plans for a 75-, 80-, and 85-story building. Only one building of a definite number of stories will be built on the available lot. The other building designs must be rejected. Almost all problems and examples in this book concern mutually exclusive alternatives.

The type of investment—mutually exclusive, mutually independent, or mutually interdependent—is important because different techniques of analysis must be used, depending on the type. In general, mutually exclusive alternative investments must be analyzed by using incremental methods, with a few exceptions. These will be explained in Chaps. 5, 6, 7, and 8.

MUTUALLY INDEPENDENT INVESTMENTS

A department of transportation of one of the western states is confronted with proposals for a number of highway projects, perhaps running into the hundreds. Building one of them does not in any technical way preclude building another. Some proposals will qualify for construction and be built. Others will not. The point is that there is no technical relation between the projects that will be built and those for which there will be insufficient funds to build. They are all *mutually independent* projects. For another example, consider a number of projects proposed by the engineering department of a large refinery. Selection of a certain project, or a number of projects, to be constructed will not in any technical way affect any of the other projects proposed. Again, these are independent projects.

Usually a list of proposed projects is presented with the supposition that not all the projects on the list will be built because available funds will not suffice to cover the cost of the whole list. A capital budget must normally be met. Therefore, certain projects on the list will be built, and others will not; but the total cost of those that are constructed must fall within the limitations of the money available—the budget. The process of selecting projects to be approved for construction is known as *capital budgeting*.

A method of rigorous analysis in capital budgeting is *zero-one integer programming*, a specialized case of linear programming. The techniques used in the analysis of alternatives form the foundation of the process, and these are incorporated into the integer programming analysis in order to arrive at an optimal set of projects to be included in a budget. All this will be explained in Chap. 14.

INTERDEPENDENT INVESTMENTS

Projects may also be mutually interdependent. For example, take machinery and the buildings to house it. The machinery and its buildings must be analyzed together, if an investment in one assumes an investment in the other. Or they may be analyzed separately, because even though the machinery cannot be properly used without buildings to house it, the buildings themselves may have other uses, as warehouses, for example. Interdependent projects will also be considered in Chap. 14.

PRIVATE INVESTMENTS

Mention has been made of the division of an economy into the public and private sectors and investments under the same classifications, depending on the source of funds of the projects concerned. Investments may be characterized by identifiers other than the source of funds for their construction, however. Private investments are controlled by private persons and private firms for private gain. A company, for example, is considering the replacement of a production lathe with another, more efficient model. No question of any public good or evil that may be brought about by the purchase of the lathe is entertained. The only point considered is the effect of the new lathe on the profits of the company. And this is as it should be, if the officers of the company are to fulfill their obligations to the stockholders. The attitude taken here is characteristic of investment in the private sector.

Along with ignoring all effects except those that affect its pocketbook, the private sector accepts prices as given by the economy. Normally, no consideration of *shadow prices*—the true societal cost of an input to an investment—is given. Even in countries where enormous labor resource unemployment exists, the cost of labor is taken as given by the labor market. The price system's perfect operation is assumed. We will see more of the meaning of these terms and concepts in Chap. 17, Sector Analysis and Viewpoint.

Private sector investments are not generally infrastructural; that is, they are not concerned with investment in roads, or education, or the judicial system, or any other area whose purpose is to give support to the operation of a society.

PUBLIC INVESTMENTS

Public investment covers such a wide range—from the U.S. space program to a county bridge—that it is difficult to define exhaustively. Its main characteristic, already mentioned, is that its funding is public. Some other points of difference between public and private investments may help to clarify their nature, however.

In public investment, the desired end result is to benefit society. The U.S. space program benefits not only U.S. citizens but also the world. A county bridge benefits the local community. The gainer from both programs is the public. The relative difficulty of estimating that gain is evident immediately as we compare public investment to private. Some space in later chapters will be devoted to a discussion of this problem.

Public investment is sometimes on a much grander scale than the private sector can manage. River development projects like the Columbia in the United States or the Nile in Egypt are far beyond the capacity of private companies—or indeed, their interest, for such grand schemes have an element of risk that private companies are unwilling to accept. Political realities often rule out participation in large projects by private companies. A change of regime can cause abandonment of support for a project with such attendant losses as only a government can stand. This is not to say that all large projects must be performed by governments. After all, the U.S. railroad systems and those of many other countries were developed by private companies. There is no doubt, however, that governments tend to take on larger projects than private firms generally do.

Infrastructural projects are almost always public. The sewer system of a city is the public's concern, and so are its streets. Roads and highways, the educational and judicial systems, and all their buildings are almost all public. Ports and airports are largely publicly owned and operated. National defense establishments and their many projects are in the public sector. Most of what provides transport, education, judicial service, and security of persons and the nation is public. No rule for judging what should be public and what should remain in the private sector will be advanced here. Although it is an interesting—even burning—question, it is beyond the scope of this text.

Analysis of public projects requires a judgment of the state of health of the price system because public projects must be judged by social costs and social benefits. When costs as given by the price system are not the same as social costs, adjustments to that system must be made if fair economic and social comparisons of alternatives and budgets are to be made. More will be said on this in a later section.

EXTERNALITIES

Economists have a word that is both descriptive and exact for an important phenomenon in socioeconomic analysis. The word is *externality*. It means that the cost or benefit in question is external—that is, not included as an item—to the books of accounts of whatever economic entity we are talking about. It is borne or enjoyed by someone else. Consider air pollution in a large city. It is a cost associated with power generation, automobile use, and bus transportation, as well as industrial production of one kind or another. Yet it never appears in dollars and cents on the account books of the power company, or the bus system, or any industrial plant, and most assuredly not on the household accounts of any of the millions of automobile operators, private and public, who drive the city streets. No one would deny that air pollution is a cost, yet it appears nowhere. It is an *external* cost, as opposed to the cost of oil for the power company, or operators for the bus company, or gasoline for the automobile users that do appear on the books of these economic entities and thus are internal to them.

Parents educate their children, let us say by sending them to an expensive university. In doing so they confer a benefit upon society, most people would agree, because educated people help the society in which they live more than uneducated people do. But the benefit is external to their household accounts. No one appears at the door, bringing a check from society that will compensate them for the benefit they have bestowed upon it.

What have externalities to do with investment? They must be included in an analysis, if the analysis is to be worthwhile; but to be included, first they must be recognized. This statement is more true of public sector studies, because in the private sector it is not expected that air pollution, for example, is going to be costed on the books of a taxicab company. However, this failure of the market system should not mean that the system remains uncorrected in cases where it can be corrected, namely, in public sector analysis. If you are studying the costs and benefits of an urban transport system, you should not ignore the decreased traffic congestion—a benefit—that will result from a subway, even though that benefit will never appear in its totality on the books of any company or household as revenue.

In sum, externalities should be included in the analysis when it is appropriate to do so, that is, when social benefits and costs are relevant.

SOCIAL BENEFITS AND COSTS

Whenever market prices diverge from the true costs of a good or service, then a comprehensive—and thus accurate—analysis of a course of action is impossible unless market prices are adjusted until they reflect true costs, that is, until they become social prices. The words *price* and *cost* are used synonymously, because in any transaction, that which a seller designates as price is the same money amount as that which the buyer calls cost. In the discussion of externalities in the previous section, we have seen that certain costs do not appear in the market at all. These must be added as new items in the analysis. Thus we have two operations to perform when making a social benefit/cost analysis—adjustment of certain prices and the addition of entirely new items with their respective prices.

Social benefits and costs are defined as the total benefits and costs of an event to society as a whole. If producing energy causes air pollution, then the cost of that air pollution must somehow be added to the kilowatt-hour cost of energy.

Not all social costs and benefits diverge from market costs and benefits because of externalities. Sometimes market prices and costs simply do not reflect true social costs because of interference in the market. For example, the common case of government-controlled prices may be the least troublesome to grasp. The U.S. government controlled the price of petroleum beginning in the early 1970s. The government of Mexico controlled the prices of food staples, such as beans, for decades of this century. The true social price—the uncontrolled or market price—would have been higher for both goods. Adjustments must be made to the controlled prices to arrive at social costs.

Many other examples exist of social costs differing from market costs. Perhaps the example of controlled prices will be sufficient to drive home the point, however.

We will not be concerned with social costs and benefits until we reach Chap. 17, Sector Analysis and Viewpoint. At that point in our study, the concepts of social benefit and cost will be expanded.

OPPORTUNITY COST

Opportunity cost is one of the most important concepts in economics. Indeed, it is one of the principal foundations on which the whole discipline rests. What does the phrase mean? To answer that question, let us establish some basic propositions.

If we have a choice to make between two courses of action, both beneficial to us, then because we cannot do both, we must give up the benefits of the course of action that we reject. But if we give up something, whatever it is, the pleasure we would have taken in it is lost to us. That lost pleasure may be thought of as a cost. A *benefit forgone* is therefore the same as a *cost*.

These two points being clear, it is now possible to define *opportunity cost.* It is the benefit forgone by choosing one mutually exclusive alternative rather than another.

Take an example. An engineer prefers to work for herself as a consultant for $55,000 per year. To do so, she gives up a job for a large corporation that pays $67,000 per year. The opportunity cost of self-employment is $12,000.

Here is another example. A couple have decided that they will own only one automobile. They presently own a 1983 station wagon, the largest of the variety offered that year. It is now fairly old with almost 100,000 miles on it. Its maintenance cost is increasing. Its mileage per gallon of gasoline is relatively low. They can sell it to a private purchaser for $500. They are contemplating buying a new car with much better fuel economy and much lower maintenance costs. Keeping their old car and buying a new car are mutually exclusive alternatives, providing they stick to their decision to keep only one automobile. If they buy the new car, they will sell the old one and benefit by having $500 to put in their pockets. However, if they keep the old car and reject the new car, they will also lose the $500 they would have received had they sold the old car. The opportunity cost of the old car is therefore $500. It is the benefit they give up by choosing to keep the old car. This idea of opportunity cost is one we automatically accept without a second thought when we say, as in the example, “We’ve got $500 tied up in that car.” (For simplicity’s sake, the example neglects all future benefits and costs.)

In questions about replacing existing equipment, the idea of opportunity cost will be used extensively. It will also appear whenever we compare accountant’s costs with those used by persons responsible for making a decision. Although the accountant’s book may say an asset is worth a certain amount—its book value—the decision maker will be interested in its opportunity cost, that is, its salvage or resale value.

COST OF CAPITAL

Opportunity cost is particularly important when the cost of capital is considered, for it is the *opportunity cost of capital* that will be one of our chief guides in making investment decisions. Some explanation of a different concept—what most people think of when the cost of capital is mentioned—is necessary here. This other concept is the *financial cost of capital.* A comparison of the two concepts of cost of capital is best seen in an example.

Suppose that a small businessman has before him several alternatives for investment in his business. He does not have the funds to finance the alternative he has selected. He borrows the money from a bank, on his signature, at 20 percent interest per annum. This bank interest rate of 20 percent is his *financial cost of capital.* The project he is considering will earn 30 percent per year. If he does not make the investment in this alternative he has selected, he will forgo earning 30 percent annually. This 30 percent is the *opportunity cost of capital.*

Of course, the opportunity cost of capital will generally exceed the financial cost of capital because one does not normally borrow money in business affairs unless one expects to make a profit on the money borrowed. One will not borrow money and pay 20 percent for it if one expects the investment of these funds to pay off at only 10 percent.

We will see that the opportunity cost of capital—also called the *discount rate*, the *minimum attractive rate of return*, and the *marginal rate of return*, among others—plays a vital role in economic analysis.

INCOME TAX

Private investments normally have income tax consequences. Because income taxes are costs to the investor, they must be included in any economic analysis where they are significant. In the major industrialized countries, they are always taken into account.

Consider the simple case of two mutually exclusive alternative investments that have different income tax consequences. There is no question that these differential tax payments will affect a choice between the alternatives and thus must be included in the economic analysis. In Chap. 10 on income tax, we will see how such analyses are handled.

In some countries there are no income taxes. In other countries, taxes are so minimal that they may be ignored. In still others, although income tax laws exist, they are not enforced and income tax considerations are of no importance. Needless to say, the ruling consideration is the situation in the country or countries where taxes on the return on investment will have to be paid.

INFLATION

Inflation is a word familiar to almost everyone in the post-World War II world. It means that prices of goods and services increase over time—in some countries almost imperceptibly, in others by leaps and bounds. *Deflation* means the opposite of inflation—prices decrease as time passes. Deflation is familiar, although not under that name, to another generation—those who lived through the Great Depression of the 1930s. Then people were confronted by falling prices, seemingly a reason for joy; but because the incomes of these same people were decreasing at the same or a greater rate than the prices they faced, the joy was mixed with sorrow.

Why must inflation—for we will confine ourselves to that aspect of price changes—be considered in the economic analysis of investments? Because investments have consequences, both benefits and costs; because those consequences are measured in money, at least in part; and because this money is paid in current dollars or francs or pesos, we must include the effect of inflation. (Current dollars are those received in any particular year.) Current dollars of 1995 buy less than current dollars of 1967 because inflation of the currency has occurred in the period between those years. Because it is ultimately the buying power of money and not its face value that concerns us, we must somehow define a method of making the buying power a constant; that is, we must remove the effect of inflation and deal only with dollars of constant buying power. How this is accomplished we will see in Chap. 11.

PROBLEMS

1.1. Robinson Crusoe is able to catch three fish per day by spear fishing. After he produces five hooks from one day's work, he is able to catch three fish in one-third of a day. He needs three fish per day to stay alive. What is his annual rate of return on his investment of one day's work producing fishhooks, assuming that the hooks will last exactly 1 year?

1.2. What is Karl Marx's labor theory of value? How can it be criticized?

1.3. What is the difference between mutually exclusive and mutually independent investments? Give an example of your own of each type of investment.

1.4. What is meant by *interdependent investments*? Give an example of your own.

1.5. Identify the following as mutually exclusive, independent, or interdependent investment situations.

(*a*) The state of Texas has before it 76 projects to be funded in its farm-to-market road program. No project has any technical effect on any other.

(*b*) Three makes of commercial sewing machines are being analyzed by Perfect Fit Company with a view to replacing the existing machines. Because maintenance and repair are facilitated by the use of a single make, only one make of those proposed will be chosen.

(*c*) The Northern Tier Power Company has before it 15 major proposals for components of its distribution system. Three of these involve alternative ways of solving a specific problem. None of the 15 proposals has a technical effect on another.

1.6. What are some of the criteria that distinguish public from private investment?

1.7. Apply some of the concepts presented in this chapter to the question of national government spending. Are those politicians who say, "Spending is spending, and calling some of it investment is just so much dust in the eyes of the public," viewing the matter correctly?

1.8. Do you have an opinion on which activities are most appropriately performed by government and which by the private sector? (A most interesting writer on this subject is Nobel Laureate Milton Friedman.) What about roads and highways? Or hydroelectric power? Or telephone service?

1.9. Metro, the publicly owned passenger transport system in the national capital area, loses money every year, yet public tax money subsidizes it year after year. Why do we not simply shut it down, as we would a losing business in the private sector?

1.10. Give five examples of each of the following.
(*a*) Private investments
(*b*) Public investments

1.11. Define *externality* as it is used by economists. Give an example of your own making (*a*) of an external benefit and (*b*) of an external cost.

1.12. An engineer may expect certain benefits to accrue to her because of the practice of her profession. What are some of these benefits? Are there benefits to society, for which she receives no remuneration, that exist because of her practice of the engineering profession?

1.13. Define *social benefits* and *social costs*. Do they differ from *government costs*? Explain.

1.14. Define *opportunity cost*. Make up your own example of a situation in which opportunity cost appears.

1.15. Give three synonyms for the opportunity cost of capital.

1.16. A photographic processing plant expects to raise money for internal improvements to its production facilities by "going public." In other words, formerly a private plant owned by three brothers, it will now be owned by stockholders. Ten million dollars in stocks will be issued in $100 shares. Each share will receive an estimated $5.00 in dividends per year. The improvements will return 23 percent over their economic life.

What is the financial cost of capital, and what is the opportunity cost of capital, for this company in these circumstances?

1.17. Why must income taxes be included in an economic analysis? Give a brief description of the income tax situation encountered by private persons—not corporations—in your city, state, and country.

1.18. Why must inflation be considered in economic analysis? In your country, what was the inflation rate last year? In general, how has that inflation rate changed over the past 10 years?

1.19. Some persons will argue that because inflation is difficult to predict, it should therefore be ignored. Would you agree or disagree?

1.20. Define *inflation* and *deflation*.

CHAPTER 2

INVESTMENT CHOICE

Choosing among investments that are competing for our funds is the action that this textbook seeks to guide. Certain general principles of logical thinking about the process of choice must be established first, in view of the tendency of newcomers to the field to concentrate on the financial mathematics involved and to give less thought to the principles that this chapter will examine. This is an error of major proportions. Forgetting the identification and proper handling of such items as sunk cost, depreciation, and differential consequences can turn a mathematically correct analysis into an absurdity. A faulty grasp of the process of investment decision making can lead to errors equally grave. These matters deserve the reader's attention during the remainder of this chapter, as much as or more than any other section of this book.

DECISION CRITERIA

A decision among alternatives must involve some delineation of optimality. If you say, "I will choose the best alternative," some definition of *best* must be forthcoming. An alternative can be best only if it carries you toward some objective better than other alternatives. Thus, optimality can be expressed only in relation to some objective. If, for example, the objective is profit, then the *best* alternative is the alternative that will produce the greatest profit. If a nation's objective is to increase the educational level of its citizens, then among alternatives such as more schools, better teachers, and television and radio courses, the best alternative will be the one that increases the educational level most.

These seemingly simple and straightforward ideas are actually not so, as evidenced by the many confusing criteria advanced. Take this criterion: That alternative is best that costs least and makes the greatest profit. Imagine three alternative ways of building a road from *A* to *B*. It is not difficult to show that the criterion suggested above can be contradictory. Consider the following table:

Alternative	Cost	Annual profit
1	6	2
2	8	4
3	10	6

The least-cost alternative, choice 1, is not the most profitable alternative, choice 3; nor is it possible to choose at all among these alternatives, given the criterion.

What, then, should the criterion of choice be? It appears that it should be as follows: That alternative is best which maximizes the difference between benefits and costs, both measured in terms of objectives. In the following section, this concept will appear in a clearer light, as we consider procedures in making decisions.

DECISION PROCEDURE

It appears obvious that to make a decision, criteria are required. Not so obvious, but equally necessary, is a decision procedure. Let us examine the steps that one might logically take in order to make an investment decision.

The first step is to set *objectives* that we wish to reach by means of our investments. In the simplest case, say, an individual investor, the objective might be maximum annual profit and nothing more. For a company, both annual profit and market expansion, implying future profit, might be joint objectives. A public entity, say, a port authority, might have as an objective the increased capacity of the port system in response to an overall national objective of increased export of internally produced goods.

The second step is to examine the *alternative ways* of reaching the objectives already set. For example, an objective of maximum annual profit set by an individual investor could be reached by investment in common stocks, or in bonds, or in some combination of the two. The investor, for personal reasons, may reject all other ways of reaching the objective except these three mutually exclusive alternatives. For the company with profit and expansion objectives, the mutually exclusive alternatives might be a new logistics system of trucks and warehouses, or a vastly expanded sales program, or a new product line. These three mutually exclusive alternatives will each accomplish the objectives. The port authority might select new docking facilities as the way to reach its objective and then be faced with the question of how many new berths to build: seven, eight, or nine—all mutually exclusive alternatives.

The third step is to predict the *consequences* of each alternative path to the desired objectives. For example, if the individual investor buys bonds, she can predict how much she will receive each year from a $10,000, 12 percent bond—$1,200—and how much she will receive at the end of the bond's term—$10,000. These are the consequences of one alternative way of reaching her objective of simple profit. In a similar fashion she can predict, with less accuracy, the consequences of an investment in common stock. The company will attempt to predict the consequences of each alternative course of action it has determined will accomplish its objectives. The expanded logistics system will result in certain consequences in terms of profit and market expansion. The trucks and warehouses all cost money, but they will also bring in money and expand the market, which will mean increased future returns. Just how much cost and how much profit will result from choosing this alternative are what must be predicted. The other two routes to the objectives—an expanded sales program and a new product line—must be studied in order to predict their consequences in relation to the objectives. Thus, for each of the three alternatives, a stream of future consequences—costs and revenues—will be predicted. The port authority is faced with a similar problem in estimating the consequences of each of its feasible alternatives. Seven, eight, or nine new berths will involve different costs and benefits in terms of the objectives. The costs of construction and operation and maintenance of the facilities and the benefits in increased capacity of the port to handle exports must be predicted for each alternative.

The fourth step is *evaluation.* Once the stream of costs and benefits for each alternative has been predicted, the next task is to apply a method of evaluation to determine which alternative should be selected, given the shortcomings of the method and the predictions. For example, the private investor interested in profit only might evaluate her cash flow for each of the alternatives—stocks, bonds, or some combination of the two—by the present-worth method, a procedure, like the others to be mentioned in this paragraph, that we will see in future chapters. [*Cash flow* means the stream of payments each year, either cash inflow (revenue or benefit) or cash outflow (cost)]. The company's financial analysts will evaluate the cash flows for each of the three alternatives perhaps by the annual-worth method, and thus determine which of the mutually exclusive alternatives is most preferable. The port authority creates a cash flow, that is, a tabulation of annual benefits and annual costs, and evaluates them by the benefit/cost ratio method. Finally each economic entity—the person, the company, the port authority—brings before the person or group that will make the decision the results of the evaluation of the alternatives to be considered in the last step of the process.

The fifth and last step of the procedure is the *decision* itself. All that has gone before in the process has led up to this step. And all that has gone before has not made the decision automatic, even though the evaluation of the alternatives has indicated the preferred one. Now all the doubts about the completeness and accuracy of the elements of the procedure must be raised: How close are the predictions of the stock dividends? Is the estimated economic life of the logistics system alternative too long? Will world demand for the products that the port will handle hold up as expected? In other words, the judgment of those making the decision must enter the process in order actually to make the decision. Elements of the alternatives, sometimes called *intangibles*, that have not been reducible to money or other commensurable terms must now be considered as well. The individual investor may ask herself, "Will I really be at ease with the uncertainty of income if I buy stocks?" The company president may inquire at the board of directors' meeting, "Will we be able to find the skilled help we will need in sufficient quantity and quickly enough to introduce our new product at the proper time?" The port authority commissioners may query, when presented with the study demonstrating that eight berths are the optimal number to build, "Is a capital-intensive loading and unloading system for the goods passing through the port really in the best interests of the country, or should a labor-intensive system be considered and studied?" In this last step, the crucial questions must be asked and answered and the decision made.

RECOGNITION OF ALTERNATIVES

An important step in the procedure just described is the examination of alternatives. But, unlike the other steps, this one deserves special attention in one respect—the recognition of alternatives. Certainly an alternative cannot be chosen if its existence has never even been imagined. Unfortunately, nonrecognition of an alternative at the beginning of a project does not mean that, at the end when the construction is complete and the project is in operation, it will not appear as the obviously correct one. Projects that are complicated and difficult to conceive often reveal themselves in the process of construction so that, when they are finished, solutions never imagined on paper appear obvious even to the most untutored eye. How can such embarrassing and, indeed, disastrous failures be prevented?

One way to reduce the chances of an alternative being overlooked is to use what is known as *brainstorming*. This method consists in assembling all the persons who might be able to contribute a solution to the problem and requesting that the group put forth and discuss all possible alternatives—even the most far-fetched. These solutions are screened with a special view to open-mindedness. Then those that appear worthy of study are selected. After this, the remainder of the procedure is carried out. Although some further assurance that all alternatives have been examined is gained by brainstorming, a definitive method of avoiding the pitfall presented here does not exist.

As an example of failure to consider a solution that turned out to be the correct one, consider the case of the design and construction engineers for Idaho Construction Company who planned a highway and railroad job on the Columbia River some years ago. Some sections of the route were opened for bidding to subcontractors. One particularly long section, which included cuts and fills in earth, was estimated by the engineers to cost $750,000, a figure determined by assuming the usual construction method in such material of tractors and scrapers. The winning bid was $600,000, considerably under Idaho Construction Company's own estimate of what that section would cost. When the job began, the successful bidder on the subcontract did not use the method foreseen by Idaho Construction Company engineers at all, but a completely different one that the engineers had not even thought of—hydraulic mining. In this method earth is moved by water flow. Its use by the subcontractor resulted in a much lower cost to him than would have been the case had he used the more usual tractor-scraper method—in fact, $300,000 less. Idaho Construction Company's engineers had to suffer the damage to their reputations that failure to recognize the proper construction method cost them, and the company had to make the $600,000 payment for work that cost only $300,000. Had the engineers realized that the hydraulic mining method was an alternative, they would have studied it, recommended it to the bidder on the subcontract, and received bids close to the actual cost of $300,000, thus saving Idaho Construction Company a considerable sum.

In any group of engineers, such stories as that just recounted can be duplicated many times, so common is the failure to recognize the existence of alternatives.

CONSEQUENCES OF ALTERNATIVES OVER TIME

In discussing the decision procedure, the consequences of alternatives were seen to play a vital role. Consequences imply a period of time—anywhere from an instant to a perpetuity—because if we decide on a course of action, a certain amount of time in which the consequences of that course of action are revealed must elapse. Consequences, therefore, are wedded to the time it takes for their occurrence. For example, should we decide that road location *A* is preferable to road location *B* or *C*, we decide on the basis of what payments—costs or benefits—will occur over the life of the road, from the costs of preliminary study to its abandonment for a new location 20 years later. Given this, it is proper to ask whether a given consequence should be somehow weighted according to how distant in the future it will come about. For example, most people view a gift of $20,000 as better occurring tomorrow than the same gift 10 years hence. Not only is the uncertainty of life involved in this feeling—who knows if we will survive into the next decade?—but also the belief that pleasure—say the extra automobile that $20,000 would buy—is better experienced sooner than later. In the same way, the unpleasant aspects of life are better deferred as long as possible, and the longer the better. Thus we tend to think of the effects of consequences in the distant future as lessened compared to consequences in the near future, whether these consequences be good or evil ones. We discount them in greater or less degree as they are farther or nearer to us in time.

Notice that in the following discussion we assume constant prices; that is, inflation or deflation is nonexistent.

A formal way exists to evaluate the same consequence experienced at different times in the future. It is known as *discounting.* To understand it, we must consider the meaning of a common term—*interest*: If a woman decides that she will not take a South American tour costing $10,000 this year but will instead wait until next year, then she is giving up pleasure this year. For this she expects, quite rightly, some recompense. She will receive it by investing the funds she would have spent on the tour in a 12-month certificate of deposit at 10 percent. The recompense for her delay is the extra 10 percent of her vacation funds, $1,000, that she will receive 1 year from now when she cashes in the certificate of deposit to pay for the trip. From her point of view as a lender, *interest* is payment for satisfaction deferred. From the bank's position as borrower, *interest* is payment for the use of borrowed money. In the more philosophical sense we have been employing, it is payment made for immediate satisfaction by the user of someone else's money.

Now suppose that she has no intention of taking the South American tour this year but prefers to take it next year. She estimates that it will cost $10,000 next year. She wishes to put money aside for it. She performs the following simple calculation: If she lets x equal the amount of funds she wishes to put aside now at 10 percent interest, then

$$x + 0.10x = \$10{,}000$$

She discovers that x equals $9,090.91. In the process she sees something else, namely, that $10,000 a year from now is worth only $9,090.91 now. In other words, she has effectively discounted her trip and its pleasures because of its 1-year distance in time from $10,000 to $9,090.91, using a discount rate of 10 percent.

This treatment of consequences over time by discounting them is most important in our study of economic analysis. It is so important that we will devote more space to it and its implications in financial mathematics in later chapters.

VIEWPOINT

By *viewpoint* is meant the institutional position an analyst takes in reference to the project under study. The word *institutional* is used in the social science sense. It refers to the entity in the society in which the analyst is located. For example, if we view the society of the United States as a whole, the analyst may be at the Brookings Institution. On the other hand, he may be in the Department of the Interior of the federal government and thus taking its viewpoint—not necessarily the viewpoint of the society but more probably that of the federal government. Or she may be on the state level in Oregon, or she may be employed by a construction company. All these analysts may be imagined as dealing with the same job, say, a dam on the Columbia River. The Brookings analyst may be concerned, because he is taking the viewpoint of society, with the effects of the purchase of foreign-made generating equipment on the balance of payments of the United States, along with all the other benefits and costs of the dam: construction costs, maintenance and operation costs, costs to Indians whose salmon fishing will be affected, and benefits in the form of energy, flood control, irrigation, and recreation. He will take the widest possible survey of the consequences of the project. The Department of the Interior analyst will ignore certain effects of the project as being outside his purview. He will not be concerned with balance-of-payments effects, for example, but all the rest of the list will occupy him. The Oregon water resource analyst will occupy herself with only those matters that will affect her state and will ignore the rest, unlike the Department of the Interior analyst who will also be concerned with the effect of the dam on other dams upriver in the state of Washington. The construction company engineering economist will be preoccupied with only the cost side, and only construction cost. Benefits are, for her, simply the construction payments from the owner (the Bureau of Reclamation) less the actual cost to the company. Thus, the same job viewed from different institutional vantage points is seen to have quite different quantities of benefits and costs. Indeed, even the lists of items to be considered are different, although each must contain the previous one, like nested boxes.

Later on when the subject of viewpoint is dealt with in depth, we will see that not only are items different relative to viewpoint, but that the same item may have a different value attached to it depending on the viewpoint. It will not seem strange, then, that one analyst can arrive at an entirely different result from another when both deal with the same project. In fact, analysis from one viewpoint may indicate approval while analysis from another may point to rejection.

SYSTEMS ANALYSIS

Systems analysis means the investigation of an interrelated entity whose interdependent parts and their effects on each other must be studied as a whole. For example, the Columbia River Authority operates a large system of dams on the Columbia River and its tributaries. An action taken at one dam—the release of water, say—can have effects on the system as a whole, its flood control capacity, for example, or its energy production potential, and on each part of the system in varying degrees.

Systems analysis is related to viewpoint in a specific way: All effects should be traced that are relevant to the viewpoint of the analyst and that are significant to the study being undertaken. The invariable question confronting the analyst at the beginning of any study is, "How far should I trace the consequences of the courses of action I am about to predict?" Take the example of a private investor in a developing country who is considering a new cement plant in his country. He divides the study into two parts, cost and revenue. The cost side he further breaks down to construction cost and operations and maintenance cost. He investigates all the labor, materials, machinery, fees, and so on connected with both kinds of cost. Because his viewpoint is a private sector one, he is not concerned with the balance-of-payments effects of the purchase of foreign machinery nor does he consider the social cost of labor—only the wages he will have to pay. On the revenue side he has to decide whether there is a significant market potential for the exportation of cement from his new plant or whether his main market exists only in his own country. He may arbitrarily decide that *significant* means over 5 percent of his cement production. He has to project demand for cement in his own country over some time span that he must also select. And if he decides that export sales are significant then he must decide what the effect of world demand will be, as far as he is concerned.

The foregoing is all systems analysis, because each element of his study—demand, plant size, machinery required, cost, revenue, and so on—is related to every other element. He must, therefore, study it as a system and not fall into the trap of analyzing parts of the problem by themselves or groups of them as subsystems. If he does so, he makes the mistake of suboptimization, that is, perfecting the operation of parts of the system without recognizing that doing so will not necessarily perfect the whole.

TIME HORIZONS AND EQUAL SERVICE LIVES

Mentioned in passing in the previous section was that the cement plant entrepreneur selects a period of time over which to predict the outcomes of his venture. It is probably obvious to the reader that consequences must happen in the future and that the analyst must try to foresee them. What is not so obvious is that the time period over which a study is conducted is of fundamental importance. It does not choose itself; that is, the time period is not necessarily the technical life of whatever investment is being considered. For example, an apartment house may have a technical life, a physical lifetime, of 50 years; yet the true horizon of the study may be only 5 years, it being envisioned that the apartment house will be sold at the end of that time.

A time horizon chosen arbitrarily for study purposes may be changed when, for example, the analysis reveals that an optimal period in which the investment should be held exists. An investment in petroleum-based bonds sold in 3 years, at a time when oil prices are predicted to be low, will appear to be a poor one. When the study period is extended to 5 years, at which time oil prices are expected to be high, the investment appears to be an excellent one. And so it is with many different kinds of investments.

It must never be forgotten in selecting a time horizon that it is not immutable. The fact that you have chosen to do a study of investment over a time horizon does not imply that the study should not be redone based on a different time horizon. For example, new electric motors for a chemical plant were studied over a time horizon of 10 years. The motors were bought and installed on the basis of this study. Six months later, in the light of new information, the study was performed again; the recently installed motors were removed and sold, and more economically efficient ones were installed. In times of swift technological change, the existence of study periods does not imply that our hands are tied until the study period has passed.

An important principle in the subject we are reviewing is that of *equal service lives.* It states that any alternative selected from a set of alternatives must provide service for the whole of the time horizon selected. It would be utterly illogical to select a time horizon of 10 years over which to compare alternatives and include in the set of alternatives one which would supply the service for only 5 years. In short, all alternatives must have the same service lives.

In later chapters we will see what assumptions must be made in order to observe the law of equal service lives where lives are different among mutually exclusive alternatives. For example, the difficulty of the previous paragraph might be resolved by assuming that the 5-year life alternative is repeated once under the same conditions as its first cycle—not that we really intend to do so, but that we observe the logical principle of equal service lives.

DIFFERENTIAL CONSEQUENCES

How are decisions among alternative courses of action really made? Is it necessary in making a decision between two mutually exclusive alternatives to include elements that are common to both? The answer is no. We need include only those elements of each course of action that are different between them. For example, take two lathes, makes *A* and *B*, whose annual costs we are examining with a view to choosing between them. If we compare total annual costs, we have

	Make A	Make B
Operation	$30,000	$50,000
Labor	20,000	20,000
Maintenance	6,000	3,000
Total	$56,000	$73,000

We will choose *A* because it is cheaper than *B* by $17,000.

If we compare only differential costs, excluding labor, we have

	Make A	Make B
Operation	$30,000	$50,000
Maintenance	6,000	3,000
Total	$36,000	$53,000
	$53,000 – $36,000 = $17,000	

We will choose *A* because its differential cost is $17,000 cheaper than *B*. The answers arrived at by the two methods are the same. In addition, we notice that even when we compared the total annual costs of the two machines, we did so by calculating the difference between them. In more complicated problems, application of the *differential-consequences principle*—only the differences between alternatives are relevant to a decision among them—may simplify our task enormously because we may not need to analyze and cost many nondifferential items. They simply disappear from the problem.

WITH/WITHOUT CRITERION

The *with/without criterion* states that the analyst must compare what will happen to the situation *with* the new investment to what will happen *without* the investment, both over a specific time horizon. Compare this correct criterion to one often used instead, the before-and-after criterion, in the analysis of a new water tank installed above a small Mexican village to purify the drinking water. The *with* portion of the with/without criterion predicted the cost of the installation and the benefit, defined as the reduction in illness attributable to pure water, over the subsequent 10 years. The *without* portion of the criterion predicted the steady increase in illness because of impure water and the increasing village population over the next 10 years. The difference between these two situations was the effect of the construction of the new tank. The before-and-after technique compared the situation before the construction of the new tank, say 50 illnesses per year, to the situation after the new tank's installation, say no illnesses per year, a benefit of 50 illnesses per year avoided, without taking into account the increased illness as the population of the town increased, thus underestimating the benefits of the new water tank.

The with/without criterion is indispensable if clear and logical reasoning in project analysis are to be maintained.

A COMMON UNIT OF MEASUREMENT

Quite often in the analysis of projects, the consequences of the alternatives being considered will not be measured in the same units. For example, a new product line for a company may have consequences on the benefit side of (1) increased profits, measured in money units such as dollars or pesos; (2) increased company size, with a resulting increase in ease of financing and in product recognition, measured in increased volume of sales expressible in money units; and (3) increased company prestige, with no obvious way of measuring it, but with certain definite indisputable benefits, such as increased ease in obtaining qualified personnel. On the cost side, the consequences are measurable in money units. Thus, some consequences are measurable in a common unit, money, but at least one other either is not measurable at all or is quantifiable only in terms of some unit to be contrived, such as the percentage increase in the recognition of this company as a desirable employer among, let us say, chemical engineers. It is evident that some way must be found to compare effects appearing in dissimilar units of measurement. In other words, the effects must be made commensurable.

The unit most used to make consequences of alternatives commensurable is money itself. The practice among economic analysts has been to reduce all consequences, or as many of them as possible, to money units which are, of course, summable algebraically. Those consequences not reducible to a money amount, called *irreducible* or *incommensurable*, have been simply listed as a benefit or cost item and thus allowed to influence a decision to the degree that the decision maker believes appropriate. In the example of the previous paragraph, all consequences appeared in terms of money units except one—company recognition by potential employees. This incommensurable consequence was listed as a benefit.

Recently, a way of putting together consequences not expressible in a common unit has been developed. It is called *multiple-goal* or *multiple-objective analysis*. We will limit ourselves, however, to the traditional method of handling consequences of alternatives, which is to reduce them to money terms. Those consequences not amenable to such reduction will be listed.

Irreducible consequences are far more likely to appear in public sector economic analysis than in private sector situations, because externalities are more likely to occur in public sector projects. A highway may have benefits that will never be evaluated in the marketplace, for example, the benefit in time saved by a new and shorter route. A dam project can generate recreational benefits as well as power, with the difference between the two types of benefits residing in the former being frequently a nonmarket item and the latter a market one.

SUNK COST

The concept of sunk cost is a subtle one, although it does not seem so at first glance. *Sunk cost* refers to a cost that has been incurred in the past; it is money or resources already spent. The sunk cost rule is that we must never allow a sunk cost to influence a decision. By definition, sunk costs belong to the past, which is unchangeable. Nothing can be done about it. It must be accepted. All unarguable statements, the reader will admit. But how is this idea related to economic analysis?

Frequently we hear the statement made, "We cannot sell this bond or that stock or that machine until we get our money out of it." Imagine that you have invested in petroleum-based bonds in an oil-rich country. The price of oil drops and with it the market value of the bonds. Is the decision with respect to the bonds limited to keeping them until they regain their former value? Clearly it is not. And not only should you not limit yourself to this course of action—it should not even be considered!

The fact is that you are allowing yourself to be limited by an action taken in the past—buying the bonds at a certain price. Nothing you do can change that; it is nondifferential among the alternatives to be considered. And what are these alternatives? They are

1. Sell the bonds at whatever you can get for them now, and invest the proceeds in something else.
2. Keep the bonds.

Both these alternatives have nothing to do with the past, in the before-tax case. They both look only to the future. If you estimate that the bonds will perform better in the future than their present market value invested in the "something else" of alternative 1, then you will hold onto them. If not, you will invest in "something else."

Notice that the bond's *value in the marketplace now* is the only value mentioned in the preceding analysis. This is the true and only value of the bonds. The original cost—the sunk cost—does not appear.

Finally, the reader is warned that the above discussion refers to analysis before income taxes are taken into account. Sunk costs do have future effects—and this is the exception to the rule—via capital gains and losses, where income tax is involved. All this will be considered fully in Chap. 10.

DEPRECIATION

Depreciation is the cost of a useful asset allocated over its estimated life. For example, you buy a residential house for rental purposes. The tax accountant who prepares your income tax return every year takes the cost of the house, not including the cost of the land it sits on, which is not depreciable, and spreads that cost over the shortest period thought to be acceptable to the Internal Revenue Service. If the cost of the house is $150,000, the allowable life is 15 years, and pre-1981 straight-line depreciation is used, the allowable depreciation will be $150,000 divided by 15, or $10,000 per year. This amount may be used as an expense of conducting a rental housing business and is therefore deducted from your taxable income. The effect is to reduce your taxes in a manner to be explained in Chap. 10.

Will your rental house actually lose value, as represented by the $10,000 per year? Perhaps yes, perhaps no. If recent experience in most areas of the United States is any guide, it probably will not. Thus depreciation of an asset—the rental house or anything else you may acquire for use in trade or business—has no necessary relation to market value. It is used only to compute income taxes.

It is too soon to go into the intricacies of computing after-tax cash flow by using various methods of depreciation and types of income tax, because all these matters will be dealt with after the fundamentals of economic analysis have been explained. But one point must be well understood: Depreciation is used only to compute income taxes when economic analysis of alternatives is the issue. (Accountants use it to compute profit or loss, balance sheets, and so on.) It has no other purpose. It may or may not represent loss in market value. The estimated life may or may not be realistic. The rate at which value is lost, if it is, may or may not be accurate. As analysts, we do not care.

But is not depreciation a cost, especially if the Internal Revenue Service recognizes it as such? It is not. It is simply the reflection of a sunk cost. In the $150,000 house example, that amount of money is sunk as soon as you buy the house. The $10,000 per year depreciation is only a reflection of that cost, spread over the depreciable life of the house. The Internal Revenue Service merely follows the legislation that has laid down the laws about how depreciation can be determined. As a reflection of a sunk cost, it must be disregarded in an economic analysis concerned with cash flow before tax effects are calculated. As economists, that is to say, as practical deciders, we will ignore it for the same reasons that we ignore the sunk cost.

To further reinforce the point, we can ask ourselves, Is depreciation a cash flow? The answer is no. It is only a number in an accountant's books. It does not represent cash outflow from the entity we are considering as, for example, do wages, material costs, rent, and so on. These reduce our bank account when checks are written to cover them. Depreciation does not. Therefore it must not be considered in a before-tax analysis.

In public sector projects, the government agency undertaking them normally pays no income taxes. Thus depreciation is disregarded because its only function—to enter into income tax calculations—is nonexistent.

INCREMENTAL ANALYSIS

Incremental analysis means that decisions about investment among competing, mutually exclusive alternatives will be made by comparing the *extra* cost of an alternative over another to the *extra* benefit to be received. *Incremental* means the same thing as *extra* or *marginal*, a word an economist would use. Some readers will remember the standard rule of classical microeconomics: *Increase production until the marginal cost of one more unit of whatever it is we are producing is just equal to its marginal revenue.* Incremental analysis is simply the application of that rule to engineering economy.

An example may serve to clarify the above definition. Imagine that a major real estate developer has requested a firm of engineers and architects to design a skyscraper. A question that must be answered is, How many floors will there be? Among the feasible heights of building from an engineering standpoint there will exist only one that will be optimal from an economic standpoint. Say that 100, 110, and 120 floors are the three designs being considered. Incremental analysis demands that the cost of going from 100 to 110 floors be more than balanced by the revenue to be gained from the extra floors. And going from 110 to 120 floors will be judged in the same way: The incremental benefits and costs will be balanced against each other and the decision made on whether the revenues are more than the costs. In other words, it is the extras that count in such a decision. On the other hand, if incremental analysis is not used, an error may be made. An individual analysis of each mutually exclusive alternative without regard to its relationship to its fellows is erroneous. We will see in later chapters many illustrations of the truth of this statement. Even when certain methods of analysis appear to deny it, we will see that these methods, too, depend on incremental analysis, although not in such an obvious way as some others. Remember, it's the extra that counts.

ANALYSIS: THE ACCOUNTANT'S AND THE MANAGER'S

Important differences exist between the professional purpose and methods of the accountant and those of the economist. These are reflected in not only the amounts of money assigned to an item in an analysis but also the inclusion of the item at all. Both may be perfectly correct in their handling of an analysis. The differences arise because of differing goals.

The accountant is responsible for producing material that will describe the condition of a company, an agency of government, or an individual. A balance sheet, a profit and loss statement, a statement of the sources and uses of funds—all these represent a financial condition. This representation is the fruit of accounting methods that are conventional; that is, the methods are arrived at by agreement among accountants as to what shall be considered good accounting practice. Whether or not these practices agree with those of any other group is immaterial.

The economist-manager, decision maker, executive, or any other title you may wish to assign, has a purpose in his or her analysis distinct from that of the accountant in that she or he wishes to arrive at a decision. This person must be guided by logic, usually that of the marketplace, where a maximum of profit is desired, not by conventions established by agreement. Thus differences between his or her view of a situation and that of a colleague in the accounting department are inevitable.

The economist will ignore sunk cost in a before-tax analysis. The accountant will include it in the cost of an asset. The economist will exclude depreciation as a cost. The accountant will include it. The economist will exclude costs allocated according to an accountant's formula. Allocated costs are those that cannot be assigned to any specific department in a company. Rent for a building housing several departments must be assigned by the accountant on the basis of the same formula, such as "The proportion of square footage occupied by each department will be used to assign the same proportion of the rent as a cost to that department." But rent is a nondifferential expense which must be paid no matter what decisions are made with respect to each department, even its termination.

Overhead expenses, which the accountant will allocate among departments in a company, the economist will ordinarily exclude. Overhead expenses are like allocated costs such as rent, because they cannot be assigned to any particular department in a company except on some arbitrary basis, such as labor costs or some other measure. They may or may not be differential, depending on how strictly a real link between department and overhead can be assigned. If no link can be established, as is ordinarily the case, overhead expenses will be excluded by the economist. Take the salary of the president of the company. There is no reason for it to enter individual departmental economic decisions. The same is true for the president's secretary's salary and the salary of many others who work and use space and supplies for the company as a whole. However, because these people and what they use must be accounted for, the accountant distributes their costs to centers—cost centers—in the company. The economist ignores such overhead costs as being nondifferential.

To sum up the matter of overhead, allocated, and similar costs: The accountant applies costs to cost centers. The economist applies costs to a decision.

The accountant disregards the value of money over time; that is, $10,000 received today and $10,000 to be received 1 year from now are viewed as the same amount. The economist recognizes that money has two values: its face value and the time it will be received or spent. (Just how the economist does this is the subject of the next few chapters.) This single difference results in major divergences in their calculations in respect to an economic analysis.

Finally, we emphasize once again that both the accountant and the economist are behaving correctly according to their viewpoints. Neither is at fault because they differ from one another.

PROBLEMS

2.1. You want to compare two possible places for spending your 3 months of leave. You prepare the following table, showing your estimate of what you and your family will spend while renting a house in Mexico or passing the time at your home in the United States. The amounts are in dollars for the 3 month period.

	Mexico	United States
Rent	$2,400	$ 0
Utilities	100	300
Travel	1,200	0
Food	750	1,500
Entertainment	300	600
Clothes	500	500
Dentist	80	240
Transportation	200	400

While in Mexico, you can rent your house in the United States for $1,500 per month. Your tenants will pay the utilities. Your mortgage payments of $1,800 per month you will continue to pay.

On the basis of all costs, neglecting the pleasures of staying in either place, what will be the difference in economic cost between staying in the United States and passing your vacation in Mexico?

Answer: $2,510 in favor of Mexico

2.2. In the table shown on page 14 at the beginning of this chapter, which of the three alternatives would be chosen by using a noncontradictory criterion such as "Maximize the difference between benefits and costs," given that the difference between benefits and costs is the profit associated with an alternative?

2.3. A correspondent in a letter to *Civil Engineering* stated, "The purpose of value engineering is to ensure that maximum value is yielded for minimum investment." Is this criterion a practical guide for choice among alternatives? *Illustrate your comments by means of a numerical example.*

2.4. Describe the five steps in making an economic decision. Imagine a situation in which an economic decision is necessary, and explain the application of the five steps to it.

2.5. It has been said that neglect of a possible alternative in the solution of a problem is a serious error. Imagine a problem from your own background, note down the alternatives available to solve it, and under each alternative write out the consequences of adopting it in terms of benefits and costs over a time horizon. Include all alternatives, even the unlikely ones.

2.6. You have an opportunity to purchase a part interest in a small firm for $3,000. You will be a silent partner, doing no work and collecting profits at the rate of $850 per year for as long as you wish to do so. Should you join the firm in this way, you intend to pass on the rights to your children. In other words, the profits may be considered to be perpetual. A friend to whom you have told your good fortune offers you $2,000 now to turn over your opportunity to him. Your wife, who has been reading your engineering economy book over your shoulder, recommends that you take this offer immediately since it represents an infinite rate of return on your investment to date (which is zero). Your personal minimum attractive rate of return, which you can receive any time for any amount, is 20 percent. Should you follow her advice? Why or why not?
Answer: Accept the $2,000.

2.7. Why is time important in an economic analysis? Can we not make our decision on the basis of first cost only? Budget-conscious analysts use the method of least first cost frequently and no doubt will continue to do so.

2.8. If you were offered $5,000 now or $5,000 in 1 year from now, which would you take? Why?

2.9. You notice a statement in a textbook on systems analysis that discounting of future consequences must be undertaken solely because of the existence of inflation and, presumably, for no other reason. How would you question this statement? Or is it true?

2.10. What is meant by *viewpoint* in an economic analysis? Give an example to illustrate your answer. Be brief.

2.11. The statement is made that "In fact, analysis from one viewpoint may indicate approval, while analysis from another may point to rejection." How is this possible? Make up an example to illustrate this important point.

2.12. What is meant by *systems analysis*?

2.13. What is meant by *suboptimization*? Is this always an error in systems analysis? Do you believe it is always possible to study a system as a whole? Or may other considerations enter to prevent such a study?

2.14. Imagine a problem in systems analysis from your own background, similar to the example offered in the text. Explain particularly how the parts are interrelated.

2.15. *Equal service lives* is a term of some importance in economic analysis. What is meant by it?

2.16. Give an example where alternatives available to solve an engineering problem have different lives. Why is it necessary to equalize the service lives of the alternatives when you are performing the economic analysis?

2.17. Is there a difference between the time horizon over which a project is to be considered and the physical life of the assets involved? Explain, using an example.

2.18. It has been stated that "Only the differences between alternatives are relevant to a decision among them." Explain and present a numerical example of your own.

2.19. Imagine a situation in which the with/without criterion has been violated. Describe the error and how it should be corrected.

2.20. "Some time ago, in an Asian country, a new high-speed freeway between two cities of over 2 million persons each was being analyzed. It was decided that the benefits exceeded the costs by no great margin.

"The method used by the analyst was to compare the stream of costs and benefits of the new freeway over the next 20 years with the costs and benefits of the present road, which had become congested. Specifically, he took the net present value of the stream of benefits from the new road and compared that amount with the annual cost of the present road *at the present time* multiplied by 20. In doing this, he violated the with-without criterion."[1] How so?

---

[1]H. M. Steiner, *Public and Private Investments: Socioeconomic Analysis*, Wiley, New York, 1980, p. 14.

2.21. *Commensurable* is a word that has a special meaning in engineering economy studies. What is this meaning? Discuss its implications.

2.22. Define *sunk cost*. Why is it important in engineering economic analysis?

2.23. A young engineer was asked by her superior in a consulting civil engineering company to put a price on the existing privately owned water system of a small northern California city. The system had been losing money for some years.

She complied by computing the original cost of each part of the system, estimating the remaining life of each part, and computing a *residual value* for all the parts. Summing up the residual values, she called the results *fair market value*.

Criticize her methods, taking into account the stable prices that had held for some time. Ignore any income tax effects for this first estimation.

2.24. Five years ago Stan Jefferson bought a residential lot as an investment. He paid \$45,000 for it. Two years ago he discovered that an urban freeway would abut his land. The freeway is now under construction. The prospective noise, air pollution, night glare, and appearance of the freeway have reduced the market value of his investment by as much as half, according to several real estate appraisers he has consulted. Jefferson insists on "getting his money out of the property" before considering other investments.

Criticize Jefferson's attitude. (Do not take into account income tax effects or prospective inflation.)

2.25. Your company has spent \$700,000 in developing a new electric toothbrush that will result in a total profit (total revenue minus total costs) to the company of \$1,100,000 over the life of the venture. (The \$700,000 has already been included in the total cost.) You are absolutely certain that the expenditure of \$600,000 more, that is, over and above the \$700,000 already spent, will ensure the predicted total income. On the other hand, if the extra money is not invested, you are equally sure that the venture will have to be abandoned. Should you recommend the expenditure of \$600,000? Defend your decision.

Answer: Yes

2.26. What is meant by *depreciation* as the word is used in this text?

2.27. Does depreciation represent the loss in market value of an asset? Briefly support your opinion.

2.28. A friend of yours, who is an accountant, maintains that depreciation is clearly a cost because

(*a*) It is universally accepted as such on all profit and loss statements.

(*b*) One of the most common errors of naive businesspeople is to overlook the cost of capital assets and take into account only operating costs, neglecting depreciation.

(*c*) The Internal Revenue Service of the United States government recognizes depreciation as a legitimate cost.

Comment on the above.

2.29. What is meant by *incremental analysis*? Make up an example similar to the skyscraper example in the text.

2.30. "An individual analysis of each mutually exclusive alternative without regard to its relationship to its fellows is erroneous." Amplify the meaning of this statement from the text. Create an example, illustrating how neglect of this dictum results in a mistaken decision.

2.31. Explain in your own words the differences between an accountant's responsibilities and a manager's.

2.32. Give examples of specific items which the accountant and the manager will handle differently.

2.33. A construction site building, used for workshops, storage, and equipment cover, is to be sold. An offer of $75,000 has been made to your company.

The building was purchased 2 years ago for $100,000. It now has a book value (first cost less accumulated depreciation) of $85,000.

A replacement building can now be bought for $70,000. It has exactly the same benefits, maintenance cost, remaining life, and salvage value as the old building.

Your accountant urges you not to sell the building. She maintains that you will lose $10,000 on the sale of the old building ($85,000 less $75,000), and this means that you will gain $65,000 ($75,000 less $10,000) and pay out $70,000, thus losing $5,000.

What decision should you make and why? Assume that tax effects are nondifferential. Assume also that *ceteris paribus* rules.

2.34. Define (*a*) opportunity cost, (*b*) allocated cost, (*c*) sunk cost, (*d*) time value of money, (*e*) differential cost, and (*f*) nondifferential cost.

CHAPTER 3

EQUIVALENCE

The concept of the time value of resources was introduced in Chap. 2 along with the definition of simple interest and discounting. Now it is time to set forth an equally important concept, that of equivalence. *Equivalence* refers to the equality of different sums considered at different times. Recall that money has two kinds of value—its quantity and the time when it is received. The further in the future that money will be received, the less is its value in the present; this relationship is expressed mathematically by means of the interest rate. In Chap. 2 we explained that $9,091 now is the same as, or equivalent to, $10,000 in 1 year from now, if interest is 10 percent per year, compounded annually. In other words, the application of the interest rate to a sum at one time converts it to an equivalent sum at another time. This relationship is true not only for single payments but also for a series of payments or, indeed, any pair of cash flow patterns, as the following examples illustrate.

EQUIVALENCE AT 0 PERCENT INTEREST RATE

In Table 3.1, interest is 0 percent, meaning that money has no time value. Now, at time 0, $10,000 is the equivalent of $2,000 per year for 5 years. Column 1 of the table is equivalent to column 2.

TABLE 3.1
Equivalence at 0 percent

Year	1	2
0	+10,000	
1		+2,000
2		+2,000
3		+2,000
4		+2,000
5		+2,000

EQUIVALENCE AT 10 PERCENT INTEREST RATE

In Table 3.2, the interest rate is 10 percent. An amount of $10,000 at year 0 is equivalent to $11,000 at year one, arrived at by adding the original $10,000 to $1,000 worth of interest. By the same token, $11,000 at year 1 is equivalent to $10,000 at year 0, with interest at 10 percent. If we continue to add interest, $11,000 at year 2 will have to be

$$11{,}000 + 11{,}000(0.10) = 12{,}100$$

at year 3

$$12{,}100 + 12{,}100(0.10) = 13{,}310$$

at year 4

$$13{,}310 + 13{,}310(0.10) = 14{,}641$$

and at year 5

$$14{,}641 + 14{,}641(0.10) = 16{,}105$$

This last amount appears in column 3. We could have said that $10,000 at year 0 is equivalent to $11,000 at year 1, or $12,100 at year 2, or $13,310 at year 3, or $14,641 at year 4, or $16,105 at year 5.

We arrived at the 16,105 figure by charging interest on the interest received. This process is called *compounding*. Thus, *compound interest* is paid on the original principal as well as the interest added to it.

It is also possible for a single amount to be equivalent to a series of amounts. For example, an investor would be satisfied, providing his 10 percent interest rate for satisfaction deferred holds good, if he replaced his $10,000 at year 0 with only the interest of $1,000 at years 1, 2, 3, 4, and 5 and deferred his $10,000 until year 5. These amounts are displayed in column 4 of Table 3.2. The series in column 4 is equivalent to the single amounts of columns 1, 2, and 3, and vice versa.

Column 5 shows yet another series. Suppose we took back $10,000 in five equal amounts of $2,000 plus the amount of interest on the amount deferred. This would also satisfy us, providing the 10 percent interest is acceptable to us. Thus $2,000 at year 1 plus the interest on $10,000 of $1,000 is $3,000. Now we have deferred $10,000 minus $2,000, or $8,000. Interest on it is $800, and we thus take $2,800, and so on. The series shown in column 5 is equivalent to the single amounts and series of amounts shown in all the other columns.

Column 6 is more difficult to explain. Evidently, if the equivalence of $10,000 to five yearly payments of $2,638 is true, then $2,638 must be composed of the interest on the amount deferred and part of the amount itself.

TABLE 3.2
Equivalence at 10 percent

Year	1	2	3	4	5	6
0	+10,000					
1		+11,000		+1,000	+3,000	+2,638
2				+1,000	+2,800	+2,638
3				+1,000	+2,600	+2,638
4				+1,000	+2,400	+2,638
5			+16,105	+11,000	+2,200	+2,638

For example, the first $2,638 must be made up of $1,000 in interest (10 percent of $10,000) and $1,638 in that part of the $10,000 we are going to take back, leaving $10,000 minus $1,638, or $8,362, still paying interest for the next year. The reader is invited to continue the calculation through the fifth year. How was the $2,638 figure determined in the first place? That explanation will be held off until Chap. 4.

It is also possible to view the year 5 as though it were the year 0 and ask, To have $16,105 now, how much would I have had to give up 5 years ago? The answer is, at 10 percent, $10,000. Or one could ask, To have $16,105 now, how much would I have had to give up 4 years ago? The answer is $11,000. Or one could ask, What equal sum for the past 5 years is equivalent to $16,105 now? The answer is $2,638. To sum it all up: Any column of Table 3.2 is equivalent to any other column at a given interest rate of 10 percent.

In back of all that has been said thus far is the idea that amounts different in quantity can be equivalent in value depending on the time at which they appear and the interest rate. The same $10,000 at time 0 is equivalent to any sum or series of sums in the subsequent columns at 10 percent interest. Equivalent value is a function of quantity and time; therefore, given an interest rate:

$$\text{Equivalence at an interest rate} = f(\text{quantity, time})$$

The same sum can be moved back and forth over time or converted to a series of one sort or another, providing we are given an interest rate to use in making the shifts. The sum or sums always have the same time value as the principal. Or, to put it another way, we can say that patterns of a cash flow are equivalent at i percent if their values at year 0 (the principal) are equal.

Exactly how this can be formulated mathematically is the subject of the next chapter.

The question of alternative financing schemes and how to compare them will be delayed until a later chapter. It is a completely separate subject from the one treated in this chapter on equivalence.

PROBLEMS

3.1. Define *equivalence*. Is it possible to have equivalence at more than one interest rate?

3.2. If you do not care whether you have $1,000 now or $1,000 a year from now, your interest rate is 0 percent. What equal sum during each of 10 years would be equivalent to $10,000 now?
Answer: $1,000

3.3. If your willingness to give up present consumption in order to have more future consumption is measured by an interest rate of 15 percent, what would an equivalent amount 3 years from now be to $8,000 now?
Answer: $12,167

3.4. If you possessed $6,000 now, and your interest rate were 7 percent, and you wished to trade off the $6,000 now for 6 years, receiving only the interest during those years and a single lump sum at the end of them, what would the equivalent amounts be for each of the 6 years?

3.5. If you had $10,000 now, what amount would be equivalent to it, 1 year ago, if your interest rate were 5 percent?

3.6. In column 6 of Table 3.2, show how much interest would be charged each year at 10 percent and how much of the $10,000 capital was returned each year.

3.7. Prepare a table similar to Table 3.2 (columns 1 and 5) showing an amount at year 0 of $15,000 in column 1. With interest at 6 percent, what equivalent amounts will appear in column 5, if interest and a capital payment of $3,000 per year are summed?

3.8. Prepare a table similar to Table 3.2 (columns 1, 2, and 3) showing an amount at year 0 of $15,000 in column 1. With interest at 6 percent, what will be the equivalent amount at the end of year 1 in column 2? At the end of year 5 in column 5?

3.9. Prepare a table similar to Table 3.2 (columns 1 and 4) showing an amount at year 0 of $15,000 in column 1. What will be the equivalent amounts in column 4 with interest at 6 percent and interest only appearing during the 5 years? The principal will be paid at the end of the fifth year.

3.10. Prepare a table similar to Table 3.2 (columns 1 through 5) showing an amount at year 0 of $25,000 in column 1. With interest at 8 percent, what equivalent amounts will appear in column 2 at year 1, column 3 at year 5, and in columns 4 and 5 for 5 years?

3.11. Prepare a table similar to Table 3.2, but change only the interest rate to 12 percent while keeping the $10,000 figure and the same number of years. What equivalent amounts will appear in columns 1 through 5?

3.12. Prepare a table similar to Table 3.2 (columns 1 through 5) showing $10,000 in column 1 at year 0. Show equivalent patterns of the cash flow in columns 1 through 5, with interest at 10 percent and the number of years increased to 10. In column 2, show the equivalent sum at 3 years; in column 3, the equivalent sum at 10 years; and in column 5, the principal paid off at $1,000 per year in addition to the interest payments.

3.13. Prepare a table similar to Table 3.2 (columns 1 through 5) showing an amount of $7,000 at year 0 in column 1. With interest at 9 percent, what equivalent sums will appear in columns 1 through 5 if the years are increased to 8? Show the equivalent amount at 3 years in column 2; in column 3 at 8 years; and in column 5, the principal paid off over 8 years in equal payments with interest.

3.14. One definition of equivalence states, "Two cash flows with completely different patterns are said to be equivalent at *k* percent interest if the single lump-sum equivalences of each at *k* percent at the same point in time are equal." Show that this definition is applicable to Table 3.2.

CHAPTER 4

INTEREST AND FINANCIAL MATHEMATICS

This chapter covers fundamental relationships concerning interest and, arising out of the concept of interest, mathematical formulas whose understanding will require of the reader no more than a knowledge of basic algebra. As explained in Chap. 2, money actually has two dimensions: one related to its face value and another, no less important than the first, related to the time when the money is received. (The term *money* is used here. The more general term is *resources*.) From this fact about money mathematical formulas will be derived. This chapter will explain gradients in cash flows, their use, and their solution. The question of nominal and effective rates of interest and a number of problems that arise out of these two definitions of *interest* will be dealt with next. Finally, the separation of interest and principal, and equations which make this matter easy, will be derived.

The above list of topics serves as a foundation for later chapters on the use of such formulas, conventions, and concepts in a more practical way. This text will deal with the standard methods of economic analysis of projects. The methods described are present worth (also called *net present value*, *present cost*, or *present benefit*), annual worth with its special case of annual cost, the benefit/cost ratio, and the internal rate of return.

The reader should be careful to understand this chapter completely. All succeeding chapters will be based upon it.

Before beginning, the reader must be aware of a convention and three assumptions. The assumptions are made so that the subject can be dealt with at this stage in a simple and understandable way. Later they will be removed.

In Chap. 2, the concept of alternatives being judged by their consequences was introduced. A widely used stipulation in regard to the treatment of those consequences is the *end-of-year convention.* It means that all consequences during the period with which we are concerned are accumulated to the end of that period. More generally it could be called the *end-of-period convention.* For example, suppose we are attempting to gather data for use in an economic analysis on the maintenance costs of a certain type of motor grader which is standard in a large construction company. The service and accounting records show costs per week for the maintenance of this type of machine. The end-of-year convention allows us to sum up all the maintenance costs for the motor grader to the end of the year. This may appear to be an imprecise way of handling the maintenance costs. The fact is, however, that for an economic analysis which uses the maintenance costs in the future for this type of motor grader, the accumulation of the costs to the end of the year is accurate enough, considering all the other inaccuracies which are sure to be present when we are trying to foresee the future. The end-of-year convention provides sufficient accuracy in the calculations connected with economic analysis for the great majority of problems encountered.

The reader may ask, Why not use the beginning-of-year convention? The answer is that we could well do so. However, the convention that has been adopted in almost all writings on the subject is the end-of-year convention. A simple example of how interest per period, by the end-of-year convention, is calculated in the single-period case will help explain the matter.

■ **Example 4.1.** A small aeronautical engineering company, Bethesda Aviation, discovered the wreckage of a classic airplane in the barn of a local farmer. It was a Fokker D7 from World War I. Bethesda Aviation paid $600 for the wreckage and, at many dates during the same year, spent a total of $5,460 on its repair. Hangar fees paid 1 month after the purchase of the D7 came to $750 for that year. After 1 year Bethesda Aviation sold the rebuilt Fokker D7 to an aeroexhibition company for $30,000. What rate of return did Bethesda Aviation make on its investment in the Fokker?

Solution. After the costs of renovating the Fokker D7 are tallied, they must be subtracted from the revenue received on the sale of the airplane. The result is then divided by the investment and multiplied by 100. The calculation is shown below.

First cost	600	(at beginning of year)
Repairs	5,460	
Hangar fees	750	
	6,810	(at end of year)

$$\text{ROR} = \frac{30{,}000 - 6{,}810}{600} \times 100 = 3{,}865\%$$

The annual rate of return (ROR) on the investment is 3,865 percent. If Bethesda Aviation had made the same deal any sooner or later than 1 year, the annual rate of return would have been different. This example shows how to compute a rate of return based on an investment that lasts only 1 year, by using the end-of-year convention. ■

What about inflation? We will not consider inflation until later in this book. This is not to say that inflation is not an important matter. It is. It is at least as important as the time value of money. But the fact is that the subject we are dealing with is much more easily understood when its components are dealt with consecutively rather than all at once. Therefore, any discussion of inflation will be postponed. This assumption of constant prices will be dropped as soon as we reach the place in the text where inflation is considered.

The second assumption that we make is that all our predictions are certain to occur. We will assume away risk. It will not be allowed to reenter the scene until it is covered in Chap. 12.

Income taxes will also be excluded from discussion until Chap. 10. The third assumption assumes them away. The existence of income tax in the United States, and in most other countries, is an important factor in the economic analysis of projects. We will ignore it until we are prepared with enough elementary knowledge to enable us to understand how its effects can be handled.

MONEY HAS A DOUBLE VALUE

In a previous chapter, one section was devoted to a discussion of consequences and their relation to time. It was said that we need some way of making events that occurred at different times commensurable, that is, able to be measured in the same units. This was to be accomplished by a method which we will now examine. Before beginning, readers will be helped if they refer to App. B, Notation, in the back of this book. The symbols follow the convention established by the Institute of Industrial Engineering, in *Industrial Engineering Terminology*. Some of the symbols are unique to this book because they have not yet been established. The symbols presented in App. B will be used throughout this book.

Simple interest means that at the end of each period a sum will be paid equal to the principal with which the account started multiplied by the interest rate on the account. Table 4.1 shows how this works. Notice that the interest payment does not vary by period. It is the same for the four periods under consideration. The principal grows from $10,000 to $18,000. *Compound interest* differs from simple interest because interest is paid on the sum of the principal plus the interest added to the principal during the previous period. (We saw this in Chap. 3.) This is shown in Table 4.2. The interest payment at the end of the first year of $2,000 is obtained by multiplying $10,000 by 20 percent. The interest payment at the end of the second year, or $2,400, is obtained by multiplying $12,000 by 20 percent.

TABLE 4.1
Simple interest at 20 percent

N	Interest payment	F_N
0	—	10,000
1	2,000	12,000
2	2,000	14,000
3	2,000	16,000
4	2,000	18,000

The result is then added to \$12,000 accumulated in the account at the end of period 1, giving \$14,400 in the account at the end of period 2. At the end of period 4 the account contains \$20,736. Contrast this to \$18,000 in the account at the end of the fourth period, as shown in the simple-interest case. Also notice that at 20 percent the amount of money invested more than doubles in four periods. This illustrates the astonishing effects of high interest rates.

It is now possible to generalize from the previous two examples so that a formula for simple interest and compound interest and their effects on capital accumulation can be calculated. In the case of simple interest,

$$\begin{aligned} F_4 &= 10{,}000 + (0.20)(10{,}000)\,(4) \\ &= 18{,}000 \end{aligned}$$

In symbols, the previous calculation is

$$F_N = P + iPN \tag{4.1}$$

where = future

F_N = future payment at period N
P = present payment
N = number of periods
i = interest rate

Following the same procedures, we can derive the formula for the accumulation of money at compound interest. It is only necessary to substitute symbols for the money sums of Table 4.2. If this is done, the following equations result:

$$\begin{aligned} F_1 &= P + Pi = P\,(1 + i) \\ F_2 &= P(1 + i) + iP\,(1 + i) = P\,(1 + i)\,(1 + i) = P\,(1 + i)^2 \\ F_3 &= P(1 + i)^2 + iP\,(1 + i)^2 = P\,(1 + i)^2(1 + i) = P\,(1 + i)^3 \\ F_4 &= P(1 + i)^3 + iP\,(1 + i)^3 = P\,(1 + i)^3(1 + i) = P\,(1 + i)^4 \end{aligned}$$

TABLE 4.2
Compounding at 20 percent

N	Interest payment	F_N
0	—	10,000
1	2,000	12,000
2	2,400	14,400
3	2,880	17,280
4	3,456	20,736

The four equations reveal that all we need do to find out how much money is accumulated in an account at the end of a certain period is to multiply the initial principal by $1 + i$ raised to a power that is the same as the number of periods one considers. In symbols this is

$$F_N = P(1+i)^N \tag{4.2}$$

The expression $(1 + i)^N$ is known as the *single-payment compound amount factor*. Rather than write this phrase, it is simpler to use a kind of shorthand to indicate this unique factor at a certain interest rate and for a certain time. This symbol is $(F/P, i, N)$. It is shown at the top of the first column of the tables of App. C, the compound interest tables. It appears for various interest rates and periods.

GRAPHICAL CONVENTIONS

Before we proceed with the examples of application of the formula, some word on the graphs to illustrate the cash flows is necessary. Generally speaking, an upward arrow indicates a cash inflow and a downward arrow an outflow. These occur along a horizontal line which represents time. The numbers on this line represent the period at which the payment, either inflow or outflow, occurs. A reference to the figures that accompany the examples following this explanation will illustrate the convention. Dates at which the payments occur may be included as points on the horizontal line. The cash flow is normally shown as starting at time 0 and proceeding to time N. According to the end-of-period convention, explained at the beginning of this chapter, each arrow represents the accumulation of funds or resources from the beginning to the end of the period in question. For example, an up arrow at the end of period 4 represents the funds or resources that will have entered the entity being considered between the end of period 3 and the end of period 4. Frequently, the periods on the horizontal line are years, but not always, as we shall see in succeeding examples.

■ **Example 4.2.** A father invests \$10,000 now in a fund paying 12 percent annually. He intends to leave the money in the fund, both interest as it accumulates and the principal. In 10 years he will withdraw the money to finance his daughter's education at a university. How much will be in the account at the end of the 10th year? See Fig. 4.1*a*.

Solution. In this problem, $P = 10{,}000$, $i = 12$ percent, and $N = 10$ years. It is solvable by using Eq. (4.2) thus:

$$F_{10} = 10{,}000\,(1 + 0.12)^{10}$$
$$= \$31{,}058$$

However, it is more easily solved by using the compound interest tables of App. C. The page showing the factors for 12 percent is consulted under the first column, entitled "Single-payment compound amount factor F/P." Opposite $N = 10$, the factor 3.1058 is found.

$$\begin{aligned} F_N &= P(F/P, i, N) \\ F_{10} &= 10{,}000(F/P, 12, 10) \\ &= 10{,}000(3.1058) \\ &= \$31{,}058 \end{aligned}$$

Figure 4.1*a* illustrates the graphical conventions outlined above. This example shows the use of the single-payment compound factor. ■

FORMULAS

Five other formulas besides the ones just presented must be derived. The *single-payment present-worth factor* is derived directly from Eq. (4.2). See Fig. 4.1*b*.

$$F_N = P(1+i)^N$$

Divide both sides by $(1 + i)^N$:

$$P = F_N \left(\frac{1}{1+i}\right)^N \tag{4.3}$$

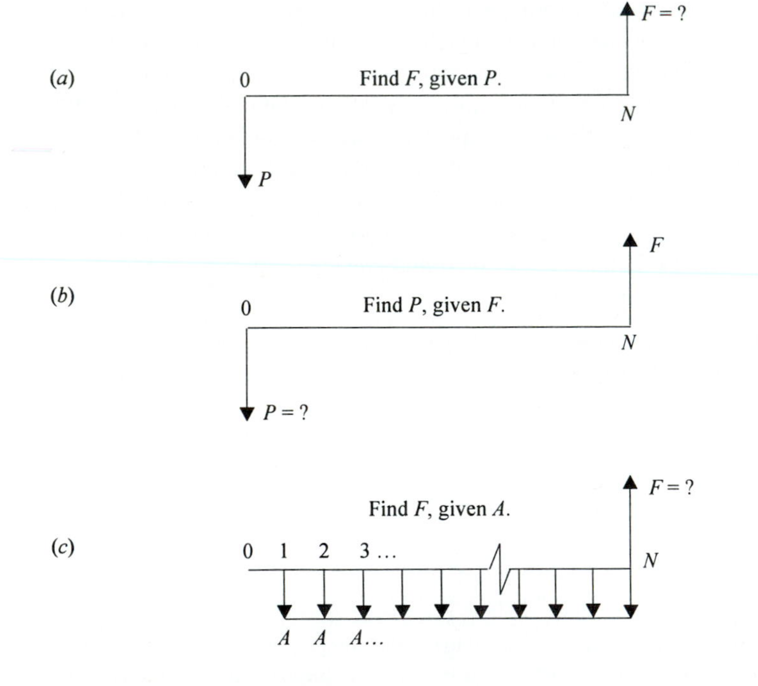

FIGURE 4.1 Typical cash flows

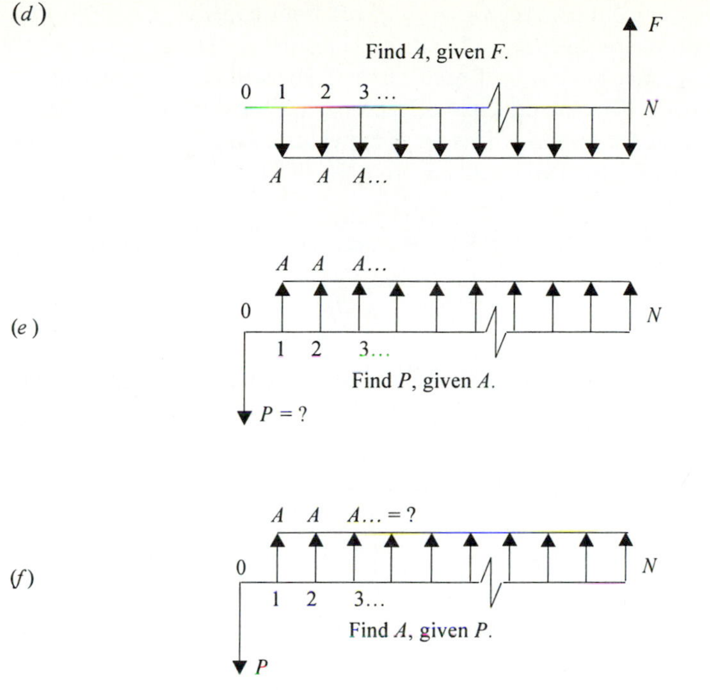

FIGURE 4.1 (continued)

The expression in parentheses is the factor needed to transform any future payment to its present worth. Its symbol is $(P/F, i, N)$, and it is found, with all other factors, in App. C for i from ¼ to 50 percent and varying N's.

■ **Example 4.3.** If you wished to accumulate \$100,000 at the end of 5 years in an account that pays 12 percent annually, how much would you need to deposit now?

Solution. Figure 4.1*b* illustrates the generalized problem, with F = \$100,000, i = 12 percent, and $N = 5$ in this specific case. As in the previous example, the question could be answered by substituting the preceding values into Eq. (4.3). It is perhaps easier, however, to use the compound interest tables of App. C, extracting the single-payment present-worth factor P/F from the column so headed, as follows:

$$\begin{aligned} P &= F(P/F, i, N) \\ &= 100{,}000\,(0.5674) \\ &= \$56{,}740 \end{aligned}$$

This sum deposited now, and left in the account along with its accrued interest, will furnish \$100,000 at the end of 5 years. The interest payments themselves will draw interest. ■

The *series compound amount factor* emerges from applying Eq. (4.2) to a series of future payments. See Fig. 4.1*c*. It measures the amount that a series of equal payments will accumulate to if each remains invested at i percent with no funds being removed. Both the principal of each payment and its periodic interest continue to be invested at i percent. Treated as individual payments into a fund, the amount accumulated F_N from all the P's is, from Eq. (4.2),

$$F_N = P_1(1+i)^{N-1} + P_2(1+i)^{N-2} + P_{N-1}(1+i)^{N-(N-1)} + P_N(1+i)^{N-N}$$
$$= P_1(1+i)^{N-1} + P_2(1+i)^{N-2} + \cdots + P_{N-1}(1+i)^1 + P_N(1+i)^0$$

If $P_1 = P_2 = \cdots = P_{N-1} = P_N = A$, then all the P's in the previous equation are replaced by A's, as follows:

$$F_N = A(1+i)^{N-1} + A(1+i)^{N-2} + \cdots + A(1+i) + A$$

and factoring out A and rewriting give

$$F_N = A\left[1 + (1+i) + \cdots + (1+i)^{N-2} + (1+i)^{N-1}\right] \quad (4.4)$$

If both sides of this equation are multiplied by $1 + i$, the result is

$$(1+i)F_N = A\left[(1+i) + (1+i)^2 + \cdots + (1+i)^{N-1} + (1+i)^N\right] \quad (4.5)$$

Subtracting Eq. (4.4) from Eq. (4.5) gives

$$(1+i)F_N - F_N = -A + A(1+i)^N$$

Factoring both sides yields

$$F_N\left[(1+i) - 1\right] = A\left[(1+i)^N - 1\right]$$
$$iF_N = A\left[(1+i)^N - 1\right]$$
$$F_N = A\left[\frac{(1+i)^N - 1}{i}\right] \quad (4.6)$$

The bracketed expression of Eq. (4.6) is the *series compound amount factor* sought and is symbolized by $(F/A, i, N)$.

■ **Example 4.4.** If you deposit \$2,000 on July 1 for each of the next 15 years in a savings and loan account paying 12 percent annually, how much money will you be able to withdraw on July 1 of the 15th year?

Solution. Substitution of A = \$2,000, N = 15, and i = 0.12 into Eq. (4.6) will furnish the correct answer. Use of the column in App. C entitled "Uniform series compound amount factor F/A" at N = 15 will give the same result.

$$
\begin{aligned}
F_N &= A(F/A, i, N) \\
&= 2{,}000\ (F/A, 12, 15) \\
&= 2{,}000\ (37.280) \\
&= \$74{,}560
\end{aligned}
$$

Figure 4.1*c* applies to this example. Because of the end-of-period convention, the withdrawal and the last payment are made at the same instant. This may seem a curious thing to do, but remember that, to balance matters, the first deposit is made only at the end of the year, not at the beginning. ■

Inverting the series compound amount factor gives the *sinking fund factor*. It is used to answer the question, "How much must be invested at interest rate i for N periods in order to arrive at a sum of F_N?" See Fig. 4.1*d*. Solving Eq. (4.6) for A gives

$$
A = F_N \left[\frac{i}{(1+i)^N - 1} \right] \tag{4.7}
$$

This sinking fund factor lies between the brackets. It is symbolized by $(A/F, i, N)$.

■ **Example 4.5.** You wish to invest in a savings and loan account paying 12 percent interest annually an amount of money that will allow you to draw out \$20,000 at the end of 4 years, in order to finance a trip around the world. How much must you deposit in the account yearly in order to accomplish this?

Solution. Once again, substitution of F = \$20,000, N = 4, and i = 0.12 in Eq. (4.7) is possible. Use of the compound interest tables for F = 20,000, N = 4, and i = 0.12 under the uniform series sinking fund factor A/F gives

$$
\begin{aligned}
A &= F(A/F, i, N) \\
&= 20{,}000\ (A/F, 12, 4) \\
&= 20{,}000\ (0.20923) \\
&= \$4{,}184.60
\end{aligned}
$$

Figure 4.1*d* reveals that for Eq. (4.7) to work, you must deposit the last payment A at the same instant that you withdraw the sum F, as was mentioned in connection with Example 4.4. ■

Imagine now that the requirement is to find the present worth of a series of payments A. See Fig. 4.1*e*.

From Eq. (4.2), $F_N = P(1 + i)^N$. Substitution for F_N in Eq. (4.6) yields

$$P(1+i)^N = A\left[\frac{(1+i)^N - 1}{i}\right]$$

Dividing both sides by $(1 + i)^N$ gives

$$P = A\left[\frac{(1+i)^N - 1}{i(1+i)^N}\right] \tag{4.8}$$

The expression within the brackets of Eq. (4.8) is the *uniform series present-worth factor*. Its symbol is $(P/A, i, N)$.

■ **Example 4.6.** What is the present worth of an annuity of $2,500 per year for the next 10 years if the appropriate interest rate is 12 percent?

Solution. Here A = $2,500, N = 10, and i = 0.12. Use of App. C under the column entitled "Uniform series present-worth factor P/A" opposite N = 10 reveals the factor 5.650. Therefore,

$$\begin{aligned} P &= A(P/A, i, N) \\ &= 2{,}500(5.650) \\ &= \$14{,}125 \\ &= \$14{,}130 \text{ (approximately)} \end{aligned}$$

This kind of calculation may be used to check the correctness of insurance company offers of annuities. Figure 4.1*e* shows the typical cash flow diagram. ■

Converting a present payment to a series of uniform future payments involves the use of the capital recovery factor. See Fig. 4.1*f*. If Eq. (4.8) is divided on both sides by

$$\frac{(1+i)^N - 1}{i(1+i)^N}$$

the result is

$$A = P\left[\frac{i(1+i)^N}{(1+i)^N - 1}\right] \tag{4.9}$$

The expression in brackets is the one desired and is symbolized by $(A/P, i, N)$.

■ **Example 4.7.** A bank is willing to finance a mortgage you need to buy a new house. Your relations with the bank are such that it will accept annual payments to pay off principal and interest on the loan. The annual rate of interest is 12 percent for a term of 30 years. The amount of the mortgage is $100,000. What is the required yearly payment?

Solution. Here P = $100,000, N = 30, and i = 0.12. We are required to find A, the yearly payment. Appendix C under the column entitled "Uniform series capital recovery factor A/P" shows opposite N = 30 the factor 0.12414. Thus

$$\begin{aligned}A &= P(A/P, i, N)\\ &= 100{,}000(A/P, 12, 30)\\ &= 100{,}000(0.12414)\\ &= 12{,}414\end{aligned}$$

The typical cash flow diagram illustrating this situation is shown in Fig. 4.1*f*. ■

Most sales of articles "on time" are financed by arrangements similar to this example—refrigerators, automobiles, boats, lots, and many more. A check of the payment amounts is possible by use of the preceding technique. How to calculate monthly payments will be explained through a later example in this chapter.

The inversions used in deriving the six-factor equation mean that

$$(F/P, i, N) = \frac{1}{(P/F, i, N)}$$

$$(A/P, i, N) = \frac{1}{(P/A, i, N)}$$

and

$$(F/A, i, N) = \frac{1}{(A/F, i, N)}$$

Note that it is unnecessary to recall the names of the six factors in order to identify them for use in solving a problem. They have been purposely symbolized in such a way that it is possible to say, "*F* given *P* must be the factor $(F/P, i, N)$; *A* given *P* is found by $(A/P, i, N)$; and so on." This is a great advantage.

GRADIENTS

In addition to the formulas presented in the previous section, another group arising out of gradients in cash flow will now be presented. Gradient cash flows appear because payments, either inflows or outflows, sometimes change each period. A piece of machinery, for example, may need more and more labor and spare parts to maintain it as it grows older, and the charge for this may increase. Cost inflation can be treated by increasing the cost each period. Similarly, a periodic increase in rents can be handled by adding an increment each time the rent is paid. Periodic changes may be *arithmetic*, that is, the change is by a constant *amount*; or the changes may be *geometric*, that is, by a constant percentage *rate*.

Arithmetic Gradient

Referring to Fig. 4.2*a*, you will observe an important point: G does not begin until the end of period 2. This convention is used so that the analyst can employ the gradient formulas with the same N that is being used to manipulate other portions of the same cash flow. The meaning of this point will become clear in the examples that follow. The analyst must be sure, however, that the gradient cash flow of concern is set up so that it follows the convention shown in Fig. 4.2*a*, namely that the first G appears at $N = 2$.

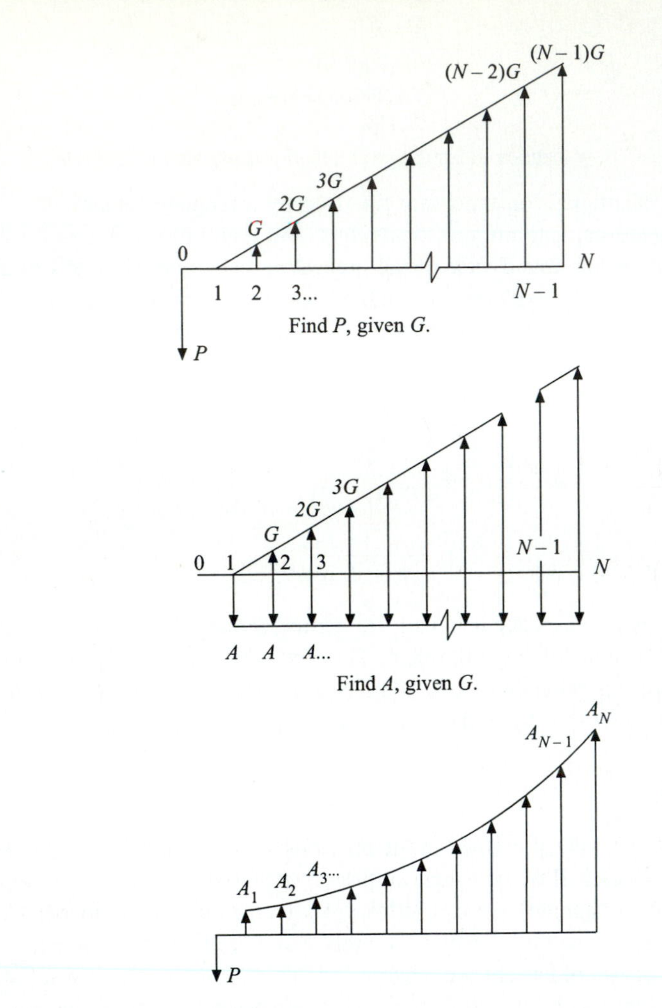

FIGURE 4.2 Typical cash flows

The most convenient way to arrive at the two most useful formulas for gradient cash flows, for converting them to a uniform annual payment and to a present worth, is to start with an expression of the future worth of the individual cash flows making up the gradient. We begin with the first G at the end of period 2:

$$F_{N_2} = G(F/A, i, N - 1)$$

Because the F/A formula was derived with the first A appearing at period 1, $N - 1$ is used. Using G in place of A and because G begins at period 2, we find $N - 1$ is the elapsed time.

The next term in the summation will be

$$F_{N_3} = G(F/A, i, N-2)$$

This term, like the preceding one and all subsequent ones, will treat G, not multiples of G, because we have broken down Fig. 4.2*a* into a pile of building blocks and are summing each block to arrive at the total thus:

$$F_N = F_{N_1} + F_{N_2} + \cdots + F_{N_{N-1}} + F_{N_N}$$

$$F_N = G\left[\frac{(1+i)^{N-1}-1}{i}\right] + G\left[\frac{(1+i)^{N-2}-1}{i}\right] + \cdots$$

$$+ G\left[\frac{(1+i)^{2}-1}{i}\right] + G\left[\frac{(1+i)^{1}-1}{i}\right]$$

$$= G\left[\frac{(1+i)^{N-1}-1}{i} + \frac{(1+i)^{N-2}-1}{i} + \cdots + \frac{(1+i)^{2}-1}{i} + \frac{(1+i)^{1}-1}{i}\right]$$

$$= \frac{G}{i}\left[(1+i)^{N-1} + (1+i)^{N-2} + \cdots + (1+i)^2 + (1+i)^1 + (N-1)(-1)\right]$$

$$= \frac{G}{i}\left[(1+i)^{N-1} + (1+i)^{N-2} + \cdots + (1+i)^1 - N + 1\right]$$

But if we wish to remove $-N$, we must multiply $-N$ by G/i and include it at the end.

$$F_N = \frac{G}{i}\left[(1+i)^{N-1} + (1+i)^{N-2} + \cdots + (1+i)^2 + (1+i)^1 + 1\right] - \frac{NG}{i}$$

The expression in brackets is the uniform series compound amount factor because we need only multiply the uniform series A by each of the terms within the brackets to arrive at the sum of each A compounded to period N. Substituting gives

$$F_N = \frac{G}{i}\left[\frac{(1+i)^N - 1}{i}\right] - \frac{NG}{i} \tag{4.10}$$

Multiplying the preceding expression by the single-payment present-worth factor gives P, the present worth of any gradient series:

$$P = F_N\left(\frac{1}{1+i}\right)^N$$

$$= \frac{G}{i(1+i)^N}\left[\frac{(1+i)^N - 1}{i}\right] - \frac{NG}{i(1+i)^N}$$

$$= G\left\{\frac{1}{i(1+i)^N}\left[\frac{(1+i)^N - 1}{i}\right] - \frac{N}{i(1+i)^N}\right\} \tag{4.11}$$

The symbol of the factor within braces is $(P/G, i, N)$.

■ **Example 4.8.** A drill press in an automobile repair shop is projected to show the following operations cost over the next 5 years:

Year	Cost
1	1,100
2	1,225
3	1,350
4	1,475
5	1,600

What is the present value of these costs if a 12 percent discount rate is used?

Figure 4.3*a* shows the situation. The gradient G is clearly \$125 per year. But the formula of Eq. (4.11) is not applicable because A must be included. Somehow the diagram must be manipulated so that the conditions of Fig. 4.2*a* appear.

Figure 4.3*b* accomplishes the necessary adjustment by splitting the cash flow of Fig. 4.3*a* into two parts, the top portion a uniform annual payment of \$1,100 and the bottom portion a gradient cash flow with G appearing for the first time at year 2, exactly as it should in order to follow the derivation assumptions of Eq. (4.10).

The solution is also in two parts, as shown by the two items.

$$P = A(P/A, i, N) + G(P/G, i, N)$$
$$= 1{,}100(P/A, 12, 5) + 125(P/G, 12, 5)$$

Substituting the factors from App. C gives

$$P = 1{,}100(3.605) + 125(6.3970)$$
$$= 3{,}966 + 800$$
$$= \$4{,}766$$

(*a*)

0 1 2 3 4 5

1,100 1,225 1,350 1,475 1,600

equals ↓

(*b*)

0 5

1,100

+

125 250 375 500

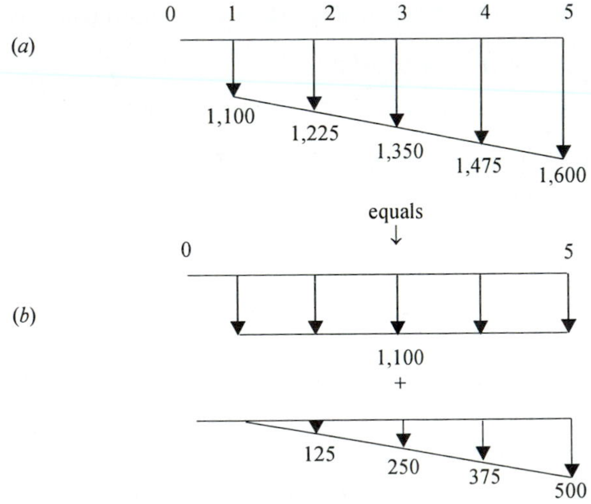

FIGURE 4.3 Gradient example 4.8

This is the present value sought.

The answer was carried out to the nearest dollar by the usual rules for rounding. The reason for doing so is that the accuracy of future projections of operational costs does not justify an answer in the cents except in contractual situations. In addition, because $(P/A, i, N)$ appears in only four significant figures, *a greater number of significant figures in an answer using it is not admissible.* Readers are advised to review their understanding of significant figures. ■

Converting a gradient series to equivalent uniform annual payments is often useful. See Fig. 4.2*b*. It can be done by multiplying the value of F_N from Eq. (4.10) by the sinking fund factor.

$$\begin{aligned} A &= F_N\left[\frac{i}{(1+i)^N - 1}\right] \\ &= \frac{G}{i}\left[\frac{(1+i)^N - 1}{i}\right]\left[\frac{i}{(1+i)^N - 1}\right] - \frac{NG}{i}\left[\frac{i}{(1+i)^N - 1}\right] \\ &= \frac{G}{i} - \frac{NG}{i}\left[\frac{i}{(1+i)^N - 1}\right] \\ &= G\left[\frac{1}{i} - \frac{N}{(1+i)^N - 1}\right] \end{aligned} \tag{4.12}$$

The symbol for the factor within the brackets is $(A/G, i, N)$.

■ **Example 4.9.** Suppose that in Example 4.8 it was required to find the equivalent uniform annual cost of the drill press operations cost rather than the present worth. Equation (4.12) could be used directly, or the appropriate factor from App. C could be used. Splitting the cash flow as shown in Fig. 4.3 fulfills the requirement that G begin at the end of the second year. The equivalent uniform annual cost is then

$$\begin{aligned} A &= 1{,}100 + 125(A/G, 12, 5) \\ &= 1{,}100 + 125(1.7746) \\ &= \$1{,}323 \end{aligned}$$

This is the answer desired. ■

The arithmetic gradient factors $(P/G, i, N)$ from Eq.(4.11) and $(A/G, i, N)$ from Eq. (4.12) are listed in App. C. And $(P/G, i, N)$ is used to convert a gradient series to a present worth and $(A/G, i, N)$ to convert a gradient series to an annual worth.

Geometric Gradient

In the geometric gradient series of cash flows, the flows change at a constant percentage rate. The change may be in the positive or negative direction. In Fig. 4.2*c*, the change is positive. In the figure A_2 is some percentage g greater than A_1, and A_3 is the same percentage g greater than A_2, and so on. That is,

$$A_N = A_{N-1}(1+g)$$

and

$$A_N = A_1(1+g)^{N-1}$$

Problem 4.9 asks the reader to derive the following formulas for the present worth of geometric gradient cash flows:

$$P = A_1\left[\frac{1-(1+g)^N(1+i)^{-N}}{i-g}\right] \quad \text{where } i \neq g \tag{4.13}$$

$$P = A_1 N(1+i)^{-1} \quad \text{where } i = g \tag{4.14}$$

The expression within the brackets of Eq. (4.13) is the *geometric series present-worth factor*. Its symbol is (P/A, i, g, N). It may also be written as

$$(P/A, i, g, N) = \left[\frac{1-(F/P, g, N)(P/F, i, N)}{i-g}\right] \tag{4.15}$$

and the factors of App. C are used.

■ **Example 4.10.** In preparing an economic analysis for a new high-rise building, Fayram and Reynolds, Architects and Engineers, estimated that revenues would be $10 million for the first year of operation of the building and would increase at the rate of 8 percent per year thereafter. The time horizon of the analysis was 10 years. The opportunity cost of capital was set at 12 percent. Taxes and inflation were ignored for this first approximation. What was the present worth of the revenues over the time horizon chosen?

Solution

$$A_1 = \$10{,}000{,}000$$
$$i = 12\%$$
$$g = 8\%$$
$$N = 10$$

From Eq. (4.15),

$$P = A_1\left[\frac{1-(F/P, g, N)(P/F, i, N)}{i-g}\right]$$

$$= 10{,}000{,}000\left[\frac{1-(F/P, 8, 10)(P/F, 12, 10)}{0.12-0.08}\right]$$

$$= 10{,}000{,}000\left[\frac{1-(2.1589)(0.3220)}{0.04}\right]$$

$$= 10{,}000{,}000\,(7.621)$$

$$= \$76{,}210{,}000$$

■

RATES OF INTEREST: NOMINAL AND EFFECTIVE

Let us begin by defining the symbols to be used in the discussion.

i_M = effective interest rate per period
i_Y = effective interest rate per year
M = number of periods per year
r = nominal interest rate—always an annual rate

The effective interest rate per period has been designated i in this chapter. The subscript M is used now to relate i more clearly to the number of periods.

The relation between the effective interest rate per period and the nominal rate is, by definition,

$$i_M = r/M \tag{4.16}$$

For example, if a 12 percent nominal rate is quoted with interest compounded monthly, r is 12 percent, M is 12, and i_M is 1 percent. However, if 12 percent had been quoted as an *effective* annual rate, a common mistake would have been to divide by M, which is 12, to find the effective monthly rate. What is the correct way to handle the question on the effective monthly rate?

Imagine that a sum of $1 is borrowed at the beginning of the year and is compounded M times during the year at interest rate i_M. At the end of the year, according to Eq. (4.2), the single dollar will have swelled to

$$F = 1(1 + i_M)^M$$

But F must also equal $1(1 + i_Y)$ given the yearly rate. This equality supplies the desired relation between i_Y and i_M.

$$1(1 + i_Y) = 1(1 + i_M)^M$$

or

$$i_Y = (1 + i_M)^M - 1 \tag{4.17}$$

and

$$i_M = (1 + i_Y)^{1/M} - 1 \tag{4.18}$$

This is the equation required to solve the problem proposed above.

■ **Example 4.11.** If the effective yearly rate is quoted at 12 percent with interest compounded monthly, what is the effective monthly rate?

Solution. From Eq. (4.18),

$$\begin{aligned} i_M &= (1 + i_Y)^{1/M} - 1 \\ &= (1 + 0.12)^{1/12} - 1 \\ &= 0.00948 \\ &= 0.95\% \end{aligned}$$

■

A single number as an interest rate can be used in at least three different ways: as related to a yearly payment, to a monthly payment with the interest rate nominal and compounding monthly, and to a monthly payment with the interest rate effective and compounding monthly. The following example illustrates the point.

■ **Example 4.12.** A ranch is offered for sale in Mexico with a 15-year mortgage at 40 percent compounded annually and a 20 percent down payment. Annual payments are to be made. The first cost of the ranch is 5 million pesos.

What yearly payment is required?

Solution

$$\begin{array}{rl} 5{,}000{,}000 & \\ -\ 1{,}000{,}000 & \text{down payment} \\ \hline 4{,}000{,}000 & \text{mortgage} \end{array}$$

$$\begin{aligned} A &= P(A/P, i, N) \\ &= 4{,}000{,}000(A/P, 40, 15) \\ &= 4{,}000{,}000(0.40259) \\ &= 1{,}610{,}400 \text{ pesos per year} \end{aligned}$$

Notice that when an annual compounding is required, the nominal and effective interest rates are equal.

Now imagine that the *nominal* interest rate is 40 percent per year and that monthly compounding and monthly payments are required. What is the monthly payment?

$$i_M = r/M = 40/12 = 3.33\%$$

$$N = 15 \text{ years} \times 12 \text{ months} = 180$$

$$A = 4{,}000{,}000(A/P, 3.33, 180)$$

$$A = P\left[\frac{i(1+i)^N}{(1+i)^N - 1}\right]$$

$$A = 4{,}000{,}000\left[\frac{0.033(1+0.0333)^{180}}{(1+0.0333)^{180} - 1}\right]$$

$$= 133{,}699$$

$$\approx 134{,}000 \text{ pesos per month}$$

If the interest rate quoted is effective and both payment and compounding are monthly, then the effective monthly rate must be computed by using Eq. (4.18):

$$\begin{aligned} i_M &= (1+i_Y)^{1/M} - 1 \\ &= (1+0.40)^{1/12} - 1 \\ &= 0.0284361 \\ &\approx 2.8\% \end{aligned}$$

$$A = P\left[\frac{i(1+i)^N}{(1+i)^N - 1}\right]$$

$$= 4{,}000{,}000\left[\frac{0.028(1.028)^{180}}{(1.028)^{180} - 1}\right]$$

$$= 4{,}000{,}000\left[\frac{4.355862}{155.5665 - 1}\right]$$

$$= 112{,}725$$

$$\approx 113{,}000 \text{ pesos per month}$$

■

SEPARATION OF INTEREST AND PRINCIPAL

Often the periodic payment on a loan must be separated into two portions, interest and principal. This separation is necessary because interest may be deducted from the taxable income of a business, while principal may not be so deducted. For all people in the United States who have borrowed money to finance a home, this is also an important consideration because home mortgage interest is deductible and only a small percentage of U.S. homes are free and clear of mortgages. Any loan made to a business—for example, a real estate developer—that requires periodic payments of principal and interest for its amortization must be split into interest and principal portions for the same reasons. In fact, continuing to own a house rather than selling it and buying another with a new mortgage may well depend on how much tax shelter—that is, benefits because of reduced income taxes—is provided by each alternative course of action. An example will illustrate the most direct procedure for separating interest and principal.

■ **Example 4.13.** You have borrowed \$100,000 for 5 years at 12 percent interest to buy a home. Because interest on a home loan is tax-deductible under present U.S. laws, you are anxious to see how much the interest on the loan will be for each of the years of its existence.

Solution. To solve this problem, the most direct approach is to compute the quantities in Table 4.3.

$$\begin{aligned} A &= P(A/P, i, N) \\ &= 100{,}000(A/P, 12, 5) \\ &= 100{,}000(0.27741) \\ &= \$27{,}741 \end{aligned}$$

This amount is the total payment, the sum of principal and interest portions, for each of the 5 years. To compute the interest portion for year 1, we must first multiply the interest rate in decimals by the remaining balance:

$$\text{Interest portion} = (0.12)(100{,}000) = \$12{,}000$$

Subtracting the \$12,000 interest portion from the total payment of \$27,741 gives the principal portion \$15,741; and subtracting it from the principal balance of the loan at the end of the previous year (year 0) results in the \$84,259 remaining balance after the first payment is made. This completes the year 1 line.

TABLE 4.3
Separation of interest and principal

Year	Payment *A*	Interest portion	Principal portion	Remaining balance
				100,000
1	27,741	12,000	15,741	84,259
2	27,741	10,111	17,630	66,629
3	27,741	7,996	19,745	46,884
4	27,741	5,626	22,115	24,769
5	27,741	2,972	24,769	0

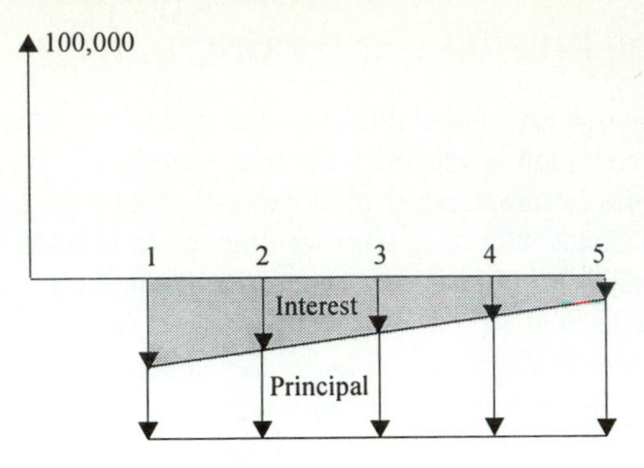

FIGURE 4.4 Separating interest and principal

Year 2 quantities are computed in the same fashion. The interest portion for the end of year 2 is

$$(0.12)(84{,}259) = 10{,}111$$

Notice that the interest is computed on the remaining balance at the end of the preceding year, not on the original principal of the loan. The rest of the calculation proceeds as before.

Table 4.3 reveals some interesting features of compound interest calculation. The interest portion of the payment decreases each year while the principal portion increases. Thus, the tax relief afforded by the loan steadily decreases. This is shown graphically in Fig. 4.4. The remaining balance declines from \$100,000 to 0. Last, the line separating the interest portion from the principal portion in the figure is not a straight line, but a curve. ■

Example 4.13 can be used to deduce the mathematical expression for the amount of interest paid in any year Y on a loan for any amount P at interest rate i for N periods.

The *remaining balance* R for any year Y is nothing more or less than the present worth at discount rate i of the remaining payments at A each:

$$R_Y = A\left(P/A, i, N-Y\right) \tag{4.19}$$

Thus, referring to Table 4.3, we see that the remaining balance at $Y = 0$ is

$$\begin{aligned} R_0 &= 27{,}741\left(P/A, 12, 5\right) \\ &= 27{,}741\left(3.6048\right) \\ &= 100{,}001 \\ &\approx 100{,}000 \end{aligned}$$

The extra \$1, really \$0.76, appears because of the lack of sufficient figures in the tables used. The series present-worth factor is the reciprocal of the capital recovery factor; this is the reason why the calculation is possible.

To illustrate further, for $Y = 3$,

$$\begin{aligned} R_3 &= 27{,}741(P/A, 12, 2) \\ &= 27{,}741(1.6901) \\ &= 46{,}885 \\ &\approx 46{,}884 \end{aligned}$$

The principal payment P, for any year Y, is the difference between the remaining balance R at the end of year Y *after* the principal payment P has been applied and the same quantity at the end of year $Y - 1$:

$$\begin{aligned} P_Y &= A\left[P/A, i, N - (Y-1)\right] - A(P/A, i, N - Y) \\ &= A\left[(P/A, i, N - Y + 1) - (P/A, i, N - Y)\right] \end{aligned} \tag{4.20}$$

However, the difference between any two entries in the series present-worth factor column of the tables of App. C is the single-payment present-worth factor for the later period because

$$(P/A, i, N) = \sum_{j=1}^{j=N} (P/F, i, j)$$

which is a description of how the series present-worth factor may be derived from the single-payment present-worth factor. Therefore,

$$(P/A, i, N-1) + (P/F, i, N) = (P/A, i, N)$$

or

$$(P/F, i, N) = (P/A, i, N) - (P/A, i, N-1)$$

In the periods set forth in Eq. (4.20), the previous equation becomes

$$(P/F, i, N - Y + 1) = (P/A, i, N - Y + 1) - (P/A, i, N - Y)$$

Substituting in Eq. (4.20) gives

$$P_Y = A(P/F, i, N - Y + 1)$$

The interest paid in any year is therefore

$$\begin{aligned} I_Y &= A - P_Y \\ &= A - A(P/F, i, N - Y + 1) \\ &= A\left[1 - (P/F, i, N - Y + 1)\right] \end{aligned} \tag{4.21}$$

Applying Eq. (4.21) to Example 4.13 for year 3, in order to find out how much interest was paid in year 3, gives

$$\begin{aligned} I_3 &= A\left[1 - (P/F, i, 5 - 3 + 1)\right] \\ &= 27{,}741\left[1 - (P/F, 12, 3)\right] \end{aligned}$$

$$= 27{,}741(1 - 0.71178)$$
$$= \$7{,}996$$

which agrees with the value in Table 4.3.

PROBLEMS

4.1. An architect friend of yours suggests an investment in a rundown house which she believes can be put into sellable condition for \$13,000. The house can be bought for \$42,000 and sold one year later for \$70,000 after repairs. You have \$30,000 in cash available, of which \$13,000 must be used for repairs and the remaining \$17,000 as a down payment. Assume that \$13,000 will be placed in a checking account as soon as the house is purchased and will earn no interest. A bank loan for the remainder of the purchase price can be negotiated at 15 percent annually for 10 years. It must be paid off when the house is sold. The architect's fee will be 10 percent of the repair costs. Taxes and insurance for the year you hold the house will be about \$800. A 6 percent commission must be paid to the real estate agent at sale. Your opportunity cost of capital, before taxes, is 20 percent. What is the annual rate of return before income taxes on this investment to the nearest whole percentage? Should you invest in this deal? (See Example 4.1.)

Answer: No, because the rate of return equals 16.5 percent.

4.2. (*a*) If you deposit \$20,000 in a bank account which pays 6 percent interest compounded annually, how much will be in the account after 8 years if you make no withdrawals? (See Example 4.2.)

Answer: \$31,876

(*b*) What is the present worth of \$100,000 which you will receive at the end of 12 years if your personal minimum attractive rate of return is 14 percent? (See Example 4.3.)

Answer: \$20,760

(*c*) A used car which your father owns will cost you \$1,500. Your father will accept five payments at the end of each of 5 years. He postulates an interest rate of 15 percent. How much will you need to pay him in five equal payments? Interest will be computed on the declining capital balance. (See Example 4.7.)

Answer: \$447.48

(*d*) You will receive an annuity of \$80,000 each year for the next 20 years. For how much would you be willing to sell the right to this annuity if you believe that your marginal attractive rate of return is 20 percent annually? (See Example 4.6.)

Answer: \$389,600

4.3. George Jenkins is promised $20,000 in 5 years if he lends $2,500 now. George's minimum acceptable return, considering inflation and income taxes, is 20 percent for an investment of this risk. Should he lend the money?

Answer: Yes

4.4. A drill press for a small machine shop has a first cost of $2,000, a prospective life of 20 years, and no salvage value. If the opportunity cost of capital for this shop is 20 percent before income taxes, what is the annual capital cost of the drill press?

Answer: $411

4.5. Fritz Simmons invests $200,000 in artistic property which he believes will appreciate at 25 percent per year, on average. How much money will he receive for the property if he sells it after 20 years—and he has guessed correctly?

4.6. Your Auntie Mame promises to give you $10,000 per year for the next 10 years out of her profits from a gambling casino she intends to establish in Atlantic City, if you will assist her with $50,000 now. She maintains that you will be making a 100 percent return on your loan because you will receive a total of $100,000 in repayment for it. On an equal-risk investment, your minimum attractive rate of return is 18 percent before account is taken of the unfavorable effects of inflation and income tax. Should you lend your aunt the money, all family feeling aside?

4.7. A garden apartment development will be constructed in four phases during 4 successive years. Each phase consists of 50 units, and each unit requires an automatic dishwasher. At the time the first phase is scheduled for completion, the dishwashers will cost $220 each. Each year thereafter their cost will increase by an estimated $70 per unit, because of inflation. The developer wishes to know the present worth of 200 dishwashers required by the development if her cost of capital is 20 percent before taxes. (See Example 4.8.)

4.8. Power shovel maintenance costs will start at $2,800 annually and increase by a constant $300 per year thereafter throughout the shovel's 10-year life. What is the present worth of these costs if the construction company purchasing the equipment uses a 20 percent before-tax cost of capital to evaluate its investments?

4.9. Derive Eqs. (4.13) and (4.14). A complete derivation of the formulas for the present worth of a geometric gradient is required. *Show all steps.*

4.10. Verify the answer of Example 4.10 by a step-by-step solution. In other words, compute the revenues by year, and use the single-payment present-worth factor to arrive at the sum of the present worth of the annual revenues.

4.11. Your small residential rental house is appraised at $240,000. It is projected to appreciate at 4.5 percent per year over the next 4 years. If your opportunity cost of capital is 10 percent, what is the present worth of the property?

4.12. In a situation similar to Example 4.12, a Peruvian real estate company tells you that the interest rate it will charge on a loan that your company intends to take out to buy a Miraflores business site will carry an effective annual rate of 24 percent. Since payments will be made monthly, what will be the effective monthly rate you should use to calculate the monthly payments?

4.13. In a situation similar to Example 4.12, prospective homeowners are considering a $350,000 house. After a down payment of 20 percent ($70,000), their mortgage will amount to $280,000. They are offered a 6.5 percent nominal rate for a term of 30 years. How much will their monthly payments be, correct to the nearest cent?

4.14. A bank quotes 3 percent per year as its interest rate on a savings account. Some questioning on your part reveals that the bank is giving you a nominal rate. The bank also states that compounding will occur monthly. What is the effective monthly rate the bank offers?

4.15. A student of Japanese history plans a trip to Kyoto. He estimates that it will cost him $10,000 for the trip and his stay in Japan. How much will he need to put monthly in a savings account paying 6 percent nominal, compounded quarterly, to make his trip 2 years from now?

4.16. A new-car buyer has computed her annual end-of-year cost of ownership—gas, oil, maintenance, etc.—at $1,535. She now wishes to convert this amount to an equivalent monthly cost if her effective opportunity cost of capital per year is 15 percent. What is her monthly cost?

4.17. A flyer entitled *Master Charge/VISA Cardholder Agreement and Disclosure Statement under Federal Law* was received by a cardholder. It read, in part:

> **Computing the finance charge.** We will figure the finance charge on your account by multiplying the balance subject to finance charge on your current monthly statement by a periodic rate of *1½ percent per month (which is a corresponding annual percentage rate of 18*

percent) on the first $1,500 thereof and a periodic rate of 1 percent per month (which is a corresponding annual percentage rate of 12 percent) on that portion from $1,500 to $2,500 thereof and a periodic rate of 0.8 percent per month (which is a corresponding annual percentage rate of 9.6 percent) on that portion over $2,500.

Are the italicized statements correct? If not, correct them numerically. You may presume that the monthly percentage rates quoted are effective interest rates.

4.18. "Caught in the middle of a bursting housing bubble, homesellers are an additional group of victims, as incredible interest costs have virtually closed down private sales of housing. The startling fact is that a 3 percent increase causes an increased monthly cost to the buyer of 23 percent. Take a $75,000 house (hardly a palace) and a $15,000 down payment; at 12 percent the monthly mortgage cost to the buyer is $617.17; at 15 percent the monthly cost is $758.67, ..."[1]

Are the statements numerically true? Show your calculations. Assume a mortgage term of 30 years. Assume also that the percentage rates given are nominal.

4.19. In how many years, to the nearest year, will money double if invested at 5, 10, 15, and 20 percent?

4.20. A 6-month bank loan of $1,000.00 required payments of $175.53 per month. After three payments were made, the borrower paid off the loan for $524.03, a sum determined by the lender. Interest was quoted at 18 percent nominal, compounded monthly.

The $1,000.00 was received on December 12, 1980. Payments of $175.53 each were made on January 12, February 12, and March 12. No payment was made during April except for the final payoff on April 12.

(*a*) Was the monthly payment correct?

(*b*) Did the bank overcharge the borrower on the payoff, and if so, by how many dollars?

4.21. A used car you are considering buying is offered for sale for $6,800. The bank will lend you $5,700 on the car at 15 percent nominal interest, compounded monthly, for 36 months. The bank manager tells you that the payments will be $276.03 per month, but he seems unsure of his figures.

Is the bank manager correct in his estimate of your monthly payments? If not, what is the correct figure?

---

[1]William J. Quirk, "The Great American Credit Collapse," *The New Republic*, April 5, 1980, p. 25.

4.22. What nominal rate of return is required to ensure an effective annual rate of 10 percent if compounding occurs quarterly?

4.23. A new Volvo you intend to buy can be financed at 18 percent nominal interest rate compounded monthly over 40 months. If you must borrow $12,000 in order to buy the car, what will be your monthly payment?

4.24. Furniture purchases for your new residence will cost about $20,000. The dealer from whom you are buying will finance the purchase for a 10 percent down payment and monthly payments over the next 2 years at an 18 percent nominal rate per year. What will your monthly payments be, to the nearest cent?

4.25. An investment program offered by a local bank advertises that funds will draw 12 percent nominal rate, compounded semiannually.

(*a*) What is the effective rate per period?

(*b*) What is the effective annual rate?

Answers must be given to two decimal places.

4.26. A $10,000 nominal 6 percent bond, bought in 1981 at par, has matured in 1991. It paid interest twice yearly. What was the effective annual rate of return for this bond to the nearest 0.01 percent?

4.27. The Soldier's and Sailor's Relief Act of 1942 allows military personnel, sent overseas in time of war, certain privileges with regard to their debts. Recently the act was invoked to assist military personnel sent to Saudi Arabia in the prelude to the Gulf War.

A certain corporal had borrowed $11,000 at a nominal 12 percent per annum to buy a car. The repayment plan consisted of 60 equal monthly payments, each comprising both principal and interest. Just after she made the 10th payment, her nominal rate was reduced to 6 percent compounded monthly.

(*a*) How much was her original monthly payment?

(*b*) What is the amount of her new monthly payment if she has 50 payments remaining?

4.28. American Security Bank offers a 2.55 percent nominal interest rate with daily compounding. Chevy Chase Bank offers a nominal interest rate of 3.55 percent with compounding quarterly. Both rates are on checking accounts. *Ceteris paribus*, which bank should you choose for your checking account?

4.29. Dr. Rogers, a university professor, is a participant in the Teacher's Insurance and Annuity Association (TIAA) and the College Retirement Equities Fund (CREF). He will reach retirement age 10 years from now.

(*a*) At that time, how much will he have in the above funds if he has $209,000 now and the fund grows at 6 percent per year?
(*b*) How much annual interest income will he receive, perpetually, if the amount in the fund at the end of year 10 is invested at 8 percent?
(*c*) Now suppose that he specifies that $50,000 of principal will be removed from the fund and given to him at the end of each year in addition to the 8 percent return on the remaining balance. How much will he receive each year, and how long will the payments to him last?

4.30. Jovita opened an account at a neighborhood bank. The bank pays a nominal 18 percent annually. The bank compounds interest monthly.
(*a*) What is the annual effective rate?
(*b*) Jovita is planning to take a vacation around the world after 5 years. How much should she deposit into her bank account monthly to have the $25,000 she needs for her vacation?

4.31. According to a credit account agreement and disclosure statement distributed by Sovran Bank, "The initial daily periodic rate applicable to your account is 0.03699 percent." (This is an annual rate of 13.5 percent divided by 365 days.) Identify or compute the
(*a*) Nominal rate of interest r
(*b*) Number of periods per year M
(*c*) Effective rate per period i_M
(*d*) Effective rate per year i_Y

4.32. Frank Keville is contemplating buying a residence larger than the one he occupies at present. The necessary loan will be $210,000. He wishes to know how much of a tax shelter—that is, how much he may deduct from his taxable income—will be provided by the interest payments 5, 10, and 15 years from now if he pays 12 percent on a 30-year mortgage. Although mortgage payments will be made monthly, sufficient accuracy will be attained by annual calculation. (See Example 4.13.)

4.33. A small photographic film processing company borrowed $1 million to finance the purchase of new equipment. The lender required an 18 percent annual rate over a loan term of 10 years. A uniform annual payment of principal and interest was required to amortize the loan.
(*a*) How much interest *each year* was paid and was therefore deductible from the company's federal income taxes as a result of this loan, over the first 3 years of the term?
(*b*) If the company had wished to pay off the loan at the end of 3 years in order to refinance at a lower rate, how much would it have had to pay after the third payment was made?

4.34. Financing for a Mercedes-Benz 300D costing $24,536 is to be considered. The purchase of the car requires a 10 percent down payment in cash. The remainder of the purchase price will be paid for by a bank loan at 15 percent annual effective rate for a term of 4 years. All costs, including transportation and dealer preparation, are included in the price.

How much of each of the four yearly payments is interest and how much principal? Assume that the loan may be treated as being paid off in one payment at the end of each of 4 years, not, as would really be the case, in 48 monthly payments.

4.35. A homeowner bought her house in June 1986 for $123,000. To do so, she put up $23,000 in cash and negotiated a mortgage of $100,000 for 30 years at 9 percent effective interest rate per annum. She will pay off the mortgage, owed to an individual, in yearly payments starting in June 1987. She wants to estimate how much of the mortgage she will still owe at her proposed retirement in June 2003, just after she has made her mortgage payment.

4.36. Repeat Prob. 4.2, using continuous compounding rather than annual compounding. See App. 4A. If you have already done Prob. 4.2, it will be interesting to compare the answers.

4.37. (*a*) Assume interest to be 15 percent before taxes and compounded annually.

- (*i*) How much must be saved annually in expenses for 8 years in order to justify spending $10,000 now?
- (*ii*) What would you accept now in exchange for a payment of $6,000 in 10 years from now?
- (*iii*) How much overhaul expenses at the end of 3 years will be justified if you do not spend $3,000 on maintenance at the end of each of 3 years?
- (*iv*) You wish to deposit in a sinking fund an amount sufficient to cover college expenses at the end of 10 years. The expenses are estimated to cost $30,000. How much must you deposit?

(*b*) Now repeat part (*a*), using continuous compounding. See App. 4A. Compare the two answers.

APPENDIX 4A: CONTINUOUS COMPOUNDING

The section beginning on page 55 explained nominal and effective rates of interest. There it was illustrated that the number of periods per year M could be expanded from semiannually to quarterly to monthly to weekly to daily. Further expansion to an infinite number of periods is possible. When M becomes infinite, the compounding is called *continuous*.

Recall that the single-payment compound amount factor is

$$F = P(1+i)^N$$

If r/M is substituted for i and the number of compounding periods in N years is recognized as MN, we have

$$F = P(1+r/M)^{MN}$$

Now, allow M to increase without limit. Then MN becomes infinitely large and r/m infinitely small:

$$F = P\left[\lim_{M\to\infty}(1+r/M)^{MN}\right]$$

Now, let $k = M/r$. Then $MN = krN$, and substituting gives

$$F = P\left[(1+1/k)^k\right]^{rN}$$

But from calculus

$$\lim_{k\to\infty}(1+1/k)^k = 2.71828 = e$$

Therefore,

$$F = Pe^{rN} \tag{4.22}$$

The remaining factors may be derived from Eq. (4.22).

The complete list of factors is shown in the following table.

To find	Given	Name of factor	Algebraic formulation
F	P	Continuous compounding compound amount (single payment)	e^{rN}
P	F	Continuous compounding present worth (single payment)	e^{-rN}
P	A	Continuous compounding present worth (uniform series)	$\frac{e^{rN}-1}{e^{rN}(e^r-1)}$
A	F	Continuous compounding sinking fund (uniform series)	$\frac{e^r-1}{e^{rN}-1}$
A	P	Continuous compounding capital recovery (uniform series)	$\frac{e^{rN}(e^r-1)}{e^{rN}-1}$
F	A	Continuous compounding compound amount (uniform series)	$\frac{e^{rN}-1}{e^r-1}$

All cash flows are discrete. Compounding occurs at nominal rate r per period—normally per year.

■ **Example 4A.1.** One thousand dollars is deposited in a bank that will pay a nominal interest rate of 10 percent per year. How much will the account contain at the end of 3 years, if compounding is continuous?

Solution

$$\begin{aligned} F &= Pe^{rN} \\ &= 1{,}000(2.71828)^{(0.10)(3)} \\ &= \$1{,}349.86 \end{aligned}$$

If compounding were annual and not continuous, then

$$\begin{aligned} F &= P(1+i)^{N} \\ &= 1{,}000(1.10)^{3} \\ &= \$1{,}331.00 \end{aligned}$$

The difference of $18.86 is significant. It becomes less and less as compounding becomes monthly, quarterly, and semiannually. ■

Continuous compounding is used in some bank contracts. Practically, its use is limited.

CHAPTER 5

PRESENT WORTH

Until we reach the chapter on capital budgeting (Chap. 14), explanations and examples will be restricted to mutually exclusive alternatives. These, as defined in Chap. 1, concern the choice among courses of action such that selecting one immediately eliminates all the others.

Four methods are standard in the economic analysis of courses of action with regard to investment: present worth (present value, net present value), annual worth (equivalent uniform annual cost, annual cost), benefit/cost ratio (cost/benefit ratio), and internal rate of return (rate of return).

The present-worth method is widely favored by analysts, particularly those in the area of operations research, because it is the most foolproof. Like the annual worth method, it does not require the incremental (marginal) analysis demanded by the benefit/cost ratio and internal rate of return methods. Its computation is straightforward and its meaning clear to persons acquainted with the method. Therein lies its disadvantage: It is often difficult to explain to someone not familiar with it. Thus its use may be restricted.

Present value, or present worth, has one very practical use—in real property appraisal. The present value of the net discounted benefits less the net discounted costs of a piece of property, over its life, is its value in the marketplace. The net-present-value (NPV) method of appraising real property is achieving greater and greater acceptance with the passage of time.

PRESENT WORTH DEFINED

The present-worth criterion states that the present worth of benefits must equal or exceed the present worth of costs if a project is to be selected. Notice that the middle ground where the analyst is indifferent between acceptance and rejection of a project is excluded from this definition. Only two results will be signaled by application of the criterion: accept the project or reject it. This rule is adopted in this text largely for the sake of convenience. We will be left with only two possible results of an analysis rather than three. Partly, however, the adoption of an accept/reject rule with no room for indifference is justifiable on economic grounds because it means that if the discounted net present value is zero, the investment will just meet the other criterion implied by the rule—that is, investment at the opportunity cost of capital in another project will return no more. In mathematical terms, we have

$$\sum_{j=0}^{N}\left(B_j - C_j\right)\left(P/F, i, j\right) \geq 0 \tag{5.1}$$

where B_j stands for benefits at the end of period j, C_j for costs at the end of period j, and $(P/F, i, j)$ for the single-payment present-worth factor for period j at discount rate i. The formula says that we must subtract the costs from the benefits at any period and then multiply the result by the single-payment present-worth factor for that period and the discount rate given. When this has been done for all the periods in the problem, we must algebraically add the result for all the periods under consideration.

■ **Example 5.1.** A friend who is trying to finance a small restaurant venture offers you payments of $2,000, $4,000, and $7,000 at the end of each of 3 years in order to repay a loan of $10,000. You are going to charge him an interest rate of 10 percent, which you would otherwise make on certificates of deposit, in order to do him this favor. This is a low rate, considering the greater risk of a restaurant in relation to a certificate of deposit and probable inflation over the next 3 years. Should you lend him the money on the repayment terms he offers?

Another way of putting this question is to ask, Does the repayment plan offer a present worth at 10 percent at least equal to $10,000? Figure 5.1 shows the cash flow.

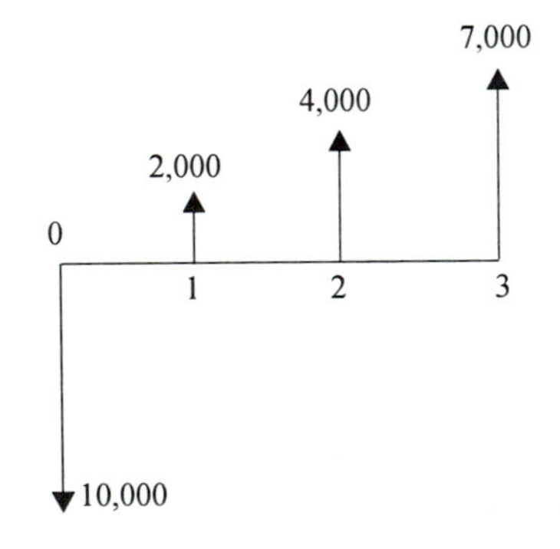

FIGURE 5.1 Repaying $10,000

Solution. Applying Eq. (5.1) yields

$$(0-10{,}000)(P/F,10,0)+(2{,}000-0)(P/F,10,1)$$
$$+(4{,}000-0)(P/F,10,2)+(7{,}000-0)(P/F,10,3)$$
$$=-10{,}000+2{,}000(0.9091)+4{,}000(0.8264)+7{,}000(0.7513)$$
$$=-10{,}000+1{,}818+3{,}306+5{,}259$$
$$=383$$

Because this amount is greater than 0, the three payments will be acceptable. ■

It often happens that $B_j - C_j$ is constant for all j's except for $j = 0$. When this occurs, Eq. (5.1) can be modified:

$$-P+(B_j-C_j)\sum_{j=1}^{N}(P/F,i,j)>0$$

where P is the cost at time 0.

But

$$\sum_{j=1}^{N}(P/F,i,j)=(P/A,i,N)$$

because it was shown in Chap. 4 that when all future payments in a series are equal, an expression for the series present-worth factor can be used in place of the single-payment factors. The reader who would like to check on the correctness of this statement can quickly do so by turning to the compound interest table in App. C for any interest rate. Select any N, say 5. Summing the P/F factors from 1 through 5 will give the P/A factor for $N = 5$. Therefore,

$$-P+(B_N-C_N)(P/A,i,N)\geq 0 \tag{5.2}$$

is a special case of Eq. (5.1).

■ **Example 5.2.** A pedestrian tunnel between two existing subway stations will cost \$800,000. Benefits over the next 50 years will be \$40,000 annually in passenger time saved. Yearly costs will be \$10,000 for lighting and janitor services. Should the tunnel be constructed, if the opportunity cost of capital is 10 percent?

Solution. Figure 5.2 shows the consequences flow. Applying Eq. (5.2) will result in

$$-800+(40-10)(P/A,10,50)=-800+30(9.915)$$
$$=-502.55$$
$$\approx -500$$

Therefore, the tunnel should not be built.

In this example, the null alternative, that is, the option of not building the tunnel at all, was selected. ■

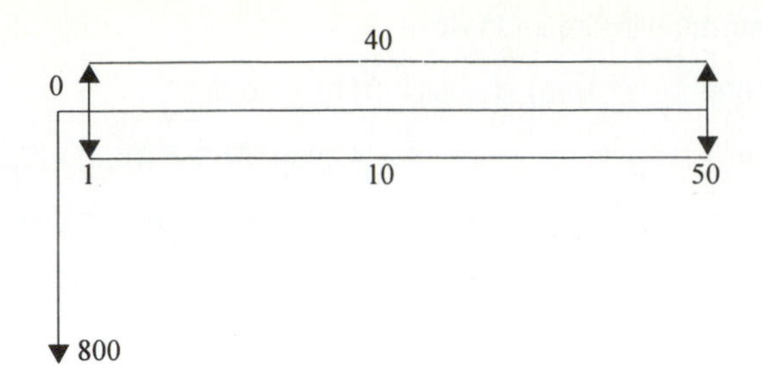

FIGURE 5.2 Pedestrian tunnel cash flow

Thus far, the examples in this chapter have dealt with two alternatives only. In the chapter on multiple alternatives (Chap. 9), we will see how more than two alternatives are handled.

Incremental analysis is conducted by comparing the *extra* benefits of a course of action to its *extra* costs. *Incremental* is synonymous with *marginal* and *extra*. If the incremental benefits exceed the incremental costs, then the course of action is justified. Incremental analysis proceeds in steps, and therefore it is necessary to order the alternatives with the lowest-cost alternative first. In the examples just seen, such ordering was automatic because the alternative of doing nothing was present as the unexpressed first alternative.

Why is incremental analysis necessary? The everyday, commonsense answer is that it is logical to measure by extras and not by totals. Is the extra time that a student spends on a course going to be compensated for by a higher grade? Is the extra time that a driver spends by taking a roundabout route to work worth the annoyance it saves by avoiding traffic congestion? Is the extra cost of an add-on room to a house going to be balanced by its extra convenience? Thus, in an industrial situation we can ask, Is the incremental cost of the new automatic drill press over retaining the old one more than balanced by the labor savings involved?

Another reason to justify incremental analysis is found in the rule enunciated early in this book that it is only the differences in alternatives that are relevant to a comparison between them. An incremental analysis makes explicit the differences between alternatives and the measurement of those differences to find out if the difference in cost is offset by the difference in benefits. Add in the idea of ordering alternatives by first cost, starting with the lower, and you have the idea of incremental analysis. (Later we will notice that incremental analysis has certain exceptions.) The equation presenting the idea of incremental analysis is, for present worth,

$$\sum_{j=0}^{N}\left[\left(B_j - C_j\right)_2 - \left(B_j - C_j\right)_1\right]\left(P/F, i, j\right) \geq 0 \tag{5.3}$$

where the symbols are the same as in Eqs. (5.1) and (5.2) and the subscripts 1 and 2 refer to the alternatives. When the incremental cash flow is composed of equal periodic costs or benefits, the series present-worth factor $(P/A, i, N)$ may be used rather than $(P/F, i, j)$.

■ **Example 5.3.** A railroad in Ecuador whose location was determined a century ago has proposed certain relocation projects to the government of the country in the hope of obtaining funds. One section can be relocated in two new alignments with characteristics as follows:

	Location 1	Location 2
Initial cost ($000,000)	102	140
Annual operation and maintenance ($000,000)	4	2
Annual benefit ($000,000)	20	26
Economic life (years)	40	40
Salvage value	0	0

The public opportunity cost of capital is 12 percent. Which of the locations is more economical? Or should the project be abandoned?

Solution. Figure 5.3 shows the cash flow for each alternative and the incremental cash flow between alternatives 1 and 2. The incremental cash flow between alternative 1 and abandoning the project—that is, continuing with the existing location—is the cash flow for alternative 1 itself.

Testing alternative 1 against the null alternative by applying Eq. (5.3) results in

$$\begin{aligned} -102 + (20 - 4)(P/A, 12, 40) &= -102 + 16(8.244) \\ &= -102 + 132 \\ &= 30 \end{aligned}$$

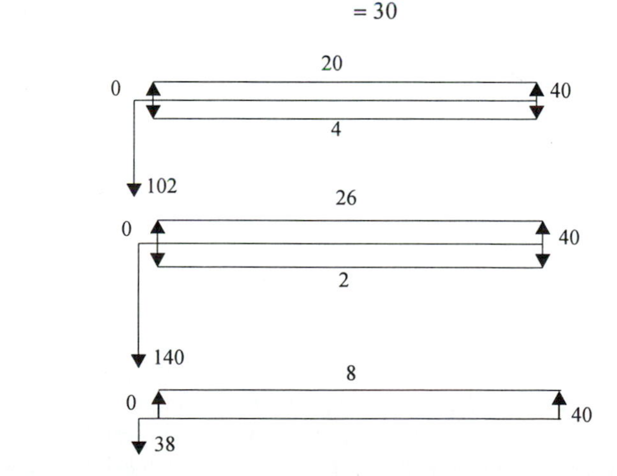

FIGURE 5.3 Railroad locations in Ecuador

This qualifies alternative 1. It shows that alternative 1 is better than the status quo.

Testing alternative 2 against alternative 1 resolves into answering the question, Is the incremental cash flow 2 – 1 economical? The incremental initial cost is

$$-140-(-102)=-38$$

The incremental benefit is

$$+26-2-(20-4)=+8$$

$$\begin{aligned}-38+8(P/A,12,40)&=-38+8(8.244)\\&=-38+66\\&=28\end{aligned}$$

This result shows that alternative 2 is better than alternative 1 because the discounted extra benefits of alternative 2 compared to alternative 1 more than balance its discounted extra costs, as shown by a positive present worth of the incremental cash flow. Said another way, the extra investment in alternative 2 yields a greater return than 12 percent.

What if this last amount had been 1 rather than 28? The answer would have been the same, although by a smaller margin.

What if the comparison between alternative 1 and the null alternative had been −5 instead of +30? Alternative 1 would not have qualified economically. The next step would have been to compare alternative 2 against the null alternative. A plus or zero answer to the consequent present-worth equation would have meant acceptance of alternative 2 rather than continuance of the status quo. A negative answer would have meant continuance of the status quo.

This question may arise from this example: Why choose alternative 2, which had a net discounted present worth of +28, rather than alternative 1, which had a higher net present worth of +30? The answer to this question is that +28 is the *incremental* net present value of alternative 2 over alternative 1, not the net benefit of alternative 2 itself.

Another question concerning why we do not analyze each alternative individually will be answered in the next section. ■

INCREMENTAL ANALYSIS VERSUS INDIVIDUAL ANALYSIS

A great deal of emphasis has been placed on incremental analysis thus far in this book, and it will continue to be emphasized. But some readers may wonder, particularly after considering Eq. (5.3), whether it might be just as correct to analyze each alternative individually, obtain its individual net present worth, compare these, and then choose the alternative with the greatest advantage—that is, the highest *net present value* (*NPV*). In the case of the present-worth method and in the annual-worth method, the same results are achieved by comparing individual worths and choosing the highest; but in the case of the benefit/cost ratio and the internal rate of return, analysis of individual alternatives can give an incorrect answer. In both present and annual worth, comparison of individual worths is fundamentally an incremental analysis and is correct because in these two cases the point in the procedure at which the comparison is made does not matter. In the benefit/cost and internal rate of return methods it does. A mathematical answer to the readers' question is as follows: Starting with Eq. (5.3):

$$\begin{aligned} \text{NPV}_{2-1} &= \sum_{j=0}^{N}\left[\left(B_j - C_j\right)_2 - \left(B_j - C_j\right)_1\right](P/F, i, j) \\ &= \sum_{j=0}^{N}\left[\left(B_j - C_j\right)_2 (P/F, i, j)\right] - \sum_{j=0}^{N}\left[\left(B_j - C_j\right)_1 (P/F, i, j)\right] \\ &= \text{NPV}_2 - \text{NPV}_1 \end{aligned}$$

Thus, the present-worth rule may also be expressed as

$$\text{NPV}_2 - \text{NPV}_1 = \sum_{j=0}^{N}\left[\left(B_j - C_j\right)_2 (P/F, i, j)\right] - \sum_{j=0}^{N}\left[\left(B_j - C_j\right)_1 (P/F, i, j)\right] \geq 0 \tag{5.4}$$

We recapitulate: If $\text{NPV}_{2-1} \geq 0$, select 2; if not, select 1. Or if $\text{NPV}_2 - \text{NPV}_1 \geq 0$, select 2; if not, select 1. Or simply compare NPV_2 with NPV_1 and select the larger. If they are equal, select 2.

■ **Example 5.4.** The circumstances of this example are the same as in Example 5.3. The individual present worth of each alternative will be determined and compared in an application of Eq. (5.4).

Solution. Testing alternative 1 against the null alternative gives the same result as in Example 5.3: a present worth of +30 that means acceptance of alternative 1 rather than doing nothing. And +30 is also the result of working out the second term of Eq. (5.4). Now we must find the present worth of alternative 2:

$$\begin{aligned} \text{NPV}_2 &= \left(B_N - C_N\right)(P/A, i, N) \\ &= -140 + (26 - 2)(P/A, 12, 40) \\ &= -140 + 24(8.244) \\ &= 58 \end{aligned}$$

Comparing the present worth of alternative 1, or 30, with the present worth of alternative 2, or 58, shows that alternative 2 is preferred and by a margin of 58 – 30, or 28, which is the same answer that we achieved by computing the incremental present worth of alternative 2 compared to alternative 1. Thus, in the case of present worth and annual worth, but only in these two methods, individual analysis and incremental analysis arrive at the same result. ■

ALTERNATIVES WITH DIFFERENT LIVES

When alternatives have different economic lives, they cannot be evaluated unless their lives are made equal, as mentioned in Chap. 2. It would be illogical to compare different lengths of service as though they were the same. The costs of an automobile which will last for 1 year cannot be compared to the costs of an automobile which will last for 10 years unless some allowance is made for the missing 9 years of service from the shorter-lived car. For example, let us say that you are comparing the costs of owning a new Nissan that costs $18,000 and will last 10 years against a used Ford that costs $1,000 and will last 1 year. Would you make the comparison by stating, "I will choose the used car because it will cost me $1,000, versus $18,000 for the new car"? Ordinary common sense dictates that you would be making the analysis on an incorrect basis—1 year of life—by ignoring the benefit of the longer life of the new car.

The device used to equalize service lives assumes that the investment in the shorter-lived investment will be repeated as many times as necessary to bring about an equality with the life of the longer-lived investment.

In the example above, you would set the Nissan purchased and held for 10 years against the Ford bought 10 times and held 1 year. You assume no salvage value for either car and equal annual costs and benefits—this is not at all likely, but is necessary to make the point simply. Now a fair comparison is possible, based on equal service lives. Remember that you have no intention of repeating the Ford purchase 10 times; you make the assumption only because you wish a logical comparison for a decision that you must make now. What you will do a year from now is an entirely separate decision to be made then. Applying the principles of this chapter, you can make a fair choice.

Benefits and costs are assumed to be the same during the successive lives. This sounds simple enough when one economic life is 10 years and another is 20 years. When one life is 7 years and the other is 8 years, however, it becomes necessary to repeat both investments for a total life of 56 years! In other words, we must use the least common multiple of the lives and repeat both investments a sufficient number of times to reach that number of years. Immediately an objection arises. How can we possibly look—using the preceding example—56 years in the future? The answer is that we are really not doing so. We are employing a device to equalize the service lives—nothing more. We are really peering into the future of only the life of the alternative finally chosen.

■ **Example 5.5.** Imagine two alternative proposals for an urban transit system. The first has a service life of 20 years, the second one of 40 years. First costs of the systems are $100 million and $150 million, respectively. Net benefits of both including externalities are equal at $30 million per year. The resource opportunity cost is 12 percent. No salvage values will be present.

Solution. To equalize the service lives, the first investment is shown repeated during another 20 years. The consequence flows appear in Fig. 5.4.

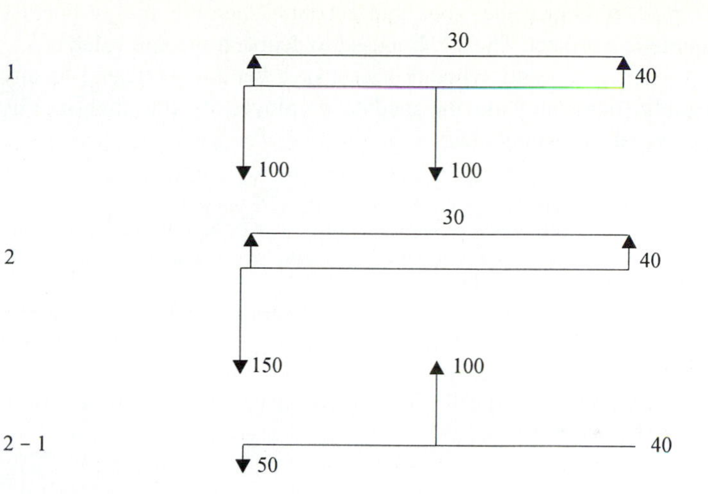

FIGURE 5.4 Two transit proposals

For alternative 1:

$$\begin{aligned} NPV_1 &= 30(P/A, 12, 40) - 100(P/F, 12, 20) - 100 \\ &= 30(8.244) - 100(0.104) - 100 \\ &= 247 - 10 - 100 \\ &= 137 \end{aligned}$$

Therefore the first alternative is acceptable.

$$\begin{aligned} \Delta NPV_{2-1} &= 100(P/F, 12, 20) - 50 \\ &= 100(0.104) - 50 \\ &= 10.4 - 50 \\ &= -39.6 \end{aligned}$$

The second alternative does not qualify. The first is chosen.

The meaning of the incremental difference between alternatives 1 and 2 is worth noting. Put in the form of a question, it is, Does investing an extra 50 in order to avoid an extra payment of 100 at the end of 20 years seem justified? The answer is no. ■

SALVAGE VALUE

Frequently, an investment retains some value at the end of its life which must be taken into account as a benefit. Less often, an investment has a negative value at the end of its life which represents the cost of its removal. An example of the latter is a highway superseded by a new alignment and construction of a more modern road. Some money must be spent to remove existing pavement and bridges. Probably in the not so distant future, society will demand that all vestiges of the existing highway be removed in order to restore the landscape to its original state.

The cost of this must, of course, be taken into account when the analyst decides whether to recommend a project. The symbol used to denote a salvage value is S.

In general, the salvage value, whether a cost or a benefit, is treated as any other cost or benefit, depending on the method employed by the analyst. The following examples make the point clear.

■ **Example 5.6.** A bridge over a river will have an initial cost of $5 million. Benefits are calculated to be $800,000 per year for 30 years. The salvage value at the end of the life of the bridge is thought to be $50,000.

Another alternative to bridging the river is to construct a ferry. The initial cost will be $200,000. Benefits will be $60,000 per year for 10 years with a $20,000 salvage value at the end of that time.

If the opportunity cost of public funds is thought to be 15 percent, which alternative should be chosen?

Solution. Solving this problem requires, first, that the salvage value be properly handled and, second, that the service lives of the two alternatives be equalized. Figure 5.5 illustrates both points.

The ferry alternative appears first because it has the lower initial cost. Its salvage value is shown as an up arrow at year 10. To provide an equal service life with the bridge alternative, two extra fictitious life cycles are added with their initial costs, benefits, and salvage values. Again notice that the fictitious cycles are included to satisfy the demands of the logic of equal service lives in a mathematical way. No repetition of the ferry investment need take place. All that is desired is a comparison that will lead to a correct choice between the two alternatives—ferry or bridge. If the ferry is selected, another analysis can be performed at the end of the ferry's life, or indeed at any time during its life, to see if it should be replaced and by what.

By the present-worth method, often called *the benefit-minus-cost method*, the ferry option is compared to doing nothing, which is the same as finding its individual *net present value* (NPV). The figures represent thousands of dollars.

$$\begin{aligned}\text{NPV} &= -200 + 60(P/A, 15, 30) + (-200 + 20)(P/F, 15, 10) \\ &\qquad + (-200 + 20)(P/F, 15, 20) + 20(P/F, 15, 30) \\ &= -200 + 60(6.566) - 180(0.2472) - 180(0.0611) + 20(0.0151) \\ &= -200 + 393.96 - 44.50 - 11.00 + 0.30 \\ &= 138.76 \\ &\approx 139\end{aligned}$$

Because $139,000 is a positive amount, the ferry project qualifies against doing nothing at all.

Analysis of the incremental cash flow of the bridge option over the ferry option yields

$$\begin{aligned}\text{NPV} &= -4{,}800 + 740(P/A, 15, 30) + 180(P/F, 15, 10) \\ &\qquad + 180(P/F, 15, 20) + 30(P/F, 15, 30)\end{aligned}$$

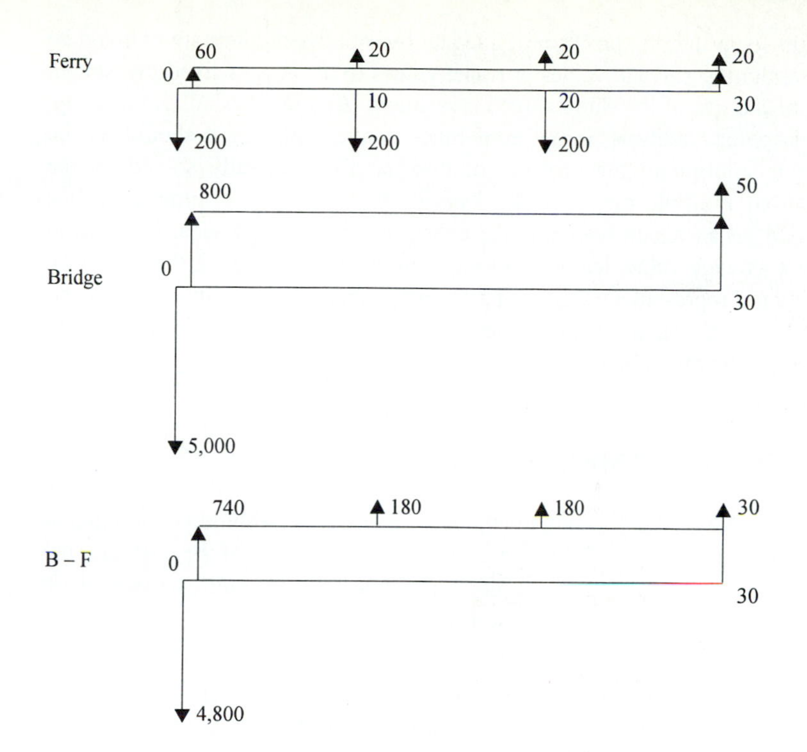

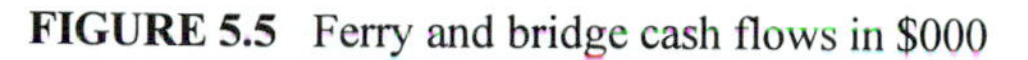
FIGURE 5.5 Ferry and bridge cash flows in $000

$$= -4{,}800 + 740(6.566) + 180(0.2472) + 180(0.0611) + 30(0.0151)$$
$$= -4{,}800 + 4{,}858.80 + 44.50 + 11.00 + 0.45$$
$$= 114.75$$
$$\approx 115$$

A positive result of approximately $115,000 indicates that the extra investment in the bridge is more than balanced by its extra benefits.

The same result is obtained if we determine the net present value of the bridge alternative individually and compare it to the individual NPV of the ferry alternative.

$$\begin{aligned} \text{NPV} &= -5{,}000 + 800(P/A, 15, 30) + 50\,(P/F, 15, 30) \\ &= -5{,}000 + 800(6.566) + 50(0.0151) \\ &= -5{,}000 + 5{,}252.80 + 0.76 \\ &= 253.56 \\ &\approx 254 \end{aligned}$$

Comparing $254,000 to $139,000 causes us to choose the higher amount and with it the bridge alternative. The difference between these two numbers is $115,000, the value of the incremental cash flow of the bridge over the ferry. ■

It might seem that the problem of unequal lives between alternatives could be solved by assigning salvage values or resale values to the longer-lived investment at the end of the life of the shorter-lived investment. And as we shall see in Chap. 16 on replacement analysis, this is sometimes convenient; but it should not be considered a solution to the problem of unequal lives, already solved by the introduction of multiple cycles in the lives of the alternatives. Indeed, such a device would be an encumbrance in the example we have just seen. How could one assign a salvage value, let alone a resale value, to the bridge at the end of 10 years that would represent a measure of 20 more years of service? In industrial and real estate projects, such an assignment may be possible sometimes, but not generally in public projects.

DEFERRED INVESTMENTS

A frequently encountered question in engineering design is, Should we design and build for full capacity now, or would it be better to defer some of the cost until the time when the larger capacity is needed? The question is readily answered by means of a present-worth analysis of the alternatives.

■ **Example 5.7.** Two alternatives are being considered by the St. Francis Water Authority for replacement of an existing water line. Alternative 1 is to install an 18-inch main now and an additional 18-inch main 10 years hence alongside it. The initial cost of each 18-inch main is $125,000. The total economic life of the installation of both mains in this manner is estimated to be 40 years. No salvage value is expected. Alternative 2 is to install a single 26-inch main now at a cost of $200,000. No salvage value is anticipated for this alternative at the end of its 40-year life. If the opportunity cost of capital of this organization is 8 percent after all considerations of inflation are taken into account, which alternative should be chosen?

Solution. Figure 5.6 shows the two alternatives. If we analyze incrementally, we will see that it is impossible to compare the lower-cost alternative to the do-nothing or null alternative, because no benefits are made explicit. In other words, the implications are that the benefits are the same for both alternatives and that the null alternative does not exist. St. Francis is determined to build one or the other of the alternatives, and we only need to decide which one shall be built. Thus we must compare, incrementally or individually, one cash flow with the other.

The net present value of the bottom cash flow diagram of Fig. 5.6 is

$$
\begin{aligned}
\Delta \text{NPV} &= -75{,}000 + 125{,}000(P/F, 8, 10) \\
&= -75{,}000 + 125{,}000(0.4632) \\
&= -75{,}000 + 57{,}900 \\
&= -17{,}100
\end{aligned}
$$

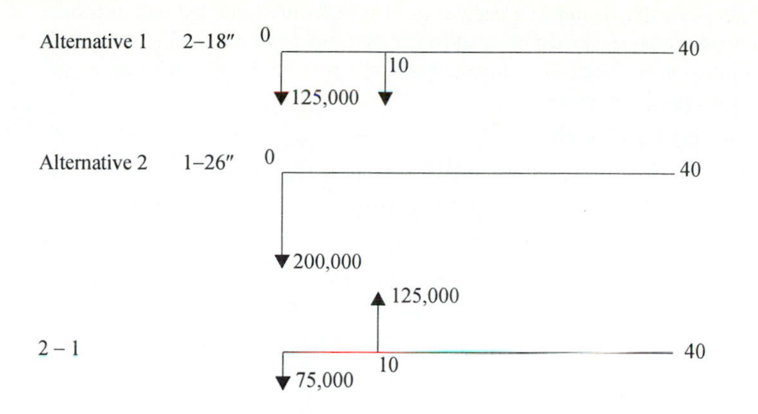

FIGURE 5.6 Deferred investment in a water main

The extra investment in the 26-inch main is not justified, therefore, and it is better to wait 10 years before installing the 18-inch second main.

Notice how two cash flows, composed exclusively of costs, resulted in an incremental cash flow where up arrows, or benefits, were found. This is usual in such situations. The commonsense interpretation is that if St. Francis chooses to build total capacity now, it will avoid a cost of $125,000 for another main 10 years in the future. A cost avoided is a benefit, hence the up arrow. ■

PERPETUAL INVESTMENTS

Investments designed to last for such a long time as to be considered perpetual deserve some special mention. If you were to invest $100,000 at 9 percent without thought of removing any of the capital, but simply enjoying the interest payments forever, you could count on an income of $9,000 per year from this source. In symbolic terms,

$$A = Pi \tag{5.5}$$

From this it follows that the capital recovery factor (A/P, i, ∞) is simply equal to i, the interest rate.

If you had asked instead, "How much money do I need to accumulate to ensure an income of $9,000 annually forever, if I can safely invest at 9 percent?" the answer would be

$$P = \frac{A}{i} = A\frac{1}{i} \tag{5.6}$$

and $9,000 divided by 0.09 is $100,000. Thus it appears that the series present-worth factor in the case of perpetual investments (P/A, i, ∞) is the inverse of the interest rate, or $1/i$.

■ **Example 5.8.** A wealthy landowner decides to set up a family fund that will distribute \$250,000 per year among her children and their descendants forever. She can invest safely, after inflation has been taken into account, at 4 percent. How much capital will she need to accomplish her purpose?

Solution. Applying Eq. (5.6) gives

$$\begin{aligned} P &= A(1/i) \\ &= 250{,}000(1/0.04) \\ &= \$6{,}250{,}000 \end{aligned}$$

■

■ **Example 5.9.** A university is given \$1,000,000 by an alumna to set up a perpetual chaired professorship in the art history of women. How much money will the holder of this chair receive annually if the grant can be invested at 6 percent per annum?

Eq. (5.5) may be used:

$$\begin{aligned} A &= Pi \\ &= (1{,}000{,}000)(0.06) \\ &= \$60{,}000 \end{aligned}$$

■

CAPITALIZED COST

Capitalized cost is the present worth of a perpetual annual payment. It is another name for P in Eq. (5.6).

Suppose, now, that a more complicated cash flow is involved, where an asset will not physically endure perpetually but must be replaced at intervals. The method is best explained by means of an example.

■ **Example 5.10.** The St. Francis Water Authority is contemplating a new main that will provide water service as far in the future as anyone can foresee. It will cost \$10 million and will need to be replaced every 50 years. If the authority requires an 8 percent rate of return on its investment, what will be the capitalized cost of the new main?

Solution. Figure 5.7 illustrates the situation. In the top diagram, \$10 million is repeated every 50 years. The middle portion of the figure shows \$10 million at time 0 converted as follows:

$$\begin{aligned} A &= P(A/P, i, N) \\ &= 10{,}000{,}000\,(A/P, 8, 50) \\ &= 10{,}000{,}000(0.08174) \\ &= \$817{,}400 \end{aligned}$$

The \$817,400 will recur within every repeated cycle of investment of \$10 million. The problem now resolves to computing the capitalized cost of \$817,400 annually forever. Equation (5.6) supplies the answer:

$$\begin{aligned} P &= A\frac{1}{i} \\ &= 817{,}400\left(\frac{1}{0.08}\right) \\ &= \$10{,}217{,}500 \end{aligned}$$

■

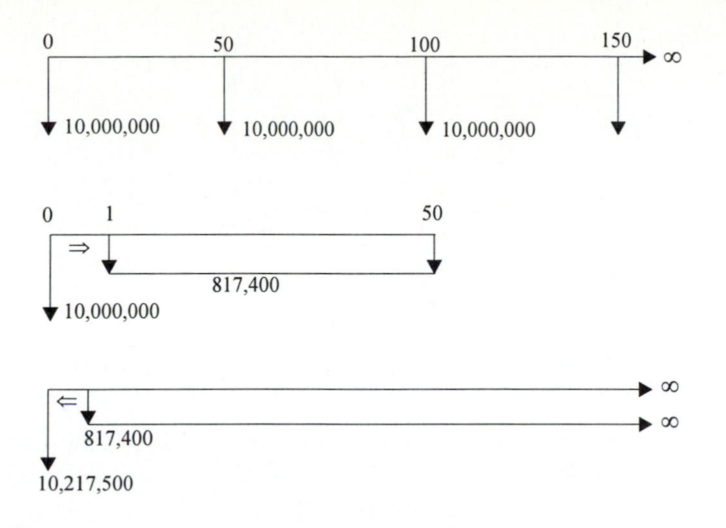

FIGURE 5.7 Capitalized cost

VALUATION AND BONDS

It is sometimes difficult to explain what present worth means to someone who has not studied it. But, as mentioned previously, one way is to use the concept of the value of future receipts, that is, *valuation.* If a stock, or a bond, or a piece of real estate investment property is estimated to have a return of money at specific points in the time you hold it, the value of the asset can be thought of as the present value of each of those sums of money. This present value is in terms of an interest rate of return that you give up. This interest rate is called the *opportunity cost of capital.* Looked at from the opposite angle, the present value is the sum of money that, invested at the opportunity cost of capital, will return money at specific points in the time you hold the asset. The point is clearly seen in the valuation of a bond.

■ **Example 5.11.** Santa Rosa Water Company issued nominal 8 percent bonds in $5,000 denominations payable in 20 years. Interest is to be paid semiannually. The first interest payment will be received 6 months after the bond is purchased. How much should an investor whose opportunity cost of capital is a nominal 10 percent compounded semiannually pay for these bonds?

Solution. The answer to this question is best seen broken down into two parts:

Part 1: Find the cash receipts of the bond. Because the bond will pay semiannually at a nominal rate of 8 percent, its effective interest rate per period is

$$\begin{aligned} i &= r/M \\ &= 0.08/2 \\ &= 0.04 \end{aligned}$$

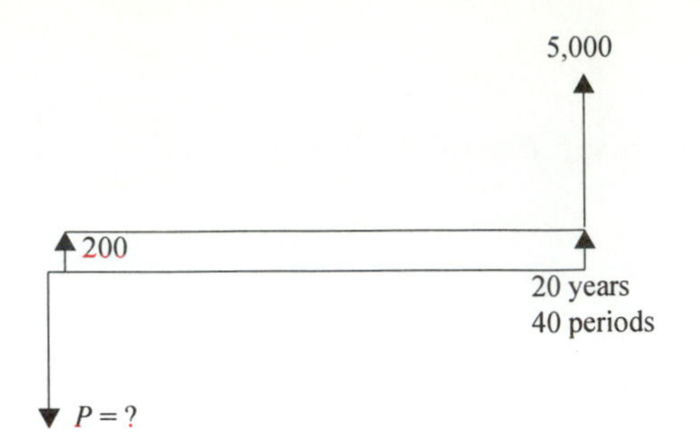

FIGURE 5.8 Bond valuation

Twenty years will equal 40 semiannual periods. At the end of each period, the bondholder will receive

$$A = (0.04)(5{,}000)$$
$$= \$200$$

At the end of 20 years, the bondholder will be repaid the face value of the bond, $5,000. The cash flow of the bond receipts is seen in Fig. 5.8, the amounts the bondholder will receive. Notice that these receipts are determined from the information on the bond itself. They represent the contractual promise to pay, the IOU, of Santa Rosa Water Company to the bondholder.

Part 2: Find the bond valuation. The bond's value, that is, its value to a specific buyer, the one whose nominal opportunity cost of capital is 10 percent, is the present worth of the bond receipts:

$$P = A(P/A, i', N,) + F(P/F, i', N)$$

Here i' is the *opportunity cost* of capital, not the interest rate on the bond.

$$i' = r'/M = 0.10/2 = 0.05$$
$$P = 200(P/A, 5, 40) + 5{,}000(P/F, 5, 40)$$
$$= 200(17.159) + 5{,}000(0.1420)$$
$$= 3{,}432 + 710 = \$4{,}142$$

The maximum value of the bond to the individual whose opportunity cost of capital is 10 percent nominal compounded semiannually is $4,142. The value would be less for someone whose minimum attractive rate of return were higher than 10 percent and more for someone else whose minimum attractive rate of return were lower than 10 percent.

Suppose that this bond is selling at $4,750 in today's bond market. Should it be bought by the investor whose nominal opportunity cost of capital is 10 percent? We have discovered that the answer is no, because $4,750, the price, is more than the value of the bond to our investor, $4,142. ■

A last example in this chapter is presented in order to familiarize the reader with more complicated problems and to introduce certain useful techniques in the present-worth method.

TABLE 5.1
Two transit proposals ($000,000)

Years	Cost	Benefit
Alternative 1: heavy rail		
0	55	—
1–9	100	—
3–7	20	30
8–10	40	35
11–∞	60	50
Alternative 2: staged rolling stock		
0	55	—
1–6	50	—
2–6	—	25
7–14	55	40
15–∞	60	50

■ **Example 5.12.** Two proposals remain for consideration from the 10 that have been put forward for a major subway system for a large city. One is certain to be chosen.

Alternative 1 envisions the installation of a heavy rail system over the next 9 years with construction costs of $100 million during each of those years, in addition to a $55 million outlay at the same time as the project is approved, that is, at time 0. Maintenance and operations costs will average $20 million annually for years 3 through 7, $40 million annually for years 8 through 10, and $60 million perpetually thereafter. Benefits will begin at year 3 at $30 million annually and remain at that figure until the end of year 7. For years 8 through 10, benefits will average $35 million per year, and for year 11 and perpetually thereafter, $50 million. These costs and benefits are shown in Table 5.1.

Alternative 2 involves stage construction of facilities. Not only extension of the system in stages is contemplated, but also that the type of rolling stock will be upgraded in three stages from electric trolley buses, to light railcars, and finally to heavy rail trains. The costs and benefits of alternative 2 are also shown in Table 5.1. Electric trolley buses will be used during the first 6 years of system operation. Benefits from this first stage will not appear until year 2 and will continue throughout year 6. During years 7 through 14, light rail will carry the system's passengers. Heavy rail will be in operation perpetually thereafter.

By using a present-worth analysis of each alternative, select the more attractive course of action.

Assume that no taxes will be paid because the enterprise is a public one. Assume also that inflation has been taken into account on the entries of Table 5.1. A 4 percent opportunity cost of capital under the preceding conditions will be used.

Solution. Figure 5.9 shows the resulting cash flows not netted out, because it is easier to handle them in blocks. Alternative 1 may be presented as follows:

Year	Cash flow ($000,000)
0	−55
1–9	−100
3–7	+30 − 20 = +10
8–10	+35 − 40 = −5
11– ∞	+50 − 60 = −10

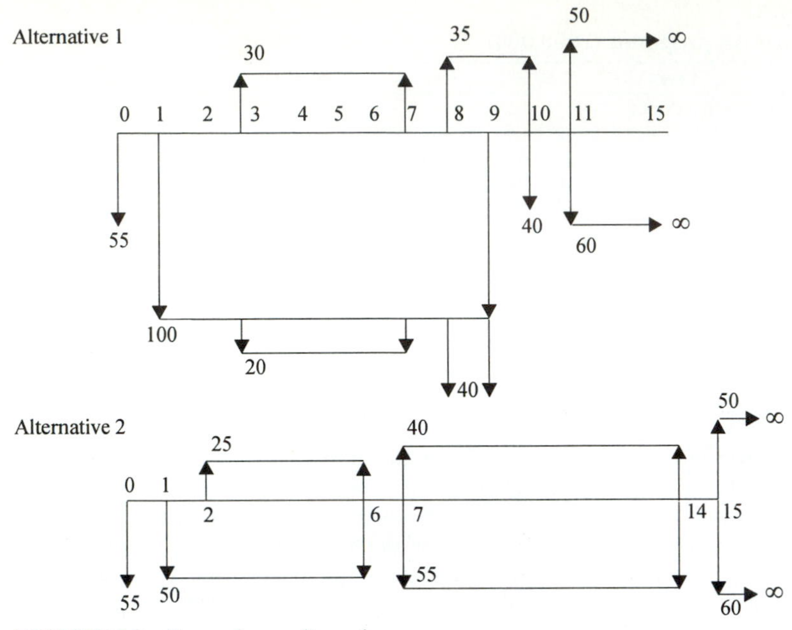

FIGURE 5.9 Two subway alternatives

The net present value (present worth) of alternative 1 is

$$\text{NPV}_1 = -55 - 100(P/A, 4, 9) + 10(P/A, 4, 5)(P/F, 4, 2) - 5(P/A, 4, 3)(P/F, 4, 7) - 10(1/0.04)(P/F, 4, 10)$$

The third term on the right-hand side of the preceding equation deserves some explanation. Five up arrows of +10 each are present at the end of years 3 through 7. Their present worth at year 2 can be found from the +10(*P*/*A*, 4, 5) portion of the calculation and then taken to time 0 by multiplying by the single-payment present-worth factor (*P*/*F*, 4, 2). The same effect could have been achieved by multiplying +10 by [(*P*/*A*, 4, 7) – (*P*/*A*, 4, 2)]. Both these techniques will give exactly the same answers. Figure 5.10 explains the manipulations graphically.

We finish the calculation:

$$\text{NPV}_1 = -55 - 100(7.435) + 10(4.452)(0.9246) - 5(2.775)(0.7599) - 10(1/0.04)(0.6756)$$

$$= -55 - 743.5 + 41.2 - 10.5 - 168.9$$

$$= -936.7$$

$$\approx -\$940$$

We present alternative 2 in the same manner as alternative 1:

Year	Cash flow ($000,000)
0	-55
1	-50
2–6	+25 - 50 = -25
7–14	+40 - 55 = -15
15– ∞	+50 - 60 = -10

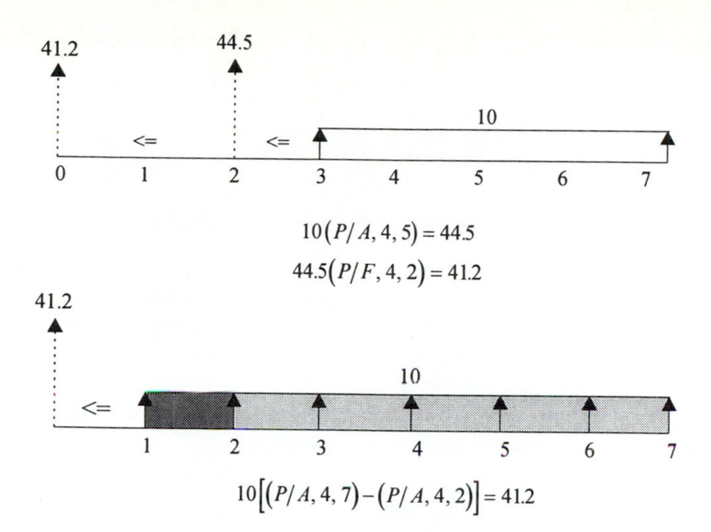

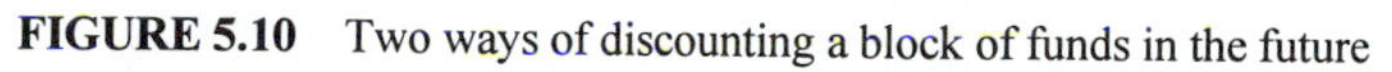

FIGURE 5.10 Two ways of discounting a block of funds in the future

The net present value of alternative 2 is

$$\begin{aligned}\text{NPV}_2 &= -55 - 50(P/F, 4, 1) - 25(P/A, 4, 5)(P/F, 4, 1)\\ &\qquad - 15(P/A, 4, 8)(P/F, 4, 6) - 10(1/0.04)(P/F, 4, 14)\\ &= -55 - 50(0.9615) - 25(4.452)(0.9615) - 15(6.733)(0.7903)\\ &\qquad - 10(1/0.04)(0.5775)\\ &= -55 - 48.1 - 107.0 - 79.8 - 144.4\\ &= -434.3\\ &\approx -430\end{aligned}$$

The lower-cost alternative is number 2, stage construction of the facilities. ■

Individual benefit minus cost analysis was chosen because it is the easiest one to use, given the pattern of the cash flows. Incremental analysis was avoided because a special problem enters which is better taken up later in Chap. 7 on the benefit/cost ratio. It is the question of how to handle equal initial investments where incremental analysis is mandatory.

FUTURE WORTH

This method of choosing among mutually exclusive projects is infrequently used, but some explanation of it is warranted. Instead of moving costs and benefits to time 0, as in the present-worth method, these are moved forward, by using compounding, to the end of the time horizon chosen. If future benefits exceed or equal future costs, then the project is accepted; if not, it is rejected. In formal terms,

$$\sum_{j=0}^{N}(B_j - C_j)(F/P,i,j) \geq 0$$

means acceptance. The symbols are the same as those in Eq. (5.1).

■ **Example 5.13.** Your friend, Mr. Arbuthnot, questions your sanity on the issue of the future worth of some washing machine stock you are about to purchase. You insist that it will appreciate to \$300,000 in 10 years. Its purchase price now is \$65,000. Expected dividends are \$3000 per year. What is the future worth, including both benefits and costs, of this transaction at the end of 10 years if the stock is sold for \$300,000? Your opportunity cost for similar-risk projects is 12 percent.

Solution. Moving benefits and costs to the end of year 10 are

$$\begin{aligned} F_{10} &= -65{,}000(F/P,12,10) + 3{,}000(F/A,12,10) + 300{,}000 \\ &= -65{,}000(3.1058) + 3{,}000(17.549) + 300{,}000 \\ &= -201{,}880 + 52{,}650 + 300{,}000 \\ &= 150{,}770 \end{aligned}$$

The net future value of the investment is about \$151,000. Because this is a positive amount, the investment is acceptable. ■

SUMMARY

Present worth, net present value (NPV), and benefit minus cost all refer to the same method which requires that the discounted value of the benefits minus the discounted value of the costs of an alternative be equal to or greater than zero for that alternative to be selected.

The consequences of alternatives over time may be judged either incrementally or individually in this method.

The lives of the alternatives must be equalized when the present-worth method is used. This is often done by selecting the least common multiple of the individual lives as the study period and repeating the payments for each cycle of life of an alternative for the number of cycles necessary to reach the end of the study period. This equalizing device ensures fairness in the comparison of alternatives. It does not imply that life cycles will in fact be repeated.

Salvage values are treated as normal payments in this method. Care should be taken in the use of salvage values as a method of equalizing lives—not always a sound procedure.

Capitalized cost is the present worth of a cost for an endless analysis period.

The future-worth method is similar to the present-worth method, but it is less frequently used.

PROBLEMS

5.1. In Example 5.1, page 70, the restaurant is discovered to cost considerably more than had been anticipated. The loan requested is now $25,000, and the payments are to be doubled, that is, to $4,000, $8,000, and $14,000, to be received at the same times as before. Show your calculations in order to decide whether you should lend the money.

5.2. An orange grove you are interested in buying will come into full bearing at the end of 3 years. The first crop and income can be expected at the end of the third year. It will have a net value of $200,000 per year for a total of 48 years. Based on this yield, how much should you pay for the grove if your opportunity cost of capital for an equal-risk investment is 15 percent before income taxes and inflation are taken into account?

5.3. In Example 5.2, page 71, passenger time benefits have been badly underestimated because of the unexpectedly heavy use of the tunnel revealed by new projections. Benefits are now believed to be at least $150,000 per year. However, the review of benefits has also revealed that construction costs will now be $1.1 million for this short tunnel. Lighting and janitor service costs remain as before. Should the tunnel be built, given the new data?

5.4. Oliver Cromwell real estate development company bought a commercially zoned lot 5 years ago on a land contract that required payments of $12,000 per year for 10 years. The company now wishes to build a shopping center on this lot, but has discovered an equally acceptable site nearby which can be purchased for cash at $38,000.

Five payments of $12,000 each have been made. If Oliver Cromwell Company reneges on its contract to buy the original lot, title to the lot will remain with the seller and Oliver Cromwell Company will lose its right to the property under the land contract. What should Oliver Cromwell Company do, if its before-tax cost of capital is 15 percent and no income tax, inflation, or any factors other than the ones mentioned above are considered?

5.5. In Example 5.3, page 73, the costs of location 1 have been revised. The initial cost will be $150 million. Annual operation and maintenance costs have also been revised upward to $6 million per year. If the rest of the data remain unchanged, what should now be the decision?

5.6. An agricultural engineer is studying a particular dairy farming operation. A major expense of this operation is feeding a herd of 25 cows and 3 bulls. On average, each animal eats 100 bales of hay per year. The dairy farmer, who is providing the data for the study, estimates that he can grow his own hay at $1 per bale for the next 10 years, the time horizon of the study. He also believes that his herd will increase by 3 head per year, beginning in the second year. His opportunity cost of farming his land for hay, rather than another use, he estimates at 20 percent per year. The engineer decides to determine the present cost of feeding this herd for the next 10 years, as part of the study. Make the calculation for him.

5.7. Elizabeth Hollister wishes to establish 20 scholarships for worthy students at the local university. Four scholarships will run concurrently for 4 years. Another group of four scholars will then be funded, and so on until 20 years have elapsed. The following table shows the outlay for the awards:

	Scholarship			
Year	**1**	**2**	**3**	**4**
1	5,000	5,000	5,000	5,000
2	5,250	5,250	5,250	5,250
3	5,500	5,500	5,500	5,500
4	5,750	5,750	5,750	5,750
5	6,000	.	.	.
6	6,250	.	.	.
7	6,500	.	.	.
8	6,750	.	.	.
9	7,000	.	.	.
.	.	.	.	.
.	.	.	.	.
.	.	.	.	.
20	9,750	9,750	9,750	9,750

The extra $250 each year is designed to offset inflation.

How much money will the donor, Elizabeth Hollister, be required to give the university if the endowment portfolio can be invested at a nominal 11.4 percent, compounded monthly, tax-free?

5.8. In Example 5.5, page 76, the resource opportunity cost is now 15 percent. The benefits of alternative 2 are increased to $40 million per year, because of the inclusion of reduced traffic congestion. Which alternative will be chosen, or will both be rejected?

5.9. Thomas Hardy is the father of three boys: John, 12; David, 8; and Peter, 4. To pay for their college educations, Hardy is considering putting a lump sum of money in the bank today. All his sons will go to college at the age of 19, which means that his oldest son, John, will start college 7 years from now, when he will need the first year's funds. The second son, David, will start college in the same year as his brother John finishes. Finally, his third son, Peter, will start college in the same year as his brother David graduates.

The cost of sending John to college for the first year will be $20,000 and will increase by $500 per year for the following 3 years. David's expenses will be $22,500 the first year and will increase by $1,000 per year for the next 3 years. Peter's first year will cost $27,000 and will increase by $1,500 annually for the subsequent 3 years. Thus the funds requirements appear as follows:

Year	Funds required
7	$20,000
8	20,500
9	21,000
10	21,500
11	22,500
12	23,500
13	24,500
14	25,500
15	27,000
16	28,500
17	30,000
18	31,500

How much should Hardy put in the bank today to pay for his sons' educations if the bank will pay 10 percent annual interest on his money for the next 18 years?

5.10. An engineering firm is considering buying new drawing equipment which will cost $34,000 and will have an estimated salvage value of $28,000 in 10 years. Special tools for the new equipment will cost $10,000 and have a resale value of $5,000 at the end of 10 years. Maintenance costs are estimated at $400 per year.

What will be the net present cost to the firm if its discount rate is 9 percent?

5.11. In Example 5.6, page 78, it is now believed that the ferry will last 15 years instead of only 10. All the other data remain unchanged. Now, however, the public road authority requires that one or the other alternative be chosen. Which one should be built?

5.12. A cattle ranch in western Nevada has been offered for sale at $6 million at the end of 1995. The owner presents books of accounts showing net after-tax profits of $1 million for 1993, $1.1 million for 1994, and $1.2 million for 1995.

If this trend in net after-tax profits continues over the next 20 years and your acceptable rate of return is 20 percent, should you buy the property now, *ceteris paribus*? (For the purpose of your calculations, assume a resale value of $12 million at the end of 20 years.)

5.13. Your company, Aviation Taxi, is attempting to decide whether to buy a Lancer aircraft. Its first cost is $29,000, used, and the projected difference between its revenues and costs is $5,500 annually. At the end of 5 years, the time horizon of the study, it can be sold for $12,000. Taxes and inflation have been included in those figures. The after-tax, after-inflation minimum attractive rate of return for your company is 10 percent.

(*a*) Should you buy the aircraft? *Use the present-worth method.*

(*b*) Whatever the results of part (*a*), your company buys the aircraft. After owning it for 2 years, you receive an offer of $20,000 for the Lancer. Assuming that your original predictions were correct, should you keep it or sell it? *Use the present-worth method.*

5.14. Exactly 4 years ago, your company, J&B Construction Leasing Associates, bought a power shovel for $95,000. At that time the projected cash flow associated with it was as follows:

Year	Cash flow	
0	−$95,000	First cost
1–10	+25,000	Net income after all taxes
10	+10,000	Resale value

The company has received an offer of $50,000 for the shovel. Should the company accept the offer, ignoring possible inflation and income tax effects for the moment? The appropriate cost of capital is 15 percent. All cash flow projections made 4 years ago still hold true for the future. *Use the NPV method.*

5.15. Two alternate plans in the design of some small industrial buildings are summarized here:

	Plan 1	Plan 2
First cost ($)	25,000	50,000
Life (years)	10	20
Salvage value ($)	5,000	0
Annual cost ($)	6,000	2,500

One of these two plans is certain to be chosen; therefore it is not necessary to consider the null alternative (status quo). Tax and inflation considerations have been included in the cash flows.

(*a*) Using a minimum attractive rate of return of 20 percent before income taxes, compare the plans on the basis of present worth. Assume that all replacements have the same first costs, lives, and salvage values and annual disbursements as the initial facilities. Which plan should be chosen?

Answer: Plan 1

(*b*) Using the present-worth method, but this time by *incremental analysis*, select the best plan.

Answer: Plan 1. ΔNPW = −4,856

(*c*) How can you check the answer given by (*a*) against that given by (*b*)?

5.16. Two air compressor makes are being compared for a large construction company with a view toward standardization on one make. The data for use in an economic analysis are as follows:

Make	First cost ($)	Annual cost ($)	Life (years)	Salvage value ($)
Fairbanks	8,000	600	4	1,000
Stromberg	10,000	500	5	1,000

The opportunity cost of capital for this company is 25 percent before taxes. Do a present-worth before-tax analysis, and make a recommendation. The tax consequences of the compressors are thought to be equal. Inflation considerations will also be the same for both makes.

Answer: $\text{NPW}_F = -15{,}079$, $\text{NPW}_S = -16{,}198$. Choose Fairbanks.

5.17. A venture in an ice cream franchise store will generate the cash flow (in thousands of dollars) shown in the table below. A similar flyer in a hamburger franchise is estimated to show the revenues and costs in the same table. The franchise will be sold at the end of the fourth year at the price shown.

Franchise and year	First cost	Revenues	Costs	Net sale income
Ice cream	20			
1		225	200	
2		280	255	
3		330	300	
4		400	350	100
Hamburger	25			
1		310	280	
2		322	290	
3		344	310	
4		370	325	125

Tax consequences will be similar for both franchises. Inflationary effects have been taken into account in the estimates. What should be the investor's decision if her minimum attractive rate of return after income taxes and before inflation is 20 percent? Should she invest at all in either franchise? Use the present-worth method.

5.18. Two types of tunneling equipment are under consideration by Borers Company. The machines are characterized as follows:

	United States	Austrian
First cost ($)	325,000	375,000
Operations and maintenance annual cost ($)	30,000	28,000
Life (years)	10	15
Resale value ($)	32,500	50,000

As an analyst for Borers, make a present-worth analysis on an *incremental* basis. Select the more economical machine if the company uses a 20 percent discount rate. No tax or inflation effects are to be considered.

5.19. Your advisers assure you that they can always get at least a 5 percent return on your funds by investing in Reputable Savings and Loan Association. However, you find that you cannot do without a car any longer. You can buy an inexpensive car for $6,000. Its useful remaining life is 5 years. The resale value at the end of this period will be about $600. Maintenance costs will start at $250 for the first year and rise at 10 percent annually.

A better car will cost you $12,000 and will have a resale value at the end of 5 years of $4,000. Maintenance will be $150 for the first year and will rise at 8 percent per year thereafter.

Which car should you buy?

5.20. John Webster wants to add a sundeck to the back of his house. He must choose between redwood or common lumber at a lower price but with less life expectancy and greater maintenance cost. Redwood will cost $4,200 initially and $30 annually thereafter in maintenance. Common lumber will cost $2,800 and $60 per year thereafter. A real estate appraiser says that, if properly maintained, the deck will add to the sales value of the house at about the deck's original cost. He expects to sell the house 10 years from now. Assume that the benefits will be the same for both alternatives. Disregarding tax effects and inflation, determine which alternative is more economical, given that Webster's opportunity cost of capital is 30 percent. *Use the NPW method.*

5.21. Perform the calculations required in Example 5.7, page 80, but analyze the alternatives *individually*. Is there a numerical relation between the individual net present values and the incremental NPV?

5.22. In a developing country, a government housing project is to be constructed in two phases. If minimal standards for streets and utilities (water, electricity, etc.) are used, the first phase will cost $27 million. If higher standards are used, the first phase will cost $40 million.

Future expansion in the second phase will cost less for the project layout with higher standards. With lower standards, the future expansion (second phase) will cost $64 million. With higher standards the future expansion will cost $38 million. Both the original project and its expansion will be demolished 50 years after the original project is constructed. No differential costs or benefits will come about by providing for expansion of the project using the higher-standard design.

In how many years, to the nearest whole year, must the second-phase expansion be constructed in order to justify the higher-standard first cost? Use a discount rate for government projects of 12 percent. Do not consider inflation effects. Because this is a government project, no taxes of any kind will be paid.

5.23. In Example 5.8, page 82, the wealthy landowner decides to be even more generous and distribute $400,000 per year but finds, to her pleasant surprise, that she can now invest at a safe 6 percent. How much money will she have to put into the trust fund?

5.24. A couple decide to provide for their old age by setting aside a sum of money each year for the next 20 years. At the end of the 21st year, they wish to withdraw $25,000 and to continue to withdraw $25,000 per year perpetually.

If their opportunity cost of capital is 10 percent, how much must they invest each year for the next 20 years? Assume that 10 percent is the return they can receive on their funds after they have allowed for inflation and after they have paid their taxes.

Answer: $4,365

5.25. In Example 5.9, page 82, the university wishes to provide the holder of the chaired professorship with $100,000 per year. How much will the donor need to give to the university if university funds can be invested at 5 percent?

5.26. A university chair is proposed in economics analysis. The prospective donor wishes to know the cost of the endowment if the chair expenses are estimated to be $40,000 per year. The endowment can be safely invested at 6 percent compounded annually. Find the cost of the endowment.

5.27. In Example 5.10, page 82, the St. Francis Water Authority becomes aware that a cheaper pipe is available. It will cost $8 million for the main, but it will last only 25 years. What will be its capitalized cost if money is worth 8 percent? Which type of main should be installed?

5.28. The interest rate on home loans has dropped from 9½ percent to 8 percent in the time you have hesitated about buying a house. On the prospective loan of $66,600 that you will need to buy the house you have in mind over a term of 30 years, how much has this drop in the interest rate saved you in terms of present worth, if your personal opportunity cost of capital is 10 percent? Do not consider tax effects.

Answer: $8,087

5.29. In Example 5.11, page 83, Santa Rosa Water Company, because of lowered interest rates on the market, has issued new bonds at a nominal 6 percent, payable quarterly, over 10 years, in $10,000 denominations. An investor whose opportunity cost of capital is 8 percent wishes to know if he should pay $9,500 for a bond. What is the top price he should pay?

5.30. A $1,000 municipal bond, offered free of state and federal taxes, is priced at $1,020. Its term is 20 years, and it pays 10 percent nominal interest in quarterly payments. Your opportunity cost of capital for an equal-risk investment is a nominal 20 percent before taxes are paid and a nominal 9 percent after taxes. Should you make this investment if inflation effects are ignored and if the first interest payment is received 3 months after you buy the bonds?

5.31. Rosita Lunny purchased a 10-year $10,000 bond exactly 2 years ago for $9,160. The bond bears interest at a nominal 8 percent compounded semiannually. Since that time, interest rates have skyrocketed, and similar bonds are now being sold at a price reflecting a nominal 18 percent. Lunny's personal opportunity cost of capital is a nominal 16 percent. She has just received an interest payment. Circumstances force her to sell the bond now.

A friend, Fred Cox, advises her to hold out for $10,000 because that is the face value of the bond. Another friend, Miguel Sanchez, believes that Lunny should sell at $8,132.26. Robert Lee, Rosita's cousin, counsels her to price the bond at $9,160 because that is what she paid for it.

Which friend is correct, if any? Or what should Lunny ask for her bond if tax effects and inflation are ignored?

5.32. A $1,000 tax-free bond offers interest at a nominal 8 percent compounded semiannually. The term of the bond is 15 years, and it is offered at $950. Your personal opportunity cost of capital for investments of equal risk is a nominal 12 percent annually after taxes. Should you purchase the bond if the first interest payment will be received 6 months after the bond is bought?

5.33. You purchased a 20-year-term bond 5 years ago for $10,469.73 with a face value of $10,000. It pays interest semiannually at 10 percent nominal rate with the next payment due in 6 months. Recently you have encountered a chance for investment in rental real estate. To invest in this real estate venture, you will need to sell the bond, which has a market value now of $8,000. It appears that your real estate investment will return a 16 percent effective annual rate for the next 5 years, and you believe that you will be able to continue to make real estate investments at that rate of return for as long as you choose to do so at the same risk as the bond investment.

A friend, Ellen Ryland, urges you not to sell the bond until you have "gotten your money out of it," that is, to hold onto it until you recover the original cost of the bond. She points out that the loss you will sustain if you sell the bond now will not be made up for almost 2 years, calculated as follows:

$$10{,}469.73 - 8{,}000 = t(0.16)(8{,}000)$$
$$t = 1.93 \text{ years}$$

On this basis, she urges you to wait 2 years and then review the situation.

Another friend, Sam Orwell, observes that you should keep the bond because interest rates are dropping. In 2 years, bond rates will return to the 10 percent level, he is certain, and therefore you will be able to sell your bond for its face value and suffer only a small loss.

Which friend's advice is sound and why? Comment on both. Or should you follow a different course of action from what either friend advises?

Do not consider tax or inflation effects in this analysis. Both the bonds and the real estate you consider equally risky in today's economic climate.

5.34. In Example 5.12, page 85, another alternative has appeared, known as alternative 3. Light rail will be used from year 2 on indefinitely. The original cost of $55 million at time 0 remains in place. The costs of light rail will continue at $55 million per year from year 1 *ad infinitum*. Benefits, at $40 million annually at the end of year 2, will continue indefinitely as well. Should alternative 3 be chosen? Do an individual NPV analysis. All other data remain the same.

5.35. Two makes of solar repeaters for telephone service to remote locations are being considered by a telephone company in the western United States. The characteristics of each make are as follows:

	A	*B*
First cost ($)	500,000	700,000
Life (years)	4	8
Salvage value ($)	50,000	70,000
Annual maintenance	10,000	40,000

The telephone company management believes that the service must be installed. Thus the null alternative is nonexistent. The opportunity cost of funds is 20 percent. Using the *net present-worth* method, choose the most attractive alternative. Either individual or incremental analysis may be used. No tax or inflation effects need be considered.

5.36. Two systems have been proposed to solve a problem of urban transport in a large city in an African country. One alternative consists of a fleet of buses and its support system. The other alternative is a subway.

The bus system will cost $100 million and will generate $18 million annually during its 15-year life. There will be no salvage value.

The subway will cost $300 million in costs accumulated to time 0 and will generate $30 million each year in net profits. It will have a salvage value of $150 million at the end of 30 years.

All tax and inflation estimates have been included in the benefits and costs cited. Using the present-worth method and assuming that the city may decide against both systems, which should you choose? The opportunity cost of resources is 12 percent.

5.37. An electric utility company is promoting the use of a new type of fluorescent lightbulb, using the following information: Existing 75-watt incandescent bulbs will need to be replaced twice per year at a cost of $1.00 each. A 22-watt fluorescent bulb will cost $20.00, but will last 5 years and save $6.72 annually in electricity costs at $0.07 per kilowatthour. Is it worth it to replace incandescent bulbs as they wear out with fluorescent, if your opportunity cost of capital is an effective annual 10 percent? *Use the incremental NPW method and the end-of-year convention.*

5.38. Three years ago, you bought 10 gasoline-powered jackhammers for use in your tool rental business at a cost of $800 each. At that time you calculated a net return of $200 per year per jackhammer, after all taxes were paid, over a life of 5 years with a $50 salvage value each.

Now you are offered $100 each for the jackhammers. Should you accept the offer if your opportunity cost of capital is 12 percent for an equal-risk investment? No capital gains or losses are expected if you accept the offer. All your estimates of 3 years ago are thought to be valid for the future. *Use the NPV method.*

5.39. A carousel at an amusement park is losing money. It will not be replaced. However, its disposal value as a piece of Americana is rising year by year. The time horizon of the study is 5 years. Its revenue, cost, and disposal values for each of the next 5 years, in thousands of dollars, are as follows:

Year	Revenue	Cost	Disposal value
0			
1	35	37	12
2	38	41	14
3	41	45	18
4	44	49	20
5	47	53	22

The relevant opportunity cost of capital is 15 percent.

If you use only the *present-worth method* and neglect taxes and inflation, when should the carousel be sold?

5.40. Repeat Prob. 5.15, using the future-worth method.

5.41. Repeat Prob. 5.31, using the future-worth method.

5.42. Two alternative plans for electrical switching gear are estimated to have the following features:

	Plan 1	**Plan 2**
First cost ($)	28,000	50,000
Life (years)	10	20
Salvage value ($)	3,000	0
Annual cost ($)	5,000	2,600

One of these plans is certain to be chosen. The opportunity cost of capital, before taxes, for this firm is 15 percent. Using the *net-present-worth method*, before taxes, choose between the plans.

CHAPTER 6

ANNUAL WORTH

The annual-worth method has the great advantage of being readily understandable by ordinary people. It simply means annual gain in the case of a positive annual worth and annual loss in the case of a negative annual worth. As with the present-worth method, its use does not require incremental methods, and thus its programming in computers is relatively easy. Because accounting information is generally in annual form, this method can use accounting data with minimal changes. Its disadvantage is its unwieldiness when annual costs and revenues are not the same each year, as will be demonstrated in this chapter.

Annual worth is the term that encompasses both *gain* and *loss*—ordinary words that everyone understands. Why, then, use such a phrase as *annual worth* when *gain* or *loss* might be used equally well? The reason is that annual worth includes within it another idea—uniformity. When we speak of annual worth, we mean *uniform* annual worth; that is, over the time period referred to, the profit or loss will be constant from year to year. Often this idea is expressed in the phrase *equivalent uniform annual benefit* (*EUAB*) or *cost* (*EUAC*).

Although annual worth may be described preliminarily in its sense of gain or loss, another meaning of annual worth also exists that does not carry the sense of favorable (gain) or unfavorable (loss) circumstances. An equivalent uniform annual cost may simply be a way of expressing the cost of a lump-sum investment in annual or periodic terms. For example, you buy a refrigerator for no money down, but you agree to pay a certain amount monthly to the seller. The lump sum, that is, the total cash cost of the refrigerator, has been replaced by a monthly payment—a periodic cost to you as a result of buying the refrigerator. Notice that this periodic cost is not thought of as "loss." This is as it should be.

As a decision tool, the annual-worth method uses the comparison between the annual worths of alternative courses of action to decide which will be preferred. If two investment possibilities exist, the choice will favor the one with the larger annual worth. To put it another way, we will accept the alternative whose difference between annual benefit (annual revenue) and annual cost is larger.

If only annual costs are involved, implying that the annual benefits of each course of action are the same, then we choose the alternative whose annual cost is the lower. Suppose, for example, you are trying to choose between two automobiles whose benefits to you will be the same, that is, no difference in appearance, riding qualities, fuel consumption, and so on. You will base your decision, then, on the difference in periodic cost between the two alternatives. You will choose the one with the lower monthly payments.

ANNUAL-WORTH ANALYSIS

Annual worth (AW) is the difference between annual benefit (revenue) and annual cost, commonly known as *gain* if annual worth turns out to be a net benefit and as *loss* if annual worth turns out to be a net loss. In symbolic terms,

$$\text{AW} = B_A - C_A$$

where B_A is the annual benefit and C_A is the annual cost. As a decision tool this becomes

$$B_A - C_A \geq 0 \tag{6.1}$$

for an alternative to be accepted, where

$$B_1 = B_2 = B_3 = \cdots = B_N = B_A$$

and

$$C_1 = C_2 = C_3 = \cdots = C_N = C_A$$

We will select the alternative where $B_A - C_A$ is larger.

To use the formula of Eq. (6.1), we must convert any lump-sum payments or benefits to equivalent uniform periodic ones by use of the capital recovery factor (A/P, i, N). The procedure is illustrated in the following example.

■ **Example 6.1.** You are considering an investment in a laundromat. The washers and dryers will cost $5,000 each and will last for 3 years with no resale value at the end of that time. Rent, labor, and maintenance will approximate $11,000 annually. Total revenue is estimated as $20,000 per year. The time horizon of the operation is 3 years. Your opportunity cost of capital for a similar-risk investment is 20 percent before taxes. For the moment, you ignore inflationary effects on your decision. Should you invest in the laundromat?

Solution. Notice that the opportunity cost of capital must be known in order to solve the problem.

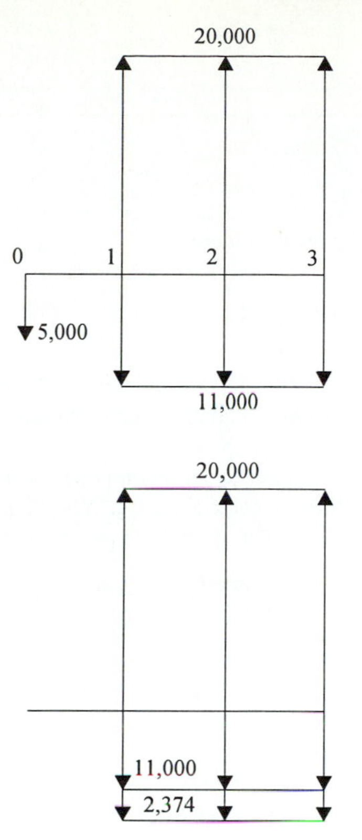

FIGURE 6.1 An investment in a laundromat

The cash flow is shown in Fig. 6.1. All revenues are already in annual terms. But one of the costs, the investment, is a single payment at time 0. The other cost, \$11,000, is an annual disbursement. The problem, then, is to convert \$5,000 from a one-time cost to an annual cost that will repay your principal with interest. This can be done by using the capital recovery factor:

Annual cost = investment cost × capital recovery factor

$$\begin{aligned} A &= P(A/P, i, N) \\ &= 5{,}000(A/P, 20, 3) \\ &= 5{,}000(0.47473) \\ &= \$2{,}374 \end{aligned}$$

The total annual revenue is \$20,000, and the total annual cost is the sum of 11,000 and 2,374, or \$13,374. The annual worth of the investment is \$6,626 and is a gain. ■

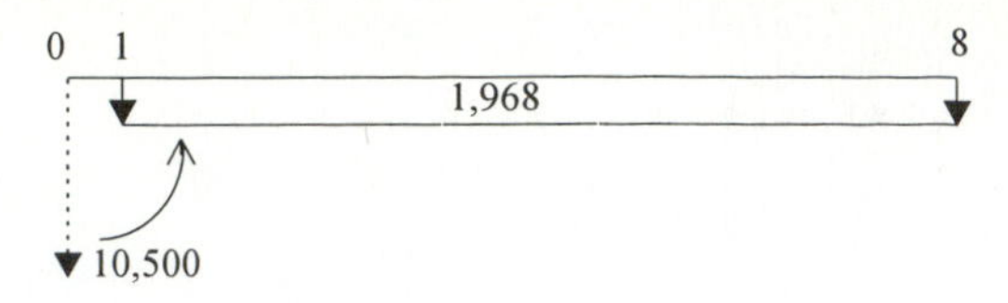

FIGURE 6.2 An EUAC for an automobile

ANNUAL-COST ANALYSIS

A special case of annual-worth analysis exists when only costs are present. The problem is solved by expressing all costs in terms of annual amounts.

■ **Example 6.2.** You have purchased an automobile for $10,500 cash. You expect to keep it for 8 years and then give it to your son. Your opportunity cost of capital is 10 percent. What will be your equivalent uniform annual cost of the car?

Solution. The $10,500 is a one-time first cost at time 0 which must be converted to a uniform annual cost. See Fig. 6.2. This is accomplished by using the capital recovery factor.

$$\begin{aligned} A &= P(A/P, i, N) \\ &= 10{,}500(A/P, 10, 8) \\ &= 10{,}500(0.18744) \\ &= \$1{,}968 \end{aligned}$$

The car will cost you $1,968 per year. Of course, you will not pay out that sum to anyone. It is a reflection of the $10,500 you have already paid. It is, indeed, the cost of the opportunity lost by tying up $10,500 in an automobile rather than investing it at 10 percent return. ■

Frequently annual-cost calculations are the only ones needed because benefits are the same among all alternatives. Take the case of a comparison between different makes of cars. If you are considering buying a Ford or a Chevrolet, you will be concerned with only costs (first cost, maintenance costs, and operations costs), not benefits, because the benefit—transportation—is the same for both makes. Alternative makes of bulldozers or drill presses or cathode-ray tubes will normally be treated from the cost side only in economic analyses, except for the salvage value which may differ among alternatives. Any asset that will perform its designed service as well as any other being considered may be treated only from the cost side because the benefits are nondifferential.

SALVAGE VALUE

The *salvage* or *resale value* is the amount an asset can be sold for when its economic life is over for the enterprise owning it. Two common computational methods exist for handling salvage value in the annual-worth method. The first requires that the salvage value S be imagined as that part of the original investment which will be returned at the end of the investment's life. Therefore the salvage value adds to the annual cost the annual interest on it. This idea can be seen most clearly in the annual-cost formula:

$$\text{Annual cost} = (P - S)(A/P, i, N) + Si \tag{6.2}$$

The portion of the investment cost that is "used up," as it were, is $P - S$, and principal and interest must be recovered on it. Notice that both members of the right-hand side of the equation are costs and that, for the purposes of this equation, we are considering costs as positive numbers.

Another way of handling salvage value in annual-worth calculations is to observe that the investment may be expressed as an annual cost simply by multiplying it by the capital recovery factor. That is, the investment cost may be spread forward over the life of the asset. The salvage value, now appearing with a negative sign because it reduces the annual cost, may be spread backward over the life of the asset by multiplying it by the sinking fund factor:

$$\text{Annual cost} = P(A/P, i, N) - S(A/F, i, N) \tag{6.3}$$

Both of these equations for annual cost will give the same answer.

The salvage value is sometimes known as the *resale value*. It is also called the *residual value*. It is not book value, defined as first cost less accumulated depreciation, because book value is simply a number appearing in the books of account that may, but more often may not, represent the market value of an asset. Book value has other uses that we will see in Chap. 10 on income tax effects. Finally, remember that *salvage value* is the term used to represent a final payment, either positive or negative, in connection with the disposal of a capital investment. In certain cases—restoring land to its original state after a surface mining operation, for example—the payment may well be a cash outflow rather than a cash inflow.

■ **Example 6.3.** What is the annual cost of a commercial sewing machine whose first cost is \$2,200, whose resale value is \$500 after an economic life of 4 years, and which is to be used in a clothing company where the opportunity cost of capital is 25 percent before taxes, without considering inflation?

Solution. Figure 6.3 shows the situation. Equation (6.2) will be used first.

$$\begin{aligned}\text{Annual cost} &= (P - S)(A/P, i, N) + Si\\ &= (2{,}200 - 500)(A/P, 25, 4) + 500(0.25)\\ &= (1{,}700)(0.42344) + 500(0.25)\\ &= 720 + 125 = \$845\end{aligned}$$

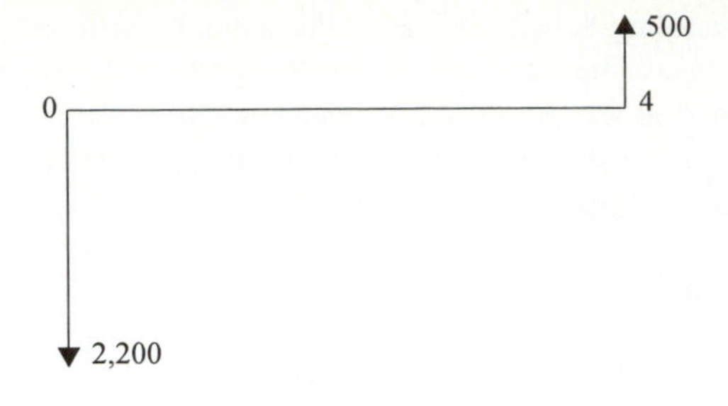

FIGURE 6.3 A commercial sewing machine

Equation (6.3) gives the same result.

$$\begin{aligned}\text{Annual cost} &= P(A/P, i, N) - S(A/F, i, N)\\ &= 2{,}200(A/P, 25, 4) - 500(A/F, 25, 4)\\ &= 2{,}200(0.42344) - 500(0.17344)\\ &= 932 - 87 = \$845\end{aligned}$$ ■

UNEQUAL LIVES: THE ADVANTAGE OF ANNUAL WORTH

Annual worth has a great advantage over other methods: It avoids the explicit repetition of life cycles. Recall that the logical foundation of comparison between mutually exclusive alternatives is equal service lives. We have seen that when lives are unequal, they must be equalized either by repeating the cash flows of both alternatives until they are equalized or by assuming a salvage value. This is done without any assumption that we are actually planning a number of repurchases of the same equipment at the same price. This would be, to say the least, a most improbable situation. Rather, we are simply setting up a condition to satisfy the logical necessities of a mathematical solution. Yet annual worth does not violate the dictum of equal service lives. It is based on the assumption of equal lives. The method cannot be disassociated from the assumption—a fact that, as we will see, makes the method unusable in capital budgeting problems. A simple example will illuminate the point.

■ **Example 6.4.** Two types of dump trucks for small jobs are being considered by a paving contractor in Mexico. Data are as follows:

	Mercedes-Benz	Ford
Cost (pesos)	150,000	100,000
Annual maintenance (pesos)	10,000	12,000
Economic life (years)	10	5
Resale value (pesos)	15,000	10,000

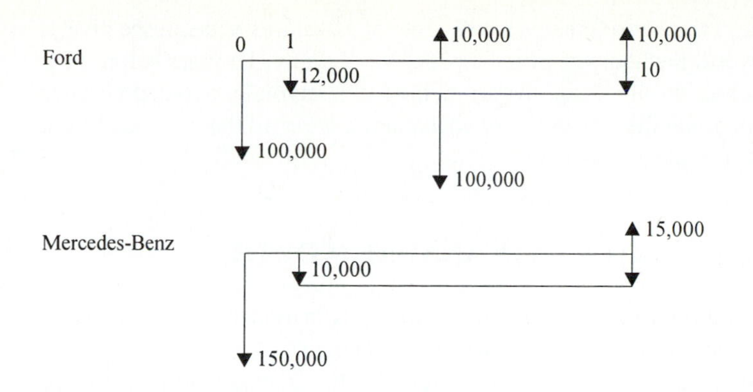

FIGURE 6.4 Two makes of dump trucks compared

The opportunity cost of capital for this contractor is 25 percent. Which make should be chosen?

Solution. Figure 6.4 shows the cash flows equalized at 10 years. The Ford cycle appears twice, the Mercedes-Benz once. Because benefits are nondifferential, that is, any truck will do the job as well as any other, they have been excluded. An annual-cost solution, therefore, is acceptable. For the first cycle of the Ford, the *annual cost* (*AC*) is

$$\begin{aligned} \mathrm{AC}_1 &= (100{,}000 - 10{,}000)(A/P, 25, 5) + 10{,}000(0.25) + 12{,}000 \\ &= 90{,}000(0.37185) + 10{,}000(0.25) + 12{,}000 \\ &= 33{,}467 + 2{,}500 + 12{,}000 \\ &= 47{,}967 \text{ pesos} \end{aligned}$$

We now calculate the annual cost of the second cycle of the Ford option:

$$\begin{aligned} \mathrm{AC}_2 &= (100{,}000 - 10{,}000)(A/P, 25, 5) + 10{,}000(0.25) + 12{,}000 \\ &= 47{,}967 \text{ pesos} \end{aligned}$$

We observe that this is *exactly the same* as for the first cycle. The costs of the first and second cycles are equal. Because of this fact, it is really unnecessary to calculate the cost of the second cycle, or write its equation, or even draw its cash flow. It would have been sufficient to concern ourselves with the first cycle only.

The annual cost of the Mercedes-Benz is

$$\begin{aligned} \mathrm{AC} &= (150{,}000 - 15{,}000)(A/P, 25, 10) + 15{,}000(0.25) + 10{,}000 \\ &= 135{,}000(0.28007) + 15{,}000(0.25) + 10{,}000 \\ &= 37{,}810 + 3{,}750 + 10{,}000 \\ &= 51{,}560 \text{ pesos} \end{aligned}$$

The Ford wins the day. ■

The point of this example is that in the annual-worth method, the lives of the alternatives need not be explicitly equalized. The method itself takes care of that point. *It is only necessary to calculate the annual worth of the first life cycle of each alternative.* Take two alternatives with lives of 10 and 11 years, respectively. In the present-worth method, we must carry out the lives to 110 years before they equalize themselves. In the annual-worth method, this is totally avoided because the annual worth of the first life cycle for each alternative is all that is needed for a comparison between the alternatives.

INCREMENTAL ANALYSIS IN ANNUAL WORTH

This book is based on the principle, among others, of incremental analysis. How does this conceptual rule hold in the annual-worth method?

The answer is that it holds equally well for the annual-worth as for all methods, but is more difficult to use than individual alternative cash flow analysis because the lives of the alternatives must be explicitly equalized before their difference is found. This removes the advantage of the annual-worth method. The following example clarifies the point.

■ **Example 6.5.** The Johnson pump is to be compared to the Silver model for use in a water treatment plant. They have the following characteristics:

	Johnson	Silver
Life (years)	15	20
Initial cost ($)	7,000	10,000
Maintenance cost (MC) per year ($)	800	500
Salvage value ($)	700	1,000

The water district uses a 10 percent opportunity cost of capital. Since it is a public nonprofit agency, it pays no income taxes.

Which model pump should be chosen if we use the annual-cost method?

Solution. The upper portion of Fig. 6.5 shows the two alternatives. Taking the individual annual worths—in this case, annual costs—of the pumps gives

$$\begin{aligned}
AC_J &= (P - S)(A/P, i, N) + Si + MC \\
&= (7{,}000 - 700)(A/P, 10, 15) + 700(0.10) + 800 \\
&= (6{,}300)(0.13147) + 70 + 800 \\
&= \$1{,}698 \\
AC_S &= (10{,}000 - 1{,}000)(A/P, 10, 20) + 1{,}000(0.10) + 500 \\
&= (9{,}000)(0.11746) + 100 + 500 \\
&= \$1{,}657
\end{aligned}$$

The Silver pump is the sound choice because its annual cost is less than that of the Johnson pump. But notice that even here we are making an incremental choice. We do not actually take the increment ($41 per year); we merely observe that there exists a difference in annual cost in favor of the Silver pump—an incremental choice in all its essentials.

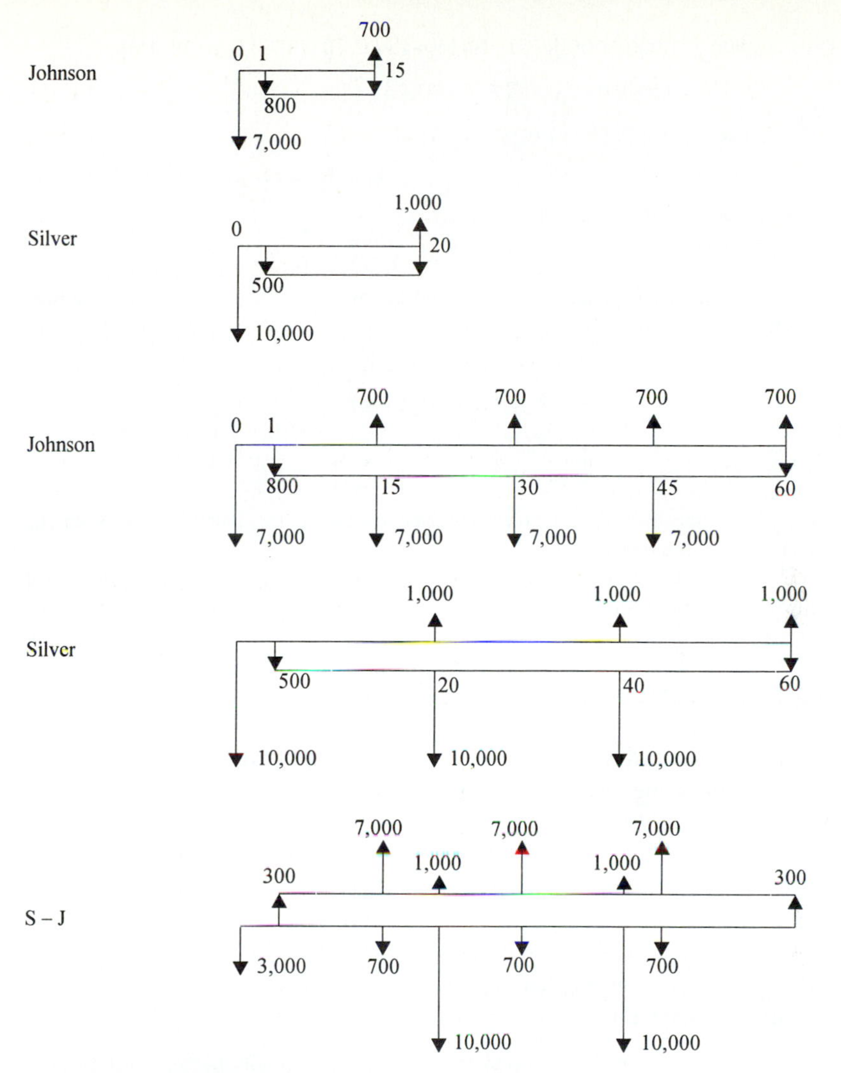

FIGURE 6.5 Water treatment pumps

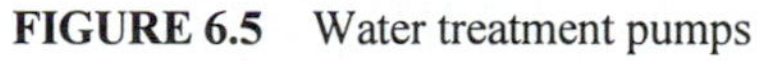

Had we taken the increment at an earlier stage in the calculation, we would have had to equalize the lives as in the bottom portion of Fig. 6.5. The incremental cash flow is complicated and difficult to compute. It is shown as the bottommost cash flow of Fig. 6.5. This difficulty causes analysts to reject incremental cash flow analysis in annual-worth calculations.

You will see just how complicated the matter is in the following calculations. It is easier to find the present worth of the single payments, sum them algebraically, and spread them forward via the capital recovery factor. This procedure, which involves finding the *present worth* (*PW*) of practically all payments, is the reason why the present-worth method is preferred for such cash flows.

$$\begin{aligned}\text{PW} &= -3{,}000 + (7{,}000 - 700)[(P/F, 10, 15) + (P/F, 10, 30) + (P/F, 10, 45)] \\ &\quad - (10{,}000 - 1{,}000)[(P/F, 10, 20) + (P/F, 10, 40)] + 300(P/F, 10, 60) \\ &= -3{,}000 + 6{,}300(0.2394 + 0.0573 + 0.0137) \\ &\quad - 9{,}000(0.1486 + 0.0221) + 300(0.0033) \\ &= -3{,}000 + 1{,}956 - 1{,}536 + 1 \\ &= -\$2{,}579\end{aligned}$$

Now it is possible to spread the present worth forward and add to it the \$300 annual benefit:

$$\begin{aligned}\text{AC} &= -2{,}579(A/P, 10, 60) + 300 \\ &= -2{,}579(0.10033) + 300 \\ &= -259 + 300 \\ &= \$41\end{aligned}$$

This is the same difference that we obtained by taking the increment between the annual costs of the first cycles.

Thus the important point is demonstrated that the individual annual-cost comparison is based on equal service lives and follows the principle of incremental analysis. ■

PERPETUAL LIVES IN ANNUAL WORTH

As in present-worth analysis, perpetual-life investments viewed from their annual-cost aspect use the same relationships expressed by

$$A = Pi$$

The annual cost of an investment with an infinite life is A, providing that P is the initial cost. The capital recovery factor (A/P, i, ∞) is i. If A is seen as the return on investment of P, the annual worth of the investment is Pi. In other words, A is the same whether it is seen as a benefit or an annual opportunity cost. The following example involves a benefit.

■ **Example 6.6.** How much can you expect to receive perpetually per year if you invest \$100,000 at 15 percent, before you take into account the effect of taxes and inflation?

Solution

$$\begin{aligned}A &= Pi \\ &= (100{,}000)(0.15) \\ &= \$15{,}000 \text{ per year}\end{aligned}$$

■

Example 6.7 concerns itself with A received as an annual cost.

■ **Example 6.7.** A Forest Service road will cost \$97,000. It is estimated that 60 percent of this will be spent on right-of-way, subgrade, and cuts and fills; 20 percent on wearing surface; and 20 percent on bridges, culverts, and other drainage structures. The right-of-way, and so forth, will have a perpetual life, the wearing surface will have a a life of 10 years, and the drainage structures will have a life of 20 years. Annual maintenance costs will be $2,500. If the Department of Transportation minimum attractive rate of return of 10 percent is used, what is the estimated annual cost of the road? No inflation effects need be considered, nor will taxes apply, because this is a public agency road.

Each element of the road will be considered to be replaced as its life ends at its original cost.

	Initial cost	N	**(A/P, 10, N)**	**Annual cost ($)**
Right-of-way, etc.	58,200	∞	0.10000	5,820
Wearing surface	19,400	10	0.16275	3,157
Drainage	19,400	20	0.11746	2,279
Maintenance				2,500
				13,756

The estimated annual cost is about $14,000. ■

SUMMARY

Annual worth may be considered as a gain or a loss when both benefits and costs are involved. When only costs are present, it is called the *annual-cost method*. The criterion of choice among alternatives is that the one with the largest annual gain or least annual cost is chosen.

Unequal lives are handled best by the annual-worth method. The assumption of equal service lives is implicitly present, which gives this method a great advantage in computational ease over the others in use.

The annual-worth method, like all others, is based on incremental analysis. It is often easier, however, to compare each individual annual worth than to find the incremental cash flow and use it to compute the incremental annual worth between alternatives.

Perpetual lives present no problems in the annual-worth method.

In conclusion, annual worth is a method easily understood by the most unsophisticated client for an economic analyst's services. It lends itself to accounting data with more facility than any other method.

PROBLEMS

6.1. In Example 6.1, page 102, you decide on a larger operation. The washers and dryers will cost $12,000. The revenue will be increased to $30,000 and the cost to $18,000 annually, with all other data remaining the same. Is it still a good deal?

6.2. A group of architectural, structural engineering, and materials consultants proposed extensive repairs to a well-known Chicago museum at a cost of $12.6 million. It was estimated that the repairs would last 50 years. If public money had an opportunity cost of 10 percent, what would be the *annual cost* of these repairs?

6.3. In Example 6.2, page 104, you consider changing the time you hold the car to 10 years, with no other changes in the data. What will be your new annual cost?

6.4. On page 105, two equations for annual cost are shown. Equalizing them should result in an identity. Does it?

6.5. A drill press in an automobile repair shop is projected to show the following operational costs over the next 5 years:

Year	Cost
1	1,100
2	1,225
3	1,350
4	1,475
5	1,600

What is the equivalent uniform annual cost if a 12 percent discount rate is used?

6.6. In Example 6.3, page 105, the life of the commercial sewing machine is extended to 5 years, which reduces its resale value to $400. What is its new annual cost if all other data remain the same?

6.7. A used Mercedes-Benz 300D costs $24,536. Its annual costs are as follows:

Fuel and maintenance	$555
Insurance	$900

Its resale value at the end of 5 years—the time horizon considered—is $7,300.

All figures are in terms of 1995 dollars. No income tax effects will be considered at this point. If the opportunity cost of capital for the purchaser is 9 percent in terms of constant 1995 dollars, what is the uniform annual cost of the car, providing the buyer pays in cash?

6.8. A car buyer paid $5,642.37 cash for a used Datsun. He wishes to know what the ownership of the car will cost him annually based on an estimated life of 5 years and $1,000 resale value, if his estimated opportunity cost of capital averages 15 percent during the next 5 years. Disregard inflationary effects. The car ownership offers no tax advantages.

Answer: $1,535 per year

6.9. A vacation home in the mountains may be bought for $80,000 in cash. Mortgage financing is available at 12 percent for 30 years with a 10 percent down payment. The upkeep and taxes will cost about $1,000 per year. The prospective buyer is an investor who regularly makes 20 percent on an investment of equal risk. Disregarding inflationary effects and possible property appreciation, how much will the property cost her each year if she holds it for 30 years? Ignore possible tax advantages, and assume that the property will have a resale value equal to its original cost.

Answer: About $11,500 per year

6.10. A small residential rental property will cost $125,000 cash. Yearly expenses will be as follows:

Item	Expense ($)
Taxes	1,822
Insurance	290
Maintenance	500
Management	696

Rents will average $725 per month. If you believe you will give up the opportunity of making 20 percent on the money tied up in this purchase on an investment of equal risk, and if you ignore the effects of inflation on all but the future sale price, what will be the annual worth of this deal? You intend to hold the property for 6 years and believe you can sell it for $260,000 net.

6.11. The accountant for your metal products firm shows you the following statement of annual projected costs over the next 5 years for a new stamping department.

Item	Cost ($)
Materials	40,000
Labor	50,000
Maintenance	4,000
Depreciation	18,000
Administration	3,000
	115,000

Your investment in depreciable assets will be $100,000. These assets will have an estimated salvage value of $10,000 after a life of 5 years. No new administrative personnel will be hired because of the new department, nor will any significant increase in administrative supply use take place.

Your minimum attractive rate of return (MARR) is 20 percent before taxes and inflation are considered. You will be satisfied with a before-tax and constant-cost analysis.

What is the annual economic cost of the new department?

6.12. Andrew Marvell, Inc., a firm of engineers, architects, and planners, must decide how to air-condition a new building with a 40-year life. Two alternatives are available, both of which have 20-year lives.

Alternative 1 has an initial cost of $180,000, annual fuel costs of $6,000, annual maintenance costs of $4,400, and a salvage value of $20,000.

Alternative 2 has an initial cost of $120,000, combined fuel and maintenance costs of $19,000, and a salvage value of $12,000. Andrew Marvell, Inc., has decided to use the *annual-cost method* to make the choice between the alternatives.

If the opportunity cost of capital for this firm is 12 percent, which alternative should be chosen?

6.13. In Example 6.4, page 106, the estimates of economic life are changed to 9 years for the Mercedes-Benz and 4 years for the Ford. If all other data remain the same, which car will we choose now?

6.14. A state highway department wishes to compare the *annual costs* of precast concrete arches and corrugated-steel arches for use in highway bridges and cut-and-cover tunnels. Data for a 57-foot span are as follows:

	Precast concrete	Corrugated steel
First cost ($)	350,000	185,000
Maintenance cost per year ($)	15,000	20,000
Life (years)	35	20

One of the arches is certain to be chosen. A 10 percent discount rate is to be used in the economic analysis. No inflation effects are to be considered. Which arch is more economical?

6.15. Repeat Prob. 5.16, using the annual-cost method.

Answer: $AC_F = \$3,814$, $AC_S = \$4,097$

(*a*) Does the annual-cost method have an advantage over the present-worth method for problems of this type? What is it?

(*b*) Does the annual-worth method compare equal service lives? Explain.

6.16. Repeat Prob. 5.17, using the annual-worth method.

(*a*) Which method is more easily performed in problems with cash flows like these, annual-worth or present-worth?

(*b*) What should be the ruling consideration for choosing between the methods?

6.17. Two types of bridges near the Pacific Ocean are to be compared. The data on them are as follows:

	Bridges	
	Steel	Concrete
First cost ($)	3,000,000	3,170,000
Salvage value ($)	30,000	—
Demolition cost ($)	—	50,000
Economic life (years)	25	50
Annual maintenance cost ($)	10,000	3,000

The discount rate is 10 percent.

Which type is more economical on the basis of (*a*) annual cost and (*b*) present worth?

6.18. A tall building is being constructed. The construction company may purchase one of two models of laser alignment equipment now on the market. Data on the two models are as follows:

	Jacobs	Continental
First cost ($)	10,000	8,500
Resale value ($)	2,500	1,000
Operating and maintenance cost per year ($)	800	1,200
Insurance per year ($)	400	350
Economic life (years)	6	5

The construction company will not consider income taxes or inflation in its analysis. Its minimum attractive rate of return is 15 percent.

(*a*) What is the annual cost of each model?

(*b*) Which should be purchased if the company must provide one model or the other?

6.19. You are considering a new addition to your fleet of for-hire limousines. The data on the two models are as follows:

	Model	
	C	*L*
First cost ($)	50,000	60,000
Annual operation and maintenance cost ($)	5,000	4,000
Resale value ($)	10,000	15,000
Economic life (years)	6	7

If your minimum attractive rate of return is 20 percent before income taxes and inflation are taken into account, which of the two should be chosen? (You are certain to choose one or the other.) *Use the annual-cost method.*

6.20. George R. Moore and Associates, Consulting Engineers, is considering the purchase of new theodolites for their property surveying practice. The instruments have the following characteristics:

	Zurich	Basel
First cost ($)	4,000	6,000
Life (years)	10	15
Annual maintenance cost ($)	300	150
Resale value ($)	400	600

This company estimates its before-tax, before-inflation rate of return at 30 percent per year.

(*a*) Using the *annual-cost method*, determine which instrument is more economical.

(*b*) Check your decision by using the *present-worth method.*

6.21. According to a recent article in *Civil Engineering*, building rehabilitation is often compared to new construction on an annual-cost basis. The following data illustrate the characteristic situation:

	Rehabilitate	Build new
First cost ($000,000)	12	18
Life (years)	15	25
Salvage value ($000,000)	1	1
Annual maintenance ($000,000)	0.5	0.2

If irreducibles, inflation, and tax effects are ignored, which of the two courses of action should be taken if the minimum attractive rate of return for the company is 25 percent before taxes? An *annual-cost* analysis is required.

6.22. We are faced with selecting one of the following two alternative projects:

Years	Project 1	Project 2
0	-100	-120
1	30	25
2	30	25
3	30	25
4	50	25
5	60	25
6		60

As shown above, project 1 has a 5-year life, and project 2 has a 6-year life. The opportunity cost of capital is 10 percent. The *annual-worth* method must be used.

Which project will be selected? Do not consider the null alternative.

6.23. In Example 6.5, page 108, it was demonstrated that the individual annual-cost comparison is based on equal service lives and follows the principle of incremental analysis. Repeat the same demonstration, but using the data of Example 6.4.

6.24. A drainage canal in Arizona will cost $105,000. Estimates show that 40 percent of this sum will be spent on right-of-way, 30 percent on concrete liner, and 30 percent on bridges, valves, and spillways. The right-of-way will have a perpetual life, the concrete liner will have a life of 20 years, and the bridges, valves, and spillways will have a life of 25 years. Annual maintenance costs will be $3,000. A Bureau of Reclamation minimum attractive rate of return of 10 percent will be used. No income taxes will be considered because this is a public project. Inflation will be ignored for the moment.

Each element of the project will be imagined as being replaced at its original cost at the end of its life.

What is the annual cost of the project?

6.25. In Example 6.6, page 110, the investment is now \$1 million, and the rate of return to be expected before taxes and inflation is 3 percent. How much can you expect to receive perpetually?

6.26. A city engineer in a small town in Nevada is considering curb installation on certain streets off Main Street. She has before her two alternatives, one of which she is certain to choose:

	Granite	Concrete
First cost (\$)	125,000	85,000
Life (years)	Infinite	25
Annual maintenance cost (\$)	1,000	5,000

The granite curbing, with proper maintenance, will last indefinitely. No salvage value is expected for either curbing.

A 10 percent discount rate is to be used.

City ordinances do not specify any particular method of economic analysis for this situation. The engineer will therefore use the most convenient correct method.

Which curbing should she choose?

6.27. In Example 6.7, page 110, the Forest Service road will now cost \$120,000. If all other data remain the same, what is the new annual cost of the road?

6.28. Because of the success of its basketball team, a university has decided to build a new sports center, in order to seat more spectators. To confirm its decision, it has requested that your company, and particularly you as an alumnus, review its analysis.

The sports center, with an expected life of 40 years, is estimated to cost \$30 million, made up of \$10 million for the land and \$20 million for the construction and landscaping. The following costs are anticipated, all in constant dollars at time 0:

Routine annual maintenance	100,000
Annual operating costs (heat, air conditioning, etc.)	250,000
Annual employee compensation	240,000
Repairs and painting every 4 years	140,000
Major overhaul of heating and air-conditioning system every 10 years	500,000
Major overhaul of floors, etc., every 20 years	1,500,000

The removal cost of the center at the end of 40 years will be \$500,000. The land will keep its value perpetually.

The revenues from the center are estimated to be $2 million for the first year, and they will increase by $100,000 each year.

If the opportunity cost of capital of the university is 15 percent and no taxes are considered, would you recommend that the center be built? *Use the annual-worth method.*

CHAPTER 7

BENEFIT/COST RATIO

The benefit/cost ratio is another method of analyzing and choosing among investments. It owes its existence and continued use to legislation. The Flood Control Act of June 22, 1936 provided that benefits should be in excess of costs in order for a federally financed project to be justified. The result was the gradual emergence of the benefit/cost ratio as a tool for judging the economy of projects. Unfortunately a ratio, which requires incremental treatment in analysis, resulted rather than a difference—benefits minus costs—which can be handled without marginal analysis. The advantage of the benefit/cost ratio is that it is widely understood by public officials. Its disadvantage is its still-prevalent misuse.

The reader may ask why it is necessary to study such a cumbersome, as some would call it, method. Why not use the net present value (NPV), for example? The benefit/cost ratio method is demanded by many government agencies in this country and others. The analyst has no real choice among methods, therefore, and must use the method the client or employer requests.

Like the two previous methods—present worth and annual worth—it relies on the time value of resources for its validity. It arranges the discounted benefits and costs as a ratio rather than as a difference. The rule is

$$\frac{B}{C} \geq 1.0$$

where B is the present discounted benefits and C is the present discounted costs.

The above equation is an outgrowth of the present-worth formula, which stipulates

$$B - C \geq 0$$

where both B and C are discounted. This means that

$$B \geq C$$

if we move C to the right-hand side of the equation. Dividing both sides by C results in

$$\frac{B}{C} \geq 1.0$$

INCREMENTAL ANALYSIS NECESSARY

But now heed an important warning on the use of the benefit/cost ratio. It cannot be used in relation to individual projects in the manner that one might expect, that is, as a measure of the worth of each individual alternative in comparison to all the other alternatives. As has been emphasized before in this book, incremental analysis of mutually exclusive alternatives must be used in the benefit/cost ratio method. An incorrect approach may lead to the wrong answer. Incorrect analysis may also lead, by accident, to the right answer, but this is no justification for it. As justification for the correctness of incremental analysis, the reader is invited to review the remarks on page 70 in Chap. 5 on the present-worth method and on page 102 in Chap. 6 on the annual-worth method. To drive the point home, an example is called for.

■ **Example 7.1.** A state highway department is attempting to choose between two highway noise barriers. Because the highway contract calls for some sort of barrier, a wood or metal barrier will be selected.

Benefits of each type of barrier differ. These benefits are the effectiveness of noise suppression, the aesthetic quality of the barrier, and an acceptance rating determined by a preference survey of adjoining property owners. By methods beyond the scope of this text, the benefits have been reduced to an equivalent annual dollar figure. Costs of each alternative are first costs and annual maintenance costs. No salvage value is expected. Net costs and benefits per mile for a life of 20 years are as follows:

	Metal	Wood
First cost ($)	152,200	183,800
Maintenance cost per year ($)	2,000	3,000
Benefits per year ($)	24,100	28,900

Because this is a public project, no taxes will be paid. Inflation effects will be practically equal between the alternatives. A 10 percent minimum attractive rate of return will be used for the analysis.

Figure 7.1 shows the two alternatives with the annual maintenance costs and benefits algebraically summed to $22,100 for alternative 1 and $25,900 for alternative 2. This algebraic summation is known as *netting out*.

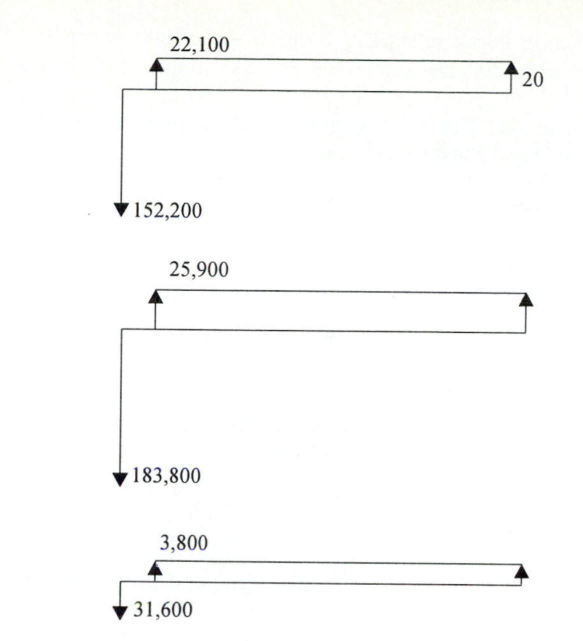

FIGURE 7.1 Highway noise barriers

Solution. Because one of the alternatives is sure to be chosen, we need not compare the lower-cost choice, alternative 1, to doing nothing. Therefore alternatives 2 and 1 must be compared with each other, but incrementally. The result is the cash flow shown as 2 − 1 in Fig. 7.1. The present worth of the benefits divided by the present worth of the costs is

$$\frac{B}{C} = \frac{3{,}800(P/A, 10, 20)}{31{,}600}$$
$$= \frac{3800(8.514)}{31{,}600}$$
$$= 1.02 > 1.00$$

This result means that the extra discounted benefits more than justify the extra discounted costs of alternative 2 compared to alternative 1. Therefore alternative 2 is selected. This is the correct answer achieved by correct methods.

Now let us see what happens if an incorrect, but widely applied, method is used, that is, if the benefit/cost ratio of each alternative is computed and the alternative with the higher benefit/cost ratio selected. The benefit/cost ratio of alternative 1 is

$$\frac{B}{C} = \frac{22{,}100(P/A, 10, 20)}{152{,}200} = 1.24$$

and of alternative 2 is

$$\frac{B}{C} = \frac{25{,}900(P/A, 10, 20)}{183{,}800} = 1.20$$

The incorrect criterion selects alternative 1 (metal) rather than alternative 2 (wood) because the metal alternative has a *B/C* ratio of 1.24 and the wood alternative a *B/C* ratio of only 1.20.

To make certain that this is the correct method, let us evaluate the net present value of alternatives 1 and 2 and the investment 2 - 1.

$$\text{NPV}_1 = -152{,}200 + 22{,}100\,(P/A, 10, 20) = 35{,}959$$
$$\text{NPV}_2 = -183{,}800 + 25{,}900\,(P/A, 10, 20) = 36{,}713$$

By this criterion, alternative 2 will be selected because it has the larger present worth.

$$\text{NPV}_{2-1} = -31{,}600 + 3{,}800\,(P/A, 10, 20) = 754$$

By this criterion, because the result of the incremental NPV is positive, alternative 2 will also be selected, confirming the result of the incremental benefit/cost ratio analysis. Notice also that 36,713 - 35,959 = 754. ■

FORMULAS

Although the definition of the benefit/cost ratio expressed in the previous section may seem definitive, it is not. The difficulty arises because of the variety of ways that exist for defining benefits and costs. This matter will be discussed at length, but first we give a more exact definition of the benefit/cost ratio as it will be used in this book. A unique benefit/cost ratio will always be obtained if the following rules are obeyed:

Step 1. Net out the cash flow whose benefit/cost ratio is desired. That is, find

$$(B - C)_j = K_j \tag{7.1}$$

This means that at each period j the benefits, represented by up arrows, must be summed algebraically with the costs, represented by down arrows. The result, an up or down arrow, is K_j, either a benefit or a cost. The process is the netting out mentioned in Example 7.1.

Step 2. Apply the equation

$$\frac{B}{C} = \frac{\sum_0^N B_j(P/F, i, j)}{\sum_0^N C_j(P/F, i, j)} \tag{7.2}$$

If the resulting *B/C* is greater than or equal to 1.0, the alternative qualifies, that is,

$$\frac{B}{C} \geq 1.0$$

When either the benefits or the costs result in a series of equal payments after netting out, $(P/A, i, N)$ may be substituted for $(P/F, i, j)$. The calculation is thus simplified for manual operations using tables. If a computer or programmed calculator is used, however, the payment at each j must be entered separately as a data input, with a plus sign if it is a benefit and a minus sign if it is a cost. Example 7.2 illustrates the process.

■ **Example 7.2.** In the central valley of California, a large canal was being investigated. Its cost was estimated at \$5,600,000. Annual maintenance costs were projected to be \$100,000. Annual benefits in irrigation for agriculture were \$360,000. No salvage value at the end of the economic life of the canal was envisioned. A time horizon of 50 years was adopted for the study. All sums were expressed in constant dollars as of project year 0, thus accounting for possible inflation or deflation over the years. A 7 percent discount rate suggested by the Federal Water Resources Council was stipulated. Compute the B/C ratio by netting out the cash flow. Will the project be accepted?

Solution. The upper cash flow of Fig. 7.2 shows the situation. First, the netting out of step 1 must be carried out. This results in the lower cash flow of Fig. 7.2. Now step 2 is performed. Equation (7.2), modified because the benefits are a series of equal cash inflows, results in

$$\frac{B}{C} = \frac{A(P/A, i, N)}{C_0}$$

$$= \frac{260{,}000(P/A, 7, 50)}{5{,}600{,}000}$$

$$= \frac{260{,}000(13.801)}{5{,}600{,}000}$$

$$= 0.64 < 1.0$$

This means that the project is rejected. ■

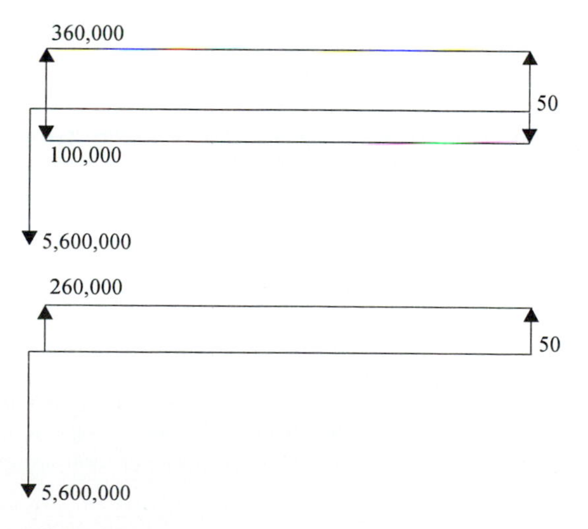

FIGURE 7.2 A California canal

As in all methods explained in this book, equal service lives of all mutually exclusive alternatives must be established before the benefit/cost ratio is calculated.

Netting out, as explained in step 1 above, is sometimes called the *rule of delta*. It must be performed before any other operation is attempted. When approaches to the benefit/cost ratio other than that of the rule of delta are described, depending on at which point in the arithmetic discounting occurs, the benefit/cost ratio will vary. This will be seen later in this chapter. Although it is an arbitrary rule, as any rule defining the benefit/cost ratio must be, the rule of delta has an intuitive appeal because it follows the logic of incremental analysis. In Example 7.2, netting out the cash flow in Fig. 7.2 is simply comparing it incrementally to the null alternative.

The other advantage of the *B/C* ratio, as has been mentioned, is that it provides a unique solution by defining benefits as up arrows after netting out and costs as down arrows. This may not appear so important at this point in the text, but the next section will underscore the advantage of uniqueness.

When two alternatives are presented for the solution of a problem, netting out follows directly from the application of incremental analysis, as demonstrated in Example 7.3.

■ **Example 7.3.** A real estate developer acting as a consultant in a Central American country is investigating two possible uses for a parcel of urban land to be redeveloped by a government agency. The first alternative is a multistory parking garage. An investment of 80 million pesos will be required. Revenues will amount to 31 million pesos annually, and costs are estimated at 12 million pesos per year over a time horizon of 20 years. At the end of 20 years, the property will be sold at 100 percent of its original cost, a typical assumption for real estate at similar sites.

The second alternative is an office building. It will cost 130 million pesos, bring in revenues of 38 million pesos annually, cost 13 million pesos per year, and be sold for 130 million pesos at the end of 20 years.

All the previous estimates are in constant pesos of year 0, the present. A constant peso rate of 10 percent will be used for discounting the future.

The government agency has requested that a benefit/cost ratio method be used for decision purposes because the proposed land development plan will be presented to the Inter-American Development Bank to support a loan request.

Solution. Figure 7.3 shows the two alternatives and their increment.

The first question to pursue is whether or not the first, the cheaper, alternative qualifies as an investment. Following the method that the consultant is requested to use leads to the question, Is the *B/C* ratio of alternative 1 greater than or equal to 1.0? In essence, this means that alternative 1 must be compared with doing nothing at all, the null alternative. We use the rules on page 122. Step 1 requires that we net out the benefits and costs at each j, resulting in a net benefit at each year of 19 million pesos for alternative 1. Step 2 applies Eq. (7.2) in its shorter form, using $(P/A, i, N)$ rather than $(P/F, i, j)$.

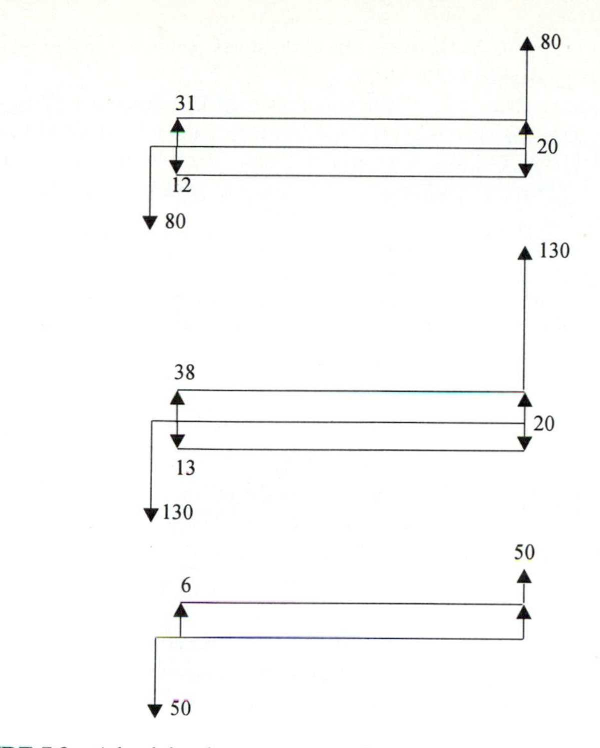

FIGURE 7.3 A land development proposal

$$\left(\frac{\Delta B}{\Delta C}\right)_{1-0} = \frac{19(P/A, 10, 20) + 80(P/F, 10, 20)}{80}$$
$$= \frac{161.77 + 11.89}{80}$$
$$= 2.2 > 1.0$$

This result indicates that alternative 1, the multistory parking garage, is an economically feasible investment. It is better than not developing the land at all. Had the benefit/cost ratio been less than 1.0, alternative 1 would have been rejected and then the comparison would have been made between alternative 2 and the null alternative, not between alternative 2 and alternative 1.

Now it is necessary to see if alternative 2 is better than alternative 1. An incremental analysis must be performed. Figure 7.3 shows the incremental cash flow between the alternatives. The calculation, again following Eq. (7.2), is

$$\left(\frac{\Delta B}{\Delta C}\right)_{2-1} = \frac{6(P/A, 10, 20) + 50(P/F, 10, 20)}{50}$$
$$= \frac{51.08 + 7.43}{50}$$
$$= 1.17 > 1.0$$

Notice that taking the increment between the alternatives performs the netting out of step 1. Step 2 is the application of Eq. (7.2).

The result justifies alternative 2 in place of alternative 1, because 1.17 is greater than 1.0. For this reason alone, completely independent of the magnitude of the ratio, alternative 2 should be selected. It would not have mattered if the incremental benefit/cost ratio had been 1.01; alternative 2 still would have been selected because 1.01 is greater than 1.0.

Imagine now that the wrong approach to the benefit/cost ratio method had been taken, with individual ratios computed for each alternative.

For alternative 1, the result would have been the same: 2.2. For alternative 2, the result would be

$$\begin{aligned}\left(\frac{B}{C}\right)_2 &= \frac{25(P/A,10,20)+130(P/F,10,20)}{130}\\ &= \frac{25(8.514)+130(0.1486)}{130}\\ &= \frac{212.85+19.32}{130}\\ &= 1.8 < 2.2\end{aligned}$$

The incorrect application of the method leads to the incorrect result: alternative 1 rather than 2. The reader should verify the correctness of the incremental approach to the benefit/cost ratio by using the NPV method, as was done in the previous example. ■

OTHER DEFINITIONS OF THE *B/C* RATIO

Up until this point the benefit/cost ratio has been uniquely defined as the present worth of the benefits of an alternative divided by the present worth of the costs of an alternative, *after* the benefit and the cost at each period have been netted out. Steps 1 and 2 of the rule of delta detailed this procedure. A number of reasons were advanced in support of the logic of this rule for determining the benefit/cost ratio. Now, however, the reader must be made aware that this procedure is by no means the only one in use. Various definitions of the benefit/cost ratio are possible, depending on what point in the arithmetic the discounting takes place, or how benefits and costs are defined. These differences in the definition and procedure for determining the ratio lead to an excellent rule for those who must concern themselves with the benefit/cost ratio, whether as a consultant to an Asian country, an analyst in the Federal Highway Administration, or an engineer for the Indian Health Service: Be sure to know how the particular agency you are working for defines costs and benefits and what procedure it uses for determining the benefit/cost ratio. If costs and benefits are defined in a rational manner and so treated, any definition of the benefit/cost ratio will result in proper decisions among competing projects, providing incremental analysis is used. The definitions and procedures must be used consistently from project to project, of course.

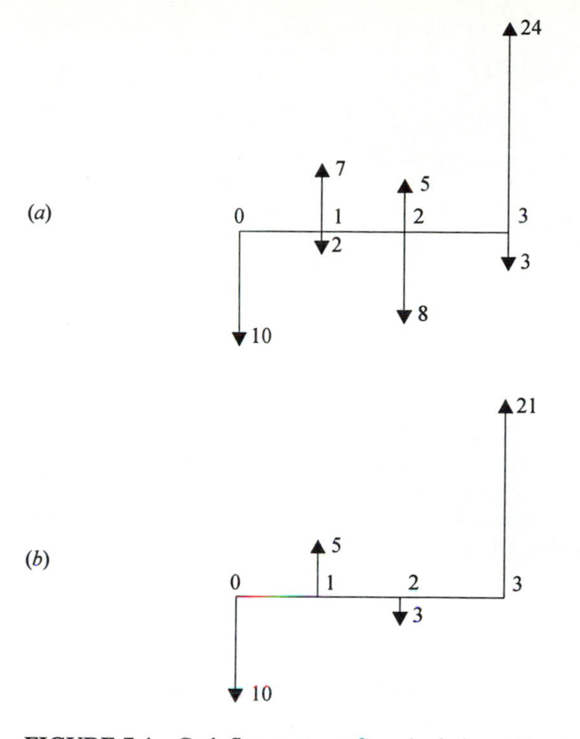

FIGURE 7.4 Cash flow patterns for calculating *B/C* ratios

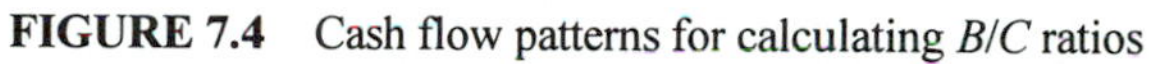

The advantage of a standard method for determining the *B/C* ratio such as the rule of delta is that it allows projects from many departments and levels of government to be compared without adjustment. Fairness in the choice of projects is thus promoted.

It goes without saying that no falsification of data in any sense is referred to here. If falsification takes place, so-called cooking the books, then any pet project can be justified and any ordering of projects achieved.

■ **Example 7.4.** Consider the consequence flow of Fig. 7.4, assuming a 10 percent discount rate. Three periods are involved. If the procedure we have developed thus far is used, the result of step 1, the netting out, is Fig. 7.4*b*. Step 2 requires us to use Eq. (7.2).

$$\frac{B}{C} = \frac{\sum_{0}^{N} B_j(P/F, i, j)}{\sum_{0}^{N} C_j(P/F, i, j)}$$

$$= \frac{5(P/F, 10, 1) + 21(P/F, 10, 3)}{10(P/F, 10, 0) + 3(P/F, 10, 2)}$$

$$= \frac{5(0.9091) + 21(0.7513)}{10(1.0000) + 3(0.8264)}$$

$$= \frac{4.546 + 15.777}{10.000 + 2.479}$$

$$= \frac{20.323}{12.479}$$

$$= 1.629$$

Now, suppose that the agency for which you are working defines the benefit/cost ratio procedure as follows: Identify each of the consequences, to whomsoever they may accrue,[1] as a benefit or as a cost. Then take the discounted present value of the benefits and divide that by the discounted present value of the costs. The result is the benefit/cost ratio. Following this rule prohibits netting out. It requires us to use Eq. (7.1) only:

$$\frac{B}{C} = \frac{7(P/F,10,1) + 5(P/F, 10, 2) + 24(P/F, 10, 3)}{10(P/F, 10, 0) + 2(P/F, 10, 1) + 8(P/F, 10, 2) + 3(P/F, 10, 3)}$$

$$= \frac{7(0.9091) + 5(0.8264) + 24(0.7513)}{10(1.0000) + 2(0.9091) + 8(0.8264) + 3(0.7513)}$$

$$= \frac{6.364 + 4.132 + 18.031}{10.000 + 1.818 + 6.611 + 2.254}$$

$$= \frac{28.527}{20.683}$$

$$= 1.379 \neq 1.629$$

A different ratio has appeared, because of the different handling of costs and benefits. Neither of the two ratios is more correct. Arithmetical procedures are different and lead to different results. Many other definitions of a benefit/cost ratio can be proposed. A common example, using a new term, is explained next. ■

The reader may wonder if annual benefits and annual costs can be used to calculate the benefit/cost ratio. The two methods will sometimes give different answers, especially for different-lived alternatives. The two methods can be reconciled. However, such bringing into agreement is of doubtful utility for the ordinary user. One method—present worth—is probably enough.

---

[1]This phrase is a famous one, taken from the aforementioned Flood Control Act of June 22, 1936. It constitutes the foundation of benefit/cost analysis in the United States.

BENEFITS, COSTS, AND DISBENEFITS

Evidently, the benefit/cost ratio can be changed by adding periodic costs into the denominator rather than, as in the netting out process embodied in the rule of delta, subtracting from the numerator of the fraction representing the benefit/cost ratio. If cost is called a *disbenefit*, this means that its discounted value is subtracted from the discounted value of the numerator rather than added to the discounted value of the denominator. Operation and maintenance costs are often treated in this way, as may be seen in the following example.

■ **Example 7.5.** Two plans for a funicular railway have been proposed in Brazil. The opportunity cost of resources is 10 percent applied to constant money units. All costs and benefits are in constant millions of cruzeiros. No income taxes are paid because this is a state project. Data on the two plans are as follows:

	Plan 1	Plan 2
Economic life (years)	40	20
Salvage value	15	12
Annual benefits	25	22
Annual operation and maintenance costs	5	3
Initial cost	300	160

Given the benefit/cost ratio method and without the assumption that the state government is sure to accept one of the plans, which, if any, should be chosen?

Treat the operation and maintenance costs first as a cost, a part of the denominator, and then as a disbenefit, by subtracting it from the numerator.

Solution. The first line of data presents the analyst with the problem of differing economic lives. He recognizes that, as stated many times in this book, the lives must be equalized. The present worths of the costs and benefits are to be compared in a ratio; but whatever the method employed in economic analysis, the alternatives must be compared over the same period of use. This rule is immutable.

Figure 7.5 illustrates the two alternatives, alternative 2 first because its initial cost is less than that of alternative 1, even though its numbering indicates the opposite. The economic life of alternative 2 has been made equal to the economic life of alternative 1 by repeating it a single time. However, for the comparison with the null alternative, no repetition is necessary.

The lower-cost alternative must be compared to the null alternative, that is, not constructing the funicular at all. Recall that this is the "without" option of the with/without criterion. The example instructions require that the operation and maintenance costs be treated first in the denominator:

$$\frac{B}{C} = \frac{22(P/A, 10, 20) + 12(P/F, 10, 20)}{160 + 3(P/A, 10, 20)}$$

$$= \frac{22(8.514) + 12(0.1486)}{160 + 3(8.514)}$$

$$= 1.019$$

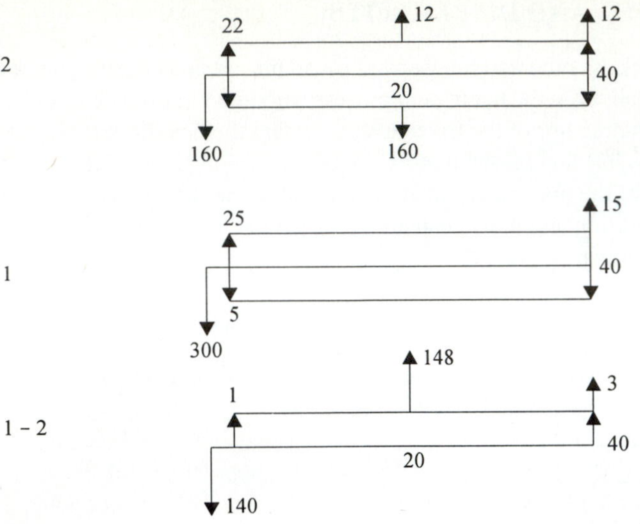

FIGURE 7.5 Funicular railway cash flows

Now, let us treat the 3 million cruzeiros per year as a disbenefit, that is, as a subtraction from the numerator.

$$\frac{B}{C} = \frac{22(P/A, 10, 20) + 12(P/F, 10, 20) - 3(P/A, 10, 20)}{160}$$

$$= \frac{22(8.514) + 12(0.1486) - 3(8.514)}{160}$$

$$= 1.022 \neq 1.019$$

The inclusion of operation and maintenance costs as a disbenefit rather than a cost has changed the benefit/cost ratio. (The reader may wonder if the treatment of a cost as a disbenefit can change an acceptable alternative to an unacceptable one. It cannot. The proof of this statement will be found in App. 7A, where it is shown that changing a cost to a disbenefit cannot cause *B/C* to fall to less than 1.0.) But the benefit/cost ratio is still greater than 1.0, and therefore alternative 2 is acceptable over the null alternative.

Now the higher-first-cost alternative must be analyzed in relation to the alternative already accepted. Incremental methods must be employed. The difference between alternatives 2 and 1 is shown in the bottom drawing of Fig. 7.5. Note that incremental analysis involves the same procedure as netting out (the rule of delta). The calculation of the incremental benefit/cost ratio follows:

$$\frac{\Delta B}{\Delta C} = \frac{1(P/A, 10, 40) + 148(P/F, 10, 20) + 3(P/F, 10, 40)}{140}$$

$$= \frac{1(9.779) + 148(0.1486) + 3(0.0221)}{140}$$

$$= 0.23 < 1.0$$

The more expensive alternative does not qualify. The extra expenditure of 140 million cruzeiros now is not justified by the extra benefits to be received in the future, namely, (1) the saving in operation and maintenance costs together represented by 1 million cruzeiros over the next 40 years, (2) the saving of 148 million cruzeiros 20 years from now, and (3) an extra 3 million cruzeiros in salvage value 40 years from now. This, in plain language, is what the incremental analysis shows us. ■

EQUAL INITIAL INVESTMENTS

Sometimes the costs of two mutually exclusive alternatives are equal. For example, consider a small investor in residential real estate who has $15,000 as a down payment on any house he buys, no matter what the cost of the house. His cash outflow at time 0 will be $15,000 on any of the alternatives he is considering. Revenues and costs may differ considerably for succeeding periods, but not for period 0. Another example: You have $100,000 to invest in stocks or bonds or some combination of the two. Whatever alternative you choose, the cost at time 0 will be essentially the same, with perhaps some minor differences depending on differing commissions paid for stocks versus bonds.

No special handling is required when the benefit/cost ratio method is used, as will be shown in the next example. However, when the internal rate of return method is employed, care must be taken in performing the incremental analysis, or else errors can be made. This will be demonstrated in the next chapter where the internal rate of return method is discussed.

■ **Example 7.6.** Two mutually exclusive alternatives have the following cash flow patterns:

	Alternative	
Year	**1**	**2**
0	-100	-100
1	75	80
2	50	40
3	25	20

The opportunity cost of resources for the public agency reviewing the proposals is 12 percent. Inflation effects have already been included in the benefits and costs. The null alternative must be considered. Using the benefit/cost ratio, select an alternative.

Solution. Figure 7.6 shows the cash flows. To use the benefit/cost ratio, the increment of costs and benefits between the two cash flows must be taken. The rule we have followed thus far is that the lower-first-cost cash flow must be subtracted from the higher-first-cost cash flow. The extra first cost of the latter must be justified by extra benefits accruing to it. In the example before us, however, such a choice is not possible because both first costs equal 100.

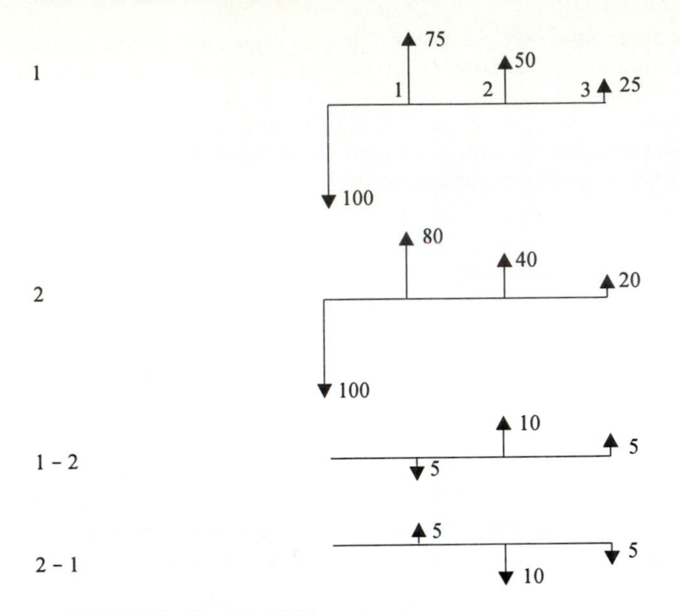

FIGURE 7.6 Equal initial investments

The alternatives may be so ordered, while we are taking the increment between them, that the first cash flow encountered is a cost. The cash flow of Fig. 7.6 marked 1 - 2 is the result of such an ordering. The first arrow is down and occurs at year 1. Therefore alternative 2 must be considered first. Employing Eq. (7.2) gives

$$\left(\frac{B}{C}\right)_2 = \frac{80(P/F, 12, 1) + 40(P/F, 12, 2) + 20(P/F, 12, 3)}{100}$$
$$= \frac{80(0.8929) + 40(0.7972) + 20(0.7118)}{100}$$
$$= \frac{71.43 + 31.89 + 14.24}{100}$$
$$= \frac{117.56}{100}$$
$$= 1.176$$

Therefore alternative 2 is acceptable. The usual reasoning may now be applied: Is the extra cost at year 1 of 5 justified by the extra benefits at year 2 of 10 and at year 3 of 5? Equation (7.2) leads to

$$\left(\frac{\Delta B}{\Delta C}\right)_{1-2} = \frac{10(P/F, 12, 2) + 5(P/F, 12, 3)}{5(P/F, 12, 1)}$$
$$= \frac{10(0.7972) + 5(0.7118)}{5(0.8929)}$$
$$= \frac{7.972 + 3.559}{4.465}$$
$$= 2.583 > 1.0$$

Therefore, alternative 1 should be selected rather than 2. What would have happened if we had chosen to take the increment 2 − 1? The result is the bottom cash flow of Fig. 7.6. It is merely the inversion of the cash flow immediately above it, 1 − 2. The implication of choosing 2 − 1 as increment is that alternative 1 must be considered first. Equation (7.2) leads to

$$\left(\frac{B}{C}\right)_1 = \frac{75(P/F, 12, 1) + 50(P/F, 12, 2) + 25(P/F, 12, 3)}{100}$$

$$= \frac{75(0.8929) + 50(0.7972) + 25(0.7118)}{100}$$

$$= \frac{66.97 + 39.86 + 17.80}{100}$$

$$= 1.25 > 1.0$$

Therefore, at least alternative 1 is acceptable.

Now we consider the increment 2 − 1; its benefit/cost ratio is the benefit/cost ratio of 1 − 2 inverted:

$$\left(\frac{\Delta B}{\Delta C}\right)_{2-1} = \frac{1}{2.583} = 0.387$$

This means that alternative 2 is not justified over 1 and that alternative 1 should be selected, as before.

In benefit/cost ratio analysis, the way the alternatives are ordered for the increment does not matter as far as the decision is concerned. However, there is something to be said for ordering the alternatives such that the customary mode of thought can be followed.

Notice that step 1 of the rule of delta was unnecessary because the alternatives required no netting out; at each period only one benefit or cost was presented. Step 2 was performed by taking the increment. ■

ZEROS IN THE RATIO

It hardly seems worthwhile to devote a separate section in this chapter to the meaning of zeros in the benefit/cost ratio, but because students raise questions about the matter, we include it here.

If a zero occurs in the denominator of the benefit/cost ratio, the rules of algebra apply; that is, the result is undefined. Division by zero is not a permissible operation in algebra. The benefit/cost ratio is definitely not infinite in such a situation. The method must be abandoned and another, such as net present worth, used in its place.

If a zero occurs in the numerator of the benefit/cost ratio, the result is zero. Because 0 is less than 1, the project is unacceptable.

COST-EFFECTIVENESS

Cost-effectiveness in general refers to economic efficiency. A project that is cost-effective is one that is economically efficient. Without going into the subject of economic efficiency, which is beyond the scope of this text, we say that its meaning in the sense used here is that a project is cost-effective if, for example, its benefit/cost ratio is greater than or equal to 1.0.

In particular, *cost-effectiveness measures* refer to an index, usually a ratio, whose numerator is an incommensurable benefit and whose denominator is the cost of achieving that benefit. It is used in the comparison and selection among alternatives where the benefits are difficult or impossible to measure in monetary units. For example, the benefits of a new road to a small village in the hinterlands of Mexico are difficult to measure in monetary terms. The benefits are (1) the lowered cost of transport for both persons and goods and (2) access to jobs, education, and medical service, all of which were available only with great effort. In general, the road brings much improved social and economic ties with the rest of Mexico, with all the implications of such a vague but important concept as improved national unity. Costs are the construction and maintenance costs of the road. If many road projects are proposed and not enough money is available to build them all—the usual economic reality—then a priority system must be set up to optimize the use of available funds. A cost-effectiveness index must be created. In the Mexican roads example, the index could be created by using as the numerator the number of people to be served by the road, which is a surrogate for all the benefits so difficult to measure, using as the denominator the present value of the costs of the road. The result is a cost-effectiveness index suitable for setting road priorities, thus facilitating the decision as to which roads to construct and which to disregard.

Cost-effectiveness indices are used in military procurement, social welfare projects, environmental pollution control planning, and many other areas. Although the preceding are public sector programs, cost-effectiveness measures can also be used in private sector planning where the same difficulty in assigning a monetary value to benefits is present. Such a benefit is company prestige—the good name of the company can be of primary importance, yet impossible to evaluate in dollars.

■ **Example 7.7.** Recently the U.S. Department of Defense analyzed a weapons procurement program. Helicopter attack squadrons, specializing in the destruction of heavy tanks, could be equipped with a number of weapons systems each of which would destroy tanks, but at a different rate per time period and at different life-cycle costs. The *life-cycle cost* is the present worth of the initial cost and of operation and maintenance costs over the economic life of the asset.

The time period was the first 3 weeks of an attack against NATO troops.

Three weapons systems were under review:

1. A pair of Gatling-type automatic cannons mounted on either side of the helicopter. The Gatling-type cannon is characterized by a high rate of fire.
2. A pair of conventional automatic cannons.
3. A set of armor-piercing, antitank rockets.

Much simplified, the projected scenario of such an attack resulted in the following table of benefits, measured in tanks destroyed in the initial attack period. The cost in billions of dollars represented the life-cycle cost of the equipment, stationed in combat-readiness in Europe. (The short combat life of the equipment was one of the factors in the determination of the benefits.) The service life of the helicopter was estimated at 10 years. A cost-effectiveness analysis was desired.

System	Benefit (tanks destroyed)	Life-cycle cost ($000,000,000)
1	1,800	9
2	1,500	7.5
3	1,000	6

Solution. Notice that the systems are numbered from highest to lowest cost. The problem must be approached from lowest to highest cost, that is, by starting with system 3.

One of the most important rules of economic analysis applies as much here as elsewhere: *It is the extra benefit as a result of extra cost that is crucial.* In other words, incremental analysis must be employed.

There is no such thing as a null alternative here because a system must be chosen. Even if it were possible to reject all alternatives, to measure the cheapest alternative against doing nothing is not feasible because of the lack of monetary benefits. Only an arbitrary cutoff index might be employed, determined by the subjective judgment of experienced officers. As will be seen, some such judgmental cutoff index cannot be avoided.

System 3 has a cost-effectiveness index of 167 tanks destroyed per $1 billion ($1{,}000 \div 6$).

The extra cost of $1.5 billion for system 2 compared to its extra benefit of 500 tanks results in an incremental cost-effectiveness index of 333.

The incremental index of system 1 is

$$\frac{1{,}800 - 1{,}500}{9 - 7.5} = 200$$

In a mutually exclusive alternative situation such as this, only judgment can prevail, buttressed by the logic of the analysis. If, say, 100 tanks per $1 billion of life-cycle cost are an acceptable figure, then alternative 1, the most expensive, should be selected because all three systems exceed that figure in an incremental sense. Discussion of what would happen if another cutoff index were chosen, say 275, must be deferred to Chap. 9 on multiple alternatives. ■

SUMMARY

The *benefit/cost ratio*, most often used to judge government projects, is defined as the discounted present worth of project benefits divided by the discounted present worth of project costs. Incremental analysis is an absolute necessity in the use of the benefit/cost ratio. The service lives of all alternatives must be equalized when incremental analysis is performed.

The magnitude of the benefit/cost ratio depends on the rules of procedure for determining it. Thus a single project may be associated with several benefit/cost ratios. A procedure that results in a unique benefit/cost ratio is the *rule of delta*:

Step 1: Net out all benefits and costs at each period.

Step 2: Find the benefit/cost ratio by dividing the present worth of all benefits by the present worth of all costs, at whatever period these occur.

Equal initial investments require no special handling in benefit/cost ratio determination, but it is recommended that the increment between alternatives be taken such that the first payment encountered in the incremental cash flow is a cost. If the internal rate of return method is used, this rule must be followed.

Zeros in the ratio are handled according to the rules of algebra.

The cost-effectiveness measure is usually used when benefits cannot be measured in monetary units. Incremental analysis must be used in this measure.

PROBLEMS

7.1. (*a*) In Example 7.1, page 120, will the result change if the life of the barriers is reduced to 15 years?

(*b*) If the null alternative were included, would the selection made in part (*a*) change?

(*c*) Do the analysis of part (*b*) by using the incorrect, though frequently used, method described in Example 7.1. Comment.

7.2. The White House solar energy system cost $28,000. It nets $1,000 annual savings in energy costs. If the U.S. government discount rate of 10 percent is used, how many years will the system need to last to achieve a benefit/cost ratio of 1.0?

7.3. In Example 7.2, page 123, a benefit of $520,000 has been suggested by higher authority as more realistic. Will the project be selected?

7.4. The state of Jalisco in Mexico is considering the investment of 1 million pesos in a rural road. The local Secretariat of Communications and Transport estimates that annual costs during the next 10 years will be 150,000 pesos. Benefits during the same period will be 400,000 pesos per year. If the opportunity cost of public funds is 12 percent, compute the benefit/cost ratio by using the rule of delta.

7.5. Repeat Prob. 5.16, using the benefit/cost ratio method. Observe the rule of delta.

7.6. Repeat Prob. 5.17, using the benefit/cost ratio method. Observe the rule of delta.

7.7. A multiple-purpose dam, to be built over 4 years, is being considered for construction. Its first cost, distributed evenly over 4 years, is $80 million. Its benefits, beginning during the first year after construction is completed, over a life of 100 years, are $6 million per year. Its annual operations cost, beginning at the same time as the benefits, is $900,000. The minimum attractive rate is 5 percent, and the salvage value is zero. Calculate its benefit/cost ratio.

(*a*) Observe the rule of delta.

Answer: $B/C = 1.17$

(*b*) Do not net out benefits and costs; that is, apply the approach of page 128.

Answer: $B/C = 1.14$

7.8. A proposed investment shows the following benefits and costs:

Year	Cost	Benefit
0	80	0
1	160	0
2	10	100
3	20	120
4	20	130
5	30	80
6	40	70

If the *minimum attractive rate of return* (MARR) is 12 percent and taxes and inflation are ignored, what is the benefit/cost ratio of this investment if netting out (that is, the rule of delta) is used? Is the project acceptable?

Answer: $B/C = 1.15$

7.9. In Example 7.3, page 124, it is suggested that the reader verify the correctness of the incremental approach to the benefit/cost ratio by use of the NPV method. Do this.

7.10. We have to select one of two projects:

Year	Project *A*	Project *B*
0	-70	-120
1	5	20
2	0	15
3	40	55
4	40	55
5	50	70

The opportunity cost of capital equals 10 percent. Do not consider the null alternative.

(*a*) Which project will be selected if the present-worth method is used?
(*b*) Perform the analysis, using the *B*/*C* method.

7.11. An architect is trying to decide between two designs, one of steel, the other of reinforced concrete, for an exposed-column, multistory building. The project is on contract to the federal government, and so a benefit/cost ratio approach must be used. The following facts are available about each design:

	First cost ($)	Life (years)	Annual maintenance cost ($)	Removal cost ($)
Concrete	3,300,000	35	125,000	50,000
Steel	2,500,000	35	140,000	40,000

No taxes will be charged since this is a government job. The MARR is 10 percent. Inflation will be ignored. Which of the two designs should be chosen?

Answer: Steel

7.12. A proposal to the Transportation Department of the state of Nevada is composed of two mutually exclusive alternatives for improvements to a certain highway overpass which has been the scene of several accidents. It is believed that the costs and benefits to the public are as follows:

Alternative	First cost ($)	Net annual benefit ($)
1	1,200,000	100,000
2	900,000	62,000

No salvage value will be forthcoming on conclusion of the 20-year life of the investment. One of the two alternatives is certain to be chosen because of public demand for improvement of the facility.

Use the benefit/cost ratio method to determine which alternative should be chosen. A 12 percent minimum attractive rate of return is used by the state of Nevada at this time for transport proposals.

7.13. In Example 7.4, page 127, another possible rule is this: *The denominator of the benefit/cost ratio will always be the first cost, that is, the cost at time 0. The numerator will be the present value of the future benefits less the present value of the future costs.* Apply the rule to the cash flow, and comment on your answer.

7.14. A government building, presently in the design stage, will have one of two possible heating and cooling systems installed. The first involves recirculation of air through electrostatic filters. The second is a makeup air system. These are the data on the systems:

Item	Recirculation	Makeup air
Installed cost ($)	52,100	20,560
Operation and maintenance cost per year ($)	10,200	15,500
Economic life (years)	30	15
Salvage value ($)	0	0
Discount rate (percent)	10	10

If the *B*/*C* ratio is used, which system should be chosen? Do not consider inflation.

7.15. (*a*) In Example 7.5, page 129, would it have been correct to test alternative 2 against the null alternative using two cycles, that is, a 40-year life, in alternative 2?

(*b*) If the analyst decides to use a 40-year life for alternative 2, that is, two cycles, will it make a difference in the benefit/cost ratio? Do the calculation and find out.

7.16. In Example 7.6, page 131, alternative 1 now has a benefit of 95 in year 1 and 30 in year 2. All the other data remain the same. Perform the *B*/*C* analysis, ordering the alternatives as recommended by the rule that the first cash flow encountered in the increment must be a cost.

7.17. In a large electric generating plant, two designs for coal car thawing sheds are being considered to replace an existing one. Each will cost the same amount—$1,090,000. The electrically heated shed will last 10 years; the gas-heated shed will last 15 years. The salvage value will approximate the removal cost for both designs. The before-tax cost of capital for this firm is 20 percent.

Maintenance and operation costs for the electric model will be $80,000 per year. The same costs for the gas shed will be $100,000 per year. Which design should be chosen if the benefit/cost ratio method is used?

7.18. In Example 7.6, page 131, alternative 1 now has a benefit of 95 in year 1. Using the benefit/cost ratio, select an alternative.

7.19. A proposed bridge on the interstate highway system is being considered at a cost of $2 million. Its life is projected at 20 years. A discount rate of 10 percent is used. Operation and maintenance costs are projected at $180,000 per year. These construction and maintenance costs will be borne by the federal and state governments. Benefits to the public are estimated at $900,000 annually. Adverse consequences to the public are projected at $250,000 per year.

(*a*) What is the *B*/*C* ratio if the rule of delta is used?

(*b*) What is the *B*/*C* ratio if all consequences to the public are counted in the numerator and all consequences to the government in the denominator?

(*c*) How will the *B*/*C* ratio change if the adverse consequences to the public are counted as cost in the denominator along with the other costs?

(*d*) What is the *B*/*C* ratio if the operation and maintenance costs are made disbenefits and the first cost and costs to the public are left in the denominator? Carry out all *B*/*C* ratios to the second decimal place.

7.20. Two highway routes were under consideration between Ras Tanura and Dhahran in Saudi Arabia. They were mutually exclusive alternatives. One was sure to be chosen. They had the following characteristics:

Route	First cost ($000,000)	Length (miles)	Annual operation and maintenance cost ($000,000)	Life (years)
A1	300	45	5.0	20
A2	200	60	6.5	20

Traffic data are as follows:

Average daily traffic	2,000 vehicles	
Vehicle composition	60 percent commercial, 40 percent private	
Average traffic speed	45 miles per hour commercial, 60 miles per hour private	
Operating costs	Commercial	$0.90 per mile
	Private	$0.30 per mile
Cost of time	Commercial	$60 per hour
	Private	$30 per hour

No data were available on accidents or aesthetics. The minimum attractive rate of return was 10 percent. Costs will include construction, operation and maintenance, and user costs. Which alternative is indicated?

7.21. In Example 7.7, page 134, what will be the result of the cost-effectiveness analysis if the cost of system 3 is raised to $8 billion?

APPENDIX 7A

In Example 7.5 we stated that the treatment of a cost as a disbenefit cannot change an acceptable alternative into an unacceptable one; that is, *B*/*C* cannot be lowered to a quantity less than 1.0. This will now be demonstrated.

Let

$$\frac{B}{C} = \frac{a-x}{b} \geq 1 \tag{7A.1}$$

Is there any x such that the following holds?

$$\frac{B}{C} = \frac{a}{b+x} < 1 \tag{7A.2}$$

Two possible situations exist:

$$\frac{a-x}{b} = 1 \qquad \text{and} \qquad \frac{a-x}{b} > 1$$

If

$$\frac{a-x}{b} = 1$$

then

$$a - x = b$$

and

$$a = b + x$$

Substitution in Eq. (7A.2) yields

$$\frac{a}{a} < 1$$

which is clearly an impossibility because

$$\frac{a}{a} = 1$$

Therefore, no such x exists.

If

$$\frac{a-x}{b} > 1$$

then

$$a - x > b$$

or

$$a > b + x$$

or

$$b + x < a$$

Substitution in Eq. (7A.2) gives

$$\frac{a}{<a} < 1$$

which is also impossible because a quantity a divided by some quantity less than itself ($< a$) must be greater than 1, not less than 1. Therefore, no such x exists.

CHAPTER 8

INTERNAL RATE OF RETURN

The *internal rate of return* (*IROR*) method for comparing benefits and costs is the last of the four that this book reviews. Present and annual worth and the benefit/cost ratio have not been without anomalies and points difficult to understand. But the IROR method is perhaps the most difficult of the methods, if practiced without a computer, and the most controversial. However, it is also widely used because it employs a percentage rate of return as the decision variable. This suits the banking community—private, public, and international. City Bank of Austin, Texas; Riggs Bank of Washington, D.C.; Federal Reserve Bank; Interamerican Development Bank; and World Bank all speak the same language of decision making. Percentage rates are understood and accepted without question.

Percentage rates are also well understood by the general public. Almost everyone pays a mortgage on a residence or makes car payments or borrows money via credit cards. Each case has a percentage interest rate attached to it that defines what the payments shall be. Almost everyone is interested in knowing what the loan interest rate is, therefore. Only those who do not care, because of wealth or ignorance, will ignore it.

Business people are familiar with rates of return because they all borrow money to finance ventures, even if the money they borrow is their own. No lengthy explanations of the meaning of a percentage rate of return need be made to them.

Consider, in contrast, the difficulty of explaining the concept of present worth to someone who has not taken the course in which this textbook is being used.

Most executives who may pay for your time, whether as a consultant or an employee, will not receive the news that they do not understand something with any great joy. In fact, it would be advisable for the consultant or employee to choose to speak in a language that the audience will understand. Because of this, in spite of much criticism, the internal rate of return method persists.

Consider, too, that so much information comes to us in the form of percentage rates. Inflation is reported as a compound interest rate, as is the gross national product (GNP). Birth and death rates and population growth and decline—those most important figures in our world—are expressed in percentages.

FORMULAS

The *internal rate of return* (IROR) is the percentage rate that causes the discounted present value of the benefits in a cash flow to be equal to the discounted present value of the costs. Thus,

$$\sum_{j=0}^{N} B_j(P/F, i^*, j) = \sum_{j=0}^{N} C_j(P/F, i^*, j) \tag{8.1}$$

where i^* is the IROR.

The internal rate of return can also be defined as the discount rate that causes the net present value (NPV) of a cash flow to equal zero.

$$\text{NPV} = \sum_{j=0}^{N} B_j(P/F, i^*, j) - \sum_{j=0}^{N} C_j(P/F, i^*, j) = 0 \tag{8.2}$$

The relationships are most easily demonstrated in the historical example that follows.

DIRECT COMPUTATION OF IROR

■ **Example 8.1.** A Merrill-Lynch Ready Assets Trust in the amount of $5,000 was purchased on November 11, 1993. It was sold on the same date 2 years later for $5,926. What was the rate of return on the investment, before taxes and inflation were taken into account?

Solution. Figure 8.1 shows the situation. If we use Eq. (8.1), then

$$\sum_{j=0}^{N} B_j(P/F, i^*, j) = \sum_{j=0}^{N} C_j(P/F, i^*, j)$$

$$5{,}926(P/F, i^*, 2) = 5{,}000(P/F, i^*, 0)$$

$$(P/F, i^*, 2) = \frac{5{,}000(1.000)}{5{,}926}$$

$$= 0.8437$$

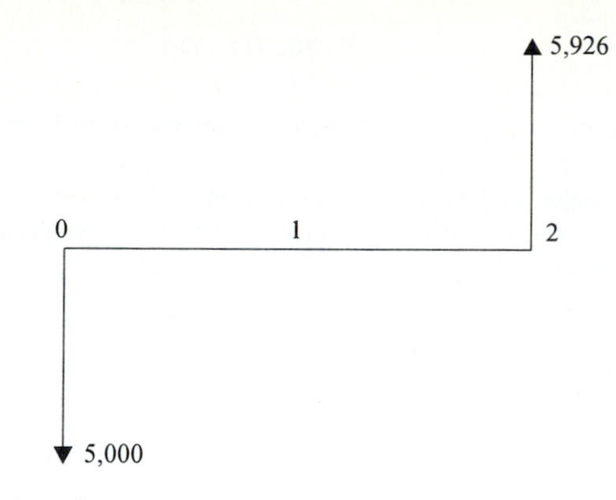

FIGURE 8.1 A ready assets trust

One way to find an approximate answer for i^* is to consult the tables in the back of the book under row 2 and the P/F column to find the P/F values that straddle 0.8437 and their corresponding interest rates. These are

0.8417 at 9%

and 0.8573 at 8%

The internal rate of return on the cash flow and thus the rate of return on the investment are about 9 percent because 0.8437 is closer to 0.8417 than to 0.8573. An interpolation will give a more exact answer:

P/F	**Percent**
0.8417	9.0
0.8437	?
0.8573	8.0

$$i^* = 9.0 - \left[\frac{0.8437 - 0.8417}{0.8573 - 0.8417} \times (9.0 - 8.0)\right]$$
$$= 8.872$$

Just how exact this answer is we cannot be sure, because interpolation assumes a linear relationship between P/F and i. However, we know that the relationship *is* really an exponential one.

$$(P/F, i, j) = \frac{1}{(1+i)^j}$$

In order to reach an exact solution, it is only necessary to solve

$$0.8437 = \frac{1}{(1+i^*)^2}$$
$$(1+i^*)^2 = \frac{1}{0.8437}$$
$$1 + i^* = \left(\frac{1}{0.8437}\right)^{0.5}$$

$$i^* = 0.0886944$$
$$= 8.869\%$$

to get an exact answer. It is slightly less than that arrived at by interpolation, with a difference of 0.003.

Interpolation becomes less accurate as the interval between the interest rates widens. For example, had we tables at intervals of 5 percent instead of 1 percent, that is, at 10 and 5 percent—the percentages straddling a (*P*/*F*, *i**, 2) of 0.8437—the interpolation would be

P*/*F	**Percent**
0.8264	10.0
0.8437	?
0.9070	5.0

$$i^* = 10.0 - \left[\frac{0.8437 - 0.8264}{0.9070 - 0.8264} \times (10.0 - 5.0)\right]$$
$$= 8.927$$

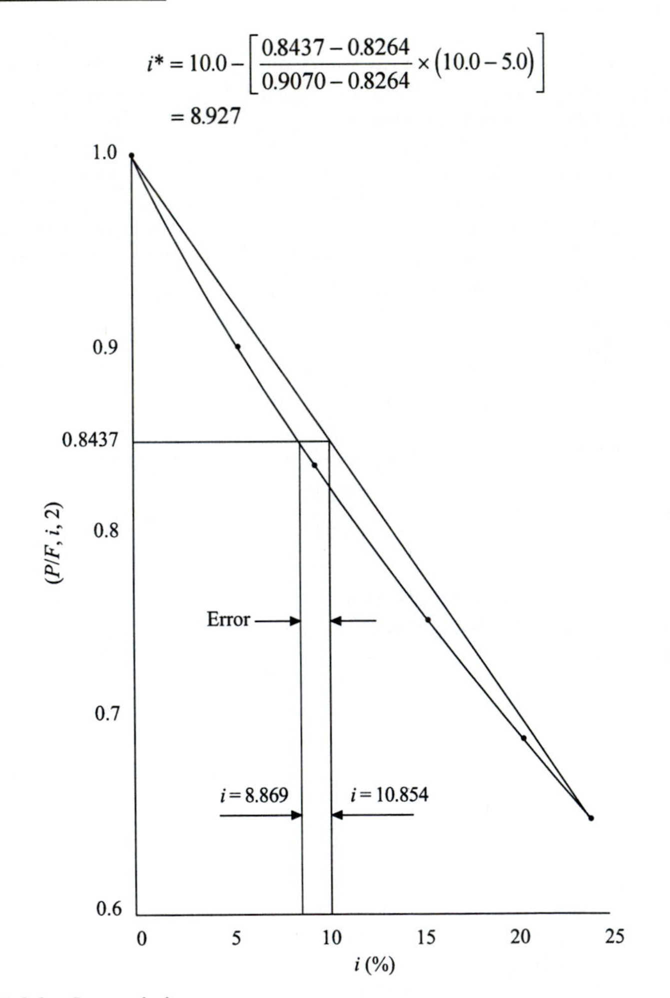

FIGURE 8.2 Interpolation error

The difference is now 8.927 − 8.869, or 0.058, some 19 times greater than the former difference between the exact and the interpolated i^*. At still larger intervals, the error grows larger. In Fig. 8.2, if the interpolation interval had been taken between 0 and 25 percent, as shown by the straight line between those two points, the error would have been almost 2 percent, as shown in the figure. By calculation it is 1.985 percent.

The lesson of this example is, in part, that if tabular interpolation is used to find IRORs, the interval between interest rates should be reduced to within the acceptable limits of error in the resulting IROR. This example has also shown two methods of finding the IROR. ■

■ **Example 8.2.** Urban University has installed a computerized energy management system with 3,500 points of audit across the campus. The $1.25 million system has achieved a 30 percent reduction in electricity use and a 50 percent savings in heating costs. This translates to yearly savings of $95,000 and $150,000, respectively. If such systems last 20 years, what rate of return to the nearest whole percentage is the university making on its investment before taxes and before considering inflation effects? No salvage value need be considered.

Solution. Figure 8.3 illustrates the situation, which is somewhat more complicated than Example 8.1. Because $245,000 is a uniform annual payment, we can use a variation on the $(P/F, i^*, j)$ portion of Eq. (8.2), as we have done before, by recalling that

$$\sum_{j=1}^{N}(P/F, i, j) = (P/A, i, N)$$

and

$$\text{NPV} = -1{,}250{,}000(P/F, i^*, 0) + 245{,}000(P/A, i^*, 20) = 0$$

$$= -1{,}250{,}000 + 245{,}000(P/A, i^*, 20) = 0$$

We will no longer bother to write in $(P/F, i^*, 0)$ in these problems, remembering that it is always equal to 1.0000.

$$(P/A, i^*, 20) = \frac{1{,}250{,}000}{245{,}000}$$

$$= 5.1020$$

Interpolation between the tables for 20 and 18 percent gives a value of 19 percent to the nearest whole percentage point. ■

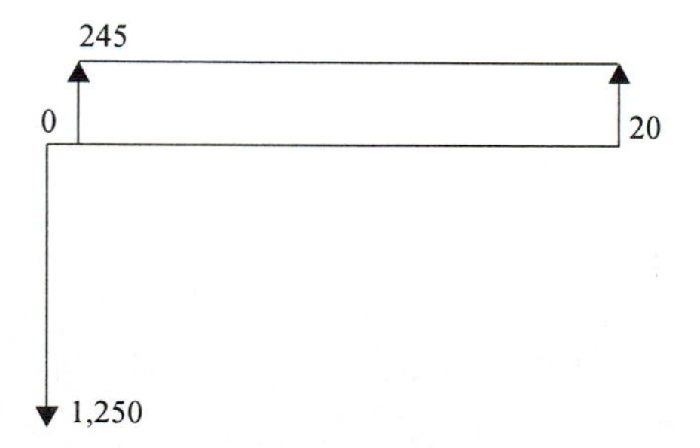

FIGURE 8.3 Energy management system

TRIAL-AND-ERROR COMPUTATION OF IROR

■ **Example 8.3.** Computing equipment was purchased 3 years ago for $200,000. Its net benefits to Development Research Company were estimated at $100,000 per year for the economic life of the equipment, thought to be 10 years. It now has a zero salvage value and will be retired. Development Research Company will purchase new computing equipment for $350,000. This new equipment is estimated to have an economic life of 5 years, no salvage value, and net benefits per year of $107,000 for the first year and $214,000 for each of the 4 succeeding years.

What is the prospective rate of return in this situation in a before-tax analysis? Do not consider inflation. Accuracy to the nearest 1 percent will be sufficient.

Solution. This example introduces the use of some concepts from previous chapters, in particular the sunk cost. The $200,000 worth of equipment purchased 3 years ago and all amounts connected with it are now sunk costs and past benefits and have no relevance whatever to the problem. We are told that the equipment has a zero salvage value and will be retired. The only relevant cash flow is that shown in Fig. 8.4. Using Eq. (8.2) gives

$$\text{NPV} = \sum_{j=0}^{N} B_j(P/F, i^*, j) - \sum_{j=0}^{N} C_j(P/F, i^*, j) = 0 \tag{8.3}$$

and adapting it to the problem at hand gives

$$107{,}000(P/F, i^*, 1) + 214{,}000(P/A, i^*, 4)(P/F, i^*, 1) - 350{,}000 = 0$$

Notice the two unknowns in the preceding equation, $(P/F, i^*, 1)$ and $(P/A, i^*, 4)$. These prevent a direct solution as in the previous two examples. A trial-and-error solution is the only possible recourse if computing equipment is not available. The procedure will

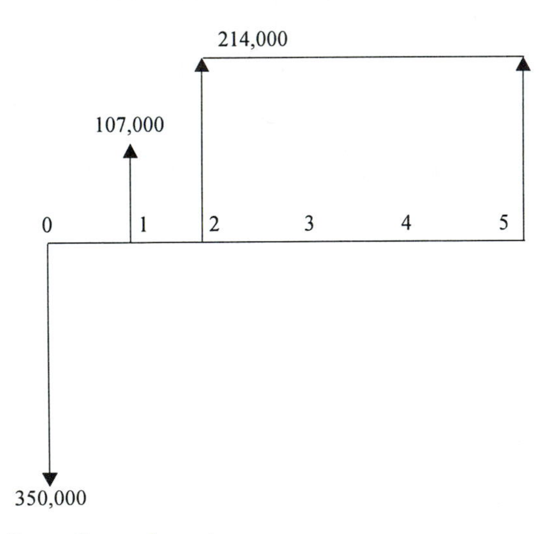

FIGURE 8.4 Computing equipment

be to try a number of interest rates to determine the two that will just straddle the IROR at an interval between them as small as necessary to achieve the desired accuracy in the IROR.

How is one to decide the discount rate for the first trial? A rough estimate can be made by imagining that the benefit cash flow is concentrated at the middle, or thereabouts, of the time stream and making a quick calculation to find what the IROR would be if that were so. Thus,

$$4(214{,}000) + 107{,}000 = 963{,}000$$

$$350{,}000 = 963{,}000(P/F, i^*, 3)$$

$$(P/F, i^*, 3) = 0.3634$$

At i = 40 percent, (P/F, 40, 3) = 0.3644. It is best to start out the trials at 40 percent, therefore. Substituting at 40 percent in the equation of the cash flow, Eq. (8.3), gives

$$\begin{aligned} \text{NPV} &= 107{,}000(P/F, 40, 1) + 214{,}000(P/A, 40, 4)(P/F, 40, 1) - 350{,}000 \overset{?}{=} 0 \\ &= 107{,}000(0.7143) + 214{,}000(1.849)(0.7143) - 350{,}000 \overset{?}{=} 0 \\ &= 76{,}430 + 282{,}639 - 350{,}000 \overset{?}{=} 0 \\ &= 9{,}069 \neq 0 \end{aligned}$$

The IROR is not exactly 40 percent because substitution of the factors associated with that percentage resulted in a net present value of 9,609, not 0.

Should our next trial be at greater or less than 40 percent? The reasoning is as follows: Because we are on the plus side of the NPV = 0 point, we want to lessen the effect of the benefits by discounting them more. This means we must go to a higher discount rate. We are attempting to straddle the true IROR. In this case, the NPV, positive at 40 percent, should become negative. Evidently, we are not far from a straddle because 9,069 is a small percentage of the sums involved. Try 45 percent, therefore.

Substitution in Eq. (8.3) gives

$$\begin{aligned} \text{NPV} &= 107{,}000(P/F, 45, 1) + 214{,}000(P/A, 45, 4)(P/F, 45, 1) - 350{,}000 \overset{?}{=} 0 \\ &= 107{,}000(0.6897) + 214{,}000(1.720)(0.6897) - 350{,}000 \overset{?}{=} 0 \\ &= -22{,}337 \neq 0 \end{aligned}$$

A straddling has occurred. The true IROR must lie between 40 and 45 percent. Graphical interpolation is shown in Fig. 8.5.

i	NPV
40%	+9,069
i^*	0
45%	−22,337

The calculation of the interpolated IROR is achieved by using similar triangles:

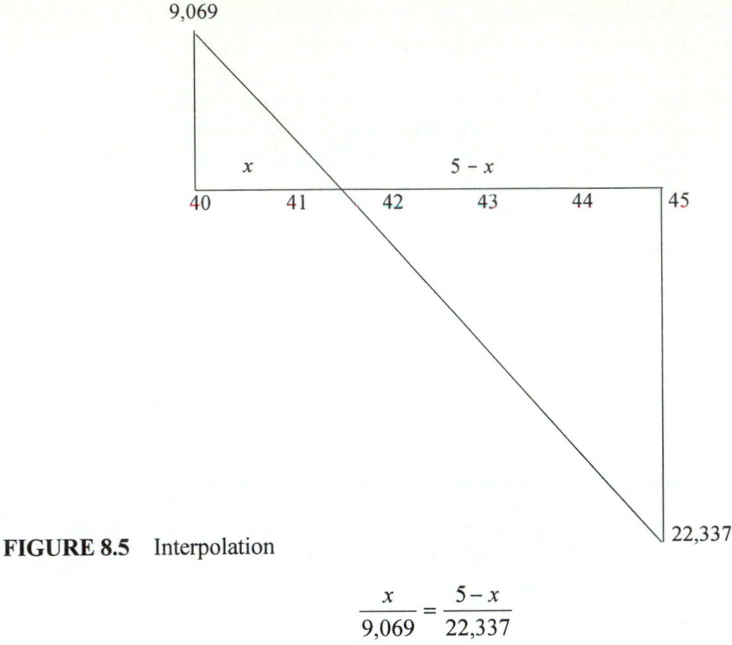

FIGURE 8.5 Interpolation

$$\frac{x}{9{,}069} = \frac{5-x}{22{,}337}$$
$$x = 1.44$$

The IROR is 41.44 percent, or 41 percent to the nearest whole percent. ■

INCREMENTAL ANALYSIS NECESSARY

The first three examples of this chapter have dealt with finding the IROR. How is the IROR method used when a decision is involved? In other words, how is it used as a decision tool with mutually exclusive alternatives? To answer this question, another relationship must be explained. When the internal rate of return of a consequences flow equals or exceeds the opportunity cost of capital, a course of action is accepted; when the internal rate of return does not, the course of action is rejected:

$$i^* \geq i \tag{8.4}$$

where i^* is the IROR and i is the opportunity cost of capital.[1] It is normal to calculate i^*; a determination that i^* is greater or less than i is not sufficient.

Equation (8.4) may be converted algebraically to

$$i^* - i \geq 0 \tag{8.5}$$

[1]Notice that only two choices are possible—acceptance or rejection—with this equation. If a third choice were allowed, when $i^* = i$, it would be called *indifference*, meaning that it would not matter whether the alternative were accepted or rejected. For this text, the indifference possibility has been excluded.

which accomplishes exactly the same purpose. Equation (8.4) states the rule in terms of a total, and Eq. (8.5) states it in terms of a difference.

For example, if i^* is 15 percent and i is 10 percent, is the project making 15 or 5 percent? Conventionally, the project is credited with a 15 percent internal rate of return. This is the way the words *internal rate of return* will be interpreted by those persons familiar with the subject—as a total, not a difference.

More than one alternative includes the null alternative. Thus if the question of Example 8.3 had been, Should Development Research Company buy the new equipment if its minimum attractive rate of return is 30 percent? we would have been comparing the alternative of buying the new equipment against not doing so. Because

$$41\% \geq 30\%$$

the decision would have been to buy the new equipment.

For consistency's sake, and because the practice follows standard microeconomic procedure, alternatives should be ordered for testing by starting with the one having the lower initial investment. This alternative's IROR is then found and compared to the opportunity cost of capital. If the first alternative qualifies under this test, then the incremental consequences flow found by subtracting the cash flows of the lower-initial-cost alternative (the first) from those of the higher-initial-cost alternative (the second) is tested by determining its *incremental* internal rate of return (ΔIROR). If ΔIROR equals or exceeds the opportunity cost of capital, the higher-cost alternative is accepted; if not, it is rejected and the lower-cost alternative accepted.

If it happens that the lower-cost alternative does not qualify, then the higher-cost alternative should be tested against the null alternative. If its IROR is greater than the opportunity cost of capital, it should be accepted; if not, it should be rejected.

It is worth pointing out that the ΔIROR need not exceed the IROR of the first alternative to be accepted. It only need exceed the opportunity cost of capital. When this is the case, the higher-cost alternative is accepted.

Incremental analysis in the IROR method is absolutely necessary when mutually exclusive alternatives are involved. In this respect it parallels the benefit/cost ratio. To do an individual analysis of each alternative in order to find each IROR, then to compare these IRORs and choose the alternative associated with the higher IROR is totally incorrect. Of course, it can sometimes produce the correct answer, as can any other incorrect method in any field.

■ **Example 8.4.** The following after-tax, after-inflation, cash flows occur for two mutually exclusive alternatives:

	Alternative	
Year(s)	**1**	**2**
0	−800	−1,500
1–5	+400	+700

The null alternative must also be considered.

(*a*) At 30 percent *minimum attractive rate of return* (*MARR*), and by using the internal rate of return method, what alternative, if any, should be selected?

(*b*) At 35 percent

(*c*) At 40 percent

(*d*) At 45 percent

All percentages must be accurate to the nearest whole percent.

Solution

(*a*) Because the null alternative is to be considered, the lower-first-cost alternative must be tested against it. Employing Eq. (8.2) on alternative 1 of Fig. 8.6 gives

$$\text{NPV} = -800 + 400(P/A, i^*, 5) = 0$$
$$(P/A, i^*, 5) = 2.000$$

From the tables of App. C, $i^* = 41$ percent. Because

$$41\% > 30\%$$

alternative 1 should be accepted according to Eq (8.4).

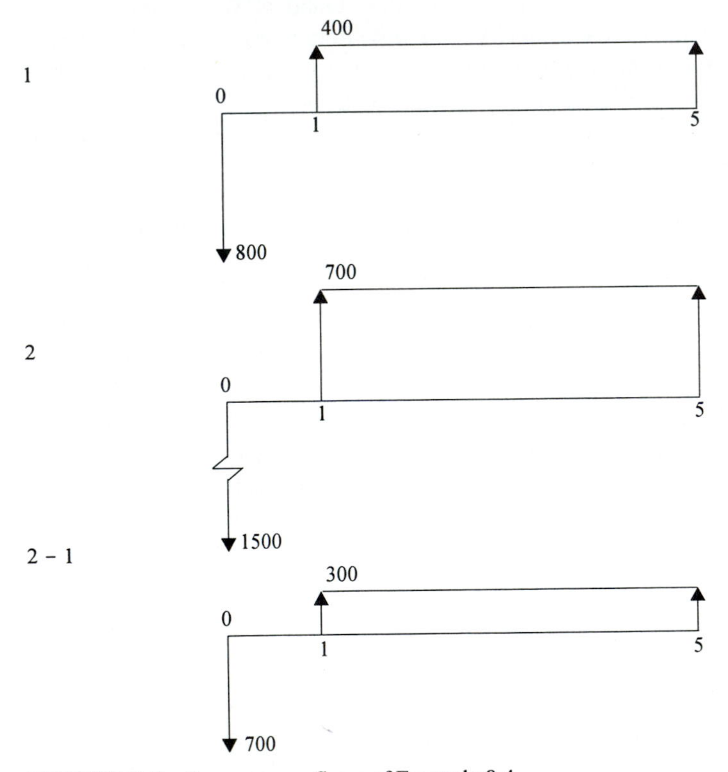

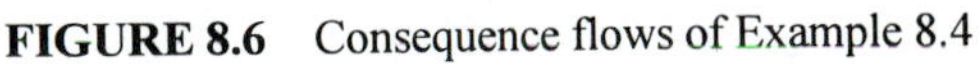

FIGURE 8.6 Consequence flows of Example 8.4

Now try alternative 2 against 1, using the incremental consequence flow 2 − 1 of Fig. 8.6:

$$NPV = -700 + 300(P/A, i^*, 5) = 0$$

$$(P/A, i^*, 5) = 2.3333$$

$$i^* = 32\%$$

Because

$$32\% > 30\%$$

alternative 2 is chosen over alternative 1.

(*b*) Now let us try the same alternatives at 35 percent. As calculated in part (*a*), the IROR of alternative 1 in comparison with the null alternative is 41 percent. Because

$$41\% > 35\%$$

alternative 1 is accepted. Alternative 2 versus 1 results as before in 32 percent. But now, because

$$32\% < 35\%$$

alternative 2 is rejected and therefore 1 is accepted.

(*c*) At 40 percent,

$$IROR_1 = 41\% > 40\%$$

Therefore alternative 1 is accepted. But

$$IROR_{2-1} = 32\% < 40\%$$

and alternative 1 is accepted.

(*d*) At 45 percent,

$$IROR_1 = 41\% < 45\%$$

Therefore alternative 1 is rejected.

Now alternative 2 must be tested against the null alternative:

$$NPV = -1{,}500 + 700(P/A, i^*, 5) = 0$$

$$(P/A, i^*, 5) = 2.143$$

Interpolation determines that i^* = 37 percent. Because

$$37\% < 45\%$$

alternative 2 is also rejected and the null alternative is accepted, that is, investment at 45 percent. Table 8.1 summarizes the process. ■

TABLE 8.1
Decision process of Example 8.4

Criterion (%)	IROR project 1 (%)	IROR 2–1 (%)	Decision
30	41	32	2
35	41	32	1
40	41	32	1
		2–0	
45	41	37	0

IROR METHOD AGREES WITH PRESENT-WORTH METHOD

For many years a controversy has overshadowed the use of the internal rate of return method. Many practitioners believed that the IROR method was flawed because, as they employed it, it did not always agree with the results obtained by the NPV (present-worth) method. The contradiction arose out of a misuse of the IROR method. This misuse consisted in determining the IROR of each alternative consequence flow of a number of mutually exclusive alternatives and then selecting the alternative with the highest IROR. (As the reader is already aware, this is incorrect procedure.) It often resulted in an incorrect solution, and when the result was checked by using the NPV method, a contradiction ensued. Over the years, however, those analysts who demonstrated the correctness of the incremental analysis procedure in the IROR method prevailed. Errors of this kind die hard, and some few analysts—and authors—still make this error. The example that follows shows the foundations of the matter.

■ **Example 8.5.** Two mutually exclusive alternatives have the cash flows shown in Fig. 8.7. Table 8.2 displays the calculation of the data necessary to draw a present-worth (net present value) profile. A present-worth (PW) profile results when the NPV of a cash flow is plotted against the discount rate. Where the profile cuts the horizontal axis, that is, where NPV = 0, is the IROR of the cash flow. Thus, creation of a present-worth profile involves the same procedure as finding the IROR of the cash flow. Figure 8.8 shows the present-worth profile of alternatives 1, 2, and 2 − 1, the incremental cash flow. The present worth of each of these cash flows was found at 0, 10, 20, 30, and 40 percent in Table 8.2, and the results were plotted. From the graph, the IROR of alternative 1 is 27 percent (27.18 exactly), of alternative 2 is 23 percent (23.11 exactly), and of the increment 2 − 1 is 20 percent (20.00 exactly).

Now notice what happens if the minimum attractive rate of return (MARR) is 10 percent. If the IROR method is misused, alternative 1 with an IROR of 27 percent will be selected over alternative 2 with an IROR of 23 percent. But at 10 percent in Table 8.2 or Fig. 8.8, the net present value of alternative 1 is +2,231 and of alternative 2 is +3,966. By the NPV method, alternative 2 should be selected—thus the contradiction. A study of Fig. 8.8 at the vertical line representing 10 percent makes the contradiction clear.

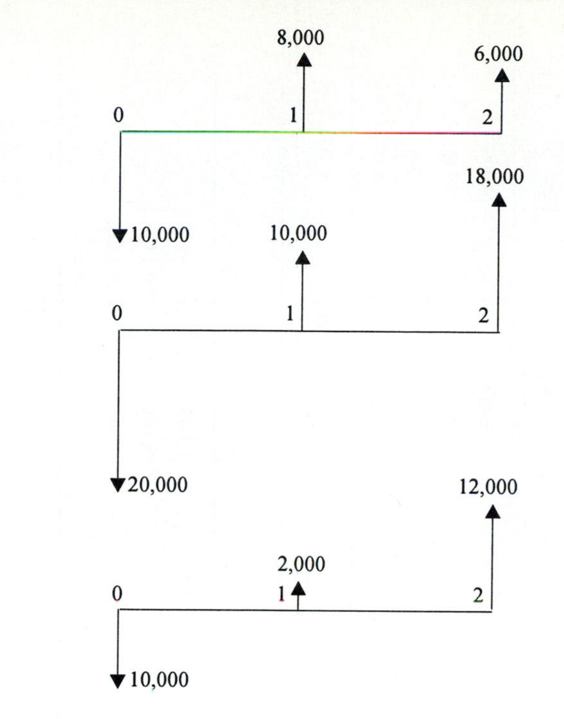

FIGURE 8.7 Present worth vs. internal rate of return: cash flows

Notice that above 20 percent MARR, no contradiction exists between the two methods. Nor would a contradiction exist if the two profiles did not intersect (at Fisher's intersection, so called).

As mentioned previously, the error in the use of the IROR method was the selection of an alternative based on individual IRORs rather than incremental IRORs.

If the correct use of the IROR method is followed, the top half of Table 8.3 appears. At 0 percent MARR at least alternative 1 will be selected because

$$27\% > 0\%$$

On investigating the ΔIROR of 2 - 1, alternative 2 is chosen over alternative 1 because

$$20\% > 0\%$$

That is, the extra investment in alternative 2 is more than justified by the extra benefits of alternative 2.

At 10 percent opportunity cost of capital, alternative 1 qualifies because

$$27\% > 10\%$$

Does alternative 2 win out over alternative 1? It does because

$$20\% > 10\%$$

TABLE 8.2

Present-worth profiles of two mutually exclusive alternatives and their incremental difference

	PW @ $i = 0\%$			PW @ $i = 10\%$				PW @ $i = 20\%$			
Year	**1**	**2**	**2 - 1**	**(*P/F*, 10, N)**	**1**	**2**	**2 - 1**	**(*P/F*, 20, N)**	**1**	**2**	**2 - 1**
0	−10,000	−20,000	−10,000	1.0000	−10,000	−20,000		1.0000	−10,000	−20,000	
1	+8,000	+10,000	+2,000	0.9091	+7,273	+9,091		0.8333	+6,666	+8,333	
2	+6,000	+18,000	+12,000	0.8264	+4,958	+14,875		0.6944	+4,166	+12,499	
	+4,000	+8,000	+4,000		+2,231	+3,966	+1,735		+832	+832	0

	PW @ $i = 30\%$				PW @ $i = 40\%$			
Year	**(*P/F*, 30, N)**	**1**	**2**	**2 - 1**	**(*P/F*, 40, N)**	**1**	**2**	**2 - 1**
0	1.0000	−10,000	−20,000		1.0000	−10,000	−20,000	
1	0.7692	+6,154	+7,692		0.7143	+5,714	+7,143	
2	0.5917	+3,550	+10,651		0.5102	+3,061	+9,184	
		−296	−1,657	−1,361		−1,225	−3,673	−2,448

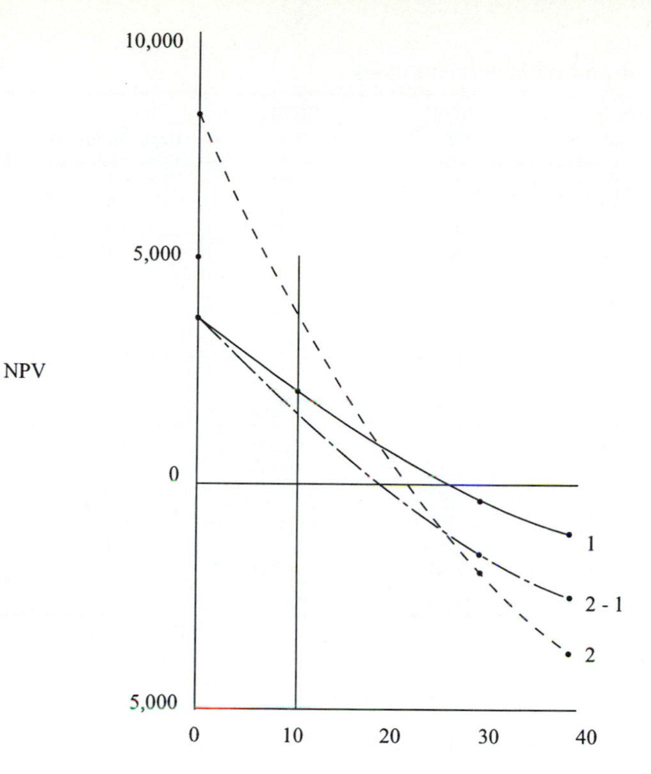

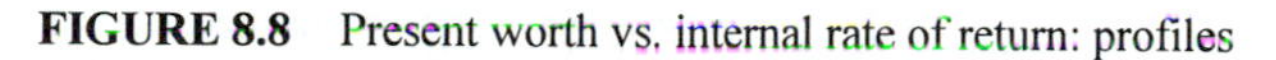

FIGURE 8.8 Present worth vs. internal rate of return: profiles

At a 20 percent discount rate, alternative 1 is better than doing nothing because

$$27\% > 20\%$$

and alternative 2 barely succeeds against alternative 1 because

$$20\% = 20\%$$

According to the rule laid down in this book, equality signifies acceptance, not indifference, between alternatives 2 and 1.

At a 25 percent minimum acceptable return, alternative 1 is accepted because

$$27\% > 25\%$$

but alternative 2 is rejected because

$$20\% < 25\%$$

and the decision favors alternative 1.

At 27 percent opportunity cost of resources, alternative 1 is accepted because

$$i^* = 27\% = i$$

and alternative 2 fails in comparison with alternative 1 because

$$20\% < 27\%$$

TABLE 8.3
Comparison of IROR and NPV decision rules

By IROR i (%)	$IROR_1$ (%)	$IROR_2$ (%)	$IROR_{2-1}$ (%)	Decision favors
0	27	NA	20	2
10	27	NA	20	2
20	27	NA	20	2
25	27	NA	20	1
27	27	NA	20	1
30	27	23	NA	0
40	27	23	NA	0
By NPV i (%)	NPV_1	NPV_2	NPV_{2-1}	**Decision favors**
0	+4,000	+8,000	+4,000	2
10	+2,231	+3,966	+1,735	2
20	+832	+832	0	2
25	+240	−480	−720	1
27	+19	−966	−985	1
30	−296	−1,657	−1,361	0
40	−1,225	−3,673	−2,448	0

At 30 percent cutoff rate, alternative 1 does not qualify because

$$27\% < 30\%$$

Now alternative 2 must be tested against the null alternative because an alternative may never be correctly compared with a rejected alternative; thus the 2 − 1 increment is not applicable. The IROR of alternative 2 is 23 percent. But

$$23\% < 30\%$$

and alternative 2 also fails. The same reasoning is used at $i = 40$ percent:

$$27\% < 40\%$$

and

$$23\% < 40\%$$

Therefore both alternatives are rejected.

The bottom half of Table 8.3 applies the NPV criterion and the incremental NPV criterion. The reader may check the results of these criteria to confirm the last column of the table and its concordance with the top portion of the table.

Finally, keep in mind that it is possible to create cash flows whose profiles will cross and where the wrong method will still give the correct answer.[2] ■

---

[2]See Henry M. Steiner, *Public and Private Investments*, Books Associates, Washington, 1980, pp. 94–97.

BENEFIT/COST RATIO AND PRESENT WORTH

The reasoning employed in the previous section could have been applied to the benefit/cost ratio method and its relation to present worth. If the individual benefit/cost ratios of cash flows are used as a decision tool, the same contradiction between the results obtained in this way and the results derived from the present-worth method will often occur. If the benefit/cost ratio method is correctly used, via incremental analysis, no contradiction will be observed between it and the present-worth method.

EQUAL INITIAL INVESTMENTS

In Chap. 7, under the section of the same title as this one (page 131), it was pointed out that although no special handling is required in the benefit/cost ratio method, special methods must be employed where the IROR method is used to decide between alternatives with equal initial investments. If the alternatives are taken in one order, the result will be correct. If they are taken in the other order, the result will be erroneous. This truth will now be demonstrated by an example.

■ **Example 8.6.** This example will use the same data and figure as Example 7.6, in order to allow you to make an instructive comparison between the two methods under equal-initial-investment conditions. Two mutually exclusive alternatives are involved, with the null alternative included. The MARR is 12 percent. Inflation has already been taken into account.

Year	1	2	1–2	2–1
0	-100	-100	0	0
1	+75	+80	-5	+5
2	+50	+40	+10	-10
3	+25	+20	+5	-5

The situation is shown in Fig. 7.6, page 132.

One incremental analysis must be performed. The rule for ordering the alternatives, used so frequently, will not serve here because both initial investments are the same. Will it make a difference if alternative 1 is taken before alternative 2, and vice versa? We shall see that indeed it does.

Dietrich R. Bergmann set forth the rule to be used in ordering alternatives with equal initial investments.[3] The projects must be ordered so that the first payment encountered in the incremental cash flow diagram is an outflow. Referring to Fig. 7.6, we see that the rule requires that the increment be 1 − 2, that is, 1 over 2 because the payment of 5 appears as an outflow in this increment. Thus alternative 2 must be considered first and alternative 1 second.

---

[3]Dietrich R. Bergmann, "Evaluating Mutually Exclusive Investment Alternatives: Rate of Return Methodology Reconciled with Net Present Worth," *Highway Research Record*, no. 437, Highway Research Board, Washington, 1973, pp. 75–82.

The top portion of Table 8.4 results from following Bergmann's rule. The discount rate shows in the first column. Alternative 2 with an IROR equal to 25 percent is considered first. At a discount rate of 0 percent

$$25\% > 0$$

therefore, alternative 2 is acceptable. Is alternative 1 accepted over 2? Yes, because[4]

$$141\% > 0\%$$

The same reasoning applies at 12 percent.

At 25 percent MARR, alternative 2 is accepted because

$$i^* = 25\% = i$$

and in this book, equality signifies acceptance. Alternative 1 is selected over alternative 2 because

$$141\% > 25\%$$

At 27 percent opportunity cost of capital, alternative 2 fails because

$$25\% < 27\%$$

The incremental cash flow is now irrelevant because no comparison can be made with a failed alternative. The IROR of alternative 1 (29 percent) is compared with the opportunity cost of capital

$$29\% > 27\%$$

and alternative 1 is accepted.

At 50 percent both alternatives fail because

$$25\% < 50\% \qquad \text{and} \qquad 29\% < 50\%$$

The null alterative is favored.

---

[4]The calculation of this rate of return with a computer or programmed calculator requires the use of algebra because 141 percent lies outside available tabular values.

The equation is

$$5\left(\frac{1}{1+i^*}\right)^1 - 10\left(\frac{1}{1+i^*}\right)^2 - 5\left(\frac{1}{1+i^*}\right)^3 = 0$$

Dividing by 5, employing a least common denominator of $(1 + i^*)^3$, and then cross-multiplying give

$$(1+i^*)^2 - 2(1+i^*) - 1 = 0$$

Expanding gives

$$1 + 2i^* + i^{*2} - 2 - 2i^* - 1 = 0$$

By eliminating we find

$$i^{*2} = 2$$

$$i^* = \pm 2^{0.5}$$

$$= 141.46\%$$

and we disregard the negative root.

TABLE 8.4
Equal initial investments

i (%)	$IROR_2$ (%)	$IROR_3$ (%)	$IROR_{1-2}$ (%)	Decision favors
0	25	NA	141	1
12	25	NA	141	1
25	25	NA	141	1
27	25	29	NA	1
50	25	29	NA	0

i (%)	$IROR_1$ (%)	$IROR_2$ (%)	$IROR_{2-1}$ (%)	Decision favors
0	29	NA	141	2
12	29	NA	141	2
25	29	NA	141	2
27	29	NA	141	2
50	29	25	141	0

i (%)	PW_2	PW_1	PW_{1-2}	PW_{2-1}	Decision favors
0	+40.00	+50.00	+10.00	−10.00	1
12	+17.55	+24.62	+7.06	−7.06	1
25	0	+4.80	+4.96	−4.96	1
27	−2.44	+2.26	+4.70	−4.70	1
50	−22.96	−20.37	+2.59	−2.59	0

The present-worth method confirms the preceding analysis. This is shown in the bottom portion of Table 8.4. For example, at 27 percent MARR, the decision favors alternative 1 because

$$+2.26 > -2.44$$

on an individual NPV basis. On an incremental NPV criterion, say 1 − 2, a +4.70 indicates acceptance of alternative 1. Conversely, by using the increment 2 − 1, a −4.70 points to rejection of alternative 2 and acceptance of alternative 1. The reader is invited to follow the reasoning at the other MARRs shown in the bottom portion of the table.

Let us see what would happen if Bergmann's rule were ignored and the only other option for incremental analysis 2 − 1 were taken. The resulting solution shows in the middle portion of Table 8.4.

At a minimum attractive return of 0 percent, alternative 1 is acceptable because

$$29\% > 0\%$$

and alternative 2 is acceptable over alternative 1 because

$$141\% > 0\%$$

Similar reasoning applies throughout the analysis, but the decision favors alternative 2—the wrong choice.

This example underlines the importance of following Bergmann's rule in the IROR analysis of equal-initial-investment alternatives. ■

■ **Example 8.7.** An investor is faced with two alternatives remaining as the best possibilities of many she has investigated. She will spend $250,000, and she estimates that she will receive the following net benefits over the lives of each, after inflation and taxes have been accounted for:

Year	Alternative 1 ($000)	Alternative 2 ($000)
0	-250	-250
1	+40	+10
2	+50	+40
3	+60	+70
4	+70	+100
5	+80	+130

If the internal rate of return method is required and her opportunity cost of capital after taxes and inflation is 4 percent, should she invest in either alternative, and, if so, which one? Rates of return must be carried out to the nearest whole percent.

Solution. The null alternative is not precluded in this example. Therefore the order of consideration of the alternatives is the first decision. The initial investment is the same in both cases. Applying Bergmann's rule, we discover that the correct increment must be 2 − 1. See Fig. 8.9. The first payment encountered in the incremental cash flow is a cost of 30 at year 1, as required for a correct incremental analysis. Thus, the first alternative to be compared against the null is 1.

$$-250 + 40(P/A, i^*, 5) + 10(P/G, i^*, 5) = 0$$

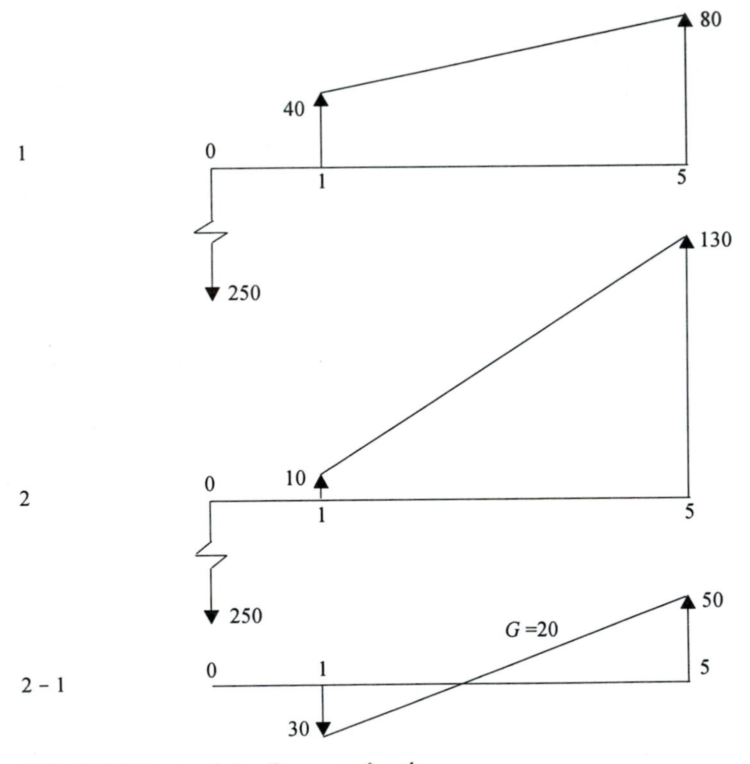

FIGURE 8.9 Applying Bergmann's rule

Dividing through by 10 gives

$$-25 + 4(P/A, i^*, 5) + (P/G, i^*, 5) = 0$$
$$i^* = 6\% > 4\%$$

Therefore, alternative 1 is accepted. The IROR of the incremental cash flow is found by using the equation

$$-30(P/A, i^*, 5) + 20(P/G, i^*, 5) = 0$$
$$i^* = 29\% > 4\%$$

Therefore, alternative 2 is accepted rather than alternative 1. ■

RATE OF RETURN ON A BOND

Bonds are frequently offered for sale at a price other than their face value. When this occurs, the percentage return on the bond is not the same as the internal rate of return of the investment in the bond. Asking the question, "Should I buy this bond?" becomes the same as asking, "How does the IROR of this bond compare with my personal opportunity cost of capital for investments of the same risk?"

■ **Example 8.8.** To illustrate the point, we will use Example 5.11. The question is now, however, What is the internal rate of return of the investment in this bond? Let us assume that the bond is selling for $4,750, at a "discount" from its $5,000 face value, to use the jargon of the bond market. This internal rate of return may then be compared with the rate mentioned in Example 5.11, a nominal 10 percent.

The procedure is the same as in Example 5.11. Part 1, finding the cash receipts of the bond, is unchanged from that example and is seen in Fig. 8.10. The $4,750 figure is the cost of the bond to you.

The IROR is found from

$$-4{,}750 + 200(P/A, i^*, 40) + 5{,}000(P/F, i^*, 40) = 0$$
$$i^* = 4.26\%$$

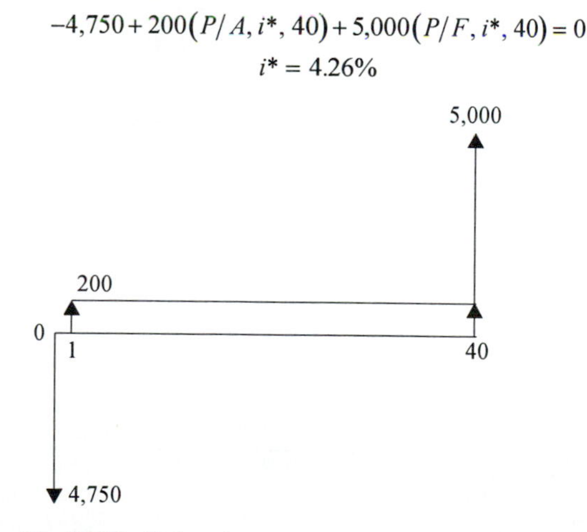

FIGURE 8.10 The IROR of a bond

This is the rate of return on the bond per period. The nominal rate of return per annum is

$$r = 2(4.26) = 8.53\% < 10\%$$

Therefore the bond investment is rejected—the same answer arrived at by use of the present-worth method.

It may occur to some readers that the equation containing i^* could simply have been tested at $i^* = 5$ percent. If the result were positive, i^* would be higher than 5 percent; if negative, lower. The decision as to the acceptability of the investment could be made on this basis. The procedure is nothing more than the NPV method in disguise. If the IROR method is specified, the NPV method will not serve. The IROR method demands that a percentage rate be found. Nothing less will do. ■

MULTIPLE RATES OF RETURN

Sometimes a cash flow can have two or more internal rates of return. The most important point about this phenomenon is how to recognize it.

Cash flows in this text have consisted of an investment, followed by net benefits or net costs but seldom by net benefits *and* net costs. Such a situation is perfectly possible, however, and often occurs in the incremental cash flow.

It appears much less frequently in primary cash flows. Figure 8.11 shows cash flows of the type concerned. Notice that more than one change in direction of the arrows distinguishes these cash flows from the usual ones we have dealt with, where a down arrow is followed by a series of up arrows or vice versa. Only one sign change—change in arrow direction—occurs in such cash flows. However, in Fig. 8.11*a*, a down arrow at time 0 is followed by an up arrow—a sign change. Up

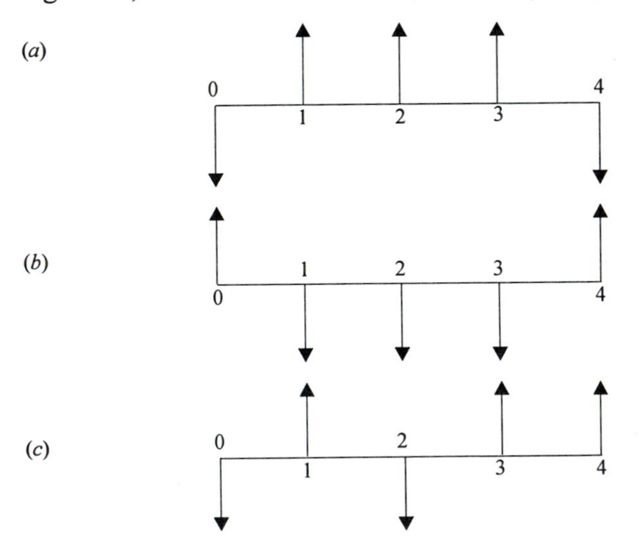

FIGURE 8.11 Cash flow with multiple rates of return

TABLE 8.5
Two building designs, increment

Discount rate (%)	NPV
0	-19,000
5	-4,530
7.85	-10
10	+1,965
15	+2,707
19.28	+2
20	-694

arrows follow until year 3. At year 4 a down arrow appears—another sign change, making two sign changes for this cash flow. The effect of two sign changes is two solving internal rates of return.

The rule is: There will be as many internal rates of return as there are sign changes in the cash flow.

In Fig. 8.11*b*, the previous cash flow inverted, two sign changes are still present and therefore two IRORs. In Fig.8.11*c*, there are three sign changes, the first between years 0 and 1, the second between years 1 and 2, and the third between years 2 and 3. Thus this cash flow will have three internal rates of return.

Figure 8.11*a* cash flows result in a *cap-shaped* present-worth profile. Figure 8.11*b* cash flows result in a *cup-shaped* present-worth profile. Figure 8.11*c* cash flows result in a *wavy* present-worth profile.

The dilemma lies in choosing among the rates. Which IROR is valid? Or are none of them valid?

Figure 8.12, drawn from Table 8.5, shows the difficulty. The two solving rates

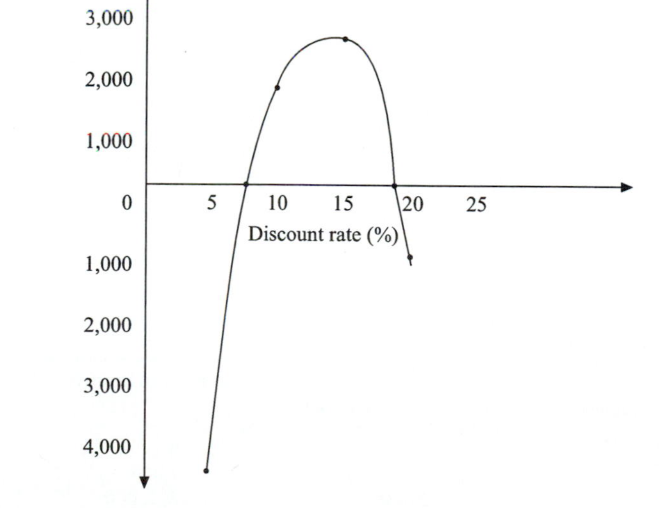

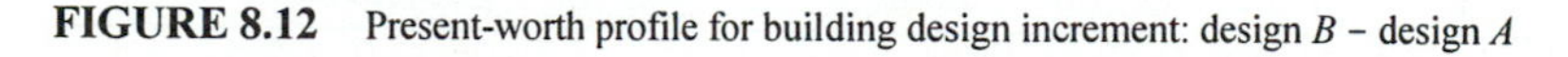
FIGURE 8.12 Present-worth profile for building design increment: design B - design A

of return are 8 and 19 percent. If the MARR is 5 percent, the project will be rejected by the NPV rule because the NPV is negative at that rate. It will, however, be accepted by the IROR rule because both IRORs are greater than the MARR. Thus, a contradiction appears.

In the following pages, we will investigate the contradiction. Some preliminary background is necessary. We begin with Descartes' rule of signs.

DESCARTES' RULE OF SIGNS

A single-variable equation has the same number of roots as the highest power to which the variable is raised. Thus

$$-P + \sum_{j=1}^{N} A_j x^j = 0 \tag{8.6}$$

has N roots. These roots can be positive, negative, or imaginary.

Descartes' rule of signs says that if a polynomial with real A's has m sign changes, then the number of positive roots will be $m - 2k$, where $k = 0, 1, 2, 3, \ldots$. Thus the first two columns of Table 8.6 appear.

Now imagine that

$$x = \frac{1}{1+i}$$

or

$$i = \frac{1-x}{x}$$

In the case where the number of sign changes of Eq. (8.6) is zero, no value of x or i appears.

Where the number of sign changes is 1, the number of positive roots is also 1. If the value of this x is between 0 and 1, then i will be positive. Variable i will be negative if the value of x is greater than 1.

If the number of sign changes is two, the number of positive roots is two or zero. If the number of positive roots is zero, the N roots of the equation will be

TABLE 8.6
Descartes' rule of signs

Number of sign changes m	Number of positive roots x	Number of positive values of i^*
0	0	0
1	1	1 or 0
2	2 or 0	2, 1, or 0
3	3 or 1	3, 2, 1, or 0

negative or imaginary. Negative x's cannot generate positive i's. Table 8.6 could be carried on further, but we will stop at two sign changes—enough to bring out the necessary points. The reader is invited to verify Table 8.6 for three sign changes.

■ **Example 8.9.** Two building designs have been developed by an architectural engineering firm for its client, a real estate developer in a major U.S. city. The developer has asked the firm to perform an economic analysis over the next 3 years for each design. The developer requires at least a 15 percent return on investment in order to accept a project. In selecting 15 percent, the developer has included the effects of inflation and risk. All money sums are expressed in terms of current dollars. The effect of income taxes has also been included in payments.

The cash flows for each design are as follows:

Year	Design *A* ($000)	Design *B* ($000)
0	−1,200	−1,815
1	+1,500	+2,258
2	+1,700	+2,360
3	+3,322	+2,500

Since both designs are based on the same lot, only one can be selected. The developer insists that the results of economic analysis be presented in rate of return form. He completely understands the idea of extra return on extra investment. Which design should the architectural engineering firm recommend? Percentages need be carried out only to the nearest whole percent.

Solution. To start out, design A must be tested against the null alternative. Its rate of return is found by the internal rate of return equation:

$$-1{,}200 + 1{,}500(P/F, i^*, 1) + 1{,}700(P/F, i^*, 2) + 3{,}322(P/F, i^*, 3) = 0$$

$$i^* = 135\%$$

This result is much greater than the 15 percent required, and thus design A is accepted rather than the null alternative.

Next, we must test the increment between designs A and B.

Year	Design *B* − design *A* ($000)
0	−615
1	+758
2	+660
3	−822

Two sign changes appear, indicating two possible IRORs or none. The equation is

$$-615 + 758(P/F, i^*, 1) + 660(P/F, i^*, 2) - 822(P/F, i^*, 3) = 0$$

$$i^* = 7.85\% \text{ or } 19.28\%$$

The form of the cash flow is shown in Fig. 8.11a. The present-worth profile of the cash flow is shown in Fig. 8.12. At 15 percent opportunity cost of capital, it appears at first glance that design A should be rejected and design B accepted. This is so if we acknowledge that the IROR is indeed 19 percent, approximately. But what if the IROR is really 8 percent? Then we reject design *B* and accept design *A*.

The contradiction cannot be resolved by use of the IROR method. We must turn to the NPV as the measure of worth. At 15 percent, the NPV of the incremental cash flow is +2,707. Design *B* should therefore be accepted. As a general rule, the present-worth profile should always be sketched and the NPV determined at the opportunity cost of capital. As before, positive values indicate acceptance and negative values indicate rejection. ■

To summarize, if we look upon Fig. 8.12 as representing a project, we will accept it at discount rates between 8 and 19 percent, where the NPV is positive. We will reject it where the discount rate is below 8 percent and above 19 percent.

■ **Example 8.10.** An international firm of real estate developers, specializing in tall buildings, is considering two mutually exclusive projects, one in Rio de Janeiro and the other in Buenos Aires. The costs and revenues, in millions of dollars, of each project are as follows:

	Year		
	0	**1**	**2**
Rio de Janeiro	−10	+15	+30
Buenos Aires	−20	+40	+14

These amounts are after-tax quantities, corrected for inflation. The MARR of the company is 12 percent under these conditions. The projects are equally risky. An IROR solution is required. Which of the two projects, if either, should be undertaken? IRORs to the nearest whole percent are acceptable.

Solution. Testing the IROR of the Rio de Janeiro alternative against the null gives

$$-10 + 15(P/F, i^*, 1) + 30(P/F, i^*, 2) = 0$$

$$i^* = 164\%$$

Thus, the Rio de Janeiro alternative will be undertaken because 164 percent exceeds 12 percent.

We are now required to test the increment between the two alternatives.

Year	Buenos Aires − Rio de Janeiro
0	−10
1	+25
2	−16

Two sign changes are evident. We immediately look for two solving rates of return or none. The cash flow is of the form of Fig. 8.11*a*. We write out the equation for finding the IROR of the increment:

$$-10 + 25\left(\frac{1}{1+i^*}\right) - 16\left(\frac{1}{1+i^*}\right)^2 = 0$$

TABLE 8.7
Incremental NPVs for Buenos Aires – Rio de Janeiro

Discount rate (%)	NPV
0	-1.00
5	-0.70
10	-0.50
15	-0.36
20	-0.28
25	-0.24
1,000	-7.86

Let $x = 1/(1 + i^*)$ and substitute:

$$-10 + 25x - 16x^2 = 0$$

Using the binomial theorem with

$$a = -16 \qquad b = 25 \qquad \text{and} \qquad c = -10$$

gives

$$\begin{aligned} x &= \frac{-b \pm \sqrt{b^2 - 4ac}}{2a} \\ &= \frac{-25 \pm \sqrt{25^2 - 4(-16)(-10)}}{2(-16)} \\ &= \frac{-25 \pm \sqrt{625 - 640}}{-32} \\ &= \frac{-25 \pm \sqrt{-15}}{-32} \end{aligned}$$

Two imaginary solutions appear. Graphing the NPV of the cash flow as shown in Table 8.7, we see Fig. 8.13. The NPV is negative for any discount rate, even 1,000

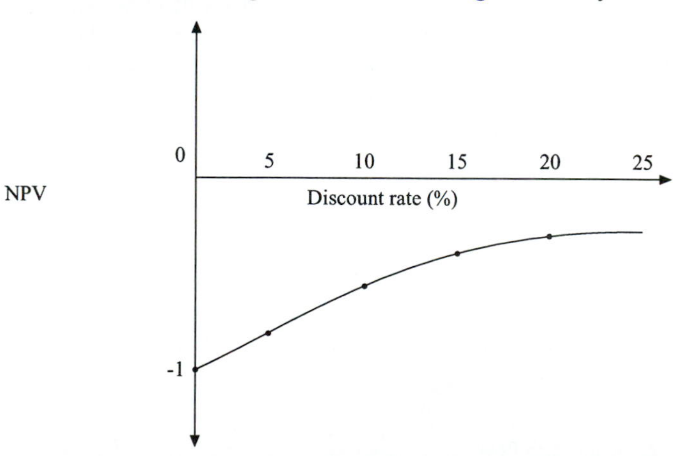

FIGURE 8.13 Present-worth profile for Buenos Aires – Rio de Janeiro

percent. Therefore, the Buenos Aires alternative is rejected, and the Rio de Janeiro alternative is accepted.

It is clear that the IROR has no meaning, indeed, existence, in the problem. We can rely on the NPV solution. This indicates that the Rio de Janeiro alternative will be accepted regardless of the opportunity cost of capital.

What if we had mistakenly taken the order of alternatives as Rio − Buenos Aires? Checking Buenos Aires against the null alternative finds that the IROR is 130 percent, indicating acceptance. The increment Rio − Buenos Aires is

Year	Rio de Janeiro − Buenos Aires
0	+10
1	−25
2	+16

Using the binomial theorem, as before, gives imaginary IRORs. Graphing the NPV, we will find Fig. 8.13 rotated 180° about the horizontal axis, indicating acceptance at all discount rates of the Rio de Janeiro alternative.

This example shows the inapplicability of the IROR criterion in a far-from-unusual problem. The individual NPV approach confirms our analysis. The present worth at 12 percent of the Rio de Janeiro alternative is 27.3 and of the Buenos Aires alternative is 26.9. ■

■ **Example 8.11.** A city is offered the ownership of a bus company now operating under franchise. The proposal requires no immediate cash from the city. The buses will make a profit of \$200,000 per year for the first 2 years of city operation. During year 3 the city will purchase a new fleet of buses and pay off the present owners at a cost of \$1,900,000. The new buses will be operated for 4 years at a net profit of \$300,000 per year. No salvage value is expected on either the present buses or the new buses. In other words, the costs of hauling them away will be exactly balanced by the revenue received from their sale. The life of the new buses is 4 years. At the end of year 7, therefore, the buses will be retired. At that time, the matter will be reinvestigated before any new investment is made. The opportunity cost of capital to the city is considered to be 10 percent. What is the true rate of return on this investment? Should the city accept the deal? Figure 8.14 shows the cash flow.

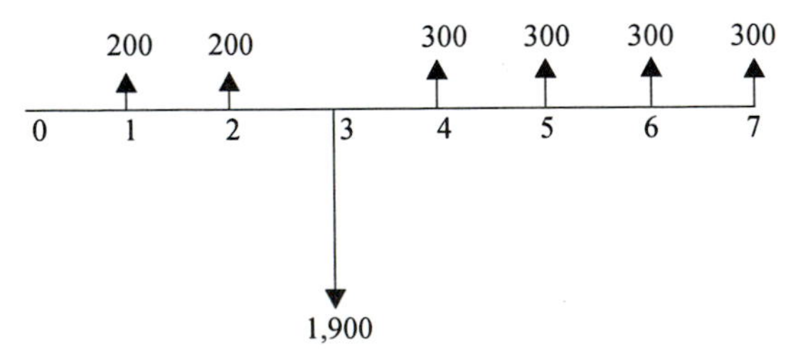

FIGURE 8.14 Bus company cash flow

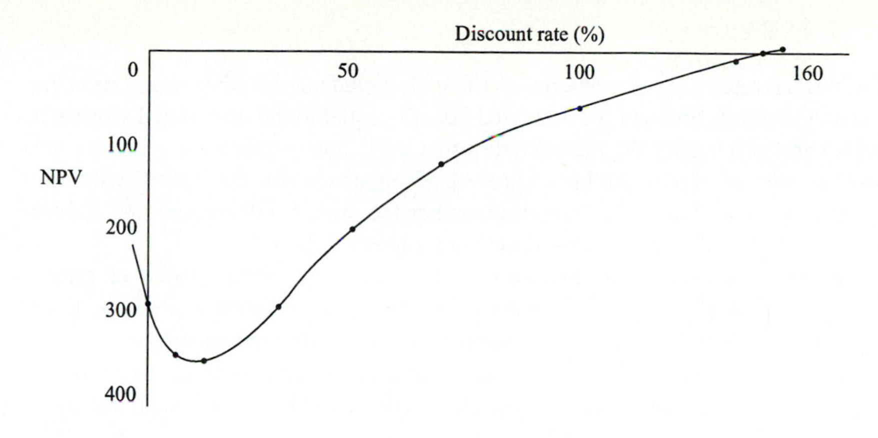

FIGURE 8.15 Bus company present-worth profile

Solution. Table 8.8 shows the net present values at various discount rates from which Fig. 8.15 was drawn. The cash flow is of the form of Fig. 8.11b. Two solving rates of return are present. One is a negative rate. The other is +146 percent. At first glance, the deal might be accepted because 146 percent exceeds the 10 percent opportunity cost of capital. But investigation of the NPV at that rate reveals −366. The deal should be rejected.

The example shows that even for a single positive solving rate of return, the IROR cannot be trusted to give the correct decision in a multiple-IROR situation. Once again it is the NPV that must be used. ■

Where a rate-of-return is demanded, the analyst has little choice with multiple IRORs but to show that such a solution is ambiguous and to recommend another method.

TABLE 8.8
Bus company present-worth profile

Discount rate (%)	NPV
0	−300
5	−350
10	−366
30	−297
50	−198
70	−123
100	−52
120	−24
140	−4
146	0
150	+3

SUMMARY

The internal rate of return method uses the widely understood percentage rate as its decision variable. Setting the discounted benefits equal to the discounted costs of a project and solving for the rate of return that will cause equality to occur are the steps in the procedure to be employed. Comparison of the IROR with the opportunity cost of capital determines whether the project will be accepted. Direct or trial-and-error methods can be used to compute the IROR.

Incremental analysis is necessary in the use of the internal rate of return method with mutually exclusive alternatives. When incremental analysis is used, no inconsistency between the IROR and the present-worth method appears.

In the case of equal initial investments, Bergmann's rule must be used to order the alternatives. This rule states that the alternatives must be ordered such that their incremental cash flow begins with a cost.

The IROR is conventionally reported as the total rate, not as the difference between it and the opportunity cost of capital.

Multiple rates of return appear when a cash flow undergoes more than one sign change. All solutions must be checked by the NPV method.

PROBLEMS

8.1. In Example 8.1, page 144, what will be the rate of return if Ready Assets Trust is sold for $7,000 after being held for 3 years? All other information in the problem remains unchanged. Find a solution by using the compound interest tables, and compare it to the exact algebraic one. How large is the difference?

8.2. The following cash flow was experienced by an investor:

Year	Cash flow
0	-1,000
1	300
2	300
3	300
4	300

(*a*) What is the internal rate of return of this cash flow?
Answer: 7.7%

(*b*) If a salvage value of $400 is received at the end of year 4, what is the IROR of the cash flow? All percentage rates should be carried out to one decimal point.
Answer: 18.6%

8.3. In Example 8.1, page 144, the Merrill-Lynch Ready Assets Trust was purchased for $4,500 and was sold on the same date 2 years later for $5,300. What was the rate of return on the investment to the nearest 0.1 percent?

8.4. A solar hot water system cost $3,000 after credits were deducted. A 3-year payback—the number of years it takes a system to pay for itself in fuel savings—meant $1,000 per year must be saved. Actually, the system is saving only $500 per year.

The owner could have invested this $3,000 at 15 percent. The life of the system is estimated to be 20 years, with no salvage value. What rate of return is the system making?

Answer: 16%

8.5. In Example 8.2, page 147, the estimated cost of the system has been revised upward after a more accurate accounting to $1,532,431. What rate of return, to the nearest whole percent, is the university making on its investment? All other data remain the same.

8.6. A Merrill-Lynch Realty Ready Assets Trust in the amount of $5,000 was purchased on November 1, 1994. It was sold on April 7, 1996, for $8,324. No taxes have been paid yet. What was the rate of return on this investment? Carry out your calculations to the nearest 0.1 percent.

8.7. In Example 8.2, page 147, Urban University will now pay $3 million for the new system, with a total savings in electricity and heating costs of $450,000 per year. With a 20-year life, what is the rate of return to the nearest whole percent? All other information in the problem remains the same.

8.8. In Example 8.3, page 148, the Development Research Company will purchase new computing equipment for $800,000. The new equipment will have an economic life of 4 years, no salvage value, and net benefits per year of $300,000 for years 1 and 2 and $325,000 for each of the remaining 2 years. What is the prospective rate of return on this investment to the nearest 1 percent?

Use the method contained in the solution to Example 8.3 to decide on the discount rate for the first trial.

8.9. A transport investment will cost 50 at time 0. Benefits are calculated to be 10 for each of 20 years. The opportunity cost of capital is 14 percent. Should the project be accepted if the only other alternative is to do nothing, with no resulting costs or benefits? Solve the problem, using

(*a*) Benefit minus cost
(*b*) Benefit/cost ratio
(*c*) Internal rate of return
(*d*) Annual profit

8.10. An investment in highways will cost 100 million pesos. Net benefits are calculated at 20 million pesos each year during an economic life of 10 years. No salvage value will be present. Inflation is ignored. The opportunity cost of resources is 15 percent. Should the government undertake the project?

Solve this problem by these methods:

(*a*) Benefit minus cost
(*b*) Benefit/cost ratio
(*c*) Internal rate of return
(*d*) Annual worth

8.11. On page 150, the following sentence appears: "It is normal to calculate i^*; a determination that i^* is greater or less than i is not sufficient." Can you cite the reasons why this remark is valid?

8.12. Georgetown Park, 33 miles due north of Georgetown, Washington, D.C., offers houses as follows: The purchase price is $58,823 with $3,000 down. The mortgage amount is $55,823 at 12½ percent interest with the following yearly principal and interest payments:

Year	Yearly payment ($)
1	5,525
2	5,939
3	6,384
4	6,863
5	7,378
6–30	7,931

Is this payment schedule really at 12½ percent?

Answer: Yes

8.13. A letter in *Money* (May, 1981) said:

> I recently paid $9,578.70 for a 3-month Treasury bill with a so-called discount rate of 16.67 percent. I figured the yield to be 20 percent; my broker says it's closer to 18 percent but doesn't know why. How do I calculate the return on my investment?
>
> C. Swan Weber
> Chevy Chase, MD

The reply said:

> Treasury bills are sold for less than the $10,000 face value you collect on maturity, and this discount, expressed as a percentage of face value and converted to an annual rate, is the discount rate. Your return, however, is the yield, or the discount expressed as a percentage of the purchase price—in this case, $421.30 divided by $9,578.70. To annualize the return, you multiply the result by 365 and divide by 91—the number of days in 3 months. Your broker was right: Your yield is 17.64 percent.

Was the broker right? What is the correct effective annual rate of return?

8.14. In Example 8.4, page 151, repeat the exercise requested there, but change the benefits received from project 2 from 700 to 775 for each of 5 years. All other data remain the same.

8.15. Evaluate the following cash flows by

(*a*) Present-worth method

(*b*) Annual-worth method

(*c*) Benefit/cost ratio

(*d*) Internal rate of return

The opportunity cost of capital is 12 percent.

Year(s)	Item	Cash flow ($000)
0	First cost	-120
1–6	Net annual receipts	+30
6	Salvage value	+40

An internal rate of return to the nearest whole percent is required.

8.16. In Example 8.5, page 154, at a discount rate of 20 percent, which alternative was chosen? Now repeat the same decision process, but apply the incorrect decision rule that the alternative with the highest IROR will be chosen. Under this erroneous rule, which alternative is chosen? Does it contradict the choice made by following the NPV method?

8.17. Two alternative designs exist for a heat-exchanger installation in a power plant. The first design will require an initial investment of $184,000, and annual net benefits will be $28,000. The second design requires an initial investment of $220,000 and will return net annual benefits of $32,900. The economic life of both options is 20 years. Which alternative should be chosen if the MARR is 12 percent? Do not consider the null alternative. There will be no salvage value. Ignore inflation.

You must use the IROR criterion to make this decision. No other method will be credited.

8.18. Consider a project for which two alternative designs are being considered. Under option 1, initial investment costs will be $184,000 and net annual benefits $28,000. With option 2, initial investment costs are $220,000 and net annual benefits $32,900. For simplicity we consider that net annual benefits are constant over the project life and that the economic life is 20 years for both options. The rates of return are

Option 1 14.1%
Option 2 13.8%

Therefore, one may think that the first option should be chosen. However, if we look at the net present value of each option, the conclusion may be different. For example, if the discount rate is 12 percent, we have

Option 1, NPV (at 12%) of 25,132
Option 2, NPV (at 12%) of 25,730

and option 2 should be chosen. Resolve the contradiction.

8.19. Two plans for a hydroelectric project in Peru have been proposed. The opportunity cost, in the monetary unit *soles*, of resources is 10 percent. Data on the two alternatives are as shown:

	Alternative	
System	**1**	**2**
First cost (S/000,000,000)	300	160
Economic life (years)	40	20
Salvage value (S/000,000,000)	15	12
Annual benefits (S/000,000,000)	25	22
Annual costs (S/000,000,000)	3	1

By the IROR method, which of the two systems should be chosen, or should either be selected?

8.20. Iranian-Arabian Petroleum Company is reviewing a proposal for a new asphalt plant at its refinery on the Persian Gulf. Two designs are on the table, with all costs in millions of dollars:

Design	First cost	Annual operation and maintenance cost	Annual revenue
1	110	15	30
2	130	18	35

The life of the plant is estimated at 30 years. The salvage value will approximate the cost of dismantling the unit.

The minimum attractive rate of return (MARR) is 8 percent, the constant-dollar opportunity cost of capital for this company. One of the designs is certain to be chosen.

No income taxes are paid by Iranian-Arabian Petroleum. All sums are in constant dollars of year 0.

Which design should be selected? The company demands that the internal rate of return method be used. The IROR must appear to the closest whole percentage point.

8.21. Repeat Prob. 5.15, using the internal rate of return method.

8.22. Repeat Prob. 5.16, using the internal rate of return method.

8.23. A small plating plant is considering a gas-only heating system versus a gas-plus-solar energy system. One of the systems is certain to be chosen. The before-tax minimum attractive rate of return is 20 percent. The bank from which funds will be borrowed demands a rate of return on investment solution based on incremental cash flow. *No other method of evaluation is acceptable to the bank.*

The gas installation will cost $3,500. It has a life of 10 years and no salvage value. It must be assumed, however, that a gas shutdown because of national shortages is considered certain to take place once per year for a 1-week duration. This shutdown will cost the company $12,900 in lost revenues per occurrence. The cost of gas will be about $12,000 annually.

The gas-plus-solar installation will cost $34,500 and will have a salvage value of $5,000, a life of 20 years, and an annual cost of $120. The cost of the gas will be an additional $6,000 annually.

Which system should be chosen for presentation to the bank's loan officer? Calculate rate of return to the nearest 0.1 percent.

8.24. Repeat Prob. 7.11, page 138, using the IROR method. What is the value of the IROR to the nearest whole percent if it is greater than zero? If the IROR is negative, the percentage need not be found, but you must indicate why you believe it is negative.

Why is ordering the alternatives correctly an important requisite in this situation? Discuss fully.

8.25. At what interest rate are the following cash flows equivalent?

Year	Option 1	Option 2
0	−12,000	−12,000
1	+1,200	+3,786
2	+1,200	+3,786
3	+1,200	+3,786
4	+13,000	+3,786

8.26. In Example 8.6, page 159, under alternative 2, the cash flow is now +70 at year 1, +55 at year 2, and +28 at year 3. Which alternative should be chosen if all the other data remain the same?

8.27. You have $100,000 to invest in portfolio *A* or portfolio *B*. The cash flows for each alternative are as shown:

	Portfolio	
Year	A	B
0	−100,000	−100,000
1	+20,000	+70,000
2	+40,000	+50,000
3	+60,000	+40,000
4	+80,000	+20,000
IROR	27%	36%

Using the internal rate of return method, which portfolio should be chosen if your opportunity cost of capital is 6 percent?

8.28. Two proposals to repair and modernize telephone facilities have been received by the national telephone company of a small African country. Each proposal has been fashioned to fit exactly within a budget of $10 million.

Proposal 1 is designed to augment urban service, while proposal 2 will concentrate on new rural installations. Their costs and net benefits in millions of dollars are as follows:

	Proposal 1		Proposal 2	
Year	**Cost**	**Net benefit**	**Cost**	**Net benefit**
0	10	—	10	—
1		1		0
2		2		1
3		2		1
4		3		2
5		3		3
6–25		4		5

No taxes need to be considered, and inflation is already included in the estimates. The applicable MARR is 15 percent. One alternative is certain to be installed.

(*a*) *A rate of return solution is required.* This means that any and all rates of return used in the solution must be expressed to the nearest whole percent.

(*b*) Check by using the NPV method.

8.29. One hundred thousand dollars is to be invested in stock *A* or stock *B*.

Stock *A* costs $20 per share. Dividends are assumed at $1 per share annually. The stock is expected to appreciate to $35 per share at the end of 5 years, when it will be sold.

Stock *B* costs $125 per share. It will return annually an expected $8 per share in dividends and will appreciate to an estimated $200 per share at the end of 5 years, when it will be sold.

A rate of return of 15 percent is the minimum acceptable for stocks of this risk. Should either stock be purchased? If so, which one? *The internal rate of return method must be used.*

8.30. Example 8.7, page 161, illustrates a solution of a gradient cash flow where the gradient line crosses the time line as well as demonstrating Bergmann's rule. Imagine now that alternative 2 begins with +5 at year 1 and continues through year 5 at an increasing gradient of 30, as before. If all other data remain unchanged, which alternative will be chosen?

8.31. In Example 8.8, page 163, the bond now sells at a "premium," that is, for more than its face value of $5,000. It sells at $5,100, with a nominal 11 percent interest paid semiannually over 20 years. The investor's nominal

10 percent opportunity cost of capital remains as before. Should the investor buy the bonds?

8.32. In 1995 a bank bought a $100,000 Treasury bond, due 20 years hence, bearing interest at a nominal 8¼ percent, payable semiannually. It paid $65,000 for the bond.

(*a*) What is the rate of return on this bond purchase, expressed as a nominal rate, compounded semiannually, to the nearest tenth of a percent?

(*b*) If the bank's opportunity cost of capital is a nominal 10 percent, should it have bought the bond? Why? Percentage rates to one decimal place are required.

8.33. A 5-year $10,000 government bond was purchased for $9,814. It paid interest at a nominal 10 percent compounded quarterly. At the end of year 4, in the expectation of a jump in interest rates, its owner sold it for $8,076.

What was the effective annual rate of return on the bond to the nearest tenth of a percent? Do not consider income taxes or possible inflation.

8.34. Accept the invitation to verify Table 8.6, page 166, for three sign changes.

8.35. How many internal rates of return will be present in each of the following cash flows?

(*a*) A mining company will invest $30 million at year 0, receive net profits for 25 years, and will spend $10 million in year 26 to restore the site to its original condition.

(*b*) For 2 years an aviation constructor will receive net income from the development and preliminary payments on its latest VO 680 passenger airplane. For the following 2 years a net cost will be experienced, and for the next 5 years after that, a net profit.

(*c*) A new printing press will replace the one you presently use in your job at a printing establishment. It will result in savings over the present press of $5,000 per year for the next 15 years.

8.36. In Example 8.9, page 167, design *B* is now estimated to cost $2,500,000. Will we still accept it if all other data in the problem remain the same?

8.37. A lease on forest lands by Wadsworth Timber Company will net $100,000 per year for the 20 years of the lease. However, extensive access road rebuilding must be undertaken by Wadsworth during the fourth year in order to continue operations. This will cost $980,000. What is the rate of return of this lease to the nearest whole percentage point?

Wadsworth is a large company with a capital budget of $50 million per year. Recently, its cutoff rate of return was 31 percent, and it is expected to remain at or near 30 percent for the life of the lease.

Should Wadsworth undertake this new lease?

8.38. A mining company is considering investment in a surface mining operation in a western state. The first cost of the project is $100 million. And $8 million net profit will be returned by the mine each year for 25 years. This profit appears after taxes have been deducted. Inflation has also been added into the figures.

At the end of the project life, an estimated $10 million is to be spent for restoring the landscape to an acceptable natural state.

The opportunity cost of capital is 2 percent.

Determine and graph the present-worth profile for this project. Would the project be accepted?

8.39. An agricultural engineer is considering leasing a cattle ranch in Montana. After taxes and inflation are taken into account, he estimates a net positive cash flow of $300,000 annually. In year 2, however, he will have to invest $2.3 million in buildings, water supply, and drainage. These are leasehold improvements that will revert to the owner at the duration of his lease 10 years hence. His opportunity cost of capital is 8 percent.

What is the rate of return, to the nearest whole percent, on his cash flow? Should he make this investment?

8.40. Repeat Prob. 8.39, but use the *net present value (NPV) method.* However, the investment in buildings, water supply, and drainage is now estimated at $3.4 million. On the basis of NPV, should the engineer make the investment? Notice that no question of cash flow sign changes arises.

8.41. An oil drilling operation is to be analyzed by a major oil company. The original drilling will cost $2,500 million. The expected annual revenue of the field is $1,000 million. The annual operations cost of the field is $500 million. After 20 years, the expected life of the field, the company will pay out $750 million to remove the equipment and restore the area to its original condition. The board of directors of the company requires that the IROR method be used to analyze such decisions. Should the company undertake the project if the opportunity cost of capital is 15 percent?

8.42. A 70-megawatt nuclear power plant in the United States experienced the following costs and benefits over its lifetime:

Years	Cash flow ($000,000)
0	0
1–3	−100
4–26	+25
27–31	−30

The negative cash flows at the beginning of the project represented construction costs. At the end of the project, they represented the costs of decommissioning the plant, including transport and burial of the reactor core.

The project is to be analyzed by the *internal rate of return method.* The existence of significant external benefits allowed negative rates of return to be considered. Should a negative rate be encountered, the fact of its existence should be indicated, but it need not be determined.

Over what range of internal rates of return, to the nearest whole percent, was the project acceptable?

8.43. A proposed outlay of $15,000 at zero date will lead to a positive after-tax cash flow of $4,000 per year for 11 years. However, there will be a negative terminal salvage value of $8,000 at the end of the 11th year.

(*a*) At what positive value or values of i^* will the present worth of the cash flow be zero?

(*b*) For the company considering the project, the opportunity cost of capital is 18 percent. At that rate, would the project be accepted?

(*c*) Between what opportunity costs of capital would the project be accepted?

8.44. The Giant Slide amusement concession is offered to an entrepreneur and investor in a small midwestern city. It will cost only $4,000 to erect, since it is merely a framework of scaffolding which supports the Giant Slide, an undulating surface which people slide down on sacks. The stairs to the top of the slide and a ticket booth complete the installation. The contract must run for 2 years. The investor foresees a $25,000 net return at the end of the first year. She wishes to calculate her rate of return for 2 years if she holds on for a second year and business falls off, since such attractions often have a short life. In the worst cases, she foresees a $25,000 loss which would include the $2,000 it will cost to remove the Giant Slide. The following table shows the Giant Slide cash flow.

Year	Cash flow
0	−4,000
1	+25,000
2	−25,000
	−4,000

First, assume that the entrepreneur's borrowing rate is 10 percent. The person who is lending the investment funds at 10 percent to the entrepreneur insists that they be used only for the Giant Slide project. The lender will be paid off at principal and interest at the end of the first year.

(*a*) Construct a present-value profile for the relevant range of rates of return for the above cash flow.

(*b*) What should be the entrepreneur's decision?

Now assume that the entrepreneur can borrow any amount of funds, for any project she cares to use them, at 15 percent. A project equal in risk to the Giant Slide will return 30 percent.

(*c*) What will her decision be on the project now? Does it agree with the NPV criterion? Explain.

8.45. A marble quarry operation in Minnesota is under consideration by a major mining company. The final cash flow estimates relative to the acquisition are as follows:

Year	Item	Cash flow ($000,000)
0	First cost	-151
1	Net profit after taxes	+205
2		+20
3		+105
4	Landscape restoration	-190

The board of directors of the company requires that the internal rate of return method be used. Should the board of directors approve the project if the company's opportunity cost of capital is 17 percent?

8.46. A group of investors have before them a proposal for the purchase of a small branch-line railroad, the Eureka and the Northern. On acquiring the railroad, the group plans to sell off some of the railroad's real estate immediately which will more than pay for the initial cost. The railroad will then be operated at a loss for 5 years. At the end of the fifth year, the railroad will be sold. The cash flow for the railroad purchase is as follows:

Year (s)	Item	Cash Flow ($000,000)
0	First cost	-25
0	Real estate assets sale	+35
1–5	Annual operating revenue	+7
1–5	Annual operating cost	-16
5	Disposal price	+37

The cash flow sums have been adjusted for inflation. Tax effects have been included in them also. A rate of return solution is required.

(*a*) If the opportunity cost of capital of this group is 3 percent after taxes and inflation, what will be the decision?

(*b*) At 20 percent, would the group purchase the railroad?

(*c*) At 75 percent, what decision would be made?

(*d*) How do you explain the variation, if any, in the answers to parts (*a*), (*b*), and (*c*)?

8.47. In Example 8.10, page 168, because of a devaluation of the Austral (the unit of currency in use in Argentina at the time), the Buenos Aires building will cost $15 million rather than $20 million. Will your decision be changed by this occurrence? All other data remain the same. Perform an algebraic solution.

8.48. In a large office machine company which considers funding many projects on its capital budget, the incremental cash flow between two mutually exclusive alternatives appears as follows:

Year	0	1	2	3	4	5	6	7
Cash flow	+6	−10	+1	+1	+1	+1	+1	+1

Compute the solving rates of return for this cash flow. Accuracy to the nearest whole percentage point in the solution will be sufficient. It will be helpful to graph the present-worth profile over the range from 0 to 50 percent.

8.49. In Prob. 8.48 (above), the opportunity cost of capital for this company is 10 percent. At that rate, would the project be accepted? Between what opportunity costs of capital would the project be accepted?

8.50. Repeat Prob. 5.35, page 97, using the IROR method.

8.51. You are considering two mutually exclusive investments. The first alternative has a first cost of $102,000. At the end of year 1 it will net $50,000, and at the end of year 2 it will net $100,000. The second alternative has a first cost of $175,000. At the end of year 1 it will net $215,000, and year 2 will end with a net of $10,000. The economic life for both alternatives is 2 years. With an opportunity cost of capital of 15 percent and without considering taxes and inflation, which investment will you select? One must be chosen. An IROR methodology is required, with IRORs to the nearest whole percentage point. Attempt an algebraic solution.

8.52. Two mutually exclusive investment opportunities display the following cash flows:

Year	*A*	*B*
0	−100	−200
1	+100	+350
2	+200	+ 50

These are both very attractive investments. Alternative *A* has an internal rate of return of 100 percent, and *B* has an internal rate of return of 88.3 percent. If your opportunity cost of capital is 50 percent, which alternative should you choose, using the IROR method? Perform an algebraic solution.

8.53. Study the following two alternatives, using the *internal rate of return method.* State what MARR conditions will govern your selection of one or the other, providing that the null alternative does not exist. Rates of return to the nearest whole percent are sufficiently accurate.

Year	*A*	*B*
0	−2,000	−1,380
1	1,700	940
2	10	10
3	1,500	850
4	400	1,210

8.54. Two types of earthmoving machines are being reconsidered by a large construction company in Panama. Both are used machines to be imported from the United States. Inspection of the machines reveals that machine type *A* will require an overhaul at the end of the first year of use which will cost an estimated $10,000 and another overhaul at the end of the third year of use costing $9,775. Machine *B* will require an overhaul at the end of the second year of use costing $20,000.

First cost, operation and maintenance costs, and salvage values are the same for both machines.

The following is a table of annual costs and revenues estimated for each machine:

First cost	$150,000
Machine life	8 years
Operation and maintenance	$12,000 per year
Labor	$10,000 per year
Salvage value	$20,000
Revenue	$50,000 per year

A *rate of return calculation is required* with rates to the nearest whole percent. The company will not now consider investments paying less than 7 percent. Taxes and inflation effects have been included in all estimates.

Should the company invest in either machine and, if so, which one?

8.55. Two fast-food franchises are available. You are certain to buy one. You intend to hold the franchise 3 years and then sell it. After-tax and after-inflation cash flows are as shown:

Year	Happy Hamburgers	Pizzazz Pizza
0	−50,000	−80,000
1	+10,000	+30,000
2	+60,000	+110,000
3	+170,000	+140,000

A rate of return calculation is required. If the appropriate cutoff rate of return is 30 percent, which franchise should you purchase?

You have established your cutoff rate of return by means of a capital budget.

8.56. Repeat Prob. 7.16, page 139, based on Example 7.6, page 131, using the internal rate of return method.

8.57. A firm of international architects and engineers has laid out two plans, out of many considered, as worthy of consideration by its client for development of her Hong Kong property. Because of the uncertain future of the colony, the client will sell at the end of 4 years.

Net cash flows after taxes and inflation have been factored in are as follows:

Year	Plan 1 ($000)	Plan 2 ($000)
0	+3,500	−4,000
1	+2,000	+2,400
2	+2,000	+2,400
3	+2,000	+2,400
4	+2,000	+1,700
4—sale	+3,500	+3,500

Should the client invest in either of these plans, and if so, in which one? A *rate of return solution* to the nearest whole percent is required. The minimum acceptable rate of return is 30 percent.

8.58. A company is considering investing in new facilities to replace existing facilities. Two possible alternatives are identified below. The funds for the project are internally generated, and the minimum acceptable rate of return (MARR) for the company is 10 percent. The life of each facility is 5 years. A rate of return method is required. What decision should the company make? (All values are in millions of dollars.) Show all work and state all assumptions.

Project	First cost	Benefits (each year, years 1–5)	Salvage value
1	20	5.0	9.1
2	25	8.0	-1.0

8.59. The offshore drilling division of a major oil company has before it two designs for drilling platforms. The first is an innovative concrete semisubmersible. The second is the traditional steel semisubmersible.

Data on the two designs are as shown:

Type	Concrete	Steel
First cost ($000,000)	50	75
Annual maintenance cost ($000,000)	10	6
Life (years)	10	10
Resale or scrap value ($000,000)	9	3.5

All tax and inflation effects have been included in these figures. The company uses a 10 percent after-tax, constant-dollar opportunity cost of capital to judge its investments. Because of financing considerations, a *rate of return* solution is demanded. Rates of return to the nearest whole percent will suffice.

Which of the two types of platform will be selected?

8.60. Repeat Prob. 7.17, page 139, using the IROR method.

8.61. The following two models of a turbine are under consideration by the Potenz Power Company:

Characteristics	Model *A*	Model *B*
Economic life (years)	15	15
First cost ($)	5,000,000	5,100,000
Annual operations and maintenance cost ($)	100,000	80,000
Major overhaul cost at year 5 ($)	0	700,000
Major overhaul cost at year 7 ($)	2,806,000	0
Major overhaul cost at year 10 ($)	0	2,894,000

Use the *rate of return method* to recommend a turbine model, given that

(*a*) No taxes are to be considered since Potenz is government owned.
(*b*) The MARR (opportunity cost of capital) of Potenz is 12 percent.
(*c*) One of the two models is certain to be selected.
(*d*) Both models are suitable in a technical sense.

APPENDIX 8A: THE REINVESTMENT FALLACY

The *reinvestment fallacy* is the idea that in order for an internal rate of return of an investment to be a valid measure of that investment's worth, the net profits must be invested, or be capable of being invested, at the internal rate of return. This somewhat confusing statement—confusing for those who have understood the correct approach to the internal rate of return presented thus far in this text—may be best explained by an example.

■ **Example 8A.1.** Consider the following cash flow, obtainable from a mining venture in a Latin American country:

Year	Cash flow ($)
0	-100,000
1	+100,000
2	+200,000

The investment required is $100,000. At the end of year 1, $100,000 over and above all expenses will be returned to the investor. The same amount will be returned at the end of year 2. Also at the end of year 2, the investor expects to sell the mining property for the amount he paid for it—$100,000. Figure 8.16 shows this.

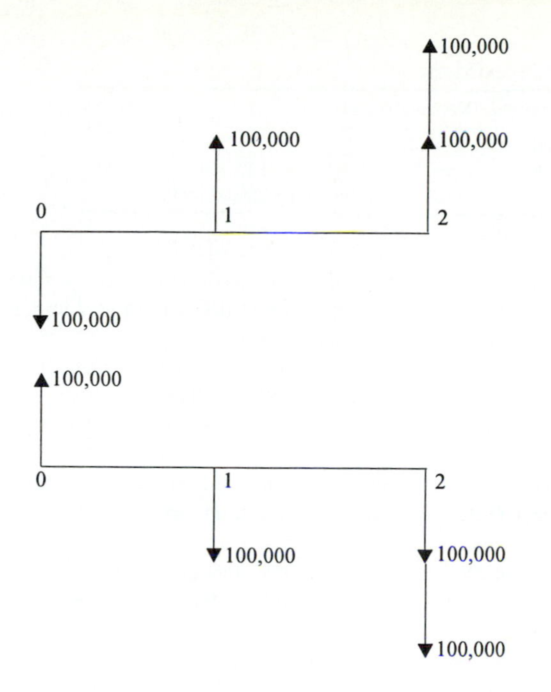

FIGURE 8.16 A mining venture

The internal rate of return of this investment is 100 percent. For the investment shown by Table 8.9 and Fig. 8.16, the IROR is 100 percent, because 100 percent of the original investment is received at the end of each year of the investment's life, and the total amount of the investment is returned at the end of the last year.

Let us see how the mechanism of the internal rate of return works. Reconsidering the cash flow of Table 8.9, we will compute the net investment in Table 8.10.

From Table 8.10, we conclude that 100 percent is the correct internal rate of return because the net investment is zero at the end of year 2. But what does the -100,000 figure at the end of year 1 mean? It appears to be the amount that must be reinvested at 100 percent in order to complete the calculation. But no further monies, beyond the original $100,000, are actually invested. Now it can be seen clearly that what we have just done in computing net investment, sometimes called—misleadingly—*unrecovered investment*, is no more than an exercise in financial mathematics. It is another way of showing that the computed IROR is correct. (In a

TABLE 8.9
A 100 percent investment

Year	Cash flow	(P/F, 100, N)	PW
0	-100,000	1.0000	-100,000
1	+100,000	0.5000	+50,000
2	+200,000	0.2500	+50,000
			0

TABLE 8.10
Computing net investment

Year	Cash flow	Net investment
0	−100,000	−100,000
1	+100,000	(−100,000) (1 + 1.00) + 100,000 = −100,000
2	+200,000	(−100,000) (1 + 1.00) + 200,000 = 0

similar fashion, we could prove that the IROR is correct by showing that the net present value of the cash flow is zero at that IROR.) But no additional investment, or retention of net revenues, is implied or occurs.

Now those who support the reinvestment fallacy maintain that, for this 100 percent IROR to be a valid measure, the $100,000 received at the end of years 1 and 2 must be reinvested at 100 percent, or must be capable of being so invested. The stipulation of reinvestment is fundamental to the fallacy.

But what if the investor wanted to spend the proceeds on a yacht? Would this fact invalidate the 100 percent return? For the proponents of the reinvestment fallacy, it would.

Suppose that another 100 percent investment were not available. Would this make nonsense of the 100 percent return? Apparently, yes, according to supporters of the fallacy.

We have said that the opportunity cost of capital (MARR) is that rate of return forgone by investing in the project under consideration. Thus any positive cash flow can be invested only at that rate, rarely or never the IROR of a project. The idea of reinvestment at another rate violates this fundamental assumption and thus is untenable.

Another of the arguments, among many, against the reinvestment fallacy is the following. If the cash flow is turned upside down, as in the bottom portion of Fig. 8.16, a loan of $100,000 is repaid at 100 percent interest. No borrower is ever required to produce further possibilities of loans at 100 percent to justify the assertion that the borrower is paying 100 percent interest on the loan. ■

The *external rate of return (ERR) method* is based on the erroneous assumption that positive and negative cash flows can be judged at different rates, thus violating the logic of opportunity cost, a fundamental principle in basic economics. It is not recommended.

Moreover, some authors have recommended the use of an auxiliary rate to solve multiple-IROR problems. In normal circumstances, the opportunity cost of capital cannot, by definition, differ from the auxiliary rate. Thus there can be no exception to the NPV method where the opportunity cost of capital exists. The borrowing-lending rate, that is, the auxiliary rate, and the opportunity cost of capital (MARR) are identical. This idea will be better understood after readers have studied Chap. 14 on capital budgeting.

Much research remains to be done before the numerous questions surrounding the IROR method are answered. Until they are, the IROR, especially as regards multiple rates of return, remains controversial.

Finally, nothing said here makes impossible a proposition where reinvestment is part and parcel of the deal. But such an arrangement is merely another cash flow to be evaluated by the methods we already know.

PROBLEMS

8A.1. In Example 8.9, page 167, test the cash flow for design *A* by computing the net investment at each period, as was done in App. 8A. What will the net investment be at year 3?

8A.2. In Example 8.9, page 167, test the incremental cash flow design *A* – design *B* by computing the net investment at each period, as was done in App. 8A.

8A.3. In Example 8.11, page 170, an analyst decides to move the $200,000 cash flow at years 1 and 2 to year 3 at the 10 percent opportunity cost of capital. Doing so results in a cash flow with one sign change and thus one positive rate of return. What is it?

The 10 percent opportunity cost of capital is called the *external interest rate* in some textbooks.

Do you think the procedure described solves the problem of multiple rates of return? Why or why not?

8A.4. In Example 8.11, page 170, show that 146 percent is a solving rate of return by using the net investment procedure described in App. 8A.

Note: The IROR will need to be carried out to three decimal places for the sake of accuracy at high rates of return like this one.

CHAPTER 9

MULTIPLE ALTERNATIVES

Thus far, we have dealt with no more than three alternatives, including the null. How are more than three—multiple—alternatives handled? The answer is, in essentially the same way, but with some variations. We will see how the four methods we have examined treat multiple alternatives via a common example, with a special additional example for each method. Then we will open the discussion to cost-effectiveness in relation to multiple alternatives.

In this chapter, multiple alternatives are all mutually exclusive; that is, as before, the choice of one excludes the choice of all the others. The problem of multiple alternatives that are not mutually exclusive, alternatives both mutually exclusive and independent, and alternatives mutually exclusive, independent, and interdependent will be discussed in Chap. 14 on capital budgeting.

INCREMENTAL ANALYSIS OF MULTIPLE ALTERNATIVES

Multiple alternatives must be analyzed incrementally, as a general rule. In the present- and annual-worth methods, individual measures of worth may be used. Figure 9.1 shows the steps to follow in the analysis. This flowchart will guide each of the methods used. The phrase *better than* will be interpreted according to the method undertaken, and *challenger* and *defender* will be explained.

Once again, the necessity for incremental analysis must be emphasized. What was true for two alternatives remains true for more than two. Each extra cost must be justified by the extra benefit accruing because of it. If the increment in cost of an alternative is not justified by an equal or greater increment in benefit, then that

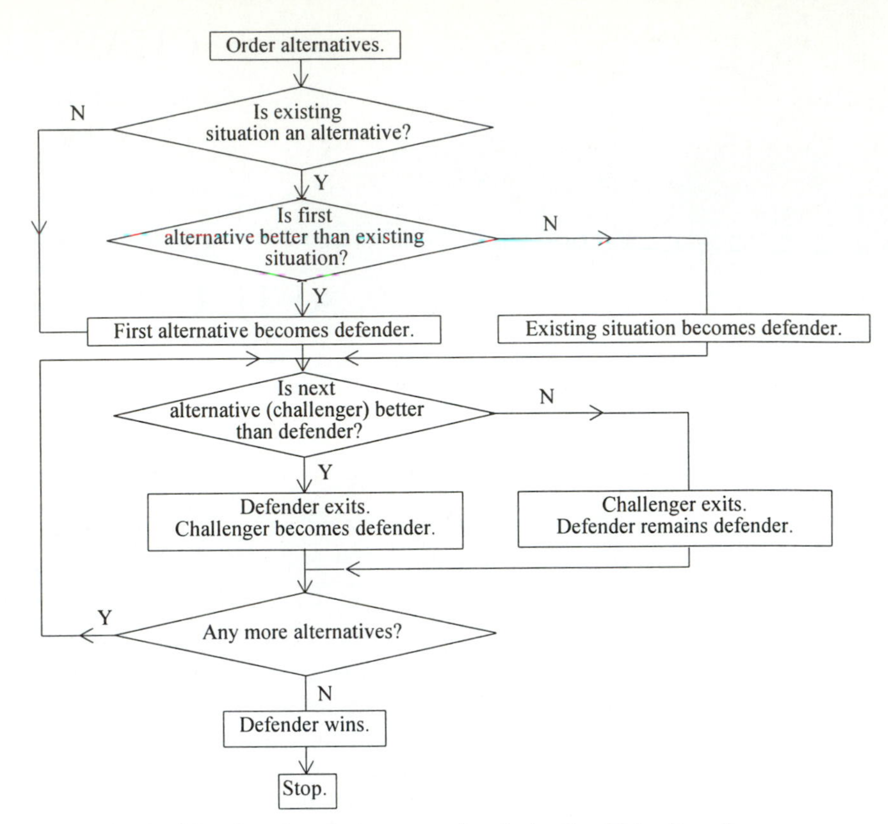

FIGURE 9.1 Decision flowchart for incremental analysis of multiple alternatives

alternative disappears from the problem. No succeeding alternative can be evaluated against a failed alternative. The failed alternative loses its status as a standard, just as an undermined benchmark in surveying can no longer be used to determine elevation. All that is necessary for a challenging alternative to be accepted over a defending alternative is that it offer a benefit at least equal to the benefit to be realized by investing the extra resources at the opportunity cost of capital.

PRESENT WORTH: INCREMENTAL AND INDIVIDUAL

The economic analysis of multiple alternatives by using the present-worth method is demonstrated in the following example.

■ **Example 9.1.** A city has decided to set up a bus transportation system. In 10 years the city council plans to sell the bus company to private interests. Four plans have been proposed for the system, involving four different schedules of cost, resale value, and net revenue. In view of the risk involved, the city council has decided that it must have a 15 percent return on its investment with no consideration given to income taxes. Inflation effects will be ignored. The four plans are as follows:

	Plan *A*	**Plan *B***	**Plan *C***	**Plan *D***
First cost	140	163	190	220
Estimated resale value	125	138	155	175
Excess of annual revenues over annual disbursements	24	28	31	38

Which plan of development, if any, should the city select?

Solution. Figure 9.2 shows the four plans. An incremental present-worth analysis is undertaken first. Following the flowchart of Fig. 9.1, we check the order of the alternatives and discover that they are already in correct order from lowest to highest first cost.

Is the existing situation (0) an alternative? It is, because the phrase *if any* appears in the example statement.

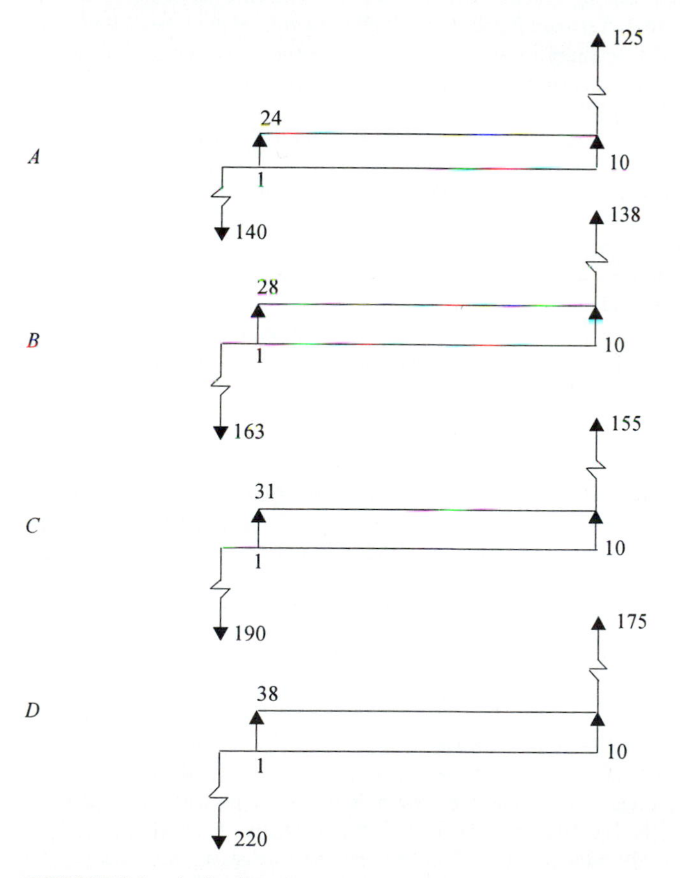

FIGURE 9.2 A city bus system

Is the first alternative better than the existing situation? To answer this question, we calculate the net present value (NPV) of the first alternative, which is A. This is the same as calculating the incremental net present value of A against 0, the null alternative.

$$\begin{aligned}\text{NPV}_A &= -140 + 24(P/A, 15, 10) + 125(P/F, 15, 10)\\ &= -140 + 24(5.019) + 125(0.2472)\\ &= +11.4\end{aligned}$$

Thus alternative A is preferred to doing nothing about the present situation. Notice that the phrase *better than* of the flowchart has been interpreted in terms of positive net present value.

The answer yes moves us to the next box. The first alternative becomes the defender. *Defender* and *challenger* are common terms in multiple-alternative analysis. The process lends itself to the analogy of the prizefighting ring where a defending champion takes on all challengers. If the defender is defeated by a challenger, the challenger becomes the defender against subsequent challengers.

Is the next alternative, B, better than the defender? Incremental analysis requires us to take the difference between B and A. Remember the rule of the earliest portion of this book: Only the differences among alternatives are relevant to a comparison among them. Figure 9.3 shows the incremental cash flow $B - A$. Therefore,

$$\text{NPV}_{B-A} = -23 + 4(P/A, 15, 10) + 13(P/F, 15, 10) = +0.3$$

This means that B is better than A because the NPV is positive. Remember that the magnitude does not matter—only whether the NPV is plus or minus. The answer to the question of this box is yes.

Moving to the next box, we see that defender A exits from the problem and previous challenger B is now the defender.

Are there any more alternatives? Yes. We move back to the question, Is the next alternative (challenger) C better than the defender B? We must now evaluate the incremental cash flow $C - B$.

$$\text{NPV}_{C-B} = -27 + 3(P/A, 15, 10) + 17(P/F, 15, 10) = -7.7$$

The minus sign indicates that the increment of investment from B to C is not justified. This time the answer to the question is no. The reply leads us to the box requesting that challenger C exit. Defender B remains the defender. We are returned to the question, Any more alternatives? Yes, there is one more.

Is the next alternative D better than defender B? Figure 9.3 illustrates the differential cash flow $D - B$. Evaluating

$$\text{NPV}_{D-B} = -57 + 10(P/A, 15, 10) + 37(P/F, 15, 10) = +2.3$$

shows a clear victory for D, and the answer is yes.

We are led to "Defender exits, challenger becomes defender." And from there to "Any more alternatives?" This time the answer is no, and this leads us to the box "Defender wins." The last defender is D, which is the alternative selected.

We have already seen in Chap. 5 on present worth that we may analyze the alternatives individually and then compare them by choosing the alternative with the

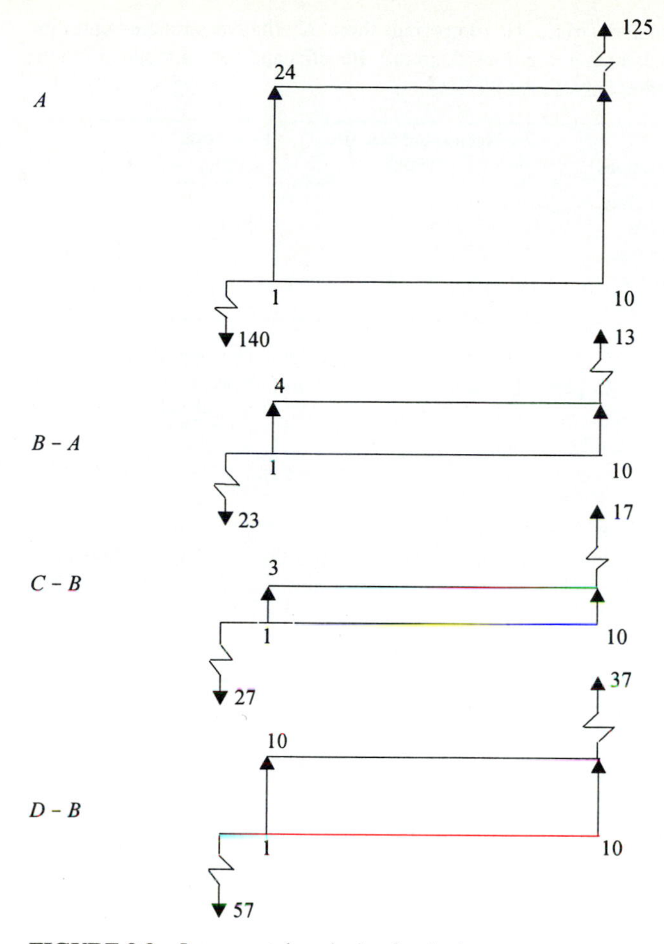

FIGURE 9.3 Incremental analysis of a city bus system

greatest present worth. This is also an incremental analysis because we are comparing by differences. The logic is the same as that used in the previous analysis, but the increment appears at a different place in the analysis.

Taking the net present value of each of the plans as seen in Fig. 9.2 gives

$$\text{NPV}_A = -140 + 24(P/A, 15, 10) + 125(P/F, 15, 10) = +11.4$$

$$\text{NPV}_B = -163 + 28(P/A, 15, 10) + 138(P/F, 15, 10) = +11.6$$

$$\text{NPV}_C = -190 + 31(P/A, 15, 10) + 155(P/F, 15, 10) = +3.9$$

$$\text{NPV}_D = -220 + 38(P/A, 15, 10) + 175(P/F, 15, 10) = +14.0$$

The choice is *D*, confirming the incremental analysis. Notice that the increments are also confirmed, within a rounding error:

$$\text{NPV}_{B-A} = +0.2 \qquad \text{NPV}_{C-B} = -7.7 \qquad \text{NPV}_{D-B} = +2.4$$ ■

■ **Example 9.2.** A major airline is comparing three alternative arrangements for passenger accommodations in a proposed aircraft. Benefits and costs for each of these mutually exclusive alternatives are as follows:

Passenger accommodations			Net annual benefits ($000)	First costs ($000)
A	250	Tourist class		
	65	First class	6,730	23,110
B	300	Tourist class		
	30	First class	6,745	23,170
C	350	Tourist class	6,771	23,260

The life of the aircraft is 15 years in average world service. The opportunity cost of capital before taxes is 25 percent. Assume that inflation and taxes will have equal effects on costs and benefits and thus may be neglected. The salvage value is $5 million, regardless of accommodations.

Should the airline invest in any of these configurations and, if so, in which one?

Solution. Referring to the flowchart of Fig. 9.1, we see that the alternatives are to be ordered. However, the order is already correct. Because the existing situation is an alternative, the first alternative, *A*, must be tested against it. Figure 9.4 shows this alternative.

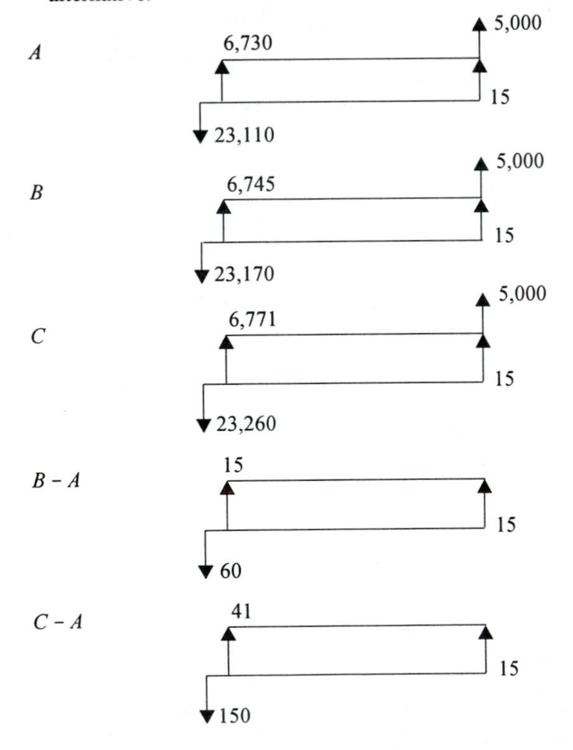

FIGURE 9.4 Airline passenger accommodations

$$\text{NPV}_A = -23{,}110 + 6{,}730(P/A, 25, 15) + 5{,}000(P/F, 25, 15) = +3{,}039$$

The positive value indicates acceptance of the course of action. The first alternative is better than the existing situation and becomes the defender.

The next alternative is B. Figure 9.4 shows the cash flow B and $B - A$.

$$\text{NPV}_{B-A} = -60 + 15(P/A, 25, 15) = -2$$

The minus sign indicates rejection of alternative B. Challenger B must exit, and defender A remains the defender. Are there any more alternatives? Yes, there are. The next alternative is C: it must challenge A incrementally. Figure 9.4 shows C and the resulting cash flow $C - A$.

$$\text{NPV}_{C-A} = -150 + 41(P/A, 25, 15) = +8$$

Alternative C is better than A. The defender exits. The challenger becomes the defender, but there are no more challengers, and the defender wins. Alternative C is accepted. ■

ANNUAL WORTH: INCREMENTAL AND INDIVIDUAL

As in present worth, it is much more usual to employ individual rather than incremental annual worth. For the sake of consistency in presentation, however, Example 9.1 will be solved first by the use of incremental annual worth and then by individual annual worths, that is, gain or loss.

■ **Example 9.1 (*continued*).** Figure 9.2 shows the cash flows of alternative courses of action with regard to a city bus system, and Fig. 9.3 shows the incremental cash flows used in determining the best alternative. The alternatives are already in correct order. In the incremental annual-worth method of analysis, we once again use the flowchart of Fig. 9.1.

The *annual worth* (AW) of A compared to the null alternative is

$$\begin{aligned}\text{AW}_A &= -140(A/P, 15, 10) + 24 + 125(A/F, 15, 10)\\ &= -140(0.19925) + 24 + 125(0.04925)\\ &= -27.90 + 24.00 + 6.16\\ &= +2.26\end{aligned}$$

The plus sign indicates a gain, and therefore A is accepted rather than the null alternative. It becomes the defender.

Alternative A is challenged by the next alternative, B. Figure 9.3 illustrates the incremental cash flow $B - A$. Its annual worth is

$$\begin{aligned}\text{AW}_{B-A} &= -23(A/P, 15, 10) + 4 + 13(A/F, 15, 10)\\ &= 23(0.19925) + 4 + 13(0.04925)\\ &= +0.06\end{aligned}$$

The plus sign shows that the challenger is successful against the defender; B is accepted over A, and B now becomes the defender.

There are more alternatives: C challenges B.

$$\text{AW}_{C-B} = -27(A/P, 15, 10) + 3 + 17(A/F, 15, 10)$$
$$= -1.54$$

Alternative C is rejected because of the minus sign and leaves the problem. Alternative B remains the defender.

The next and last course of action is D, which now challenges B:

$$\text{AW}_{D-B} = -57(A/P, 15, 10) + 10 + 37(A/F, 15, 10)$$
$$= +0.46$$

The plus sign signals acceptance of D rather than B. Now D becomes the defender. Because there are no more challengers, D wins, thereby checking the present-worth solutions.

The same example handled in the more usual manner consists of computing the individual annual worths, comparing them, and choosing the alternative with the highest gain.

Figure 9.2 shows the cash flows. The annual worth of A was computed above:

$$\text{AW}_A = +2.26$$

The annual worth of B is

$$\text{AW}_B = -163(A/P, 15, 10) + 28 + 138(A/F, 15, 10)$$
$$= +2.32$$

In a similar manner, the remaining annual worths are

$$\text{AW}_C = -190(A/P, 15, 10) + 31 + 155(A/F, 15, 10) = +0.78$$
$$\text{AW}_D = -220\,(A/P, 15, 10) + 38 + 175(A/F, 15, 10) = +2.78$$

The alternative with the greatest annual worth is D.

Notice that, as in present worth, it is possible to calculate the incremental differences between the alternatives:

$$\text{AW}_{A-0} = +2.26 \qquad \text{AW}_{C-B} = -1.54$$
$$\text{AW}_{B-A} = +0.06 \qquad \text{AW}_{D-B} = +0.46$$

Within rounding errors, the answers are identical. ■

It should be obvious to the reader by now why individual annual worths are preferred to incremental annual worths in solutions of multiple-alternative problems.

■ **Example 9.3.** The passenger bus system of the city of Guadalajara, Mexico, used three types of engines in the buses: gasoline, natural gas, and diesel. Otherwise the buses were identical. All motors, regardless of type, had to be changed every 5 years. Table 9.1 shows the buses by type of motor. The first cost of the bus included the cost

TABLE 9.1
Guadalajara buses

Type	First cost (pesos)	Economic life (years)	Motor cost (pesos)	Motor life (years)	Resale value (pesos)
1. Gasoline	145,000	10	17,000	5	18,000
2. Natural gas	146,400	9	15,000	5	18,500
3. Diesel	153,000	10	47,000	5	19,000

of the motor. The resale value was that of the bus with motor included, whatever the motor's age.

If the opportunity cost of capital was 18 percent, find the annual cost of each type of bus. Which type was cheapest?

Solution. Figure 9.5 illustrates the cash flows for each of the three bus types. The usual method of annual-cost analysis will be used. We compute the annual cost of each type:

$$\mathrm{AC}_1 = -145{,}000(A/P, 18, 10) - 17{,}000(P/F, 18, 5)(A/P, 18, 10) + 18{,}000(A/F, 18, 10)$$

$$= -145{,}000(0.22251) - 17{,}000(0.4371)(0.22251) + 18{,}000(0.04251)$$

$$= -33{,}152$$

$$\mathrm{AC}_2 = -146{,}400(A/P, 18, 9) - 15{,}000(P/F, 18, 5)(A/P, 18, 9) + 18{,}500(A/F, 18, 9)$$

$$= -146{,}400(0.23239) - 15{,}000(0.4371)(0.23239) + 18{,}500(0.05239)$$

$$= -34{,}576$$

$$\mathrm{AC}_3 = -153{,}000(A/P, 18, 10) - 47{,}000(P/F, 18, 5)(A/P, 18, 10) + 19{,}000(A/F, 18, 10)$$

$$= -37{,}808$$

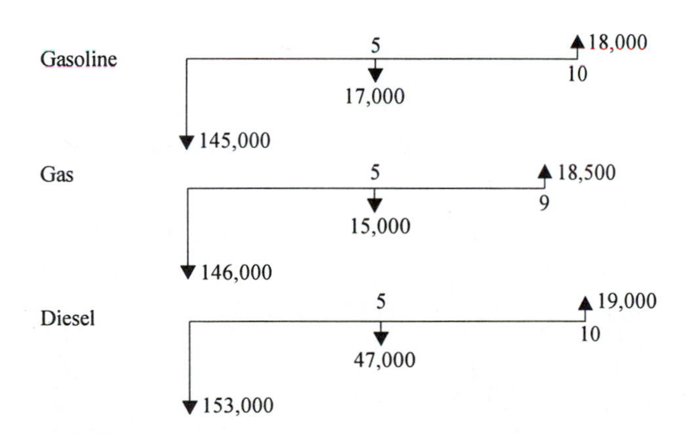

FIGURE 9.5 Three bus types

The gasoline engine type has the lowest annual cost. Notice that the advantage of annual cost holds in multiple-alternative even more than in double-alternative problems. Had other methods been requested in this example, the analyst would have had to equalize the lives to make comparisons between the alternatives. For example, if present worth had been required, the equalizing life would have been 90 years! ■

BENEFIT/COST RATIO: INCREMENTAL

The benefit/cost ratio method must employ incremental analysis. Individual benefit/cost ratios lack meaning in economic analysis. To demonstrate the use of the benefit/cost ratio method in multiple-alternative situations, two examples are offered.

■ **Example 9.1 (*continued*).** As before, Figs. 9.2 and 9.3 show the individual and incremental cash flows, respectively. The analysis follows the flowchart of Fig. 9.1.

The courses of action are already in correct order. The existing situation is an alternative. Therefore alternative A must be compared to the null alternative. Using the top drawing of Fig. 9.3, we find

$$\left(\frac{B}{C}\right)_{A-0} = \frac{24(P/A, 15, 10) + 125(P/F, 15, 10)}{140}$$
$$= \frac{151.35}{140}$$
$$= 1.08 > 1.0$$

Thus alternative A is better than the existing situation because its B/C exceeds 1.0. The first alternative now becomes the defender.

Is the next alternative better than the defender? Incremental analysis between alternatives B and A must be used. The cash flow $B - A$ is shown in Fig. 9.3. Its benefit/cost ratio is

$$\left(\frac{\Delta B}{\Delta C}\right)_{B-A} = \frac{4(P/A, 15, 10) + 13(P/F, 15, 10)}{23}$$
$$= \frac{23.29}{23}$$
$$= 1.01 > 1.0$$

Challenger B beats out defender A. Alternative A exits, and B becomes the defender. There are two more alternatives.

Is the next alternative C better than defender B? To answer this question, incremental analysis between C and B must be employed. The differential cash flow $C - B$ is shown in Fig. 9.3.

$$\left(\frac{\Delta B}{\Delta C}\right)_{C-B} = \frac{3(P/A, 15, 10) + 17(P/F, 15, 10)}{27}$$
$$= \frac{19.26}{27}$$
$$= 0.71 < 1.0$$

The challenge fails, and B remains the defender while C exits the example.

One alternative remains, and it becomes challenger D. An incremental analysis will answer the question of its superiority over the defender. See Fig. 9.3.

$$\left(\frac{\Delta B}{\Delta C}\right)_{D-B} = \frac{10(P/A, 15, 10) + 37(P/F, 15, 10)}{57}$$
$$= \frac{59.33}{57}$$
$$= 1.04 > 1.0$$

Because 1.04 exceeds 1.0, the challenger wins, the defender exits, and the challenger becomes the defender against new challengers. There are none. Defender D wins the contest.

Let us now observe what happens if the *wrong* method is used. From Fig. 9.2 we compute the individual benefit/cost ratios and discover

$$\left(\frac{B}{C}\right)_A = 1.08 \qquad \left(\frac{B}{C}\right)_C = 1.02$$
$$\left(\frac{B}{C}\right)_B = 1.07 \qquad \left(\frac{B}{C}\right)_D = 1.06$$

Using the erroneous method results in an incorrect answer. The highest B/C ratio belongs to alternative A, and the next to B. The correct answer—D—appears in third place. ■

■ **Example 9.4.** Three proposals exist for the improvement of highway communication between two cities in a developing country. The proposals are mutually exclusive. Project 1 maintains the existing highway over the next 10 years. Project 2 improves the alignment of the existing highway and maintains it. Project 3 builds an entirely new highway. The stream of net costs and benefits for the three alternatives appears in Table 9.2.

TABLE 9.2
Three highway projects

	Alternative		
Year	**1**	**2**	**3**
0	–50	–90	–100
1	2	5	2
2	8	15	10
3	12	25	15
4	15	30	20
5	20	34	30
6	22	30	35
7	18	22	38
8	10	15	35
9	8	10	25
10	5	5	15

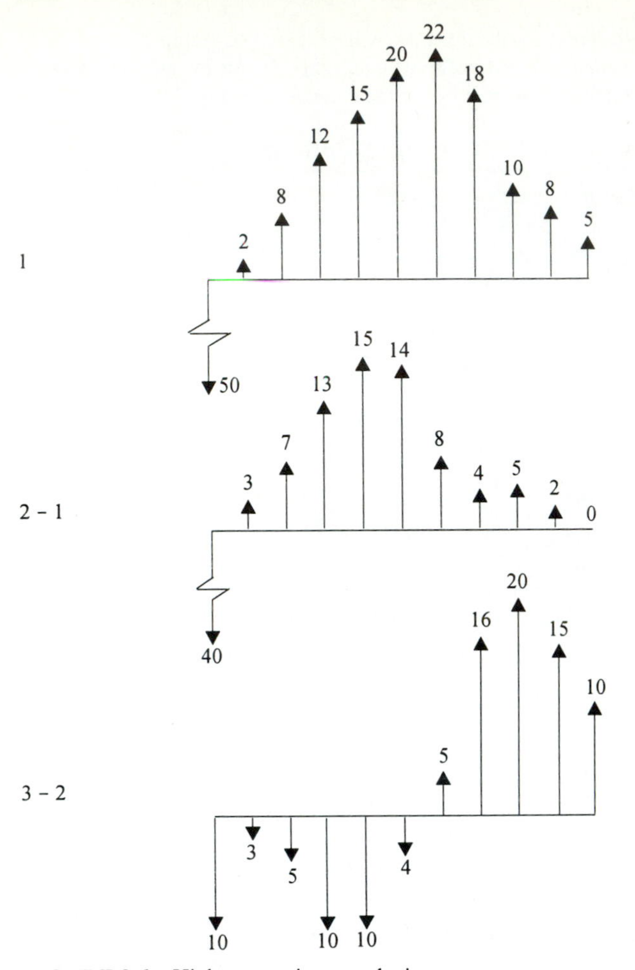

FIGURE 9.6 Highway projects analysis

Select one or none of these alternatives, using the benefit/cost ratio method. The opportunity cost of capital for the public sector in this country is 10 percent after inflation has been accounted for. The figures of Table 9.2 are in constant money units of year 0. No salvage value will be considered for this project.

Solution. Because all the projects may be rejected, we must test alternative 1, shown in Fig. 9.6, against the null alternative. Table 9.3 demonstrates an easy way to set up the calculation of the present worths necessary to compute the benefit/cost ratio. Therefore,

$$\left(\frac{B}{C}\right)_{1-0} = \frac{71.7}{50} = 1.43$$

Alternative 1 has successfully defended itself against the null alternative. It now becomes the defender, as we follow the instructions of Fig. 9.1.

TABLE 9.3
Benefit and cost of project 1

Year	Cash flow	$(P/F, 10, N)$	Discounted C	Discounted B
0	-50	1.0000	-50	
1	2	0.9091		1.8
2	8	0.8264		6.6
3	12	0.7513		9.0
4	15	0.6830		10.3
5	20	0.6209		12.4
6	22	0.5645		12.4
7	18	0.5132		9.2
8	10	0.4665		4.7
9	8	0.4241		3.4
10	5	0.3855		1.9
			Σ-50	71.7

Is the challenger, alternative 2, better than the defender, alternative 1? We must now test the incremental consequences flow between alternatives 2 and 1. Figure 9.6 shows this flow. Table 9.4 demonstrates the calculation of the discounted benefits and costs. The benefit/cost ratio is

$$\left(\frac{\Delta B}{\Delta C}\right)_{2-1} = \frac{47.0}{40}$$

$$= 1.17 > 1.0$$

Alternative 2 wins over alternative 1. Alternative 1 exits the problem, and alternative 2 becomes the new defender. One more alternative exists. Alternative 3 becomes the challenger. The incremental cash flow 3 − 2 is the bottom display of Fig. 9.6. Table 9.5 facilitates the calculation of the incremental benefit/cost ratio.

TABLE 9.4
Benefit and cost of 2 − 1

Year	Cash flow (2 −1)	$(P/F, 10, N)$	Discounted C	Discounted B
0	-40	1.0000	-40	
1	3	0.9091		2.7
2	7	0.8264		5.8
3	13	0.7513		9.8
4	15	0.6830		10.2
5	14	0.6209		8.7
6	8	0.5645		4.5
7	4	0.5132		2.1
8	5	0.4665		2.3
9	2	0.4241		0.9
10	0	0.3855		0
			Σ -40	47.0

TABLE 9.5
Benefit and cost of 3 − 2

Year	Cash flow (3 − 2)	(*P*/*F*, 10, *N*)	Discounted *C*	Discounted *B*
0	−10	1.0000	−10	
1	−3	0.9091	−2.7	
2	−5	0.8264	−4.1	
3	−10	0.7513	−7.5	
4	−10	0.6830	−6.8	
5	−4	0.6209	−2.5	
6	5	0.5645		2.8
7	16	0.5132		8.2
8	20	0.4665		9.3
9	15	0.4241		6.4
10	10	0.3855		3.9
			Σ − 33.6	30.6

$$\left(\frac{\Delta B}{\Delta C}\right)_{3-2} = \frac{30.6}{33.6}$$
$$= 0.91 < 1.0$$

The challenge fails. Alternative 2 remains the defending champion. There are no more alternatives, and therefore 2 is chosen.

Had incorrect methods been followed and individual rather than incremental benefit/cost ratios been calculated, the result would have been

$$\left(\frac{B}{C}\right)_1 = 1.43 \qquad \left(\frac{B}{C}\right)_2 = 1.32 \qquad \left(\frac{B}{C}\right)_3 = 1.26$$

The highest *B*/*C* ratio belongs to alternative 1. Its choice, rather than the correct alternative 2, would mean a heavy loss in comparison to what would have been gained by the choice of 2. ■

INTERNAL RATE OF RETURN: INCREMENTAL

To illustrate the internal rate of return method with regard to multiple-alternative problems, Example 9.1 will be reused. Incremental analysis *must* be employed.

■ **Example 9.1 (*continued*).** Figure 9.3 shows the incremental analysis of the four plans. If we follow the flowchart of Fig. 9.1, we recognize immediately that the plans are ordered correctly and that the existing situation is an alternative. Alternative *A* is tested against the existing situation by finding its internal rate of return:

$$0 = -140 + 24(P/A, i^*, 10) + 125(P/F, i^*, 10)$$
$$i^* = 16.66\% > 15\%$$

Therefore alternative A is accepted rather than the existing situation. It now becomes the defender. The challenger is alternative B. To decide between A and B, incremental analysis must be used. Figure 9.3 shows the incremental cash flow $B - A$. Its internal rate of return is

$$0 = -23 + 4(P/A, i^*, 10) + 13(P/F, i^*, 10)$$

$$i^* = 15.28\% > 15\%$$

Thus alternative B is better than alternative A by a narrow margin. Alternative B becomes the defender against C, the challenger. The incremental cash flow $C - B$ appears in Fig. 9.3. Its internal rate of return (IROR) is found from

$$0 = -27 + 3(P/A, i^*, 10) + 17(P/F, i^*, 10)$$

$$i^* = 8.63\% < 15\%$$

Therefore, alternative C loses out to B and exits the example. Defender B remains the defender. Now B is challenged by D. The incremental cash flow of Fig. 9.3 results in the equation

$$0 = -57 + 10(P/A, i^*, 10) + 37(P/F, i^*, 10)$$

$$i^* = 15.89\% > 15\%$$

and D is successful over B. It becomes the surviving defender and the choice.

Had incorrect methods been followed, usually using the criterion of favoring the highest-IROR project, the selection would have been A, as shown:

$$\text{IROR}_A = 16.66\% \qquad \text{IROR}_C = 15.43\%$$
$$\text{IROR}_B = 16.48\% \qquad \text{IROR}_D = 16.33\%$$

■

■ **Example 9.5.** Five projects have been put forward as alternative means of correcting a technical problem in the Number One Crude Unit of a refinery in Saudi Arabia. The five are shown below:

Alternative	First cost (000,000 riyals)	Net yearly benefit (000,000 riyals)
1	38	4
2	30	2
3	45	9
4	50	10
5	42	8

The economic life of all alternatives is 12 years. The opportunity cost of capital after taxes and inflation have been accounted for is 15 percent. All estimates of cost and benefit may be used without adjustment for inflation or taxes. Use the IROR method.

Solution. Once again, the decision flowchart of Fig. 9.1 is followed. To order the alternatives requires that we change the order of the preceding table thus:

Alternative	First cost (000,000 riyals)	New yearly benefit (000,000 riyals)
A	30	2
B	38	4
C	42	8
D	45	9
E	50	10

The alternatives are now in correct order. Figure 9.7 illustrates them. Assuming that all alternatives may be rejected, the existing situation becomes an alternative. We must test the first alternative *A* against the existing situation by finding its incremental rate of return against the null alternative.

$$0 = -30 + 2(P/A, i^*, 12)$$

$$(P/A, i^*, 12) = 15.0000$$

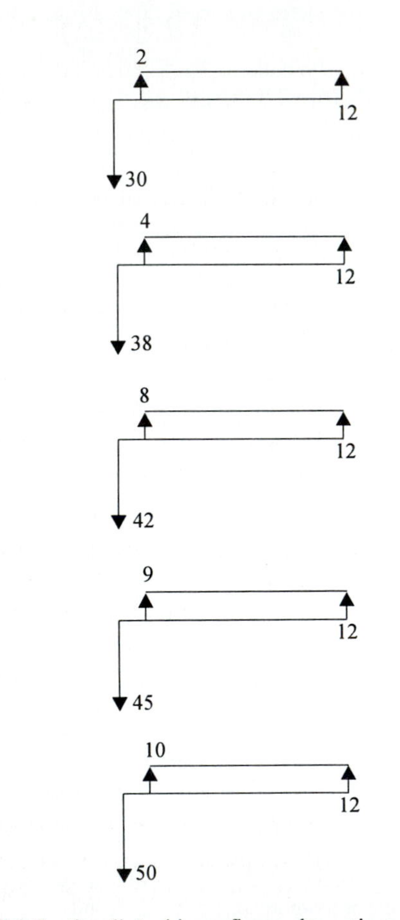

FIGURE 9.7 Saudi Arabian refinery alternatives

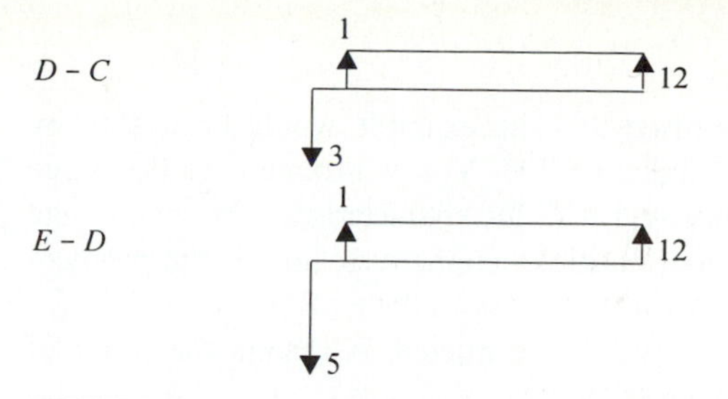

FIGURE 9.8 Incremental cash flow of Example 9.5

Investigation in the compound-interest tables of the appendix reveals that at no interest rate where N equals 12 under the P/A column can we arrive at 15.0000. The internal rate of return appears to be negative, and this is confirmed when we realize that at 0 percent, the cost of −30 exceeds the benefits of 24 (2 times 12 years). Thus alternative A fails. The existing situation becomes the defender.

Is the next alternative B better than the defender? To find out, we set

$$0 = -38 + 4(P/A, i^*, 12)$$
$$(P/A, i^*, 12) = 9.5000$$
$$i^* = 3.8\% < 15\%$$

Alternative B is not better than the null alternative. Now B exits, joining A, and the existing situation remains the defender. Now C challenges the existing situation.

$$0 = -42 + 8(P/A, i^*, 12)$$
$$(P/A, i^*, 12) = 5.2500$$
$$i^* = 15.8\% > 15\%$$

Thus, C is successful and becomes the new defender.

Any more alternatives? Yes, there are. We now must test incrementally D against C. Figure 9.8 shows the resulting incremental cash flow. The testing equation is

$$0 = -3 + 1(P/A, i^*, 12)$$
$$(P/A, i^*, 12) = 3.0000$$
$$i^* = 32.2\% > 15\%$$

So D wins out over C.

The last alternative E enters to challenge the new defender D. The incremental cash flow $E - D$ is shown in Fig. 9.8. Testing gives

$$0 = -5 + 1(P/A, i^*, 12)$$
$$(P/A, i^*, 12) = 5.0000$$
$$i^* = 16.9\% > 15\%$$

and E wins out over D and becomes the defender and the choice among the alternatives.

No demonstration of incorrect methods will be presented here. The reader will discover that if individual IRORs are calculated, E and D tie for winner—once again the incorrect answer. ■

COST-EFFECTIVENESS

In Example 7.7, page 134, we promised that the example would be treated by using a minimum acceptable cutoff index of 275. You will remember that three alternative weapon systems were proposed with the characteristics shown on page 135. Given what you now know about multiple alternatives, how is the problem attacked?

Using the flowchart of Fig. 9.1, we are instructed to change the order of alternatives, starting with 3, then 2, then 1.

The existing situation is not an alternative. One of the systems will be chosen. The first alternative, number 3, becomes the defender, because we start by choosing the minimum-cost alternative. The challenger is number 2. The incremental analysis reveals that an extra (\$7.5 – \$6) \$1.5 billion spent will result in an extra 500 tanks destroyed or 333 tanks per \$1 billion.

$$333 > 275$$

Therefore system 2 is better than system 3. The defender exits, and challenger 2 becomes the defender. One more challenger, system 1, exists. The incremental cost-effectiveness index is

$$\frac{1{,}800 - 1{,}500}{9 - 7.5} = \frac{300}{1.5}$$
$$= 200 < 275$$

The challenger fails, and system 2 is selected.

SUMMARY

Multiple alternatives are defined as more than two excluding the null alternative. They require a special decision analysis which is presented in essence in Fig. 9.1.

Multiple alternatives must be analyzed incrementally. In the present-worth (net present-value) method, both individual and incremental analysis can be used. Both are based on incremental judgments. The comparison of the present worth of individual alternatives is more common. In the annual-worth method, the same statements hold true.

In both the benefit/cost ratio method and the internal rate of return method, incremental analysis must be used. Individual analysis often yields incorrect solutions.

Cost-effectiveness of multiple alternatives follows the same rules as the more usual methods. The major difference is that the minimum acceptable cost-effectiveness index must be determined by the judgment of knowledgeable persons in the area concerned.

PROBLEMS

9.1. In Example 9.1, page 192, the excess of annual revenues over annual disbursements in plan B is now 30 and in plan C is now 35. Which plan will be selected if all other inputs remain unchanged? Use the *incremental present-worth method.*

9.2. Five plans have been proposed to perform a certain task. The time horizon of the analysis is 5 years. The minimum attractive rate of return for this government agency is 15 percent. The total investment necessary for each plan, the annual net receipts, and the salvage value at the end of 5 years are shown below:

	Plan				
	1	**2**	**3**	**4**	**5**
Investment	50	60	70	80	100
Annual net receipts	15	18	20	25	28
Salvage value	10	12	15	18	20

Which of the five plans should be chosen, using (*a*) the *incremental* NPV method and (*b*) the *individual* NPV method?

Answer: Plan 4

9.3. Suppose you can choose only one of the mutually exclusive plans *A*, *B*, *C*, and *D*. Their present values in dollars at various interest rates are as follows:

	Interest rate (percent)				
Plan	**0**	**5**	**10**	**15**	**20**
A	500	280	50	–150	–400
B	350	280	200	140	80
C	250	200	170	140	100
D	180	80	–50	–140	–280

(*a*) If your MARR = 20 percent, what plan will you choose?
(*b*) To choose plan *B*, how much must your MARR be?

9.4. Alternative *A* in Example 9.2, page 196, is now estimated to cost $23,185. All other elements of the situation remain the same. Will the selection change? If so, to which alternative? Use the *incremental net present-worth method.*

9.5. Three makes of jet engines are to be analyzed for use in a new short-range, high-capacity aircraft. Their characteristics are as follows:

	Make		
	1	2	3
First cost (000,000 francs)	2.6	3.0	2.3
Fuel cost per year (000,000 francs)	0.25	0.16	0.29
Maintenance cost per year (000,000 francs)	0.1	0.1	0.15
Life (years)	4	4	3
Salvage value (000,000 francs)	0.2	0.2	0.2

The opportunity cost of capital is 25 percent before taxes and inflation are considered. Perform an analysis, using this discount rate, by the NPV method. Which make is most economical?

9.6. Five alternative high-rise complexes have been proposed for 331 Bush Street in San Francisco. These are asymmetric configurations with numerous structural irregularities. The economic characteristics of each complex are as follows. (All sums are in millions of dollars.)

Complex	First cost	Net income per year	Resale value in 50 years
1	35	6	10
2	32	5	10
3	37	7	10
4	43	8	10
5	34	6	10

All complexes will have a 50-year life. Inflation and taxes, including income taxes, have been included in the estimates. The opportunity cost of capital is 8 percent. Which alternative, *if any*, should be chosen?

Use the *incremental* NPV method.

9.7. In Example 9.1, page 192, which plan will be selected by the *individual annual-worth method*, if the excess of annual revenues over annual costs is 30 in plan *B* and 35 in plan *C*? All other data remain the same.

9.8. Regal Printing Company is in the market for new hand pallet trucks. Three makes, of 2-ton capacity, are under consideration with data as follows:

	Make		
	White	Yellow	Green
First cost ($)	2,000	1,800	1,500
Maintenance cost per year ($)	220	250	300
Life (years)	10	9	7
Salvage value ($)	200	180	150

A before-tax, before-inflation method will be used. You, as the analyst, are required to use the annual-cost method. No other method will be acceptable by management.

An opportunity cost of capital of 15 percent should be employed. Remember, no tax or inflation considerations need be included.

One truck is certain to be chosen. Which one?

9.9. In Example 9.3, page 198, the motor powered by natural gas is now, by means of a government subsidy, to cost 7,500 pesos. The subsidy is designed to encourage the use of such low-pollution motors. Will this change the decision if no other data are altered? Use the individual annual-worth method.

9.10. In an oil refinery, a proposal is received by the engineer in charge from the refining division for evaluation. The opportunity cost of capital is 35 percent. All installations have a zero salvage value. Since this operation is outside the U.S. and since no local taxes are charged, a before-tax analysis will suffice. Inflation did not exist in the country.

Which of the four alternative makes of high-speed pump should be chosen? It is unnecessary to investigate the null alternative because one make of pump is sure to be chosen. Make your choice via the *annual-worth method.*

Pump	First cost	Annual cost	Economic life
A	450,000	70,000	5
B	560,000	60,000	7
C	590,000	45,000	9
D	640,000	40,000	10

9.11. It is often said that an expensive car is justified by its high resale value. To test this supposition, a prospective buyer decides to investigate three cars, all used, made in 1992, four-door models with automatic transmission, power steering, power brakes, and air conditioning. She selects a time horizon of 5 years. After expending some effort, she determines the facts contained in the table at the top of page 212.

The buyer intends to pay cash for the automobile selected from the table. She has no preference vis-à-vis ownership, and all three cars are assumed equally safe. The maintenance cost and all other fixed and operating costs are assumed to be the same among the cars.

Her opportunity cost of capital, after inflation has been taken into account, is 9 percent. All costs have been deflated to 1995 dollars. Thus inflation need not be handled further.

Assume that the new automobile will be driven half its yearly mileage of 10,000 miles in the city and half in the country. Using the *annual-cost method*, determine which car should be purchased. No tax effects will be considered.

Car facts

	Mercedes-Benz 300D	Chevrolet Cavalier	Volvo 244
First cost ($)	24,536	6,168	12,985
Mileage per gallon of fuel			
City	17.1	15.1	10.6
Highway	28.7	32.4	19.9
Gasoline cost per gallon ($)	—	1.20	1.20
Diesel fuel cost per gallon ($)	1.27		
Maintenance expectation	Very good	Unknown	Good
Resale value after 5 years ($)	7,300	740	3,225
Insurance cost per year ($)			
Liability	190	190	190
Collision and comprehensive	412	182	296
	602	372	486

9.12. In Example 9.1, page 192, the null alternative is now excluded. The excess of annual revenues over annual cost is 30 in plan *B* and 35 in plan *C*. No other data changes occur. If the benefit/cost ratio method is used, which alternative will be chosen?

9.13. In Example 9.1, page 192, the first costs are revised to 150, 160, 175, and 200 for plans *A*, *B*, *C*, and *D*, respectively. All other data remain the same. Use the benefit/cost ratio method and the rule of delta.

9.14. Consider the case of the optimal number of berths in a projected port. The benefits and costs of each mutually exclusive alternative discounted to its present value are shown below, in millions.

Number of berths	Total benefits	Total costs
1	2.5	1.0
2	4.5	1.5
3	6.0	2.0
4	6.9	2.5
5	7.5	3.0
6	7.8	3.5
7	8.0	4.0

Select the optimal number of berths to be built by

(*a*) Net present-worth method

(*b*) Benefit/cost ratio method

9.15. The Bureau of Reclamation, Department of the Interior, had under consideration four possible dam designs for a site on the Snake River. All bureau projects have to be justified by a *B/C* ratio computed at 7 percent. The costs and benefits over a life of 100 years of these dams were as shown:

Dam	Height (feet)	First cost ($000,000)	Net benefits per year ($000,000)
1	250	45	3.80
2	300	54	4.40
3	350	60	5.00
4	400	63	5.20

No salvage value was included, nor was inflation considered. If none of the designs is acceptable, the site will be abandoned. Which design, if any, should be chosen?

9.16. In Example 9.4, page 201, the first costs of alternatives 1, 2, and 3 have been raised to 55, 98, and 110, respectively. In addition, the first-year benefits are now estimated at 4, 9, and 7, respectively, for alternatives 1, 2, and 3. If all other data remain the same, which alternative should be chosen, if one option is certain to be required? The benefit/cost ratio method must be used.

9.17. One of three types of electrical (12,000-volt) transmission line supports must be chosen for a new urban route. The data on the poles are as shown:

	Wooden poles	Steel structures	Steel poles
Initial cost ($000,000)	−10	−25	−20
Annual maintenance ($000,000)	−1	−2	−3
Life (years)	15	30	30
Salvage value or removal cost ($000,000)	−1	+0.9	+1.2

No income tax or inflation effects will be taken into account at this time. The minimum attractive rate of return is 9 percent.

(*a*) By the annual-worth method, which support should be selected?

(*b*) By the benefit/cost ratio method, which structure should be selected? Assume that all costs appear in the denominator and all revenues in the numerator of the B/C ratio.

9.18. Perform a benefit/cost ratio calculation on three alternative designs for an underwater tunnel project. The opportunity cost of capital is 10 percent over a life of 20 years. No salvage value is to be considered. Inflation is to be ignored for the moment. Data, in millions of dollars, are as follows:

	Alternative 1	Alternative 2	Alternative 3
Investment	130	152	184
Annual net benefits	16	22	26

9.19. Three types of buses are under consideration for a new bus system. The first cost, annual cost, life, and salvage value for each type are given:

Type	First cost ($000,000)	Annual operation and maintenance cost ($000,000)	Life (years)	Salvage value ($000,000)
1	60	2.0	20	6
2	50	2.5	20	5
3	70	0.75	40	10

One of the three types must be chosen. No inflation or taxes need to be considered. Which type should be selected if the minimum acceptable rate of return is 10 percent? A *B*/*C* ratio analysis is required.

9.20. Pima County is considering three mutually exclusive projects which will add to its equipment maintenance facilities. Characteristics are as follows, with money figures in thousands of dollars:

	Project 1	Project 2	Project 3
Initial cost	500	600	700
Life (years)	10	15	30
Annual cost savings	100	125	160
Annual maintenance cost	50	45	60
Salvage value	200	250	310

If the opportunity cost of capital of the county is 11 percent, which alternative, if any, should be chosen? Ignore inflation for the moment. Use the *B*/*C* ratio method.

9.21. The Mexican Secretaria de Comunicaciones y Transportes is considering three mutually exclusive alternatives for a ferry across a river in the state of Chiapas. Data, in millions of pesos, on the three options are as follows:

	Option 1	Option 2	Option 3
Initial cost	550	650	750
Projected life (years)	10	15	30
Annual revenue	110	130	165
Annual cost	55	50	65
Salvage value	200	250	310

If the social opportunity cost of capital for Mexico is 12 percent, which alternative should be chosen, if any? It is possible to reject all alternatives in this situation. Use the *B*/*C* ratio method. Do not consider income tax or inflation.

9.22. Three patented methods to control erosion of a riverbank are being considered by the Corps of Engineers. The first is a fabric foam filled with pumped concrete. The second consists of hand-placed precast concrete blocks. The third method is a system composed of concrete blocks and interconnecting cables installed in factory-preassembled mattresses. Data, with costs in constant dollars, and the methods for a certain job are as follows:

Method	First cost	Life (years)	Maintenance cost per year
1 Fabric foam	4,300,000	20	100,000
2 Concrete block	6,100,000	30	200,000
3 Blocks and cables	7,200,000	60	100,000

The opportunity cost of capital is 9 percent for constant dollars. One method is sure to be chosen. No salvage value or removal cost is estimated. Which system should be chosen if the benefit/cost ratio method is used? Because the project is a public one, no taxes need be considered. Use any reasonable definition of the B/C ratio.

9.23. Five different locations are proposed for a rural road between two small mountain villages in Mexico. Of course, only one location will be chosen. The opportunity cost of public capital is 15 percent. The economic life of the road is 12 years. No salvage value is contemplated. The road locations are estimated to have the following net annual benefits and costs, in millions of pesos:

	Alternative				
	1	2	3	4	5
Investment	30	38	42	45	50
Net annual benefit	2	4	8	9	10

By the *benefit/cost ratio method,* which location should be chosen?

9.24. In the refinery mentioned in Prob. 9.10, a proposal is received to replace valves on the hydroformer unit. Which of the valves should be chosen? Again, it is unnecessary to investigate the null alternative. All other information is the same as in Prob. 9.10. Use the *benefit/cost ratio method.* You may use any definition of benefits and costs you like, but you must describe in words how these will be determined and treated.

Proposal	First cost	Annual cost	Economic life
A	120,000	1,000	4
B	135,000	800	5
C	140,000	1,200	7

9.25. A southern state recently instituted a new highway safety program in order to reduce the shocking fatality rate on its roadways. Four programs were considered, shown in the accompanying table. Also shown are the annual costs of the programs and the projected fatality rates associated with the implementation of each one. Only the three years 1996, 1997, and 1998 are to be considered; that is, the time horizon of the projects is 3 years.

	Program countermeasures	Annual cost ($000,000)	Projected fatalities			
			1995	1996	1997	1998
1	Bridge improvements	16	540	450	357	312
2	Railroad-grade crossing gates	35	210	24	12	4
3	High-hazard illumination	12	284	216	152	110
4	Hazardous-spot improvements	14	246	191	146	120

The state department of transportation uses a 10 percent opportunity cost of capital. Since this is a state program, no income taxes need to be considered. Possible inflation of program costs will also be ignored.

The National Safety Council's estimate of $170,000 as the value of a human life will be used to estimate benefits. The program is scheduled to begin on January 1, 1996. Assume that without the program the 1995 fatality level would be reached in 1996, 1997, and 1998.

By the *benefit/cost ratio method,* which program would be chosen? The null alternative does not exist. Remember, only one of the four programs may be chosen. Use the *rule of Δ* for *B/C* ratios.

9.26. Three types of subway cars are under consideration for a new system in a major Asian city. *A solution based on the B/C ratio is required.* One of the three types must be chosen. No inflation need be considered. Which type should be selected, if the opportunity cost of capital is 15 percent?

Type	First cost (millions)	Annual operation and maintenance cost (millions)	Life (years)	Salvage value (millions)
1	50	2.6	20	5
2	60	1.6	20	6
3	65	0.5	20	10

9.27. Three road locations are under consideration for construction in the Jicarilla Apache Indian reservation in New Mexico. The Bureau of Indian Affairs analyst believes costs and benefits over the 20-year life of the road will be as shown:

Location	First cost ($000)	Annual operation and maintenance costs ($000)	Annual benefits ($000)
1	1,200	90	250
2	1,750	75	295
3	2,000	70	335

The analyst must use the *benefit/cost ratio method* with a 10 percent discount rate. The road will have no salvage value or removal cost. The possibility of the road's not qualifying for construction exists.

(*a*) Which road location would be chosen if the rule of delta were not used to determine the *B/C* ratios?

(*b*) Which road would be chosen if operation and maintenance costs were considered "disbenefits" and were included in the numerator of the *B*/*C* ratio, rather than in the denominator?

Carry out the *B*/*C* ratios to three decimal places.

9.28. In Example 9.1, page 192, the excess of annual revenues over annual costs is 30 in plan *B* and 35 in plan *C*. All other data remain the same. Use the IROR method to repeat the analysis.

9.29. Five plans have been proposed to accomplish the same purpose. The minimum attractive rate of return (MARR) in this situation is 10 percent. All plans have 100 percent resale value at the end of 20 years. Capital available is 1,300,000. One of the plans will be selected as the solution. The following table summarizes the plans.

	Plan				
	1	**2**	**3**	**4**	**5**
Total investment ($000)	800	1,000	1,100	1,150	1,300
Annual net receipts ($000)	200	210	230	240	250

(*a*) Use the benefit/cost ratio method.
(*b*) Use the internal rate of return (IROR) method.

9.30. An automobile motor-rebuilding company is analyzing the addition of new computerized cylinder-boring equipment. It may or may not choose one of the three alternatives described in the following table. Money figures are in thousands of dollars.

	Alpha	Beta	Gamma
First cost	20	35	50
Net annual profit	8	7	12
Life (years)	4	8	6
Salvage value	3	0	0

The opportunity cost of capital of this company is 30 percent. No income taxes or inflation will be included at this time. The internal rate of return method must be used. Which alternative, if any, should be chosen?

9.31. Repeat the analysis in Example 9.5 using the *incorrect individual IRORs*. What is your answer? Why is it incorrect?

9.32. A real estate developer is considering investment in a skyscraper in a major U.S. city. Designs, cost estimates, and net profit figures have been developed for three heights of building as follows:

Number of floors	First cost ($000,000)	Net annual benefit ($000,000)
96	360	53
100	380	58
108	410	62

The economic life of the building will be 50 years, and the salvage value plus land at the end of that period is believed to be $40 million.

A before-tax calculation is desired. The net annual benefit does not include any allowance for depreciation or capital recovery; it is annual revenue minus annual cost.

By the *internal rate of return method*, which height of building should be chosen? The opportunity cost of capital is 15 percent before taxes. Interpolate to one decimal point on solving rates of return.

Answer: 100 floors

9.33. Three plans are being considered for extensions to the Washington, D.C., Metro. First costs and benefits are in millions.

	Plan 1	Plan 2	Plan 3
Year(s)		**Costs**	
1	−100	−105	−115
2	−100	−105	−115
3	−100	−105	−115
		Benefits	
4–10	+30	+40	+42
11–20	+20	+10	+15

One of the plans is sure to be chosen. An internal rate of return solution is required. Carry out the rates of return to the nearest percent. All payments are in constant dollars of year 0. The opportunity cost of capital is 12 percent. Which plan will be selected?

9.34. Repeat Prob. 9.29, using the annual-worth method.

9.35. The Oregon state highway department's value engineering team was presented with four likely alternatives for proposed asphaltic concrete overlay for a 9.7-mile section of Interstate 5. Costs were as follows for each alternative:

Method	First cost ($)	Life (years)	Annual maintenance cost ($)
1 Heater scarification	2,200,000	12	180,000
2 Stress-absorbing membrane	1,800,000	10	200,000
3 Cold-planing (1.5-inch depth)	1,500,000	12	300,000
4 Cold-planing (3.5-inch depth)	2,300,000	15	170,000

No salvage value was expected. If the minimum attractive rate of return was 10 percent, which alternative should be chosen? No inflation effects were considered. Select the best analysis method.

9.36. The Port of New Orleans is trying to decide which of three types of cranes to purchase as part of the expansion of its container terminal. The port determined the annual benefit of each crane based on its production capacity. Initial cost, benefits, maintenance cost, all in dollars, and life in years were estimated as follows:

Crane	First cost	Annual benefit	Annual maintenance cost	Life
A	1,500,000	300,000	81,000	12
B	1,100,000	220,000	59,400	10
C	1,750,000	350,000	94,500	15

No salvage value was expected. If the minimum attractive rate of return was 10 percent, which alternative should be chosen? The null alternative is nonexistent. No taxes or inflation will be considered. Select the best analysis method.

9.37. Four makes of elevators are under consideration for use in a high-rise building. Their economic characteristics are as follows:

	Elevator make			
	A	*B*	*C*	*D*
First cost ($000)	300	380	460	500
Operation and maintenance cost per year ($000)	50	35	28	22
Economic life (years)	15	20	28	40

The building owners require a 15 percent rate of return before taxes and inflation are considered. Which make of elevator should be chosen? Select the best analysis method.

CHAPTER 10

EFFECT OF INCOME TAX ON ECONOMIC ANALYSIS

Until this point in our study of engineering economy, we have assumed away several important aspects of reality in capital investment. This was done so that the fundamental points of discounted cash flow analysis could be presented in as simple a form as possible. We have assumed that all alternatives are subject to the same amount of risk, that inflation is nonexistent, and that income tax exerts a negligible effect on choices among mutually exclusive alternatives. Now, the first of three steps back into reality will be taken: We readmit income tax and its effect on decisions. In Chap. 11, inflation will be allowed to assume its proper role in economic analysis. Chapter 12 will return this discussion to complete reality once again as we undertake the study of risk. The complicated problems of the real world, where income taxes, inflation, and risks are present, will then be addressed and solutions proposed.

Why is the influence of income tax matters on economic decisions so important? The answer is straightforward. Income tax is always a cost, and often a large part of the cost, of a private investment. Not to include it is to ignore a vital consequence of an alternative, so vital that the income tax effects of choosing the alternative may well make the difference between acceptance and rejection. Thus, before-tax analyses as bases for decisions are, in the great majority of cases, worse than useless, for they mislead one into believing that proper analytic support exists where it does not.

If income tax is a cost, why may it not be treated as other costs have, such as labor costs or maintenance costs? The reason is that income tax cost is dependent on the amount of all the other costs and on the revenues as well. Both revenues and costs of an alternative must be determined before the consequent income tax can be figured.

In general, income tax is a disbursement (cost) that must be computed based on a cash flow that includes all other payments, both costs and benefits.

Income tax is borne, in the United States, by persons and private corporations. No income tax is paid by public agencies at the federal, state, or local level. No public sector investment will be subject to income tax consequences, and thus income taxation of public investment is nonexistent. Regulated private companies, such as railroads and electric power suppliers, pay income tax, and all investments made by them must consider the tax consequences of an investment.

This is not to say that public investment has no effect on income taxes paid. The public investor does not pay taxes, true, but the investment may have indirect but important effects on the amount of income tax that citizens pay. Consider a highway investment made by a state. The users of the highway experience a lowering of their automobile and truck operating and maintenance costs. Because costs are lowered, net income is increased and more income taxes are paid. In ordinary benefit/cost analysis, however, such secondary effects are not taken into account.

You should understand that what follows in this chapter will not be an exhaustive treatment of income tax laws and their interpretation. The constantly changing income tax law is a lifelong study for many lawyers and accountants. Libraries of books have been written on income tax law, court cases, and accounting aspects and conventions. This chapter will devote itself to enough commentary and illustration of how income tax and economic decisions are related to bring about an appreciation of that relation's importance and to develop a methodology for the quantitative analysis of that relation. No more than this can be hoped for in a text on essentials. Questions on the finer points of tax law and tax accounting will be left where they belong—with the experts in the field. And it is to those experts that you are referred.

INCOME TAX

Before we describe the complications of income tax, it seems appropriate to ask, What is meant by income? "Income is the money we make," might be a quick answer. It would also be a true one except for capital gains, to be covered later. Income is divided into personal income and business and professional income. *Personal income* is the sum of wages, salaries, fees paid to us, tips if we occupy a position where we receive them, and so on. *Business income* is the total revenue we receive from a business or profession minus the total cost of conducting it. It is on the sum of personal and business or professional income that we are taxed.

Income tax is the amount of our income that we are required by law to turn over to the government. After certain deductions are made from total income, a figure is arrived at called *taxable income*. A percentage of this taxable income is our income tax. It is not a property tax, which depends on the value of the real or personal property one owns. It is not an inheritance tax, which is based on property and money received from deceased persons. It is not a gift tax. It is not an excise tax, based on the value of purchased goods such as cigarettes and liquor.

Income and the tax on it can be categorized in a number of ways. As mentioned, it can be personal, business, or professional income. If it is business income, it can be derived from an incorporated or unincorporated business. An incorporated business is called a *corporation*, which is a legal person brought into being under the laws of one of the 50 states of the United States. In other countries, a corporation goes by other names: a *limited company* in Great Britain and a *sociedad anonima* in Mexico. Corporations are taxed under different rules than persons.

Another way income tax can be categorized is by the government agency collecting it. In the United States, the most important agency is the Internal Revenue Service (IRS) of the Department of the Treasury of the federal government. Some states of the United States collect income tax. California and Maryland do, for example; Texas does not. Cities can also levy taxes based on income. San Francisco, California, and New York City do. Thus, two or more tax rates may apply to the same income at the same time. Sometimes the applicable tax rate is not the sum of the rates involved. This occurs when one tax is a deductible expense from another jurisdiction's taxable income.

TAX BRACKETS

Income tax is levied by means of percentages, not of total income, but of increments of income or brackets of income. One whose income falls within a bracket is said to belong to a certain percentage tax bracket. As taxable income rises, the percentages attached to each bracket rise also, but at an increasing rate. Table 10.1 shows the tax rates and taxes for corporations for tax year 1994. The first increment of income, or bracket, is $50,000. The next bracket is $25,000.

TABLE 10.1
Federal income tax rates for corporations, 1994[a]

Taxable income	Tax
Not over $50,000	15%
Over $50,000 but not over $75,000	$7,500 plus 25% of excess over $50,000
Over $75,000 but not over $100,000	$13,750 plus 34% of excess over $75,000
Over $100,000 but not over $335,000	$22,250 plus 39% of excess over $100,000
Over $335,000 but not over $10,000,000	$113,900 plus 34% of excess over $335,000
Over $10,000,000 but not over $15,000,000	$3,400,000 plus 35% of excess over $10,000,000
Over $15,000,000 but not over $18,333,333	$5,150,000 plus 38% of excess over $15,000,000
Over $18,333,333	35% of the amount over 0

[a]*Source*: U.S. Department of the Treasury, *Instructions for Forms 1120 and 1120-A*, Washington: Government Printing Office, 1993, p. 15.

Associated with each bracket is a fixed tax plus what an economist would call a *marginal tax rate* to be applied to all amounts within that bracket. For example, if a corporation had an income of $85,000 in 1994, it paid $13,750 plus $3,400, or $17,150, in federal taxes. The corporation was, at that time, in the 34 percent tax bracket. Incomes of over $100,000 put a corporation in the 39 percent tax bracket, where by far the greatest majority of corporations found themselves.

Individual income tax schedules are arranged in a similar manner. They are irregular in their jumps as they rise from bracket to bracket. The marginal tax rate rises, as in the corporate tax structure. This is why income tax is called *graduated.*

Evidently two rates are present in the graduated income tax structure—an average rate and a marginal rate. In the example cited above, the average rate is $17,150 divided by $85,000 times 100, or 20.2 percent. The marginal rate, or bracket rate, is 34 percent. Which is the correct rate to use in after-tax analyses of proposed investments? Surely, the marginal rate because this is the rate at which additional profits will be taxed. Failure to appreciate this point of fundamental economic logic has led to many an error.

DEDUCTIONS

Evidently, any sum that can be subtracted from income will cause a reduction in the amount of income tax one pays. For example, if your business borrows money from a bank and pays interest of $500 per year, under certain conditions that $500 can be subtracted from total income under present IRS rules. The result is that your taxes are reduced by an amount that depends on the tax bracket in which total income places the taxpayer. If that tax bracket rate is, say, 37 percent, then $500 of income will escape taxation at that rate. The interest deduction results in a saving in taxes of 37 percent of $500, or $185. Put another way, the cost of your loan before its effect on income taxes is considered is $500 per year. When the effect on income taxes is taken into account, the loan costs $500 less $185, or $315. The implications of this simple idea are truly earthshaking, as we will see.

All subtractions from total income reduce income tax that must be paid. The IRS recognizes many kinds of subtractions. Subtractions that are not the result of business operations are called *deductions*. Some of these deductions are medical and dental expenses, sales tax, interest on loans and mortgages, property tax on one's residence, charitable contributions, losses due to theft, and many more. Just the deductions allowable because homeowners pay interest on a mortgage and property taxes on their houses result in a tremendous encouragement for home ownership. It is essentially a subsidy by the federal government to support a hallowed U.S. institution, one of the cornerstones of the American dream—a home of one's own.

Some of the items mentioned in the previous paragraph are unrelated to capital investment—medical expenses, for example. Others are related, if one considers home ownership an investment—mortgage interest and residential taxes, for example.

■ **Example 10.1.** Guy Gibson is considering purchasing a residence. It will cost $200,000. He will put 20 percent down. The remaining $160,000 he will finance with a 13 percent, 30-year mortgage. Property taxes are estimated at $2,000 per year.

(*a*) What will Gibson be able to deduct from his taxable income during his first year of ownership as a result of the house purchase?

(*b*) What amount of money will Gibson save in income tax if his applicable tax bracket is 43 percent?

Solution

(*a*) The annual interest amount will be calculated as though Gibson's house payments occurred only once a year. Normally, such payments would be made monthly.

$$\begin{array}{rl} (160{,}000)(0.13) = \$20{,}800 & \text{interest} \\ \underline{+2{,}000} & \text{property taxes} \\ \$22{,}800 & \end{array}$$

(*b*) Gibson's income tax bill will be reduced by

$$(\$22{,}800)(0.43) = \$9{,}804$$ ■

In business and professional income tax accounting, however, subtractions from income are simply expenses. Wages and salaries, rent, utility payments, interest on debt, overhead, real estate taxes, insurance, and so forth are all expenses that can be deducted from revenue, that is, income. Giving workers a raise results in a higher tax deduction and thus lower income tax for your business.

But there is another class of expenses that is not simply deducted from your annual revenue; these are capital expenditures. Investment in a piece of machinery that will last several years cannot be accounted for under IRS rules by deducting its cost from the revenue of the year in which it is purchased.[1] Its cost must be spread over several years. This process of allocating the cost of capital goods over their lives is called *depreciation*. Depreciation may be deducted from income, and thus the tax effects of capital investment enter into economic analysis. These are of great importance.

The IRS uses another word, *amortization*, to describe similar deductions for capital investment such as business start-up expenses, reforestation, and others. Discussion of such matters is left for more advanced texts. The principles are unchanged.

All the expenses mentioned above can have an effect on a firm's income tax bill, and thus all need to be considered. A new piece of machinery can affect wages, insurance, utility payments, maintenance, and so forth. It can result in depreciation expense as well. All these are deductions from income, and all affect income tax. A simple example will show how all these business expenses are handled in conventional accounting practice.

---

[1]The cost of certain qualifying property, called by the IRS *section 179 property*, may be treated as an expense rather than a capital expenditure. See page 227 for further discussion.

■ **Example 10.2.** In 1993, Jovita Fuentes began a small landscaping business after her graduation as a landscape architect. At the end of the first year, her profit and loss statement was as follows:

Income		$56,000
Less operating expenses		
Disbursements for rent, wages, maintenance, etc.	$14,000	
Depreciation on truck, trailer, mowers, tools	$10,000	
		$24,000
Profit		$32,000

The $56,000 figure needs no explanation. It is simply the total of the monies clients paid. The $14,000 is the sum of $6,000 in rent for a small building used exclusively for the business, $4,000 for part-time help by high school students in mowing lawns, $3,200 for repairs and maintenance of a pickup truck and trailer, and $800 for other expenses. All these are cash outflows from her business, no more subject to change at her discretion than the receipts of $56,000. All are deductible. But the $10,000 depreciation charge is another matter.

Fuentes must use the *modified accelerated cost recovery system* (MACRS) for depreciation. She was able to claim a 5-year depreciation period for all assets she had acquired, according to the advice of her tax preparer. This choice resulted in the $10,000 depreciation charge. On looking into the matter a little further, she discovered that she could have chosen an alternate MACRS method, using a 5-year life, that would have changed her depreciation from $10,000 to $5,000, thus increasing the profit to $37,000. Had she chosen a 15-year life for her truck, mowers, and tools, the depreciation charge would have lessened and her profits increased correspondingly. Her taxes would also have increased.

It was now clear to Fuentes that, unlike the expenses for rent, wages, etc., the depreciation expense could be changed as she wished, as long as she stayed within IRS rules. It also was clear that she should choose the depreciation so as to reduce profit to a minimum because that reduced her taxes to a minimum. Fuentes saw that the money in her bank balance was paramount in importance, not profits on an accountant's profit and loss statement. Before she paid her taxes, that amount was $42,000, her before-tax cash flow, calculated by subtracting operating expenses ($14,000) from income ($56,000). Her taxes were minimized by choosing the depreciation method that gave the largest depreciation. She therefore stayed with the MACRS 5-year life depreciation ($10,000). We shall go into the mechanics of the depreciation methods and resulting charges later in this chapter.

Fuentes realized a number of important business lessons from this first encounter with income taxes as applied to business operations:

1. Depreciation is not a cash flow.
2. To minimize taxes, which are a cash flow, assets should be depreciated by using the largest depreciation possible as early in the life of the asset as possible under IRS rules.
3. Her choice of a depreciable life had nothing whatever to do with the real useful life of the truck, mowers, and tools. All these would be in use far longer than 5 years because Fuentes took personal pride in the things she owned and used.
4. Therefore, the depreciable life was simply the time over which she recovered the capital costs. The depreciation charges had nothing to do with the market value of her truck, mowers, or tools. ■

"Section 179 of the Internal Revenue Code permits certain taxpayers to *elect* to deduct all or part of the cost of certain qualifying property in the year they place it in service, instead of taking depreciation deductions over a specified recovery period."[2] In other words, an asset can be expensed in 1 year, rather than depreciated over a number of years. This, of course, can be much to the taxpayer's advantage. The option is limited in a number of ways; for example, real property is excluded. What can and cannot be deducted may be ascertained by referring to the publication cited.

VALUE: BOOK AND MARKET

We will not spend a great deal of time with the philosophical concept of value. Value is so subjective, so changeable with time, indeed, such a capricious and fleeting wraith, that we will leave it to the real estate appraisers, art dealers, rare book sellers, and all like them to ponder. We will be concerned with ideas as hard as concrete.

The *book value* of an asset is its first cost minus the total depreciation taken against it to date. Immediately it can be seen that we are dealing with accounting allocation of cost—numbers in books of accounts or in computer storage files. In economic analysis, the only use of book value is to compute taxes; just how, we will see later in this chapter.

The *market value* is the amount of money that will be received if the asset is sold. It is the opportunity cost of the asset. In other words, it is the amount of money the owner gives up by keeping the asset rather than selling it. Here, the value is in real dollars and cents, or pesos and centavos. In contrast to book value, it is an amount that will directly influence the cash flow.

It has been suggested that a much more descriptive phrase than *book value* is the term *unamortized cost*. The amount of the first cost of an asset not yet "dead" is all it is. (*Unamortized* has its roots in the same place as *mortal*, *mortician*, and *moribund*. Most authors, however, use *book value* for the concept.)

DEPRECIATION

Depreciation is the recovery of capital costs over statutory recovery periods. This is the definition that arises out of the Economic Recovery Tax Act of 1981. It underlines the fact that depreciation is not concerned with the useful life of an asset. It is not involved with salvage value. Both these variables, useful life and salvage value, are difficult to predict. Consider the errors in useful life that were

---

[2]U.S. Department of the Treasury, *Depreciation*, IRS Publication 534. Washington: Government Printing Office, 1994, pp. 70–80.

evident when handheld electronic calculators replaced cumbersome mechanical ones. And what is the "salvage value" of a piece of residential real estate in a good neighborhood whose resale value has risen since the Second World War? Methods of depreciation in use before 1981 retained vestiges of the idea that it was a method of valuation. A salvage value had to be assigned that, on the basis of experience, would be the market value of the asset at the end of its useful life. An attempt was made to estimate useful life as well. All this was done with the notion in mind that book value ought to approximate market value. Many were the attempts made to accomplish this. As we will see, a number of depreciation methods, now happily disappearing, were developed, and several attempts were made at classifying assets according to useful life. Yet continually intruding and supported by real-world experience was the idea that depreciation ought to be only an allocation of capital cost over a period of time. These two opposing points of view might be called the economist's and the accountant's. This is not the place to describe the pros and cons of each. The argument is, and remains, absorbing.

At the moment of writing, the attempt to change depreciation accounting has complicated an already complicated situation. At least seven depreciation methods are now in use—two post-1986, two 1981–1986 methods, and three pre-1981 methods. All these will be examined in subsequent pages.

DEPRECIATION: MODIFIED ACCELERATED COST RECOVERY SYSTEM

The *modified accelerated cost recovery system* (*MACRS*) came into being as a result of the Tax Reform Act of 1986. It replaced the *accelerated cost recovery system* (*ACRS*), which was mandatory for property placed in service after 1980 and before 1987. The provisions of the Tax Reform Act of 1986 were further changed by the Technical and Miscellaneous Revenue Act of 1988. Further changes took place up to the present (1995). Remember, however, that once a depreciation method is chosen, a change to another method may not be allowed or may be subject to special rules.

The Tax Reform Act of 1986 provided a modified accelerated cost recovery system for all tangible property placed in service after December 31, 1986. MACRS consists of two systems: the *general depreciation system* (*GDS*) and the *alternative depreciation system* (*ADS*). GDS is the one generally used. Under it, most property, via class life, is assigned to eight property classes: 3-year, 5-year, 7-year, 10-year, 15-year, 20-year, nonresidential real property, and residential real property. Some 3-, 5-, and 7-year assets are shown in Table 10.2. The recovery period under the MACRS column determines the property class.

ADS generally provides longer recovery periods. It uses a straight-line method of depreciation that excludes salvage value.

TABLE 10.2
MACRS class lives and recovery periods: Selected items

Asset class	Description of assets included	Class life	Recovery periods	
			GDS	ADS
00.11	Office furniture	10	7	10
00.12	Information systems: computers	6	5	5
00.22	Automobile, taxis	3	5	5
00.241	Light general-purpose trucks	4	5	5
00.242	Heavy general-purpose trucks	6	5	6
15.0	Construction equipment	6	5	6
20.5	Returnable pallets	4	3	4
34.0	Manufacture of fabricated metal products	12	7	12
37.31	Ship and boat building machinery and equipment	12	7	12
45.0	Air transport	12	7	12
	Personal property with no class life	—	7	12

Source: U.S. Department of the Treasury, *Depreciation*, IRS Publication 534, Washington: Government Printing Office, 1994, pp. 70–80.

Personal Property

Personal property includes all business property, tangible or intangible, that is not real estate. Tangible property can be seen or touched; intangible property, such as a franchise, cannot be seen or touched.

The first step in computing MACRS depreciation, once you have decided it can be used and whether it falls under GDS or ADS—presumably with the assistance of a tax expert—is to gather the following information:

1. *Basis*. This is usually the original cost of the asset. However, if the property was inherited, received as a gift, or received in some other way, a tax expert should be consulted.
2. *Property class and recovery period*. This is done by referring to the Table of Class Lives and Recovery Periods in the latest IRS Publication 534, entitled *Depreciation*. In that publication will be found more than 130 asset classes, along with a description of the assets included, the class life, and the recovery periods in years under the MACRS general depreciation system (GDS) and the alternative depreciation system (ADS). Once the asset class is determined, use the MACRS recovery period assigned to that asset class. (See Table 10.2. The asset selection was partially determined by its use in the examples and problems of this chapter.)

   The depreciation deduction is determined by using the applicable depreciation method and convention: For property in the 3-, 5-, 7-, and 10-year classes, you may use the double (200 percent) declining-balance method over 3, 5, 7, or 10 years. For property in the 15- or 20-year class, you may use the 150 percent declining-balance method.

   You may use the 150 percent declining-balance method for property, other than residential rental and nonresidential real property, that is placed in

TABLE 10.3
GDS half-year depreciation rates (percent)

	Property class (GDS recovery period)					
Year	**3-Year**	**5-Year**	**7-Year**	**10-Year**	**15-Year**	**20-Year**
1	33.33	20.00	14.29	10.00	5.00	3.750
2	44.45	32.00	24.49	18.00	9.50	7.219
3	14.81	19.20	17.49	14.40	8.55	6.677
4	7.41	11.52	12.49	11.52	7.70	6.177
5		11.52	8.93	9.22	6.93	5.713
6		5.76	8.92	7.37	6.23	5.285
7			8.93	6.55	5.90	4.888
8			4.46	6.55	5.90	4.522
9				6.56	5.91	4.462
10				6.55	5.90	4.461
11				3.28	5.91	4.462
12					5.90	4.461
13					5.91	4.462
14					5.90	4.461
15					5.91	4.462
16					2.95	4.461
17						4.462
18						4.461
19						4.462
20						4.461
21						2.231

Source: Department of the Treasury, *Depreciation*, IRS Publication 534. Washington: Government Printing Office, 1994, p. 46.

service after 1986. The recovery period is the property's ADS period. (See Table 10.2.) If the property does not have a class life assigned to it, the recovery period is 12 years. For 15- and 20-year property, the 150 percent declining-balance method may be used with the GDS recovery period determining the number of years over which depreciation may be taken. For the 150 percent tables, see IRS Publication 534.

In any of the above methods, a switch to the straight-line method is ordinarily made for the first tax year for which the switch gives an equal or larger depreciation than the declining-balance method. If the MACRS percentage tables are used (Tables 10.3, 10.4, and 10.5), there is no need to determine in what year your deduction is greater by switching to the straight-line method. The tables have the switch to the straight-line method built into their percentages.

ADS must be used for certain property, as was implied above. However, if the property comes under GDS, one can elect to use ADS. We will not cover all depreciation possibilities in this section, only the most frequently used. This will be sufficient for our purposes. Once again, the services of a tax expert are recommended.

3. *Placed-in-service date*. Depreciation begins, in general, when property is placed in service in a trade or business or for the production of revenue. If

property is for personal use, depreciation is not allowed. The important point is that depreciation begins not when the property is purchased, but when its use begins. These dates may be quite different.

4. *Convention to be used.* For both GDS and ADS, the half-year, midquarter, or midmonth convention must be employed.

   The half-year convention is normally used. (Under certain conditions, to be explained in the next section, the midquarter convention must be used.) Only half the allowable depreciation may be taken for the first year, and half may be taken during the year following the end of the recovery period. Table 10.3 contains the depreciation percentages resulting from this convention and the recovery periods. These percentages are applied to the original basis, usually the first cost of the asset. The midmonth convention is used for real estate. It will be illustrated in a later section.

5. *Depreciation method.* Under MACRS, there are five methods to depreciate property. We will concern ourselves with the two oldest as exemplary: the 200 percent declining-balance method over the GDS recovery period that switches to the straight-line method, and the straight-line method over fixed ADS recovery periods.

   For the rest, the reader is once again referred to IRS Publication 534.

■ **Example 10.3.** Office furniture, costing $10,000, bought by Steinhoff Corporation on December 21, 1992, was not placed in service until January 1, 1993, because of installation and delivery delays. What maximum depreciation may Steinhoff take on its income tax returns?

Solution. The property is asset class 00.11, with a class life of 10 years, a GDS recovery period of 7 years, and an ADS recovery period of 10 years. See Table 10.2. GDS is selected. Thus the GDS recovery period makes this 7-year property.

The date placed in service governs the first year that depreciation may be taken, not the date purchased. Thus the depreciation starts in 1993 and runs through 2000, with a half-year in 1993 and 2000.

The declining-balance rate for 7-year property is 200 percent divided by 7, or 28.57 percent. The calculation of the yearly depreciation over 8 years is as follows:

Year	Depreciation	Book value	Half-year convention
0		10,000	—
1	(10,000)(0.2857)(0.5) = 1,429	8,571	Yes
2	(8,571)(0.2857) = 2,449	6,122	No
3	(6,122)(0.2857) = 1,749	4,373	No
4	(4,373)(0.2857) = 1,249	3,124	No
5	(3,124)(0.2857) = 893 or 3,124/3.5 years = 893	2,231	No
6	892	1,339	No
7	893	446	No
8	446	0	Yes

Notice that the depreciation checks with Table 10.3. ■

Midquarter Convention

The use of this convention, under the Technical and Miscellaneous Revenue Act of 1988, supersedes the use of the half-year convention when the total basis of depreciable property placed in service during the last 3 months of the tax year exceeds 40 percent of the total basis of *all* depreciable property placed in service during that year. Real property is excluded from the calculation. For tax years beginning after March 31, 1988, property placed in service and disposed of in the same tax year is not counted in determining whether the midquarter convention must be used rather than the half-year convention.

To calculate the GDS deduction for property subject to the midquarter convention:

1. Calculate the depreciation for a full year.
2. Multiply that amount by the following percentages according to the quarter of the tax year in which the property is placed in service.

Quarter	Percentage
1	87.5
2	62.5
3	37.5
4	12.5

In this way, the percentages of Table 10.3, based on the half-year convention, may be used to convert to the midquarter convention for the years where the double-declining balance method is used. The straight-line portion of the table must be computed separately, as in the next example.

TABLE 10.4
GDS midquarter depreciation rates: Selected property classes (percent)

Year	3-Year	5-Year	7-Year	Year	3-Year	5-Year	7-Year
	First quarter				Second quarter		
1	58.33	35.00	25.00	1	41.67	25.00	17.85
2	27.78	26.00	21.43	2	38.89	30.00	23.47
3	12.35	15.60	15.31	3	14.14	18.00	16.76
4	1.54	11.01	10.93	4	5.30	11.37	11.97
5		11.01	8.75	5		11.37	8.87
6		1.38	8.74	6		4.26	8.87
7			8.75	7			8.87
8			1.09	8			3.33
	Third quarter				Fourth quarter		
1	25.00	15.00	10.71	1	8.33	5.00	3.57
2	50.00	34.00	25.51	2	61.11	38.00	27.55
3	16.67	20.40	18.22	3	20.37	22.80	19.68
4	8.33	12.24	13.02	4	10.19	13.68	14.06
5		11.30	9.30	5		10.94	10.04
6		7.06	8.85	6		9.38	8.73
7			8.86	7			8.73
8			5.53	8			7.64

Source: Department of the Treasury, *Depreciation*, IRS Publication 534. Washington: Government Printing Office, 1994, pp. 46–48.

■ **Example 10.4.** Your company has put in service in 1993 the following personal, as opposed to real, property: a milling machine, costing $40,000, in January; office furniture, costing $10,000, in September; personal computers, costing $50,000, in October. Compute the depreciation.

Solution. The total basis of all property placed in service is $100,000. The computers' basis for depreciation is over 40 percent of the total basis of all property placed in service that tax year:

$$\frac{50{,}000}{10{,}000+40{,}000+50{,}000}=0.50>0.40$$

The depreciation rate for a full year for the milling machine, which is asset class 34.0, a 7-year property, is

$$\frac{200\%}{7}=28.57\%$$

Because it was placed in service in the first-quarter, its first-year depreciation rate is

$$28.57\times 87.5\%=25\%$$

The first-year depreciation is

$$(0.25)(40{,}000)=\$10{,}000$$

The remaining percentages can be calculated from Table 10.3 and straight-line depreciation by using the midquarter convention. The same rates also appear in Table 10.4 in the first quarter.

Year	Depreciation rate (percent)	Depreciation
1	25.00	10,000
2	24.49 × 87.5% = 21.43	8,572
3	17.49 × 87.5% = 15.31	6,124
4	12.49 × 87.5% = 10.93	4,372
5	27.33 / 3.125% = 8.75	3,500
6	8.74	3,496
7	8.75	3,500
8	100.00 − 98.91% = 1.09	436
	100.00	40,000

The office furniture, a 7-year property, will be depreciated using the percentages in Table 10.4, third quarter.

Year	Depreciation rate (percent)	Depreciation
1	10.71	1,071
2	25.51	2,551
3	18.22	1,822
4	13.02	1,302
5	9.30	930
6	8.85	885
7	8.86	886
8	5.53	553
	100.00	10,000

The personal computers will be assigned the following depreciations by using Table 10.4, fourth quarter, 5-year property.

Year	Depreciation rate (percent)	Depreciation
1	5.00	2,500
2	38.00	19,000
3	22.80	11,400
4	13.68	6,840
5	10.94	5,470
6	9.58	4,790
	100.00	50,000

■

Real Property

Residential real property is depreciated over 27.5 years. Nonresidential real property is depreciated over 31.5 years if placed in service before May 13, 1993, or over 39 years if placed in service after May 12, 1993. See Table 10.5 for residential real property and nonresidential real property with a 31.5-year depreciation schedule. Nonresidential real property with a 39-year depreciation schedule has a separate schedule; see IRS Publication 534.

■ **Example 10.5.** Werner Molders placed in service an apartment building costing $250,000 on March 15, 1993. What depreciation amounts can Molders take for the MACRS recovery period?

Solution. March is the third month of Molders' tax year. Table 10.5 shows that he is able to take 2.879 percent for 1993, or $7,198.

$$\text{For 1994 to 2019,} \quad (250{,}000)(0.03636) = \$9{,}090$$
$$\text{For 2020,} \quad (250{,}000)(0.02576) = \$6{,}440$$
$$\text{For 2021,} \quad (250{,}000)(0.00000) = 0$$

The total depreciation sums to $249,978. Because of the rounding error, $22 worth of depreciation has been lost. ■

DEPRECIATION: ADS METHOD

For most property, the ADS method may be used at the taxpayer's discretion. Depreciation is calculated by using the straight-line method with no salvage value under the applicable convention. See Table 10.2 for recovery periods under the ADS method.

The midmonth convention is used for nonresidential real estate and residential property. As already stated, the half-year or midquarter convention is used for all other property.

TABLE 10.5

MACRS real property depreciation rates (percent)

Year(s)	Month placed in service											
	1	2	3	4	5	6	7	8	9	10	11	12
Residential—27.5 years												
1	3.485	3.182	2.879	2.576	2.273	1.970	1.667	1.364	1.061	0.758	0.455	0.152
2–27	3.636	3.636	3.636	3.636	3.636	3.636	3.636	3.636	3.636	3.636	3.636	3.636
28	1.970	2.273	2.576	2.879	3.182	3.485	3.636	3.636	3.636	3.636	3.636	3.636
29	0.000	0.000	0.000	0.000	0.000	0.000	0.152	0.455	0.758	1.061	1.364	1.667
Nonresidential—31.5 years												
1	3.042	2.778	2.513	2.249	1.984	1.720	1.455	1.190	0.926	0.661	0.397	0.132
2–31	3.175	3.175	3.175	3.175	3.175	3.175	3.175	3.175	3.175	3.175	3.175	3.175
32	1.720	1.984	2.249	2.513	2.778	3.042	3.175	3.174	3.175	3.174	3.175	3.174
33	0.000	0.000	0.000	0.000	0.000	0.000	0.132	0.397	0.661	0.926	1.190	1.455

Source: Department of the Treasury, *Depreciation*, IRS Publication 534. Washington: Government Printing Office, 1994, pp. 48–49. The table is not exact because the original tables make-up for rounding errors by changes in the third decimal place in the body of the tables. The error is very small, however.

ADS is required for some property, for example, any tangible property used predominantly outside the United States during the year, any tax-exempt use property, and some others.

■ **Example 10.6.** Linda Tse buys a barge, costing $60,000, for use in her ports and harbors construction business. She decides to depreciate it by the ADS. How much depreciation can she take per year over what years?

Solution. Table 10.2 shows, under asset class 15.0, the applicable recovery period to be 6 years. The straight-line depreciation rate is 100 percent/6 years, which equals 16.67 percent. Because she qualifies to use the half-year convention, her depreciation is as follows:

Year(s)	Depreciation
1	(0.5)(0.1667)(60,000) = 5,000
2–6	(0.1667)(60,000) = 10,000
7	(0.5)(0.1667)(60,000) = 5,000
	Σ = 60,000

■

MACRS DISPOSITIONS

Abandonment of a property allows its book value to be deducted as a loss at the time of abandonment.

Disposal of a property before the end of its recovery period, that is, early disposition, permits depreciation to be taken in the year of disposal. The amount of depreciation depends on the convention used when the property was placed in service.

For property depreciated according to the half-year convention, only half the depreciation is allowed for the year of disposition.

For property depreciated according to the midquarter convention, a percentage of the full-year depreciation is allowed depending on the quarter in which disposal occurs. See Table 10.6.

Residential rental property and nonresidential real estate use a midmonth convention. For example, real property sold in March will take 2.5/12 of that year's depreciation.

■ **Example 10.7.** On November 15, 1993, J. B. McCudden placed in service a piece of 5-year property, costing $10,000. Because this was the only new asset acquired during 1993, he had to use the midquarter convention for the fourth quarter. He sells the property on April 12, 1995. What is his allowable depreciation?

Solution. Using Table 10.4 and the percentage for second-quarter disposal from Table 10.6, we get the following:

Year	Depreciation
1993	(0.05)(10,000) = 500
1994	(0.38)(10,000) = 3,800
1995	(0.375)(0.228)(10,000) = 855

The 1995 deduction took 37.5 percent of the full year's depreciation. ■

TABLE 10.6
Midquarter convention, allowable year-of-disposition depreciation

Quarter	Percentage
1	12.5
2	37.5
3	62.5
4	87.5

Source: Department of the Treasury, *Depreciation*, IRS Publication 534. Washington: Government Printing Office, 1994, p. 25.

AFTER-TAX CASH FLOW ACCORDING TO MACRS

The foregoing pages of this chapter have led up to this section. The effect of income taxes on economic analysis was the subject of our investigation. Now it is time to observe that effect.

Returnable pallets used in the food and beverage industry may be imagined; see Table 10.2, asset class 20.5. The percentages of Table 10.3 are employed. No salvage value is anticipated. Table 10.7 illustrates the situation.

Profit equals revenues minus costs. Included in the costs, and thus already deducted, were such items as labor, operation and maintenance, interest, all taxes except income tax, and all other costs attributable to this investment.

Depreciation has been taken according to MACRS GDS rules. See Table 10.7. Columns 1 and 2 are added algebraically to obtain column 3, the taxable income. A tax rate of forty-eight percent of column 3 results in column 4, the income tax payable. To obtain the after-tax cash flow, columns 1 and 4 are added. Notice that these are cash flows. Columns 2 and 3 are not. They serve only as numerical stepping stones to arrive at cash flows.

The effect of income tax is to reduce the positive cash flow and thus to make all the indicators of economic worth less favorable. How much less favorable is the question, however, and column 5 answers it.

TABLE 10.7
Income tax effect according to MACRS

Year	Cash flow (1)	Depreciation (2)	Taxable income (3) = (1) + (2)	Income tax payment (4) = −0.48 × (3)	After-tax cash flow (5) = (1) + (4)
0	−1,000				−1,000
1	+500	−333	+167	−80	+420
2	+500	−445	+55	−26	+474
3	+500	−148	+352	−169	+331
4	+500	−74	+426	−204	+296

DEPRECIATION: ACCELERATED COST RECOVERY SYSTEM

The *accelerated cost recovery system* (*ACRS*) was used for many assets placed in service after December 31, 1980, and before January 1, 1987.

MACRS supplanted ACRS. However, taxpayers must continue to use whatever depreciation method they started with, including ACRS, declining balance, etc.

For decisions, its use is almost nil. Although it has some historical interest, a knowledge of ACRS may be useful only in certain situations such as replacement analysis where the defender was depreciated by ACRS. It is therefore excluded from further discussion here. The reader who is interested in more information on ACRS is referred to the first edition of the present textbook.

PRE-1981 DEPRECIATION METHODS

You may wonder why it is necessary to explain pre-1981 depreciation methods. If the law now requires depreciation to be performed by MACRS, then what is the use of studying former methods? Here are five reasons:

1. All property acquired before December 31, 1980, must continue to be depreciated according to pre-1981 rules for depreciation.
2. Many states have not switched to the new rules. Thus the federal tax return must be prepared according to the new rules, while the state tax return must be made out according to the old ones.
3. Some foreign countries still may require the use of methods that will be explained in the following pages.
4. An appreciation of what depreciation is and the implications of the different methods cannot be reached unless, at least, the most important, widely-used methods are understood.
5. Some assets must still be depreciated by pre-1981 methods. Patents and copyrights must use the straight-line method, for example.

A description of the three principal pre-1981 methods—straight-line, declining-balance, and sum-of-the-years'-digits—follows. In these methods the basis, useful life, and salvage value must be estimated. The *basis* is normally the first cost of the asset. The *useful life* is an estimate of the number of years the property will be of use to its owner. This life may vary from user to user. The *salvage value* is the owner's estimate of its value on the market when it is no longer of use. In all methods, yearly depreciation must be prorated for the first and last years of use according to the month in which the asset is bought or sold. The examples that follow assume that all assets are bought in January and sold in December.

DEPRECIATION: STRAIGHT-LINE

The simplest of the pre-1981 methods is straight-line depreciation. (It is not the same as MACRS straight-line depreciation, which does not take into account salvage value.) It *must* still be used for certain classes of property, a franchise, for example, or computer software developed before August 11, 1994, where MACRS does not apply under the present (1993) laws. To calculate the yearly depreciation charge, the analyst subtracts the estimated salvage value from the basis, usually the first cost, and divides the result by the estimated years of useful life.

■ **Example 10.8.** In 1994, Wanda Hershey purchased an ice cream franchise. It cost $6,000. She believed that she could sell it for $2,000 in 3 years. She must use straight-line depreciation. How much could she deduct from her taxable income each year as a result of depreciation on this franchise?

Solution

$$\frac{6{,}000 - 2{,}000}{3} = \$1{,}333 \text{ per year}$$

As a percentage rate per year this is

$$\frac{1{,}333}{6{,}000} \times 100 = 22.2\%$$

The depreciation and the book value that results are shown in Table 10.8. The book value, you will recall, is the first cost of an asset less accumulated depreciation. ■

DEPRECIATION: DECLINING-BALANCE

Another pre-1981 method still required by Internal Revenue Service regulations is the declining-balance method. Like the pre-1981 straight-line method, it can be used to depreciate intangible property and certain other property that the taxpayer can elect to exclude from MACRS.

In the declining-balance method, the annual depreciation charge is a fixed percentage of the previous year's book value, an amount which declines year by year, hence the name. The salvage value is not considered, except in the sense that the asset cannot be depreciated to a book value below the expected salvage value.

TABLE 10.8
Straight-line depreciation

Year	Depreciation	Book value
0		6,000
1	1,333	4,667
2	1,333	3,334
3	1,334	2,000

The fixed percentage is tied to the straight-line rate, or rather, what the straight-line rate would be if the salvage value were neglected. There are three main rates possible under this method: 200, 150, and 125 percent. Which percentage may be selected for a certain asset is best left to the reader's tax lawyer or accountant. To determine the fixed percentage in the double-declining-balance method, divide 200 percent by the estimated life. For example, if the life of the asset is 5 years,

$$\frac{200\%}{5} = 40\%$$

If the 150 percent declining-balance method is selected for a 5-year life, the yearly percentage is

$$\frac{150\%}{5} = 30\%$$

■ **Example 10.9.** If Wanda Hershey of the previous example used the double-declining-balance method to compute depreciation, what was her annual depreciation on the $6,000 franchise with a 3-year life?

Solution. Her yearly fixed percentage is

$$\frac{200\%}{3} = 66.667\%$$

For the first year, her depreciation was

$$(6{,}000)(0.66667) = \$4{,}000$$

The book value at the end of the first year is

$$\$6{,}000 - \$4{,}000 = \$2{,}000$$

The second-year depreciation is

$$(2{,}000)(0.66667) = \$1{,}333$$

but none of this was taken because it would have reduced the book value below $2,000. For the same reason, no depreciation was taken in the third year. These results are shown in Table 10.9. ■

TABLE 10.9
Double-declining-balance depreciation

Year	Depreciation	Book value
0		6,000
1	4,000	2,000
2	0	2,000
3	0	2,000

DEPRECIATION: SUM-OF-THE-YEARS'-DIGITS

The *sum-of-the-years'-digits* (*SYD*) method is now obsolete. It is not mentioned as a possible method for any new-asset depreciation in the latest (1994) Publication 534 of the Internal Revenue Service. It is included here because it still may be in use for assets placed in service before 1981, and thus it will still appear on the books of some firms. It may also be used outside the United States.

Like the declining-balance method, the SYD method allows larger deductions in early years and smaller deductions in later years. It, too, is an accelerated depreciation method. But unlike the declining-balance method, it uses an estimated salvage value to compute the depreciable value, that is, the total amount that may be depreciated. The basis for computing depreciation is the first cost less the salvage value, as in the straight-line method.

The sum of the years' digits is exactly that. For example, a 3-year life gives a sum of the years of that life of 1 + 2 + 3, or 6. The formula is

$$\text{SYD} = \frac{N^2 + N}{2}$$

where N = estimated life.

The depreciation charge for any year is obtained by multiplying the basis by a fraction, the numerator of which is the remaining life of the asset and the denominator is the SYD.

■ **Example 10.10.** Hershey's franchise of the previous two examples shows the depreciation charges of Table 10.10, computed as follows for the SYD method:

Solution

$$\begin{aligned}\text{SYD} &= \frac{N^2 + N}{2} \\ &= \frac{3^2 + 3}{2} \\ &= 6 \\ \text{Depreciable value} &= 6{,}000 - 2{,}000 = 4{,}000\end{aligned}$$

First year's depreciation is

$$\frac{3}{6}(4{,}000) = 2{,}000$$

TABLE 10.10

Sum-of-the-years'-digits depreciation

Year	Depreciation	Book value
0		6,000
1	2,000	4,000
2	1,333	2,667
3	667	2,000

Second year's depreciation is

$$\frac{2}{6}(4{,}000) = 1{,}333$$

Third year's depreciation is

$$\frac{1}{6}(4{,}000) = 667$$

Notice that the total amount of depreciation taken is $4,000, the depreciable value, just as in the straight-line method. ■

You will not be surprised to learn that still other depreciation methods exist: the sinking fund method and the units-of-production method among others. As of 1994, you must use MACRS if the asset is eligible for MACRS. None of these less-used methods will be considered here. Once again, consult your tax lawyer or accountant.

The prospective taxpayer should be aware, however, that it is possible to change from one depreciation method to another. It has always been possible to change from the declining-balance method to the straight-line method without the permission of the IRS, and often this combination results in greater after-tax economies, as we have seen in MACRS. How such decisions are made will be discussed in the section entitled "Selection of a Depreciation Method." However, "The various elections made under MACRS, once made cannot be changed."[3]

AFTER-TAX CASH FLOW USING PRE-1981 STRAIGHT-LINE DEPRECIATION

After-tax cash flow using straight-line depreciation is best explained by means of an example—that of Table 10.7, page 237. However, a salvage value will now be included. Let us estimate the salvage value at $200 at the end of 4 years. The before-tax cash flow then appears as in Table 10.11, with the $200 salvage value shown on a separate line. The reason for a separate line will be clear shortly.

The straight-line depreciation is

$$\frac{1{,}000 - 200}{4} = \$200 \text{ annually}$$

This is shown in column 2. Taxable income, column 3, is the algebraic sum of columns 1 and 2. The tax, at 48 percent of taxable income, appears in column 4. Cash flow after taxes is the algebraic sum of the cash flow columns 1 and 4.

---

[3]*Source*: Department of the Treasury, *Depreciation*, IRS Publication 534. Washington: Government Printing Office, 1994, p. 43.

TABLE 10.11
Income tax effect using pre-1981 straight-line depreciation

Year	Cash flow (1)	Depreciation (2)	Taxable income (3)	Income tax payment (4)	Cash flow after taxes (5)
0	-1,000				-1,000
1	+500	-200	+300	-144	+356
2	+500	-200	+300	-144	+356
3	+500	-200	+300	-144	+356
4	+500	-200	+300	-144	+356
4	+200				+200

The salvage value is not taxed, but is brought directly from column 1 to column 5. The reason it is not taxed is that the salvage value represents what remains of the capital portion of the investment. It is not income and, therefore, is not subject to income tax. To emphasize this, it is shown on a separate line.

AFTER-TAX CASH FLOW USING DOUBLE-DECLINING-BALANCE DEPRECIATION

The same example used in the foregoing section will now be employed to illustrate the income tax effect of using double-declining-balance depreciation. Table 10.12 shows the situation, with column 1 as before.

The fixed percentage for double-declining-balance depreciation is

$$\frac{200\%}{4} = 50\%$$

Columns 2 and 3 show the depreciation and the resulting book value after the depreciation change is made. Column 2 is subtracted from the previous year's entry in column 3 in order to write in the current year's entry in column 3. Notice that the asset cannot be depreciated below its salvage value of $200.

The income tax payment is computed by taking 48 percent of column 4, the taxable income. Cash flow after taxes is the algebraic sum of columns 1 and 5, the cash flows.

TABLE 10.12
Income tax effect using double-declining-balance depreciation

Year	Cash flow (1)	Depreciation (2)	Book value (3)	Taxable income (4)	Income tax payment (5)	Cash flow after taxes (6)
0	-1,000		1,000			-1,000
1	+500	-500	500	0	0	+500
2	+500	-250	250	+250	-120	+380
3	+500	-50	200	+450	-216	+284
4	+500	0	200	+500	-240	+260
4	+200					+200

AFTER-TAX CASH FLOW USING THE SUM-OF-THE-YEARS'-DIGITS DEPRECIATION

Again, the same example is used as in the previous two sections. Table 10.13 shows the before-tax cash flow in column 1. The sum of the years' digits is

$$\text{SYD} = \frac{4^2 + 4}{2}$$
$$= 10$$

The first year's depreciation is

$$\frac{4}{10}(800) = 320$$

The second year's depreciation is

$$\frac{3}{10}(800) = 240$$

The third year's depreciation is

$$\frac{2}{10}(800) = 160$$

The fourth year's depreciation is

$$\frac{1}{10}(800) = 80$$

The taxable income, tax at 48 percent, and the cash flow after taxes are calculated as before.

This concludes the explanation of the principal depreciation methods and their effect on after-tax cash flows. We will now use this information in a practical way.

TABLE 10.13
Income tax effect using sum-of-the-years'-digits depreciation

Year	Cash flow (1)	Depreciation (2)	Taxable income (3)	Income tax payment (4)	Cash flow after taxes (5)
0	−1,000				−1,000
1	+500	−320	+180	−86	+414
2	+500	−240	+260	−125	+375
3	+500	−160	+340	−163	+337
4	+500	−80	+420	−202	+298
4	+200				+200

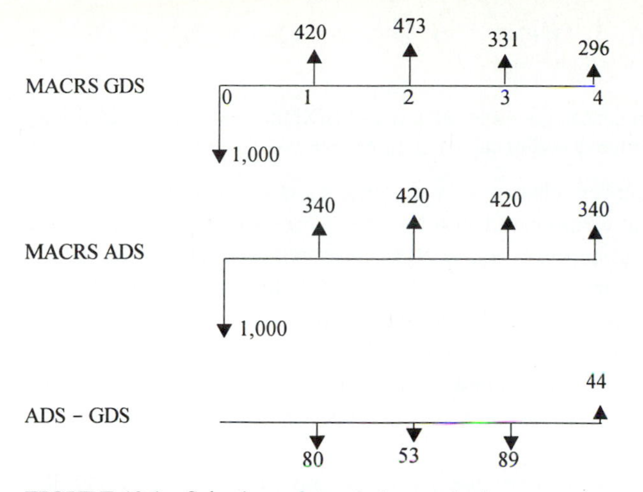

FIGURE 10.1 Selecting a depreciation method

SELECTION OF DEPRECIATION METHOD

On what basis is one to choose one depreciation method rather than another, when both are possible choices under the law? Two points should be noted:

1. These are mutually exclusive alternatives. Therefore, it would be incorrect to calculate an after-tax rate of return from a cash flow obtained by using one depreciation method and compare it to a rate of return based on an alternative depreciation method. Incremental analysis is the correct approach in such a problem.
2. The advantage, in terms of net present worth of choosing one method or another, depends on the taxpayer's opportunity cost of capital.

Let us see how the choice is made via an example.

■ **Example 10.11.** Townsend Company has been attempting to decide on a depreciation method for an asset acquired in 1994. The asset falls under the 3-year property category by IRS rules. It must be depreciated by the MACRS GDS or the ADS. The half-year convention will be used. The asset will be retained in service for 4 years and sold at the end of that time. The timing of the sale will be such that 4 years of depreciation may be taken under both methods. Its first cost is $10 million, its salvage value is zero, and its acquisition will result in profits for Townsend Company of $5 million per year. Townsend Company has an after-tax opportunity cost of capital of 15 percent and is in the 48 percent tax bracket.

Solution. The cash flows after taxes are illustrated in Fig. 10.1. (The reader is invited to verify them.)

The analyst decides to use the incremental net present-value method. The incremental cash flow (ADS − GDS) is shown in the bottommost diagram of Fig. 10.1. The incremental NPV at 15 percent of that cash flow is

$$\text{NPV} = -80(P/F, 15, 1) - 53(P/F, 15, 2) - 89(P/F, 15, 3) + 44(P/F, 15, 4)$$
$$= -143$$

Thus—the minus sign indicates it—it is better to use GDS than ADS for this cash flow with an after-tax opportunity cost of capital of 15 percent. ■

What may not be entirely clear is why a company would decide to choose a depreciation method that would result in a lower net present worth rather than a higher one. Consider that because depreciation is shown as a cost on the profit and loss statement of a company, large depreciation deductions in earlier years can cause the company to make less accountant's profits. This could result in difficulties in the sale of stock to the public, in raising more capital through loans, and so forth. Or, for a conglomerate, lower accountant's profits for one of the companies in the conglomerate may result in higher after-tax profits for the conglomerate as a whole.

You are now aware that the more quickly depreciation is taken, the greater the after-tax cash flow earlier in time. Thus, greater net present worth follows from accelerated depreciation, *ceteris paribus*. The criterion is to select the depreciation method that results in the greater net present worth of the after-tax cash flow. The method thus determined will always be selected no matter what the opportunity cost of capital, as long as it is greater than or equal to zero, again *ceteris paribus*.

LIVES

The *technical life* of an asset is the period of time during which it can provide service. A piece of production machinery can continue to punch out blanks for 10 years with only routine maintenance required. A DC-8 airliner can continue to operate, hauling passengers on the South American service, for 20 years before wearing out. But *economic* life can be different from technical life. *Economic life* is the period during which an asset is kept in service because of the superior economic benefits it is believed to furnish over any other feasible alternative. The punch press continues in service only 5 years, at which time it is replaced by another make that uses less labor. The DC-8, although it could easily continue carrying passengers for another 10 years, is replaced by the more economical DC-10. Economic life can be the same as technical life or less than technical life. It can *never* be greater than technical life.

MACRS recovery periods are the years allowed to recover the investment in the asset via depreciation. They have some relation to economic life, but it is a tenuous one. Everyone knows that a residential rental property, in the majority of cases, will last more than 27.5 years, for example; yet 27.5 years is the recovery period allowed.

Economic life is usually the basis for the time horizon over which an economic analysis is made. It may be longer than, shorter than, or the same as the MACRS recovery period. It is perfectly possible to retain a piece of 5-year recovery property for 10 years. We may, if we wish, take all the depreciation in the first 6 years and zero depreciation for the next 4 years. Or we may spread the depreciation over 10 years by using the ADS. What we may not do is depreciate the asset after it has left our service.

CAPITAL GAINS OR LOSSES ON DISPOSITION

The laws on taxation of capital gains or losses on the disposition of assets are complex and changeable. They involve such matters as these:

1. What assets are capital assets—and thus subject to rules on capital gains or losses; and what assets are not capital assets—and thus to be treated in a different way?
2. What constitutes a long-term capital gain? For businesses, taxation of a long-term capital gain is managed differently from that of a short-term capital gain.[4] As of this writing, a long-term capital gain was one arising from disposition of a capital asset held over one year. A short-term capital gain arose from a capital asset held less than 1 year.
3. What percentage of long-term capital gains is to be taxed and at what rate?.
4. How are long- and short-term capital losses to be treated? When the sale price of an asset exceeds its book value, the excess is taxed. When the reverse is true, a tax benefit occurs. When the sale price equals book value at the time of sale, there are no tax effects. The IRS rules for taxation on disposition of an asset must be determined by consultation with a tax expert—usually a lawyer or accountant specializing in taxation. When the prevailing IRS rules have been determined and the accounting information has been collected, the economic analysis can proceed. How such an analysis is performed is shown in Example 10.21.

For corporations, the Tax Reform Act of 1986 stipulated that all long-term capital gains are to be treated as ordinary income for the year of disposal. The rule continued for 1994 tax filings.

Capital losses for corporations were deductible only to the extent of capital gains. For 1994, any amount of capital loss may be carried forward until it is completely used up.[5]

■ **Example 10.12.** Bishop Cab Company is contemplating the acquisition of 300 new taxis estimated to cost $3,000,000. These cabs will bring in $800,000 annually in net profit to the company. It is expected that they will be sold at the end of their third year of use for $2,100,000.

Depreciation will be according to the MACRS, with a 5-year recovery period and the half-year convention.

---

[4]U.S. Department of the Treasury, *Tax Guide for Small Business*, IRS Publication 334. Washington: Government Printing Office, 1994, p. 112.
[5]*Ibid.* p. 113.

One half-year's depreciation will be allowed during the third year of use because the cabs will be disposed of during that year. Table 10.3 shows the allowable depreciation percentages.

Bishop Cab Company is in the 45 percent income tax bracket and requires a 15 percent after-tax rate of return on investments.

Capital gains for the year of disposal will be taxed, if Bishop Cab's tax attorney's guess as to the law 3 years hence is correct, at 40 percent of the gain taxed as ordinary income. Should Bishop Cab buy the new taxis? Calculate the internal rate of return before and after taxes and the NPV after taxes at 15 percent.

Solution. Table 10.14 shows the results of the analyst's calculations.

The second-year calculations deserve some comment. The $960,000 depreciation for that year is 32 percent of the original investment of $3,000,000. The book value at the end of the second year is $3,000,000, less the accumulated depreciation of $1,560,000, or $1,440,000. The taxable income is a negative $160,000 (800,000 − 960,000), indicating that this amount may be subtracted from the net profits of the remainder of Bishop Cab's operations. It is believed that this $160,000 will be exceeded by Bishop Cab's net profits so that the tax advantage will be realized. The tax advantage is $72,000. The total after-tax cash flow for the second year is $800,000 net profit plus $72,000 tax advantage, resulting in $872,000.

$$\text{Capital gains } = 2{,}100{,}000 - 1{,}152{,}000 = +948{,}000 \times 0.40 = 379{,}200$$

The capital gain on disposition of the assets is $2,100,000 less $1,152,000, or $948,000. Only 40 percent of this amount is taxable—$379,200, shown in auxiliary line 3. Taxing it at 45 percent gives a tax of $170,640.

The internal rate of return of the before-tax cash flow is

$$0 = -3{,}000{,}000 + 800{,}000(P/A, i^*, 3) + 2{,}100{,}000(P/F, i^*, 3)$$
$$i^* = 18.3\%$$

The internal rate of return of the after-tax cash flow is

$$0 = -3{,}000{,}000 + 710{,}000(P/F, i^*, 1) + 872{,}000(P/F, i^*, 2) + (569{,}600 + 1{,}929{,}360)(P/F, i^*, 3)$$
$$i^* = 13.7\% < 15\%$$
$$\text{NPV}_{15\%} = -80{,}145$$

Bishop Cab Company should not purchase the new cab fleet as offered because the rate of return of the investment is less than the required 15 percent.

This example makes evident the important point that there is no general relationship between the before- and after-tax rates of return. Each determination of after-tax rate of return is a specific one. ■

INVESTMENT TAX CREDIT

Once again, the reader is cautioned to determine the prevailing IRS rules before attempting an analysis. In general, an investment tax credit means that taxpayers

TABLE 10.14
Bishop Cab Company's investment

Year	Cash flow	Depreciation (percent)	Depreciation ($)	Book value at end of year	Taxable income	Income tax payment @ −0.45	After-tax cash flow
0	−3,000,000						−3,000,000
1	+800,000	20.00	−600,000	2,400,000	+200,000	−90,000	+710,000
2	+800,000	32.00	−960,000	1,440,000	−160,000	+72,000	+872,000
3	+800,000	$\frac{19.20}{2}$	−288,000	1,152,000	+512,000	−230,000	+569,600
3	+2,100,000	—	—	—	+379,200	−170,640	+1,929,360
							+1,080,960

may subtract some percentage of the first cost of an asset from their tax payments for the year in which the asset is purchased. This is equivalent to a reduction in the cost of the asset by the amount of the investment tax credit.

A question immediately arises as to how the investment tax credit affects the depreciable value of an asset. This has been answered by the IRS in various ways at different times. Current rules on the matter must be ascertained by the analyst before the analysis can be started.

Although the Tax Reform Act of 1986 removed the investment tax credit, it may well reappear whenever Congress so ordains. As of this writing (1995), it has not done so.

ALTERNATIVE MINIMUM TAX

Completely apart from what has been said so far in this chapter, another way of computing tax liability is possible for taxpayers. It is the *alternative minimum tax*.

"The tax laws give special treatment to some kinds of income and allow special deductions and credits for some kinds of expenses. So that taxpayers (including corporations) who benefit from these laws will pay at least a minimum amount of tax, a special tax has been enacted—the 'alternative minimum tax' for corporations and individuals."[6]

Of interest to students of engineering economy is that taxpayers are required to use a special application of depreciation rules and conventions when computing the alternative minimum tax. Suffice it to say that the principles explained in this text are still valid for the computation. The reader is referred to the publication *Tax Guide for Small Business*. (See References at the chapter's end.)

INCOME TAX IN OTHER COUNTRIES

The discussion and examples in this book are based on the income tax laws in the United States at the time of writing, 1995. The income tax laws of other countries are different from U.S. laws and thus must be ascertained before an analysis can be begun that involves them. This seems obvious. What is often not so obvious is the difference between laws as written and laws as enforced. In many countries, the body of income tax laws presents as formidable a facade as that of the United States. But their enforcement is almost nonexistent because of lack of resources or will or some other reason. It behooves anyone doing an analysis of projects in other countries to inquire into the customs of those countries, particularly with regard to tax enforcement. This means establishing a relationship with those persons likely to be thoroughly acquainted with such matters within the country.

---

[6]U.S. Department of the Treasury, *Tax Guide for Small Business*, IRS Publication 334. Washington: 1994, p. 180.

The graduated income tax appears to be a feature of most countries. Sometimes, however, the tax may be determined not by the bracket system but by a precise point on a curve. For each income a specific tax can be cited, according to the equation of the curve relating income and income tax. The principle remains the same, however. The higher the income, the greater the percentage of that income is paid to the government as tax.

SUMMARY

The most important point made in this chapter is that income tax is a cost, particularly in developed countries, that must be incorporated into economic decisions. The amount of income tax that an individual or corporation pays depends on the tax bracket associated with the income.

Deductions from income reduce the taxpayer's liability. Business expenses are tax-deductible, as are interest payments and depreciation. The last is important because it is connected with capital investment.

Two general depreciation systems are now possible in the United States: post-1986 and pre-1981. The categories include the modified accelerated cost recovery system (MACRS) GDS, ADS, and, among others, straight-line, and declining-balance methods. Given a before-tax cash flow and a depreciation method, the after-tax cash flow can be obtained. It is this after-tax cash flow that determines the decision on investment.

Capital gain or loss on the disposition of an asset is important because income tax to be paid is affected. The same is true of investment tax credits.

Individual countries will have different rules for computing income tax. These rules must be well understood before an analysis is undertaken.

REFERENCES

U.S. Department of the Treasury. *Your Federal Income Tax*, IRS Publication 17. Washington: Government Printing Office. (See latest revision.)

———. *Tax Guide for Small Business*, IRS Publication 334. Washington: Government Printing Office. (See latest revision.)

———. *Depreciation*, IRS Publication 534. Washington: Government Printing Office. (See latest revision.)

———. *Business Expenses*, IRS Publication 535. Washington: Government Printing Office. (See latest revision.)

———. *Sales and Other Dispositions of Assets*, IRS Publication 544. Washington: Government Printing Office. (See latest revision.)

———. *Investment Income and Expenses*, IRS Publication 550. Washington: Government Printing Office. (See latest revision.)

———. *General Business Credit*, IRS Publication 572. Washington: Government Printing Office. (See latest revision.)

PROBLEMS

10.1. Using Table 10.1, page 223, for a tax year starting January 1, 1994, compute the corporate income tax owed to the federal government for an income of $400,000.

10.2. In Example 10.1, page 225, Gibson's new house will cost $450,000. Property taxes will be $3,200. His tax bracket is 35 percent. All other information remains the same, as are the questions.

10.3. Two plans are available for the purchase of a house. One involves a cash payment of $80,000, which sum includes all fees and expenses of purchase. Another plan involves financing the purchase on the following terms:

Down payment	$20,000	
Fees	$4,000	
Yearly payment	$3,000	at end of each year for 20 years with interest at 6%

Interest payments are deductible for the purposes of income tax. The prospective buyer estimates her tax bracket to be 25 percent. In addition, she assumes that she will be able to sell the house at the end of 20 years for $100,000. What decision should be made if the after-tax MARR of this buyer is 10 percent? Assume that the buyer will be able to postpone paying tax on capital gain from the sale. Thus, capital gain is excluded from the problem.

10.4. The Mercedes-Benz 300D of Prob. 6.7 will now be financed via a bank loan. The terms of the loan are a 10 percent down payment, with the remainder of the purchase price paid for via a 15 percent loan for a term of 4 years. The 48 payments normally required will be assumed to be paid in four payments at the end of each year.

Each year, interest and principal will be paid as follows:

Year	Payment	Interest	Principal
1	7,735	3,312	4,423
2	7,735	2,649	5,086
3	7,735	1,887	5,848
4	7,735	1,009	6,726
			22,083

The buyer is in the 43 percent tax bracket. Compute the tax relief for each of 4 years for the buyer. Assume that the car is used for business driving only and that interest is tax-deductible.

10.5. Your business loan of $3,825, originally to be paid off in 36 payments, requires 13 more monthly payments of $132.59 to be paid off completely. Your next payment is due soon. The bank states in the payment book on the stub for the next payment, "Your balance after this payment will be $1,594.20." If the interest on your loan is a nominal 15 percent:

(*a*) Is $1,594.20 a correct figure? If not, what is?

(*b*) Your MARR is a nominal 12 percent. Should you pay off the loan after the next payment, tax and inflation considerations aside?

Answer: Yes

(*c*) You are in the 32 percent tax bracket. Should you pay off the loan? Your MARR before taxes is a nominal 18 percent and after taxes is a nominal 12 percent. Assume that the 12 remaining payments after your next one will occur during a single taxable year and that your taxes will be paid exactly at the end of that year. Ignore inflation effects.

Answer: No

10.6. In Example 10.2, page 226, Jovita Fuentes decides to determine the effect of using a 15-year life for her depreciable property. The result is a depreciation of $2,500. By how much does this affect her profit? Is not the increase in profit sufficient reason to use the longer depreciable life?

10.7. According to the Tax Reform Act of 1986, the MACRS GDS depreciation for 3-, 5-, 7-, and 10-year classes of depreciable property is based on double-declining-balance depreciation rates with conversion to straight-line depreciation at the year when straight-line depreciation gives the greater deductible amount. In 5-year class property, show that the percentages in Table 10.3 follow this rule.

10.8. In Example 10.3, page 231, Steinhoff Corporation places the furniture in service the day after it is received. Its basis is its cost of $10,000. The corporation's tax lawyer advises that the MACRS GDS, using the half-year convention, is appropriate for this asset. What yearly depreciation may now be taken on the furniture?

10.9. Systems Engineering, Inc., is about to buy new office furniture for $30,000. The company will use MACRS GDS, with the half-year convention, to compute depreciation.

(*a*) How much depreciation, in dollars, will the company be able to take on this investment each year if the furniture remains in service for 9 years?

(*b*) If the company is in the 36 percent tax bracket, how much yearly tax relief will it receive if it purchases the office furniture?

(*c*) What is the present worth of this tax relief, if the company's opportunity cost of capital is expected to be 20 percent after taxes are considered and 35 percent before taxes are considered?

10.10. The Boelcke Shipping Company is buying new returnable pallets. It must use the MACRS GDS depreciation. It qualifies to use the half-year convention.

The pallets cost $5,250. Maintenance and storage costs are estimated at $500 per year. The pallets will be disposed of at the end of the year, *after* the ultimate allowable depreciation is taken. The cost of disposal will be $450.

The company is in the 37 percent bracket for income taxes. If the after-tax opportunity cost of capital for the company is 15 percent, what is the uniform annual after-tax cost of the pallets for each year of their use?

10.11. R. Bong and Associates propose to invest $40,000 in new computer hardware to augment the memory of existing mainframe equipment. It is estimated that $15,000 annually will be saved during the 6-year life of the new hardware. MACRS GDS with the half-year convention will be employed for depreciation calculations. The resale value will be zero. The depreciable life will be exactly the same as the economic life. If R. Bong and Associates finds itself in the 34 percent corporate income tax bracket, calculate the after-tax rate of return on this investment to the nearest whole percentage. (Compare to Prob. 10.12.)

10.12. A proposal to invest $40,000 in a piece of construction equipment was to be analyzed. Annual net benefits were $15,000 per year for 4 years. The machine was depreciated by the sum-of-the-years'-digits method. The salvage value was zero. The economic life and the depreciable life were assumed to be the same. The company was in the 48 percent tax bracket for income tax purposes. Calculate the before- and after-tax rates of return on this investment to the nearest 1 percent. (Compare to Prob. 10.11.)

10.13. Repeat Prob. 10.11, but instead of computing the internal rate of return of the after-tax cash flow, calculate the net present value of the after-tax cash flow if the MARR (opportunity cost of capital) of the firm is 18 percent. Will the project be accepted?

10.14. In Example 10.4, page 233, suppose that your company's cost for the computers turned out to be only $30,000 when all discounts came in. You must file a correction to your company's income tax return. You will, of course, take the most advantageous depreciation policy available to your company. What will be your new depreciation amounts year by year for the three assets? All other information remains the same.

10.15. Using the midquarter convention, compute the depreciation rates with MACRS GDS for 3-year property placed in service during the first quarter of the tax year.

10.16. In Prob. 10.11, assume that the investment in new computer equipment is made during the last quarter of the year, with no other investments made during that year. All other information in the problem remains the same.

Under this new condition, make appropriate changes to the allowable depreciation. Calculate the after-tax rate of return to the nearest whole percentage.

10.17. In Example 10.5, page 234, Werner Molders changed his mind and bought an office building rather than an apartment house. What annual depreciation amounts can he now take?

10.18. On July 14, 1991, Guenther Rall purchased and placed in service a residential apartment building costing $1 million, not including the cost of the land. He used Table 10.5 to compute his depreciation. He sold the property on March 26, 1993. What depreciation per year was he able to take during his ownership of the property?

10.19. As the outright owner of a small residential house, Frances Allen Cleveland wants to decide whether to sell it or keep it. It can be sold for $240,000. Selling it would mean paying a real estate agent's commission of 6 percent of the sale price. Because she has depreciated the property fully, capital gains taxes would also have to be paid on the total receipts, after the agent's commission, at 28 percent federal tax, 5 percent state tax, and 2.5 percent county tax.

Keeping the house means the following consequences:

Rent	$1,100 month
Vacancy rate	5%
Maintenance	$2,000 per year
Property taxes	$2,650 per year
Insurance	$237 per year

Cleveland believes that timed mutual fund investments can yield 10 percent on her investment at about the same risk.

(*a*) What rate of return is she making on the rental house, assuming she holds it perpetually?

(*b*) If she retains the present house, with a time horizon of 4 years, how much will she have to sell it for at the end of 4 years to make a 10 percent return? Assume no increase in income or expenses. Be sure to include the real estate agent's commission and the effect of capital gains taxes on 100 percent of her sale receipts.

(*c*) What appreciation rate on the property per year over the next 4 years is implied by your calculation in (*b*)?

10.20. Refer to Prob. 10.19. Will it pay Frances Cleveland to sell the fully depreciated rental house and buy another so that she can start depreciating again? Assume that she invests the full after-tax proceeds of her sale without further financing. Assume also that her income and expenses on the new house will be reduced exactly in proportion to the reduction in the sale price between the present house and the new one. Finally, assume that she uses the second calendar year's MACRS depreciation percentage for her calculations.

10.21. Courtney and Scott, Engineers and Architects, have before them three designs for a new shopping mall. The costs and revenues in millions of current dollars are shown here:

Design	First cost	Annual operation and maintenance	Annual revenues	Resale value
1	15	3	6	22
2	18	4	8	27
3	20	4	10	30

Courtney and Scott are required to present the designs, with an economic analysis, to the owners.

MACRS depreciation percentages are 3.042 percent for the first year of operation, reflecting a January purchase, and 3.175 percent for the following years up to the beginning of the seventh year, when the property will be sold. During the seventh year, 0.132 percent depreciation may be taken, reflecting a January sale. No costs or revenues will be assigned to the seventh year.

Capital gains will be treated as ordinary income for the year of disposal.

The mall owners are in the 40 percent bracket for all income taxes. The opportunity cost of capital is 15 percent on current dollars after income taxes have been taken into account. Assume that inflation is not considered.

According to the *net present-value method*, what design, if any, should Courtney and Scott recommend to the owners?

10.22. In Example 10.6, page 236, Linda Tse decides to examine the results of using MACRS GDS rather than ADS on her new barge. Do this for her.

10.23. Rob Rickenbacker puts a new car in service in his electrical engineering consulting business. It costs $8,000. He intends to hold it for 6 years. Compute the depreciation he can take for each of those years under the following methods:

(*a*) MACRS GDS with the half-year convention
(*b*) MACRS ADS with the half-year convention
(*c*) MACRS GDS with the midquarter convention for property placed in service on April 15 if he uses the calendar year as his tax year

10.24. In a problem similar to Example 10.11, page 245, compare the ADS and GDS depreciation under MACRS by using the data of Example 10.6, page 236. Linda Tse's barge is depreciated by ADS in this example. Use exactly the same data under GDS. Which method is more profitable for Tse if her opportunity cost of capital is 15 percent after taxes?

10.25. Table 10.7, page 237, shows the income tax effect of using MACRS depreciation. Make up a similar table for the following cash flow:

Year	Cash flow ($000)
0	-98
1	+24
2	+24
3	+24
4	+24
5	+24
6	+24

The cash flow will be the estimated result of investing in a front-end loader for your construction business. It will be put on standby after the sixth year. Its precise fate after that will be decided at that time. Therefore no capital gain or loss should be considered now. Use 5-year GDS and the half-year convention. The company's tax bracket remains at 48 percent.

10.26. In Example 10.7, page 236, J. B. McCudden places the property, a light truck, in service on May 10, 1993. He sells the truck on September 1, 1995. What is his allowable depreciation per year, using GDS and the half-year convention? All other information remains the same.

10.27. You buy a new car for $3,500 for use in your business. You decide to use MACRS GDS depreciation for income tax purposes. The life of the car is 4 years, but it may be depreciated as 3-year property.

(*a*) Show the amount of depreciation you may take for each of these years.

(*b*) What will be the book value of the car at the end of the fourth year?

10.28. You buy a new car for $8,000 for use in your business. Show the amount of depreciation you may take for the 4 years you intend to use the car and its book value at the end of each of those years under the following methods:

(*a*) MACRS GDS 3-year property

(*b*) ADS for 3-year property

(*c*) Straight-line depreciation with a $2,000 salvage value

(*d*) Double-declining-balance depreciation with a $2,000 salvage value

(*e*) Sum-of-the-years'-digits depreciation with a $2,000 salvage value

10.29. (*a*) At what rate of return, after taxes, would the taxpayer prefer the MACRS GDS method to ADS for 3-year property? Carry out your ROR calculations to the nearest 0.1 percent.

(*b*) Would the taxpayer choose MACRS GDS above or below this rate?

10.30. A major transatlantic airline is considering the purchase of a new passenger aircraft. This aircraft has a capacity of 400 passengers at an expected value of load factor of 0.95. A fare of $275 one way will be charged, and the airplane will make an expected 300 one-way crossings per year.

The aircraft will cost $25,000,000 each. Operations and maintenance expenses will be $21,275,000 per year. The resale value of the aircraft will be $5,000,000. All amounts are in current dollars.

The IRS will allow, and the airline elects, ADS, with the half-year convention over a life of 12 years. The aircraft will be sold for $5,000,000 at the end of the 13th year. The $5,000,000 will be considered ordinary income. The airline is in the 46 percent tax bracket.

If the airline uses an after-tax, current dollar rate of 18 percent, should this aircraft be purchased? All calculated percentages must be correct to the nearest whole percent.

10.31. (*a*) An after-tax analysis of the proposed purchase of a Lancer aircraft is needed. The Lancer will cost your company $29,000, will result in a profit of $5,500 per year for 6 years—the analysis period—and will have a resale value of $12,000.

The MACRS GDS is used with the half-year convention. Gain or loss on disposal is counted as income or loss for the year during which resale occurs. The income tax rate for your company, Aviation Taxi, is 45 percent. Your after-tax MARR is 12 percent. Using the *NPV method*, determine whether your company should buy the airplane. Do not consider inflation.

(*b*) Your company buys the airplane of part (*a*), whatever your recommendation. After operating the aircraft for 2 years, the company receives an offer of $20,000 for it. Should Aviation Taxi accept or reject the offer, providing that estimates of original cost, profit, and resale value remain the same? Do not consider inflation.

10.32. A small, family-owned power company in California, organized under the Public Utilities Regulatory Powers Act of 1978, a part of the National Energy Act, had under review a plan to build a 600-kilowatt Pelton wheel plant for $210,000. The plant would return $80,000 per year in profit throughout its physical life of about 50 years.

Under the MACRS, required for most assets acquired after 1986, such public utility assets can be depreciated over 20 years. No salvage value will be realized at the end of the 50-year life.

Construction would start on January 1, 1987, and would terminate on December 31, 1987.

Do not consider inflation. The income tax bracket of the owners was 45 percent for all state and federal income taxes. The plant would begin operation on January 1, 1998.

What rate of return, after taxes, to the nearest whole percentage, will the venture achieve if held for the full 50 years of physical life?

10.33. Five years *after* the plant begins operation, the company in Prob. 10.32 receives an offer of $400,000 for the plant. Should the family accept the offer?

In making its decision, the company must adhere to the capital gain and loss provisions of the current tax law, which state that all capital gains or losses for such companies must be considered as income or loss in the year sold. The sale will take place at midnight on December 31 of the fifth year of operation.

Assume that all original projections of physical life, tax rate, profit, and acceptability of depreciation method have been correct. Assume also that the fifth year's depreciation has been legally taken. The minimum attractive rate of return for a similar-risk project is 11 percent after taxes. The NPV method must be used.

10.34. An asset is purchased for $250,000. At the time of purchase, the taxing agency allows an 8 percent investment tax credit. It will also allow depreciation to be taken according to the following schedule:

Year	Depreciation percent
1	15
2	22
3	21
4	21
5	21
	100

The depreciation will be allowed on the full first cost of the asset, that is, the cost before the investment tax credit is deducted. What will be the tax effect (that is, the cash flow after taxes) of the purchase of this asset on the company, assuming that the benefits provided by, and the operations cost of, the asset are not ascertainable because it is a unit of a general production system. There will be no net salvage value. The company is in the 48 percent tax bracket.

10.35. In Example 10.8, page 239, Wanda Hershey acquired computer software on May 1, 1995, for $6,000. She decides not to expense the cost. She must therefore depreciate it over 5 years. It will have no salvage value at the end of this time. If she uses MACRS GDS and the half-year convention, how much can she deduct from her taxable income each year as a result of depreciation on this equipment?

10.36. A drill press, if purchased, will cost $30,000, and will have a life of 10 years and a salvage value of $3,000. An investment tax credit of 7 percent will be permitted under this year's regulations. Pre-1981 straight-line depreciation will be used for accounting purposes. Annual operation and maintenance costs will be $4,000. The income tax rate for this company is 47 percent, and the after-tax marginal rate of return is 15 percent. Compute the after-tax cash flow for this machine and its annual cost after taxes.

10.37. In Example 10.9, page 240, Hershey decides to use the double-declining-balance method to depreciate the cost of her franchise, which she now is told will cost $10,000. What is her annual depreciation allowance if she uses a life of 10 years with no salvage value?

10.38. In 1993, a large company with many investments had a choice of depreciation methods for income tax purposes related to a certain investment in intangible property. The cash flow was composed of a $100,000 first cost and a net revenue before taxes of $30,000 per year for 10 years in constant dollars. No further inflation effects need be considered, therefore.

The company was trying to decide which depreciation rate to use. It was able to use pre-1981 straight-line or double-declining-balance method. Salvage value at the end of the 10-year life of the asset was $10,000.

The company was in the 45 percent tax bracket. Its minimum attractive rate of return was 10 percent after taxes and inflation.

Which type of depreciation schedule was better for the company? *Use the NPV method to answer this question.* Show all calculations on which you base your answer.

10.39. In Table 10.11, page 243, the cash flow is now as shown:

Year	Cash flow
0	−3,500
1	600
2	600
3	600
4	600
5	600
5	350

Compute the after-tax cash flow, using the pre-1981 straight-line method and a 48 percent tax bracket.

10.40. Make up a table similar to Table 10.12, page 243, but for 150 percent declining-balance depreciation.

10.41. The 1994 IRS regulations allowed the 150 percent declining-balance method to be used for certain property Smith Company acquired during that year. The property had a class life of 3 years. The half-year convention is to be employed. What was the allowable depreciation each year for a property costing $10,000?

10.42. In Example 10.12, page 247, a book disk example, the cost of the taxis is reduced to $2,900,000. Will this make a sufficient difference to cause Bishop Cab Company to invest? All other information remains as before.

10.43. Frank Luke Construction Company, Inc., a small family business, has been renting a backhoe whenever necessary. Frank Luke Construction is wondering whether it would save money after income tax payments to buy a new backhoe. Savings in costs between buying and renting would amount to $12,000 per year, considering all costs.

The backhoe would cost $54,000, have a life of 8 years, and have a salvage value at the end of 8 years of $8,000.

After-tax opportunity cost of capital for Frank Luke Construction is 12 percent.

The MACRS GDS depreciation method must be used, under the half-year convention. A 40 percent tax bracket is estimated for the next 8 years. Capital gains treatment is assumed to follow the 1993 tax law.

By the NPV method, should the backhoe be bought?

10.44. A floating dry dock for power boats has a first cost of $310,000, a salvage value of $10,000, and an economic life of 30 years. It can be depreciated by the MACRS GDS or ADS with the half-year convention.

Estimated revenues will be $500,000 per year. Operating and maintenance costs will be $378,000 per year. The proprietor, a power boat company, is in the 34 percent tax bracket. No gain or loss on the dry dock's disposal is anticipated.

The company employs a 20 percent rate of return after taxes to decide on its investments. Using the *rate of return method*, determine which of the two depreciation methods should be used. (Compare with Prob. 10.45.)

10.45. A barge-mounted crane with a first cost of $310,000, a salvage value of $10,000, and an economic life of 30 years may be depreciated by the pre-1981 straight-line method or the sum-of-the-years'-digits method.

It will earn estimated revenue of $500,000 per year. Operating and maintenance costs for the crane will be $378,000 per year. The proprietors, a large ship salvage company, are in the 48 percent tax bracket. No gain or loss on the equipment's disposal is anticipated.

The company uses a 20 percent rate of return after taxes to decide on its investments. Using the *rate of return method*, decide which of the two depreciation methods should be used. (Compare to Prob. 10.44.)

10.46. In 1980, two electric motor models were being considered for use in the lifting cranes for a building construction company. The data on the motors were as shown:

	Model *A*	Model *B*
Life (years)	8	10
Initial cost	35,000	42,000
Annual maintenance cost	1,200	1,000
Salvage value	3,000	4,200

If the construction company had an after-tax opportunity cost of capital of 15 percent and elected to use pre-1981 straight-line depreciation, which model would be chosen? The company was in the 45 percent tax bracket. (Compare with Prob. 10.47.)

10.47. In Prob. 10.46, the date is now 1994. Rather than pre-1981 straight-line depreciation, ADS with the half-year convention and a 10-year life will be used. All other data remain the same. Choose one of the models.

10.48. A four-wheel-drive, specially equipped surveyor's pickup is to be purchased for your consulting engineering business. It will cost \$15,000, with an estimated resale value of \$6,500. Operation and maintenance costs will run about \$12,000 annually. Assume the truck is held for 6 years.

Your after-tax opportunity cost of capital is 12 percent. You are in the 39 percent income tax bracket. The ADS will be used for depreciation purposes, with the half-year convention.

What is the annual, after-tax cost of the pickup? (Compare to Prob. 10.49.)

10.49. A small desk-top computer to be used in your consulting engineering business will cost \$15,000. Its economic life will be 4 years with an estimated resale value of \$5,000 at the end of that time. Expenses—power, paper, etc.—will be \$100 per month. Use \$1,200 per year.

Your company is a corporation in the 45 percent income tax bracket. The after-tax MARR is 12 percent. Depreciation will be computed for both income tax and economic analysis by the sum-of-the-years'-digits method.

Compute the equivalent uniform annual after-tax cost of the computer. (Compare to Prob. 10.48.)

10.50. Interstate Trucking Company will be purchasing new trucks on December 16, 1996, the only capital investment it will make during its calendar tax year. Each truck has the following economic characteristics:

Annual revenue	\$40,000
First cost	\$34,200
Operation, maintenance, etc., costs per year	\$24,800
Life	5 years
Resale value	\$4,200

The company is in the 39 percent tax bracket. The truck will be depreciated by MACRS GDS (5-year), midquarter convention. What is the prospective rate of return on this equipment after income taxes are paid? (Compare to Prob. 10.51.)

10.51. Ready-Mix Concrete Company purchased new trucks in 1980. Each truck had the following economic characteristics:

Annual revenue	\$40,000
First cost	\$34,200
Operation, maintenance, etc., costs per year	\$24,800
Life	5 years
Resale value	\$4,200

The company is in the 45 percent tax bracket. The trucks were depreciated by the pre-1981 straight-line method for 5 years. What was the prospective rate of return on this equipment after income taxes were paid? (Compare to Prob. 10.50.)

10.52. A turret lathe, acquired on June 30, 1980, was depreciated by the double-declining-balance method. It had a first cost of \$90,000 and an estimated salvage value of \$18,000. Its estimated life was 8 years. Operation and maintenance costs were projected at \$25,000 per year.

The before-tax cost of capital was 20 percent, and the after-tax cost of capital was 12 percent. No inflation effects were considered. The tax rate was 48 percent. Capital gains were taxed at 40 percent of the gain considered as income.

The lathe was sold on June 30, 1986, for \$25,000. What was its *present worth*, as of 1980, after taxes? (See Prob. 10.53.)

10.53. Repeat Prob. 10.52 by MACRS GDS. Imagine that the lathe was put in service on June 30, 1993, and will be sold, according to best estimates, on June 30, 1999. The GDS recovery period is 7 years. Use the half-year convention. The remaining information is as before.

10.54. Two machines were to be analyzed for a certain manufacturing task. One was certain to be purchased. Their salvage value was zero at the end of an estimated life of 3 years. The sum-of-the-years'-digits depreciation method was used in determining depreciation charges. The after-tax marginal cost of capital was 10 percent. The income tax rate for this company was 40 percent.

Use the net present-worth method for your solution. The cash flows were as follows:

Year	Machine *A*	Machine *B*
0	−120,000	−132,000
1	+70,000	+80,000
2	+70,000	+80,000
3	+70,000	+80,000

(*a*) Which machine should be purchased?

(*b*) Is your decision changed by using the double-declining-balance depreciation? (See Prob. 10.55.)

10.55. In Prob. 10.54, the machines are now to be analyzed by using MACRS GDS and the half-year convention. They will be sold at the end of the third year. All other information remains the same. Answer question (*a*) only.

10.56. Clostermann Construction Company will be acquiring new trenching equipment this year. It will cost $2,500,000. Maintenance and operations costs will be $300,000 per year. The equipment will be depreciated by the MACRS GDS with the half-year convention. The equipment will be used for 8 years. At the end of the eighth year, it will be sold for an estimated $500,000.

The company is in the 34 percent tax bracket. Capital gains are to be taxed as income of the disposal year.

An opportunity cost of capital of 15 percent after taxes will be used in the analysis. Do not consider inflation.

What is the after-tax *evaluation uniform annual cost* (*EUAC*) of this investment?

10.57. Patricia Sawyer, vice president of Jack Currie Company, believed the company needed two new trucks. As of October 1, 1993, the price per truck was $50,000, implying a total expenditure of $100,000. The trucks were to be put into service immediately and thus would earn $5,000 profit per month ($2,500 per truck) for the remainder of 1993 and for the rest of their estimated life through 1998.

The company made no other depreciable investments during 1993. Thus the MACRS GDS midquarter convention must be used for depreciation purposes.

Sawyer assumed that the trucks would be sold for 10 percent of their first cost on January 1, 1999. Because the trucks will have been fully depreciated at that time, capital gains will be treated as income for the year of disposal, at 100 percent of the gains.

The company's opportunity cost of capital is 12 percent after taxes have been accounted for. It is in the 34 percent tax bracket.

Use the NPV method for your solution and the end-of-year convention. Should Sawyer have recommended the purchase of the trucks?

10.58. A $150,000 investment in special materials handling equipment to be placed in service May 1, 1994, was proposed to management by the warehousing division of a large food processing company. The equipment will last 5 years and will be depreciated by the MACRS GDS (3-year) method. The midquarter convention is to be used. The salvage value of the equipment is estimated at 10 percent of the initial cost. It will save about $40,000 annually in labor costs. The company is in the 34 percent tax bracket. What are the before- and after-tax rates of return on this investment to the nearest whole percent? (Compare to Prob. 10.59.)

10.59. A $150,000 investment in materials handling equipment is proposed to management by the warehousing division of a large food processing company. The equipment will last 5 years and will be depreciated by the sum-of-the-years'-digits method. The salvage value of the equipment is estimated at 10 percent of the initial cost. It will save about $40,000 annually in labor costs. The company is in the 45 percent tax bracket. What are the before- and after-tax rates of return on this investment to the nearest whole percent? Compare to Prob. 10.58.

10.60. A transport company was considering purchasing a new truck and placing it in service on January 1, 1993. The truck would cost $40,000. The annual revenue attributable to the truck was estimated at $20,000. The plan was for the truck to be used for 8 years. At the end of the eighth year, it would be sold for $5,000.

The company must use the MACRS GDS depreciation. The midquarter convention is required for 1993 for this company. The company is in the 35 percent tax bracket. Capital gains will be treated as income earned during the year of disposal. If the after-tax opportunity cost of capital is 12 percent, should the company have purchased the truck?

10.61. You are contemplating buying a house. You wish to know whether it will be better to pay $100,000 cash with all fees paid or to take out a loan at 12 percent for 30 years with a 10 percent down payment. A mortgage loan fee of $2,700 must also be paid on purchase. Interest is deductible from your income tax. You are in the 30 percent bracket. You plan to hold the house 4 years and then sell it for an estimated $150,000. Your after-tax cost of capital is 15 percent. A capital gains tax will be paid on 40 percent of the gain counted as income.

Answer: Mortgage is preferred

10.62. An industrial bond with a face value of $10,000 is available for sale at $9,200. The bond will pay a nominal 12 percent semiannually for a term of 10 years.

The buyer is in the 33 percent tax bracket and expects to remain there for the term of the bond. He assumes that capital gains will be treated for tax purposes on the basis of 40 percent of the capital gain taxed as ordinary income for the year of sale.

For an investment of similar risk, the buyer expects to make 8 percent nominal interest compounded semiannually, after taxes. For the moment, the prospective buyer will ignore the effect of inflation. Should the bond be bought by this investor?

Answer: Yes

10.63. A Sears Roebuck Company bond has a face value of $10,000. It will pay interest at a nominal 6 percent, compounded semiannually. Its term is 14 years. It is offered at $7,500.

A $10,000 municipal bond, for sewers in Milwaukee, offered free of federal, state, and local taxes is being sold at par, that is, at face value. It will pay interest at a nominal 7 percent compounded semiannually, also for 14 years.

The prospective buyer is in the 37 percent bracket for all federal, state, and local taxes. Her opportunity cost of capital is a nominal 6 percent for semiannual compounding and equal-risk investments.

Income tax must be paid on the Sears bond but not on the municipal bond. Capital gains will be taxed, she assumes, at 40 percent of the gain counted as current income.

If she ignores possible inflation, should she spend $750,000 to buy either bond, and if so, which one? Use the *NPV method.*

CHAPTER 11

INFLATION

Earlier in this text, income taxes, inflation, and risk were assumed away in order to make the presentation of the basic concepts simpler and thus easier for the reader to understand. In Chap. 10, the first step on the return to reality was made when income tax was included in the analysis. Now it is time to make the second step toward reality by including inflation, that is, the effect of price changes on analysis.

The importance of inflation in economic analysis cannot be minimized. It would be absurd to make an analysis in a country where inflation is occurring and not include this effect. Russia in 1993 had an inflation rate of 20 percent per month. Were the effect of inflation ignored in an analysis whose time horizon was that year, an error of almost 800 percent would be possible. Even in countries where inflation rates are much lower, inflation exerts a significant influence on economic decisions.

PRICE CHANGES

Inflation describes an upward change in prices; *deflation*, a downward change. When the Argentinean radio reports a 20 percent monthly rate of inflation during July 1993, the meaning is that the price level of goods and services in Argentina has risen by 20 percent over the level at the end of June. When the Bureau of Labor Statistics of the U.S. government estimates a consumer price index of 51 for the year 1800 and 25 for the year 1900, it indicates that the general level of prices fell from 1800 to 1900 by more than half. Deflation occurred. This change in the general level of prices does not mean that all prices of all goods and services changed by the same amount. It is possible that the prices of some goods and some

TABLE 11.1
Consumer price index (1967 = 100)

Year	Index	Year	Index	Year	Index
1913	29.7	1940	42.0	1967	100.0
1914	30.1	1941	44.1	1968	104.2
1915	30.4	1942	48.8	1969	109.8
1916	32.7	1943	51.8	1970	116.3
1917	38.4	1944	52.7	1971	121.3
1918	45.1	1945	53.9	1972	125.3
1919	51.8	1946	58.5	1973	133.1
1920	60.1	1947	66.9	1974	147.7
1921	53.6	1948	72.1	1975	161.2
1922	50.2	1949	71.4	1976	170.5
1923	51.1	1950	72.1	1977	181.5
1924	51.2	1951	77.8	1978	195.4
1925	52.5	1952	79.5	1979	217.5
1926	53.0	1953	80.1	1980	246.8
1927	52.0	1954	80.5	1981	272.4
1928	51.3	1955	80.2	1982	289.1
1929	51.3	1956	81.4	1983	298.4
1930	50.0	1957	84.3	1984	311.1
1931	45.6	1958	86.6	1985	322.2
1932	40.9	1959	87.3	1986	328.4
1933	38.8	1960	88.7	1987	340.4
1934	40.1	1961	89.6	1988	354.3
1935	41.1	1962	90.6	1989	371.3
1936	41.5	1963	91.7	1990	391.4
1937	43.0	1964	92.9	1991	408.0
1938	42.2	1965	94.5	1992	420.3
1939	41.6	1966	97.2	1993	432.7
				1994	444.0

Source: Department of Labor, Bureau of Labor Statistics, *CPI Detailed Report*. Washington: Government Printing Office, 1995.

services rose or fell at a different rate, or remained stable, or showed an opposite trend to the general price movement. What is meant is that the weighted-average level of prices fell over this period. What this implies will be explained now.

PRICE INDEXES

"The consumer price index (CPI) measures the average change in prices of goods and services in day to day living."[1] This index is the one most often cited in discussions of inflation in the United States. It is the cost of living that this index measures, based on a weighted average of a "market basket" of goods and services. The quantity and quality of items making up the market basket remain essentially the same between pricing periods. Thus the index considers only the change in prices between two dates. Table 11.1 shows the *consumer price index*

[1]U.S. Department of Labor, Bureau of Labor Statistics, *Handbook of Labor Statistics*, Washington: Government Printing Office, 1989.

(*CPI*) from 1913 to 1994 for all items. Notice that the base of the CPI in this table is 1967, where the CPI is equal to 100.0. A new base has been established for the CPI in which 1982–1984 equals 100.00. In this scale, no single year equals exactly 100.00. Such goods as automobiles, washing machines, dishes, and kitchen knives; and such services as haircuts, gardening, and bus transportation are included in the market basket whose changes the CPI records.

Because the rate of inflation for a year depends on the amount of change in prices over that year compared to the price of the market basket at the beginning of the year, the phenomenon of inflation is similar to the phenomenon of capital accumulation as a result of interest. If an investor's account begins the year at $10,000 and ends it at $11,000 because of interest being added to the account, then the amount of change in capital is $1,000 and the interest rate is $1,000 divided by $10,000 times 100, or 10 percent.

Similarly, if the market basket of goods and services costs $10,000 at the beginning of the year and $11,000 at the end of the year, then the rise in prices is $1,000 and the inflation rate is 10 percent. Therefore, inflation is a compound-interest phenomenon, because each year's rise is measured on the base of the previous year's end-of-year cost. Thus it may be treated with the same formulas and methods that have been developed thus far in this text.

■ **Example 11.1.** The market basket of goods and services in New York City for a certain year costs $12,120, a year later costs $13,240, and a year after that $14,100. What is the inflation rate f for the second and third years?

Solution. For year 2

$$f_2 = \frac{13{,}240 - 12{,}120}{12{,}120} \times 100 = 9.2\%$$

For year 3

$$f_3 = \frac{14{,}100 - 13{,}240}{13{,}240} \times 100 = 6.5\%$$

■

The CPI itself is calculated by arbitrarily choosing a certain year and calling its price index 100.0. The market basket is priced for that base year. A year later the market basket is priced again. The latter price is divided by the former price and multiplied by the previous year's CPI. The result is the CPI for the subsequent year. The same procedure is followed to determine the CPI for later years.

■ **Example 11.2.** The market basket of goods and services cost $12,120, $13,240, and $14,100 at the end of each of 3 consecutive years. If the first year is designated as the base year with a CPI of 100.0, calculate the CPI for the following 2 years.

Solution. For year 2

$$\text{CPI} = 100.0 \times \frac{13{,}240}{12{,}120} = 109.2$$

For year 3

$$\text{CPI} = 109.2 \times \frac{14{,}100}{13{,}240} = 116.3$$

The inflation rate for year 3 as calculated in Example 11.1 (6.5 percent) is not added to the CPI for year 2 (109.2) to obtain the CPI for year 3. Rather, the CPI for year 3 can be calculated from the CPI for year 2 by multiplying the latter by 1 plus the inflation rate in decimals. ■

The CPI allows us to convert any year's current money to money of the base year, that is, to money of constant buying power. This is done by dividing the current cash flow by the CPI for that year and multiplying by 100, a procedure mandated by the definition of the CPI. It is known as *deflating*.

■ **Example 11.3.** Given the years, CPIs, and cash flow in current money shown below, convert the cash flow to constant dollars of year 0.

Year	CPI	Cash flow (current $)
1	100.0	1,000
2	109.2	3,000
3	116.3	5,000

Solution. For year 1

$$\frac{1{,}000}{100.0} \times 100 = 1{,}000$$

For year 2

$$\frac{3{,}000}{109.2} \times 100 = 2{,}747$$

For year 3

$$\frac{5{,}000}{116.3} \times 100 = 4{,}299$$

■

Conversely, any year's constant dollars can be inflated into current dollars by multiplying by the CPI and dividing by 100. The base-year dollars are both constant and current.

■ **Example 11.4.** Given the years, CPIs, and constant-dollar cash flows shown below, calculate the corresponding current-dollar cash flow.

Year	CPI	Cash flow (constant $)
1	100.0	1,000
2	109.2	2,747
3	116.3	4,299

Solution. For year 1

$$1{,}000 \times \frac{100.0}{100} = 1{,}000$$

For year 2

$$2{,}747 \times \frac{109.2}{100} = 3{,}000$$

For year 3

$$4{,}299 \times \frac{116.3}{100} = 5{,}000$$

■

In some problems the base year of the CPI, or any other index similar to it, must be changed to some other year. This is easily done by dividing the CPI of the new base year by itself and multiplying the result by 100. All the other CPIs in the list are converted to the new base year by dividing them by the same number, that is, the original CPI of the new base year, and multiplying the result by 100. The next two examples illustrate the process.

■ **Example 11.5.** Given the data shown below, change the base year to year 2.

Year	CPI (base year 1)
1	100.0
2	109.2
3	116.3

Solution. For year 2

$$\text{CPI} = \frac{109.2}{109.2} \times 100 = 100.0$$

For year 1

$$\text{CPI} = \frac{100.0}{109.2} \times 100 = 91.6$$

For year 3

$$\text{CPI} = \frac{116.3}{109.2} \times 100 = 106.5$$

We summarize below:

Year	CPI (base year 2)
1	91.6
2	100.0
3	106.5

■

■ **Example 11.6.** Using the CPIs of Table 11.1, convert the table to a new base year of 1975 and compute the CPIs for 1975 through 1979.

Solution

Year	CPI (1967 = 100)	CPI (1975 = 100)
1975	161.2	100.0
1976	170.5	105.8
1977	181.5	112.6
1978	195.4	121.2
1979	217.5	134.9

For 1975

$$\frac{161.2}{161.2} \times 100 = 100.0$$

For 1976

$$\frac{170.5}{161.2} \times 100 = 105.8$$

For 1977

$$\frac{181.5}{161.2} \times 100 = 112.6$$

For 1978

$$\frac{195.4}{161.2} \times 100 = 121.2$$

For 1979

$$\frac{217.5}{161.2} \times 100 = 134.9$$

■

TABLE 11.2

Producer price indexes by commodity group, selected years from 1926 to 1993 (1982 = 100)

Year	All commodities	Construction machinery and equipment
1926	17.2	—
1930	14.9	—
1935	13.8	—
1940	13.5	9.5
1945	18.2	10.4
1950	27.3	15.8
1955	29.3	19.5
1960	31.7	25.0
1965	32.3	27.2
1970	36.9	33.7
1975	58.4	53.8
1980	89.8	84.2
1981	98.0	93.3
1982	100.0	100.0
1983	101.3	102.3
1984	103.7	103.8
1985	103.2	105.4
1986	100.2	106.7
1987	102.8	108.9
1988	106.9	111.8
1989	112.2	117.2
1990	116.3	121.6
1991	116.5	125.2
1992	117.2	128.7
1993	118.9	132.0
1994	120.4	133.7

Source: Department of Labor, Bureau of Labor Statistics, *Producer Price Indexes*, Washington: Government Printing Office, 1995.

Other price indexes than the CPI are compiled by the Bureau of Labor Statistics. Producer price indexes are a measure of average changes in prices received by commodity producers. For example, Table 11.2 shows the *producer price index* (*PPI*) for all commodities for selected years from 1926 to 1994. It also shows the PPI for construction machinery and for equipment for the same years. They are different.

A question arises as to which index should be used to analyze a particular problem. It would not be appropriate to use the PPI for flat glass to analyze a number of alternatives for the replacement of a bulldozer. The construction machinery and equipment price index would be called for if predictions of price movements of such machinery in the future or an after-the-fact analysis of the results of a historical decision were to be made. An investment in a certificate of deposit whose interest payments will be spent on consumption would properly use the CPI. An analysis of an investment in construction machinery will use the PPI for construction machinery, the construction labor index, and the CPI for the profits realized by the investment if these are judged to be destined for consumption expenditures. Some examples of the use of different indexes will be seen later in this chapter.

Many indexes are available from other sources than the Bureau of Labor Statistics. The United Nations publishes indexes related to the interaction of nations. Trade associations, individual countries, and some trade magazines publish indexes. The analyst need only remember that choosing the proper index is worth consideration.

PRICE CHANGES IN THE UNITED STATES AND OTHER COUNTRIES

Price changes in the United States have been upward during all the wars in which the country fought—the War of 1812, the Civil War, the Spanish-American War, the First and Second World Wars, the Korean war, and the Vietnamese war. After the surge of wartime inflation, a deflation followed immediately for all the wars up to the Second World War. The expected deflation never came after that war, or after any of the wars that followed it. The inflation rate slowed, but that was all, except for 1955 when a minuscule deflation (−0.04 percent) occurred.

The highest single yearly inflation rate since 1945 (14.4 percent) was experienced during 1947 and the lowest (−0.04 percent) during the aforementioned 1955.

In the industrialized nations, a sampling of the four countries in the top portion of Table 11.3 shows inflation to be present in all of them in varying degrees. A relatively new arrival to the so-called first world, Japan has been more successful at controlling inflation than Sweden, a socialized member of long standing. The giant United States has done about as well as tiny Switzerland, banker to the world.

TABLE 11.3
Consumer price indexes for selected countries (1980 = 100)

Country	1988	1989	1990	1991
Japan	116	119	122	126
Sweden	177	188	208	227
Switzerland	128	133	140	148
United States	144	151	159	165
Belarus	110	111	116	227
Hungary	184	215	277	374
Yugoslavia	8,517	115,142	783,401	1,399,241*
Algeria	198	216	252	310
Brazil	364	4,733	142,023	723,218
Korea	160	169	184	201
Mexico	9,907	11,889	15,058	18,470

Source: United Nations, *Statistical Yearbook, 1990/91*, New York: United Nations, 1993, p. 347.
*Average of less than 12 months.

Former eastern bloc countries form the second group, and so-called third world countries, the third. See Table 11.3. They display a wide variation in inflation rates.

The reasons for varying rates of inflation are many and beyond the scope of this book. Examples of price rises are included here to make the reader understand that economic analysis that neglects the effect of inflation is naive indeed. Imagine conducting a study in Brazil that assumed constant prices!

Inflation is an evil because it makes predictions of future consequences of a decision, difficult in any case, far more difficult. A completely new dimension has been added to the problem of economic analysis which is impossible to ignore. Inflation causes contracts to become more risky for lenders, who then are less willing to lend at all except at higher interest rates to protect themselves. Interest rates rise, therefore, further contributing to inflation. Some countries, such as Brazil, have introduced a system of tying contractual payments to a government index, thus ensuring with more or less success that the contract payments maintain the same buying power. Finally, inflation results in grave harm to those persons with fixed incomes, such as pensioners, while it favors those with incomes that can be varied to match the progress of inflation, such as dentists. Thus it introduces another element of injustice into a society to add to those already present.

RATES: CONSTANT, CURRENT, AND INFLATION

We have been using the constant discount rate i up until this point because we have assumed away inflation. Now that we are including inflation as a phenomenon that must be considered, we must also define i more exactly, as well as two more rates. The *inflation rate* f is the rate of change in the price of a market basket of goods and services over 1 year. (See Example 11.1.) The *constant rate* i

TABLE 11.4
Rates

Symbol	Rate	Synonyms
f	Inflation	—
i	Constant $	Deflated
u	Current $	Inflated, composite, inclusive

is the opportunity cost of capital when price changes are not expected or when they are assumed away. It is the rate associated with constant-currency cash flows. The *current rate u* is the opportunity cost of capital when payments will be made in inflated currency; it is the rate associated with the current cash flows.[2] All compound-interest formulas will use u in place of i when current cash flows are dealt with.

Constant dollars are dollars of equal buying power. They are expressed in terms of base-year dollars. As we have seen, one of the United States' CPIs, with a magnitude of 100.0, is presently based on the year 1967. This base year can be changed to any new base year we choose. (See Example 11.5.) Or it can be simply year 0 in project analysis.

Current money is money of any particular year. Actual payments made in any year are always made in current money—whence *currency*. No one ever pays out or receives anything but current money; for actual payments, constant dollars are an impossibility. This is not to say that enough current dollars cannot be paid to match the buying power of a certain number of constant dollars.

Table 11.4 shows the symbols described above with their names and synonyms.

Rates of change are sometimes projected for other than inflation or deflation as we have defined them. For example, in the case of labor increasing at 6 percent per year while materials increase at 8 percent annually, neither of these rates is f. They can, however, be included in problems by showing labor and material costs at their current prices for each year of the analysis. They may then be handled mathematically by revisions to the standard factors or by treating each year individually and netting-out.

One rule is paramount. *A constant rate is always associated with a constant-dollar cash flow, and a current rate is always associated with a current-dollar cash flow.* Never discount a current cash flow with the constant rate or a constant cash flow with the current rate.

■ **Example 11.7**

(*a*) A bank quotes a return of 11 percent annually on its 2-year certificates of deposit. You wish to use this rate as your opportunity cost of capital. Should you call it i or u?

---

[2]As of 1995, no official body has established symbols for these variables. Because it is the second letter of the word "current," u is used.

(*b*) A proposed investment in Mexican petrobonds shows an internal rate of return of 120 percent, based on prospective peso payments. You wish to compare this result to an opportunity cost of capital for Mexican investments of 150 percent for current cash flows and 50 percent for constant-peso cash flows. Should you invest in petrobonds?

Solution

(*a*) Payments to you on your certificate of deposit will be made in current dollars. Because u is associated with current-dollar cash flows, your opportunity cost of capital, 11 percent, is u, not i.

(*b*) Because the petrobonds will pay in current pesos, u is 120 percent, which is less than 150 percent. Therefore reject the investment. ■

MATHEMATICS OF INFLATION

For the discussion that follows, these symbols are used:

P = amount at year 0
N = year
F_I = current (inflated) dollars in year N
F_D = constant (deflated) dollars in year N
i = constant rate
f = inflation rate
u = current rate

As we have seen in Example 11.3, current dollars of any year N can be converted to constant dollars of the same year by dividing the current dollars by the price index for that year and multiplying by 100. In Example 11.3, each year had a different price index. Moreover, the price index did not change at a constant rate, which is the same as saying that f varied year by year. What is the situation when the inflation is considered to remain constant? This is an assumed condition that occurs in economic analysis of future consequences. If, at year 0, the price index is taken as 100.0, then the price index for year 1 must be $100.0(1 + f)$.

To deflate a current dollar of year 1 to a constant dollar of year 1, it must be divided by $100(1 + f)$ and multiplied by 100, or simply divided by $1 + f$. Because f is constant, a current dollar of year 2 must be divided by $(1 + f)^2$ to deflate it to a constant dollar of year 2. A current dollar of year N can be deflated to a constant dollar of year N by dividing it by $(1 + f)^N$:

$$F_D = F_I\left(1+f\right)^{-N} \tag{11.1}$$

But this is no more than multiplying F_I by the single-payment present-worth factor, at interest rate f.

$$F_D = F_I\left(P/F, f, N\right) \tag{11.2}$$

For the sake of convenience, we may use the tables.

■ **Example 11.8.** José Anguelas is to receive a legacy, willed to him by his grandfather, when he reaches 30 years of age. He is now 20 years old. The inflation rate in his country is 1,000 percent per year. If the legacy is 1 billion pesos, what will be its buying power 10 years from now in terms of today's pesos, if inflation continues at its present rate?

Solution. The inflation of 1,000 percent per year must be included in Eq. (11.1) as a decimal.

$$F_D = 1{,}000{,}000{,}000\left(\frac{1}{1+10.00}\right)^{10} = 0.04$$

The billion pesos will have a buying power of 4 centavos 10 years from now. ■

Constant dollars in any future year can be inflated to current dollars by dividing both sides of Eq. (11.1) by F_D and cross-multiplying.

$$F_I = F_D(1+f)^N \tag{11.3}$$

or

$$F_I = F_D(F/P, f, N) \tag{11.4}$$

■ **Example 11.9.** Wages for common labor are expected to increase by 6 percent per year in the area of a dam job in the western United States. What will the payroll for common labor amount to for the fourth year of the job, if this item is estimated at $15 million annually?

Solution. The wage rate has been estimated for time 0, the beginning of the job. If wages did not increase at all, the bill for common labor during the fourth year of the job would be $15 million, that is,

$$F_D = 15{,}000{,}000$$

From Eq. (11.3)

$$F_I = 15{,}000{,}000(1+0.06)^4 = 18{,}937{,}154$$

or about $19 million.

In this example f represents a constant rate of increase in a single item, not the increase in all items. ■

The present worth of constant dollars of year N is

$$P = F_D(1+i)^{-N} \tag{11.5}$$

If we substitute the right side of Eq. (11.1) for F_D in Eq. (11.5), we have

$$P = F_I\left(\frac{1}{1+f}\right)^N\left(\frac{1}{1+i}\right)^N$$

$$= F_I \left[\frac{1}{(1+f)(1+i)} \right]^N$$

But $[(1+f)(1+i)]^{-N}$ is simply the present-worth factor with $(1+f)(1+i)$ present instead of $1+i$. Subtracting 1 from the denominator reveals the rate u to be applied to current dollars F_I:

$$u = (1+f)(1+i) - 1 \tag{11.6}$$

or

$$u = i + f + if \tag{11.7}$$

This is the fundamental relationship among the rates discussed previously.

■ **Example 11.10.** A couple is able to invest in residential real estate that will return 26 percent. They estimate that inflation will average about 5 percent over the life of the investment. What is their constant-dollar rate of return on the investment?

Solution. It may appear that the answer to this question is 26 percent. But you should remember that all returns on the investment will be paid in current money. The rate associated with current cash flows is the current rate u. Therefore,

$$u = 26\% \qquad f = 5\%$$

Substituting in Eq. (11.7) gives

$$0.26 = i + 0.05 + i(0.05)$$

$$i = \frac{0.26 - 0.05}{1 + 0.05} = 0.20 \text{ or } 20\%$$ ■

Thus i is seen as the difference between u and f deflated, or

$$i = \frac{u - f}{1 + f} \tag{11.8}$$

■ **Example 11.11.** If you placed $2,000 in an account paying 9½ percent compounded annually, and inflation was 9 percent during the same year, what is the value of your investment at the end of the year in terms of constant dollars at the beginning of the year?

Solution. The rates must be identified first. Clearly, f equals 9 percent. But what, then, is i, and what is u? You must recognize that the amount paid back in interest and principal at the end of the year will be in current money. Therefore, because u is the rate associated with current-dollar cash flows, u equals 9½ percent. Substituting in Eq. (11.8) gives

$$i = \frac{0.095 - 0.09}{1 + 0.09} = 0.0045871$$

or about one-half percent.

This answer suits our commonsense perception of the situation: While the interest rate of 9½ percent moves us upward, the 9 percent inflation rate drags us downward. The result is that we move upward by about 0.5 percent. It is the 18½ percent answers that are to be avoided.

$$
\begin{aligned}
F_D &= P(1+i)^N \\
&= 2{,}000(1+0.0045871)^1 \\
&= \$2{,}009.17
\end{aligned}
$$

This is the buying power of the principal and interest at the end of 1 year, that is, the value of the investment. ■

We could have attacked the problem in another way by first asking, What is the amount of principal and interest that will appear in your account after 1 year? This is

$$F_I = P(1+u)^N = 2{,}000(1+0.095)^1 = 2{,}190.00$$

Deflation to constant dollars by using Eq. (11.1) yields:

$$F_D = \frac{2{,}190.00}{1+0.09} = \$2{,}009.17$$

The answer is exactly the same by either method.

Nominal and Effective Rates Related to Inflation

Because inflation (or deflation) is measured as the percentage rise (or fall) in a market basket of goods and services at the end of the year, it is an effective annual rate, not a nominal rate. How is a problem to be handled in which either i or u is given as a nominal rate? The answer is that i or u must be converted to an effective annual rate before it can be used in conjunction with f, the inflation rate. (Conceivably, f could be converted to a nominal rate, depending on the compounding period, but a more laborious solution would result.)

■ **Example 11.12.** Mr. and Mrs. James Longstreet invest in a savings and loan association that guarantees a nominal 12 percent return, compounded monthly. Inflation is projected at 6 percent annually. What is their true (deflated) rate of return on this use of their money?

Solution

$$f = 6\% \qquad u = 12\% \text{ nominal} \qquad i = ?$$

The 12 percent nominal rate must be converted to an effective annual rate.

$$M = 12$$

$$u_M = \frac{12}{12} = 1\% \text{ effective monthly rate}$$

$$u_Y = (1 + u_M)^M - 1 = (1 + 0.01)^{12} - 1$$
$$= 12.68\% \text{ effective annual rate}$$

From Eq. (11.8)

$$i = \frac{u - f}{1 + f}$$
$$= \frac{0.1268 - 0.06}{1.06} = 0.0630 = 6.30\%$$

■

It does not matter how inflation is actually occurring—in spurts, at an increasing rate during the year, or continuously. The fact is that its measurement at the end of the year ensures that it is an effective annual rate.

INCORPORATING INFLATION IN AN ECONOMIC DECISION

How does inflation affect an economic decision among alternatives? The higher the inflation rate, the more profound the effect will be. Imagine an economic decision made in a country suffering an inflation rate of 1,000 percent per year. The exclusion of any inflation considerations from the analysis would render the decision useless and misleading. The examples that follow show how inflation may be included in a before-tax economic analysis.

■ **Example 11.13.** Given a constant inflation rate f of 20 percent and a constant-dollar opportunity cost of capital i of 15 percent, find the present worth (PW) in constant dollars in terms of year 0 of the following cash flow:

Year	Current-year cash flow
0	−10,000
1–5	+5,000

Solution. The problem can be solved in two different ways. The first applies the current rate u to the current-dollar cash flow. See Fig. 11.1.

$$u = i + f + if$$
$$= 0.15 + 0.20 + 0.03 = 0.38 \text{ or } 38\%$$
$$\text{PW} = -10.000 + 5.000(P/A, 38, 5)$$

Because the factor is not shown in the tables, it must be calculated.

$$(P/A, 38, 5) = \frac{(1 + 0.38)^5 - 1}{0.38(1 + 0.38)^5} = 2.1058$$
$$\text{PW} = -10{,}000 + 5{,}000(2.1058)$$
$$= -10{,}000 + 10{,}529 = +529$$

A plus sign indicates that the project is acceptable after inflation has been taken into account.

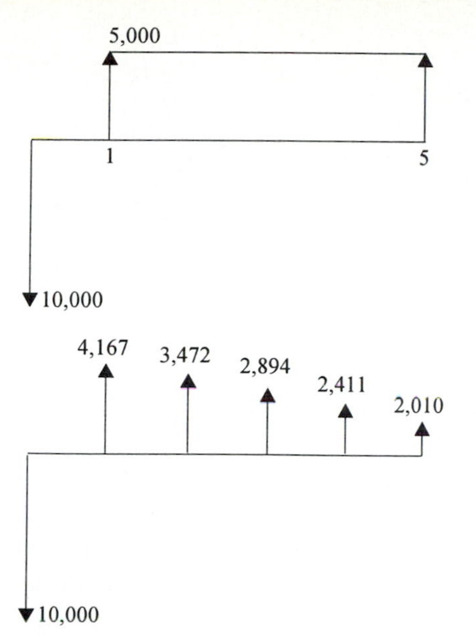

FIGURE 11.1 Current and constant dollars

Another method, equally acceptable, is to calculate the price indexes corresponding to an inflation rate of 20 percent. Although this is no onerous task for 5 years, it would certainly be so for 100 years.

Year	Current $	Price index	Constant $	(*P*/*F*, 15, *N*)	PW
0	−10,000	100.0	−10,000	1.0000	−10,000
1	+5,000	120.0	+4,167	0.8696	+3,623
2	+5,000	144.0	+3,472	0.7561	+2,625
3	+5,000	172.8	+2,894	0.6575	+1,903
4	+5,000	207.4	+2,411	0.5718	+1,379
5	+5,000	248.8	+2,010	0.4972	999
					+529

The answers are the same by both methods. ■

■ **Example 11.14.** A small investor in a developing country estimates that inflation in her country will average about 10 percent per year over the next 30 years. She wishes to purchase a residential house for $100,000 cash and rent it as a hedge against such an inflation. She is willing to accept a negative cash flow of $1,000 annually, after taxes, in constant (year 0) dollars, providing she receives at the end of 30 years a sum, after taxes, in dollars of that year—that is, current dollars—such that her return on all her cash outflow is 21 percent. This 21 percent is her estimate of the rate of return she would make, after taxes, from yearly payments to her if she invested in an international trust company under conditions not relevant here. She considers the risks to be about equal between the two investments.

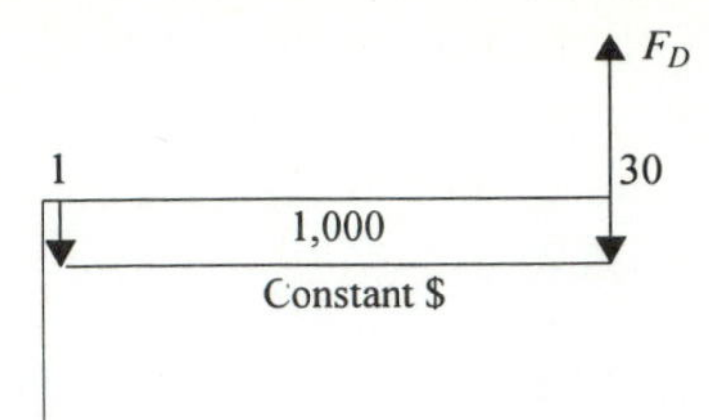

FIGURE 11.2 A house investment

Annual compounding is to be assumed.

What must be the sale price of the property 30 years from now in current dollars, after taxes, such that the rate of return is equal between the two possible uses of her investment cash? See Fig. 11.2.

Solution. First, we must decide on the values of i, u, and f. The rate of return the investor makes on the international trust company payments is a return on current dollars because, needless to say, the trust company will pay in current money. The 21 percent is her opportunity cost of capital based on current cash flow and is therefore u. The inflation rate of 10 percent is f. According to Eq. (11.8),

$$\begin{aligned} i &= \frac{u-f}{1+f} \\ &= \frac{0.21-0.10}{1.10} \\ &= 0.10 \text{ or } 10\% \end{aligned}$$

Figure 11.2 shows the constant-dollar cash flow.

$$\begin{aligned} F_D &= P(F/P, i, N) + A(F/A, i, N) \\ &= 100{,}000(F/P, 10, 30) + 1{,}000(F/A, 10, 30) \\ &= 100{,}000(17.4494) + 1{,}000(164.494) \\ &= 1{,}909{,}434 \end{aligned}$$

From Eq. (11.3),

$$\begin{aligned} F_I &= F_D(1+f)^N \\ &= 1{,}909{,}434(1.10)^{30} \\ &= 1{,}909{,}434(17.4494) \\ &= \$33{,}318{,}000 \end{aligned}$$

This large amount is the sale price needed 30 years from now to equal a 21 percent current-money return on the investment. The effect of a 10 percent inflation is appreciable. ■

■ **Example 11.15.** A now-concluded investment in handling equipment at a coal port facility is being investigated with regard to its economic effectiveness in the light of inflation. The equipment's original cost was $10,250,000 in 1981. It saved, over the previous method, $2,750,000 annually during the first 5 years, starting in 1982, and

$3,400,000 annually during the succeeding 6 years, all in current after-tax dollars. It has just been removed, with removal costs approximating its scrap value.

During each of these years, the appropriate inflation index was as follows:

(1980 = 100)		
1981 - 104.2		1987 - 147.7
1982 - 109.8		1988 - 161.2
1983 - 116.3		1989 - 170.5
1984 - 121.3		1990 - 181.5
1985 - 125.3		1991 - 195.3
1986 - 133.1		1992 - 217.7
		1993 - 247.0

What was the net present value of this investment in 1981, in terms of 1981 dollars, after inflation has been considered? Use a constant-dollar discount rate of 10 percent.

Solution. The following table shows the calculations.

Year	Current \$ (000)	*PI*	Constant \$ (1980 = 100)	(*P*/*F*, 10, *N*)	Present worth in constant \$ (1980 = 100)
1980	—	100.0			
1981	-10,250	104.2	-9,837	1.0000	-9,837
1982	2,750	109.8	2,505	0.9091	2,277
1983	2,750	116.3	2,365	0.8264	1,954
1984	2,750	121.3	2,267	0.7513	1,703
1985	2,750	125.3	2,195	0.6830	1,499
1986	2,750	133.1	2,066	0.6209	1,283
1987	3,400	147.7	2,302	0.5645	1,299
1988	3,400	161.2	2,109	0.5132	1,082
1989	3,400	170.5	1,994	0.4665	930
1990	3,400	181.5	1,873	0.4241	794
1991	3,400	195.3	1,741	0.3855	671
1992	3,400	217.7	1,562	0.3505	548
1993	—	247.0			+4,203

The $4,203,000 in constant (1980) dollars is converted to 1981 dollars:

$$4{,}203{,}000 \times \frac{104.2}{100} = +4{,}379{,}520 \quad \text{in 1981 dollars}$$

$$\approx +4.380 \text{ million}$$

The figure sought is +4.38 million. It shows that the investment was an excellent one even after inflation is taken into account. The problem could also have been solved by converting all the 1980-based price indexes to 1981-based indexes. The calculation would have taken longer. ■

VARIABLE INFLATION RATES

What will be the effect of varying the inflation rate? How can this common situation be handled? The following example will explain.

■ **Example 11.16.** You intend to invest $10,000 in a savings and loan company account. It will return 12 percent compounded annually over 3 years. You expect to spend the yearly receipts, not reinvest them, and to withdraw $10,000 at the end of 3 years. You estimate that inflation will remain at 4 percent for the coming year, will increase to 8 percent for the following year, and will go to 10 percent for the last year.

(a) Calculate the rate of return in constant dollars, before taxes, for this investment to the nearest whole percent.

(b) Is this rate i, u, or f?

Solution. The most direct way to handle this situation is to calculate the consumer price indexes for the time horizon of the investment. The consumer price indexes, if the present moment is considered to be time 0 at an index of 100, will be as follows:

Year	CPI
0	100.0
1	104.0 = 100.0 × 1.04
2	112.3 = 104.0 × 1.08
3	123.5 = 112.3 × 1.10

The following table shows the conversion from current to constant dollars.

Year	Current $	CPI	Constant $
0	−10,000	100.0	−10,000
1	+1,200	104.0	+1,154
2	+1,200	112.3	+1,069
3	+11,200	123.5	+9,069

To calculate the IROR of the constant-dollar cash flow, we solve for i^* in the following equation:

$$0 = -10{,}000 + 1{,}154(P/F, i^*, 1) + 1{,}069(P/F, i^*, 2) + 9{,}069(P/F, i^*, 3)$$

Trial and error results in

$$i^* = 5\%$$

(b) It is the constant-dollar rate of return and is therefore i. ■

The annual-worth method can also be used in an inflationary situation. Some care needs to be taken, however, as the comments and the following example explain.

■ **Example 11.17.** A city engineer in Uruguay is trying to decide between *corrugated steel pipe* (*CSP*) and *reinforced-concrete pipe* (*RCP*) for extensions to a storm sewer system. These data are available:

	CSP	RCP
Life (years)	30	60
First cost per meter (pesos)	195,000	260,000
Current-peso cost of capital (percent)	30	30
Inflation rate (percent)	4	4
Maintenance cost (pesos)	0	0

Which type of pipe should be chosen?

Solution

$$f = 4\% \qquad u = 30\%$$

The constant-peso cost of capital is

$$\begin{aligned} i &= \frac{u-f}{1+f} \\ &= \frac{0.30-0.04}{1+0.04} \\ &= 0.25 \text{ or } 25\% \end{aligned}$$

The *annual cost* (AC) of each alternative is as follows:

$$\begin{aligned} \text{AC}_{\text{CSP}} &= 195{,}000(A/P, 25, 30) \\ &= 195{,}000(0.25031) = \$48{,}811 \\ \text{AC}_{\text{RCP}} &= 260{,}000(A/P, 25, 60) \\ &= 260{,}000(0.25000) = \$65{,}000 \end{aligned}$$

The choice is CSP by a wide margin.

The assumption behind the preceding calculation is that the service lives are equalized by an investment of $195,000 at year 30 of the CSP alternative.

You will remember that the annual-worth method postulates a repetition of the cycles of investment in order to equalize service lives. You may not have noticed that the same amount of investment in money must be made at the beginning of each cycle for the annual cost to be equal over all the cycles. (See Fig. 11.3.) This fact precludes expressing the investment at year 30 in current money for the CSP alternative. Therefore, the deflated cost of capital must be found first, and constant money compared using it, as was done above.

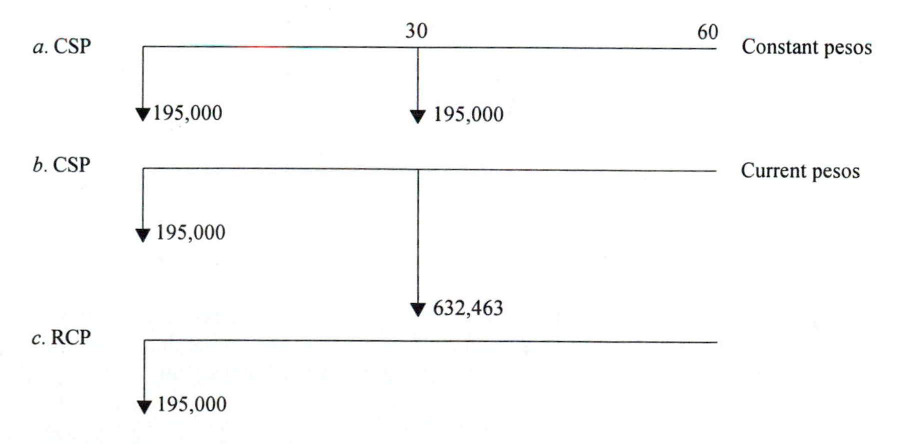

FIGURE 11.3 Storm sewers

In a present-worth solution, we can use the current-money cash flow and the current cost of capital to evaluate it, as follows. (See Fig. 11.3*b*.)

$$\begin{aligned} F_I &= F_D(1+f)^N \\ &= 195{,}000(1.04)^{30} \\ &= 195{,}000(F/P, 4, 30) \\ &= 195{,}000(3.2434) = 632{,}463 \end{aligned}$$

$$\begin{aligned} \text{PW}_{\text{CSP}} &= 195{,}000 + 632{,}463(P/F, 30, 30) \\ &= 195{,}000 + 632{,}463(0.0004) \\ &= 195{,}000 + 253 = 195{,}253 \\ \text{PW}_{\text{RCP}} &= 260{,}000 \end{aligned}$$

Again, the choice is CSP. Further resolution of the two methods is impossible. ■

What will happen when inflation and income tax effects are considered simultaneously? This, of course, is the normal form of the problem in deciding among mutually exclusive alternatives. One important point must be noted: Income tax is levied against current-money cash flow, not against constant-money cash flow. This means that all cash flows must be converted to current money before evaluation begins. By following this procedure, the after-tax cash flow in current money can be analyzed by whatever method is desired.

■ **Example 11.18.** On January 1, 1993, Aviation Taxi, your company, was considering the purchase of a used Lancer aircraft for its air service. As of that date, you estimated the following constant-dollar (as of January 1, 1993) figures:

First cost	\$29,000
Annual profit	\$ 5,500
Resale value (at end of study period)	\$12,000

The study period was to be 5 years.

You also estimated that profit from the operation of the aircraft will rise at the rate of 5 percent per year, while the resale value will rise at 8 percent per year.

A pre-1981 straight-line depreciation method is to be used based on the constant-dollar values given above. The income tax rate for your company is 45 percent. Your deflated opportunity cost of capital is 6 percent, the constant-dollar rate. The current-dollar after-tax opportunity cost of capital is 15 percent. Gain or loss on disposal of assets are to be counted as income or loss, respectively, for the year during which resale occurs. Use the *net present-value* method to determine whether the aircraft should have been purchased.

Solution. Because taxes are based on current income, the before-tax cash flow is expressed by applying the 5 percent annual increase in profits to the time 0 profit of \$5,500. See Table 11.5. The resale value is raised at 8 percent annually to

$$\begin{aligned} F_I &= F_D(F/P, 8, 5) \\ &= 12{,}000(1.4693) = \$17{,}632 \end{aligned}$$

TABLE 11.5
Lancer aircraft cash flows

Year	Before-tax cash flow	Depreciation	Taxable income	Tax @ −0.45	After-tax cash flow
0	−29,000				−29,000
1	+5,775	−3,400	+2,375	−1,069	+4,706
2	+6,064	−3,400	+2,664	−1,199	+4,865
3	+6,367	−3,400	+2,967	−1,335	+5,032
4	+6,685	−3,400	+3,285	−1,478	+5,207
5	+7,020	−3,400	+3,620	−1,629	+5,391
5	+17,632		(17,632 − 12,000)	−2,534	+15,098

The annual depreciation is calculated as

$$\frac{29{,}000 - 12{,}000}{5} = \$3{,}400$$

The rest of the table is calculated in the same way as in Chap. 10. Finally, the net present value (NPV) of the after-tax cash flow is computed by using the current-dollar rate u of 15 percent because the cash flow is expressed in current dollars. The constant-dollar rate of 6 percent is ignored.

$$\begin{aligned}\text{NPV} &= -29{,}000 + 4{,}706(P/F, 15, 1) + 4{,}865(P/F, 15, 2) + 5{,}032(P/F, 15, 3)\\ &\qquad + 5{,}207(P/F, 15, 4) + 5{,}391(P/F, 15, 5) + 15{,}098(P/F, 15, 5)\\ &= -29{,}000 + 24{,}244 = -4{,}756\end{aligned}$$

Therefore, because of the minus sign, the aircraft should not have been purchased. (See Prob. 11.53.) ■

■ Example 11.19

(*a*) In 1976, a Latin American property was bought for 297,500 pesos cash. The property was used as a second residence; therefore, no depreciation was charged. In 1993, it was sold for 20 million pesos cash. What rate of return, to the nearest whole percent, was made on this transaction? Ignore taxes and inflation for the moment.

(*b*) A capital gains tax had to be paid under the laws of the Latin American country in which the property was located. It was based on 40 percent of the gain taxed as ordinary income. The owner was in the 37 percent tax bracket. What was the after-tax return, to the nearest whole percent, on the sale of the property? Remember that no depreciation was allowed. Ignore inflation for the moment.

(*c*) The consumer price indexes within the country were

1978	=	100
1976		93.1
1993		1,045.0

After taxes, what was the rate of return on the transaction in terms of the buying power of 1976 pesos?

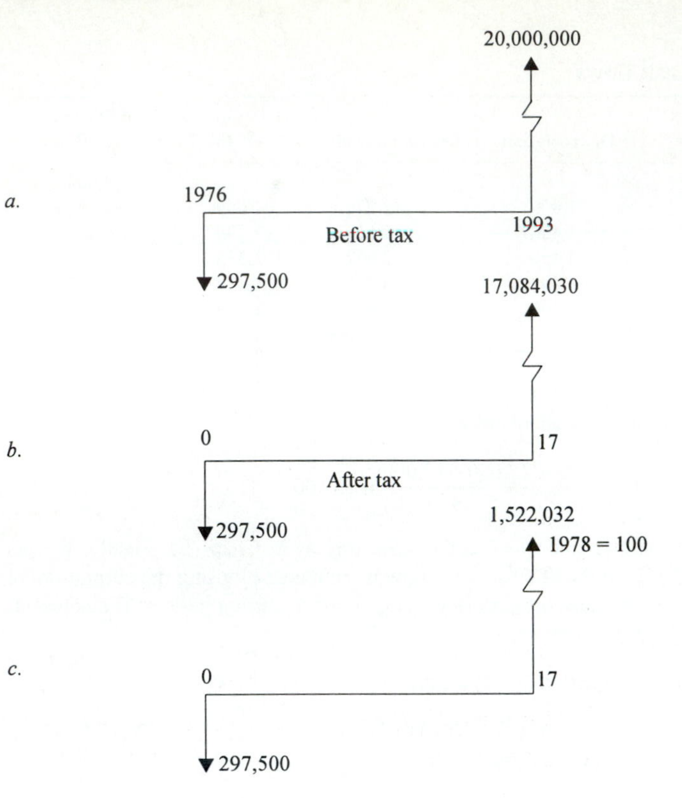

FIGURE 11.4 A Latin American residence

Solution

(*a*) See Fig. 11.4*a*.

$$297{,}500 = 20{,}000{,}000(P/F, i^*, 17)$$

$$(P/F, i^*, 17) = \frac{297{,}500}{20{,}000{,}000}$$

$$= 0.0149$$

Interpolation between 25 percent and 30 percent results in

$$i^* = 28\%$$

(*b*) The capital gains were

$$\begin{array}{r} 20{,}000{,}000 \\ -\ 297{,}500 \\ \hline 19{,}702{,}500 \end{array}$$

But only 40 percent of the gain was taxable. Therefore, the taxable portion was

$$(19{,}702{,}500)(0.40) = 7{,}881{,}000$$

The after-tax receipts on the sale of the property were

Sale price	20,000,000
Less tax	2,915,970
	17,084,030

Figure 11.4*b* leads to

$$297{,}500 = 17{,}084{,}030(P/F, i^*, 17)$$

$$(P/F, i^*, 17) = \frac{297{,}500}{17{,}084{,}030}$$

$$= 0.0174$$

Interpolation between 25 and 30 percent gives

$$i^* = 27\%$$

(*c*) The cash flow after taxes must be converted first to 1978 pesos

$$\frac{17{,}084{,}030}{10.450} = 1{,}634{,}835$$

and then to 1976 pesos:

$$(1{,}634{,}835)(0.931) = 1{,}522{,}032$$

Figure 11.4*c* leads to the relationship

$$297{,}500 = 1{,}522{,}032(P/F, i^*, 17)$$

$$(P/F, i^*, 17) = \frac{297{,}500}{1{,}522{,}032}$$

$$= 0.1955$$

Interpolation between 10 and 11 percent gives

$$i^* = 10\%$$

■

Problems that combine inflation, taxes, and some of the more complicated aspects of material covered in earlier sections of this text can be difficult to solve. The following example illustrates.

■ **Example 11.20.** You are in the process of paying annually on an automobile loan of $3,800 at 15 percent effective annual rate over a 4-year period. The car is used for business only, and thus deductions from taxes resulting from interest payments may be taken. The loan agreement allows you to pay off the balance of the loan at any time. You are considering this course of action just after making the first annual payment.

You estimate that the inflation rate based on the consumer price index will be 10 percent per year. Your income tax bracket is 43 percent. Your marginal cost of capital after taxes and inflation are considered is 2 percent.

Is paying the remaining balance of the loan an attractive alternative at this time, given the tax and inflation effects? You will, of course, ignore depreciation effects as irrelevant to the question.

Solution. The annual payment on the loan must be calculated first:

$$A = 3{,}800(A/P, 15, 4) = \$1{,}331$$

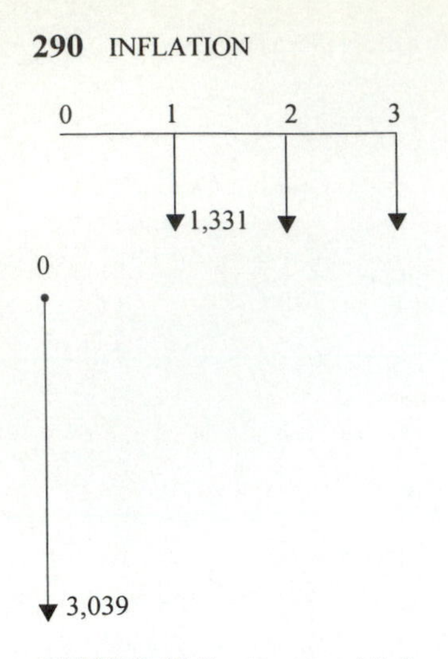

FIGURE 11.5 An automobile loan

Table 11.6 shows the split between interest and principal. It would be possible to pay off the loan at the end of year 1 for $3,039. Figure 11.5 shows the alternatives.

Table 11.7 shows the after-tax cash flow, the effects of inflation, and the present worth after taxes and inflation have been considered. The present cost of keeping the loan is $2,847. Paying off the loan requires $3,039. After taxes and inflation, it is better to keep the loan, therefore. ■

TABLE 11.6
An automobile loan

Year	*A*	*P*	*I*	Balance
0				3,800
1	1,331	761	570	3,039
2	1,331	875	456	2,164
3	1,331	1,006	325	1,158
4	1,331	1,157	174	1

TABLE 11.7
Automobile loan after taxes and inflation

Year	Before-tax cash flow	Interest	Deductions from taxes @ 0.43	After-tax cash flow	Price index	Deflated cash flow	*P/F*, 2, *N*)	PW
0	—				100.0	—		
1	−1,331	−456	+196	−1,135	110.0	−1,032	0.9804	−1,012
2	−1,331	−325	+140	−1,191	121.0	−984	0.9612	−946
3	−1,331	−174	+75	−1,256	133.1	−944	0.9423	−889
								−2,847

NEGATIVE RATES

Negative rates may appear in inflation problems. They are substituted into the compound-interest formulas with the minus sign attached, and the formulas are used as each problem dictates. The factors are *not* those of the compound-interest tables, negatively signed.

■ **Example 11.21.** If $2,000 of your money is invested in a savings account at a commercial bank at 5¼ percent compounded annually and inflation is 9 percent during the same year, what is your investment worth at the end of the year?

Solution

$$f = 9\% \qquad u = 5\,{}^1\!/\!_4\,\%$$

From Eq. (11.8):

$$\begin{aligned} i &= \frac{u-f}{1+f} \\ &= \frac{0.0525-0.09}{1+0.09} \\ &= -0.0344036 \end{aligned}$$

The deflated interest rate has passed into the negative range, thus illustrating that as compound interest has increased the amount actually in the account, inflation has removed its buying power even more. Its buying power after 1 year is

$$\begin{aligned} F_D &= P(1+i)^N \\ &= 2{,}000(1-0.0344036)^1 \\ &= \$1{,}931.19 \end{aligned}$$

The problem could have been solved in a different way:

$$\begin{aligned} F_I &= P(1+u)^N \\ &= 2{,}000(1.0525)^1 \\ &= \$2{,}105.00 \\ F_D &= \frac{2{,}105.00}{1.09} \\ &= \$1{,}931.19 \end{aligned}$$

■

Negative rates are by no means a rare phenomenon. Many millions of people are forced to invest at negative rates. In many countries, the inflation rates are higher than the interest rates available to investors. Their choice is to invest at the local bank, where interest rates are usually less than the inflation rate, or not to invest at all. In other words, they can invest at a negative rate or hold cash. Holding cash is equivalent to investing at $-f$, which results in an even greater loss than investing at the bank interest rate. Thus, inflation tends to further impoverish those with few opportunities for investment—a poor farmer in Brazil, for example, who wishes to safeguard his crop receipts until needed for household expenses, fertilizer, etc. Equally true is that inflation tends to further the fortunes of those with more opportunities for investment—the rich. Thus inflation is destabilizing, and the larger it is, the more it destabilizes. Apart from such social considerations, the negative investment rates it fosters cannot be ignored.

Another way that negative rates can enter analysis is not via i as in the previous example, but via f itself. Thus far, f has been treated as positive, that is, as an inflation rate. If deflation occurs—a general drop in the level of prices—f is negative. For example, if \$10,000 were held in cash for 1 year and a deflation of 2 percent occurred during that year, the buying power of \$10,000 at the end of the year would be

$$\frac{P}{1+f} = \frac{10{,}000}{1-0.02}$$
$$= \$10{,}204$$

The buying power of the money has increased by \$204, but the amount of cash in hand still remains \$10,000. Table 11.1 reveals that deflation took place from 1926 through 1933 in the United States. It may happen again. By substituting the word *deflation* for *inflation* and making the proper sign changes in the formulas, all that has been explained in this chapter with regard to inflation applies as well to deflation.

IS INFLATION EVER NONDIFFERENTIAL?

It has often been argued that because inflation will have the same effect on prices for any of the alternatives under consideration, it may be excluded because it is a nondifferential element among them. Consider two road projects that use the same materials. After construction at time 0, maintenance costs and user costs will rise at the same rate for each alternative. Does this mean that inflation can be disregarded? As the following example demonstrates, it cannot be ignored. The pattern of the cash flows becomes decisive.

■ **Example 11.22.** Consider the following two mutually exclusive alternatives A and B. The deflated opportunity cost of capital for the organization is 15 percent. Inflation is expected to maintain a constant rate of 10 percent per year. Using the current-dollar alternative cash flows below, comment on whether inflation is differential.

Y	*A*	*B*	*B* – *A*
0	–1,000	–1,200	–200
1	600	550	–50
2	500	550	+50
3	400	550	+150
4	300	550	+250

Solution. Disregarding inflation, the net present value of each alternative at i = 15 percent is

$$\begin{aligned}\text{NPV}_A &= -1{,}000 + 600(P/A, 15, 4) - 100(P/G, 15, 4) \\ &= -1{,}000 + 1{,}713 - 379 \\ &= +334\end{aligned}$$

and

$$\begin{aligned}\text{NPV}_B &= -1{,}200 + 550(P/A, 15, 4) \\ &= -1{,}200 + 1{,}570 \\ &= +370\end{aligned}$$

Therefore, select B.

Now, let us include inflation. This means evaluating the alternatives at u, the inflated rate, rather than i. That is, we will treat the cash flow as current dollars rather than constant ones, which they become if we exclude all consideration of inflation.

$$\begin{aligned}u &= i + f + if \\ &= 0.15 + 0.10 + 0.015 \\ &= 0.265 \text{ or } 26.5\%\end{aligned}$$

$$\begin{aligned}\text{NPV}_{B-A} &= -200 - 50\left(\frac{1}{1.265}\right)^1 + 50\left(\frac{1}{1.265}\right)^2 + 150\left(\frac{1}{1.265}\right)^3 + 250\left(\frac{1}{1.265}\right)^4 \\ &= -200 - 39.5 + 31.2 + 74.1 + 97.6 \\ &= -36.6\end{aligned}$$

Therefore, select A. This is the reverse of the answer arrived at when inflation is neglected. Therefore, inflation must be taken into account. ■

SUMMARY

Inflation becomes more important in economic analysis as its rate increases.

Inflation is an upward change in prices, and deflation is a downward change.

Because of the way inflation is measured—as an annual change in prices based on the previous year's prices—it is a compound-interest phenomenon. The most important measure of inflation in the United States is the *consumer price index* (CPI). Other indexes may be more appropriate for a particular problem.

One must distinguish among the following rates and cash flows:

1. Constant-dollar cash flows use the constant-dollar rate i.
2. Current-dollar cash flows use the current-dollar rate u.
3. The inflation (or deflation) rate is f. One must never use the constant (deflated) rate with a current-dollar cash flow, or vice versa.

The fundamental relation among these rates is

$$u = i + f + if$$

Inflation is measured as an effective annual rate. The other rate(s), *i* or *u*, must be changed to annual effective rates as well before they can be used in an analysis where *f* enters.

Inflation rates can be varied over the time horizon of the analysis with ease by computing yearly price indexes.

Any of the methods we have used thus far may be employed in inflation problems.

Income tax and inflation effects can be analyzed simultaneously. The important point to remember is that income taxes are paid on current-dollar cash flow.

Negative rates can be substituted into the compound-interest formulas.

Inflation should always be considered differential, even among alternatives that will experience the same rates of price increases.

PROBLEMS

11.1. On page 267 it was stated that, in late 1993, inflation in Russia was averaging 20 percent *per month.* What *yearly* inflation rate does this imply?

11.2. One billion (1×10^9) percent inflation rates have been reported in countries of the former Soviet Union and its satellites. What must be the daily rate of inflation to arrive at this yearly figure?

11.3. According to Table 11.1, page 268, 1994 prices in the urban areas of the United States have increased by how many times since these dates?
(*a*) 1980
(*b*) 1967
(*c*) 1950
(*d*) 1913

11.4. In Example 11.1, page 269, what is the inflation rate between the first and third years? Can the correct answer be obtained by adding the inflation rates for year 2 and year 3?

11.5. In Example 11.2, page 269, the cost of the market basket of goods and services for the third year has been recalculated to be $14,710. Compute the new CPI for year 3. All other information in the example remains valid.

11.6. In Example 11.3, page 270, the CPI for year 3 is now recalculated and becomes 121.3. Convert the current $5,000 cash flow to constant dollars of year 1.

11.7. In Example 11.4, page 270, the CPIs remain the same, but the cash flow in constant dollars is now 1,500, 1,800, and 2,200, respectively, for each of 3 years. Convert the new cash flow to current dollars.

11.8. Repeat Example 11.5, page 271, but change the base year to year 3. All other information remains the same.

11.9. In Example 11.6, page 271, convert Table 11.1, page 268, to a new base year of 1982 for the years 1982, 1983, and 1984.

11.10. In the latest CPI scale, no single year, or even month, is exactly 100.0 during the years 1982, 1983, and 1984. The exact CPIs are as shown:

Year	CPI (1982–1984 = 100)
1980	82.4
1981	90.9
1982	96.5
1983	99.6
1984	103.9
1985	107.6

Can you guess how the Bureau of Labor Statistics arrived at 1982–1984 = 100?

11.11. Current dollars are shown by years, along with a price index. Convert the cash flow to constant dollars of year 0.

Year	Price index	Current-dollar cash flow
0	100.0	–10,000
1	106.2	+1,500
2	108.4	+1,500
3	109.7	+1,500
4	111.8	+1,500
5	114.3	+11,500

11.12. An estimate is shown below for a project in which the costs and benefits have been projected in constant dollars based on year 0. The price index for each year is also shown. Calculate the cash flow in current money.

Year	Price index	Constant-dollar cash flow
0	100.0	–20,000
1	106.0	+8,000
2	113.0	+10,000
3	118.5	+12,000
4	126.1	+14,000

11.13. In Prob. 11.12, change the base year to 3. Calculate the price index based on year 3.

11.14. The *producer price index* (*PPI*) base year was changed from 1967 to 1982. See Table 11.2, page 272. For "all commodities," the situation was as shown:

Year	PPI (1967 = 100)
1967	100.0
1982	299.3

Changing the base year to 1982 caused 1967 to change to what index?

11.15. Read Prob. 5.17, page 93. A more precise account of inflationary effects will be included by inflating the *costs only*. The revenues will remain as they appear in the problem. The base year of the analysis is year 0. The producer price index (PPI) (all commodities) for the years shown below will be used to adjust all costs to that year. See Table 11.2, page 272.

Problem year	PPI year
0	1989
1	1990
2	1991
3	1992
4	1993

The minimum attractive rate of return will be changed to 25 percent under these new circumstances. All other data and information on the situation remain the same. Continue to use the *present-worth method.*

Should either franchise be chosen?

11.16. Over the years from 1988 through 1991, shown in Table 11.3, page 274, what was the average annual compound rate of inflation for the countries listed?

11.17. A comparison of certain bonds, of equal risk, and inflation rates within countries results in the following table:

Country	Bond interest rate (percent)	Inflation rate (percent)
Italy	7	5
Sweden	13	4
United States	3	3

If you were financial vice president for a major corporation, how would you make, among the countries, a decision to invest if the inflation rates and the bond interest rates were the only factors considered?

11.18. An investment in Germany will make a rate of return of 12 percent. The German inflation rate is projected at 4 percent for the life of the investment.

An equal amount may be invested in the United States at 16½ percent. Inflation is likely to be 10 percent over the life of the investment.

Where should you invest in order to realize the maximum deflated rate of return on your investment? Ignore the possible effects of exchange rates between Deutsche marks and dollars. Assume that *ceteris paribus* holds for all other aspects of the question.

Answer: Germany

11.19. As in Example 11.7, page 275, a U.S. government 30-year bond carries an interest rate of 5 percent. One of the most common mistakes in the solution of inflation problems is to misname the symbol. What symbol goes with this percentage?

11.20. In Example 11.8, page 277, use Brazil's average annual compound inflation rate for the years 1988 through 1991 to compute the amount of monetary buying power in today's money that José Anguelas will receive 10 years from now.

11.21. In Example 11.9, page 277, concrete for the dam at this moment, time 0, is estimated at $35 million per year for the next 4 years of construction. What will the concrete cost each year for these 4 years, in current dollars, if its cost increases at 5 percent per year?

11.22. In Example 11.10, page 278, imagine that the couple's estimated rate of return is 38 percent, along with an estimated inflation rate of 20 percent. What is their constant-dollar rate of return?

11.23. Barry Sarnoff, aged 30, is examining an investment in an individual retirement account. If he decides to invest in it, he will make the first payment of $2,000.00 at the end of this year. Twenty-nine similar payments in the same amount will follow for a total of 30 payments over the next 30 years. Sarnoff is promised an annual interest rate of 10 percent. On that basis he is told by his banker that he should be able to accumulate $328,988 by the end of the 30-year investment period.

Sarnoff wants to "inflation-proof" his investment, stating that he wishes to be able to realize a rate of 4 percent on deflated dollars. What is the maximum uniform annual inflation rate that Sarnoff can tolerate and still be able to fulfill his investment expectations? (Verify all figures provided by Sarnoff's banker.) Present your answer accurate to one decimal place.

11.24. A couple discover that they must invest $4,365 in constant dollars annually for the next 20 years to accumulate a sum they consider to be sufficient to guarantee an acceptable lifestyle when they retire. However, they have calculated the $4,365 based on a 10 percent return. That is, $4,365 is in deflated dollars in terms of year 0.

At what annual rate of return on current money must they invest if they estimate that the constant annual inflation over the next 20 years will be 15 percent?

Answer: 26.5 percent

11.25. Repeat Example 11.11, page 278, except the interest rate on your account has been increased to 15 percent. The rest of the information and the question remain unchanged.

11.26. A mutual fund account began with $100,000 on August 29, 1988. When it was cashed in on April 4, 1994, it was worth $224,538. The owners claim an annualized gain of 15.41 percent. Are they correct?

With inflation at 4 percent annually, what was the constant-dollar rate of return, based on 1988 dollars?

11.27. The airline of Prob. 10.30 estimates an inflation of revenues, costs, and resale values of 9 percent annually over the next 13 years. All other information remains the same.

If the airline requires an after-tax, constant-dollar rate of return of at least 6 percent, should the aircraft be purchased?

11.28. If you had placed $1,000 in a savings account on January 1, 1994, at a Mexican bank paying an 8½ percent effective annual rate and if the inflation rate during the same year was 9 percent, what would have been the value of your investment, in January 1, 1994, dollars, on January 1, 1995?

11.29. In *The New Republic* of April 5, 1980, the following statement[3] was made:

> ... The long-term fixed interest bond is totally dependent upon a stable currency. For example, take a rosy view and assume that inflation goes along at a steady and cool rate of 10 percent per year. Assume a $10,000 twenty-year bond due in the year 2000. Simply to preserve the purchasing power of his original capital the lender would have to be paid back $67,276 in the year 2000; it does not include any return for the use of his money and related credit risk.

Is the statement numerically true or false? Show your calculations.

Answer: True

11.30. (*a*) One million dollars is invested at 10 percent in the Voss Savings and Loan Association. If the investor wishes to withdraw an equal

---

[3]William J. Quirk, "The Great American Credit Collapse," *The New Republic*, April 5, 1980, p. 24.

amount at the end of each of 2 years, how much can be withdrawn so as to leave nothing in the account? Ignore inflation for the moment.

(*b*) During this period, a 5 percent inflation occurs for each of 2 years. Will the *amount* the investor can withdraw at the end of each of 2 years change because of this occurrence?

(*c*) Will the *value*, that is, the buying power, of the two equal amounts change because of inflation? If so, what will be the buying power in constant dollars of year 0 of the two equal amounts?

(*d*) In this situation, identify or compute f, u, and i.

(*e*) What amount can be withdrawn from the account at the end of 2 years to close it out, if nothing is withdrawn before then?

(*f*) What will be the value in constant dollars of year 0 of the closing withdrawal of part (*e*)?

11.31. Now imagine that the Voss Savings and Loan Association of Prob. 11.30 establishes a branch in Rio de Janeiro, Brazil. Because of a high rate of inflation for many years, savings institutions' payments to investors have been tied to an inflation index so that investors will receive a current *cruzeiro* (the monetary unity of Brazil) payment equal in buying power to the constant payment they would have received had there been no inflation.

(*a*) Given the same dollar amounts and the same interest and inflation rates as in Prob. 11.30, what amount in dollars would be received at the end of each of 2 years if the payments were indexed to the inflation? Assume that the dollars are converted to cruzeiros and back again at a constant rate without gain, loss, or commission.

(*b*) Identify or compute f, i, and u for the cash flow of part (*a*).

(*c*) What amount can be withdrawn at the end of 2 years if no funds are removed? Under Brazilian law, the amount accumulated in the account at the end of each year must be adjusted upward to retain its buying power.

11.32. In Example 11.12, page 279, the Longstreets shift their investment to another savings and loan association that guarantees the same nominal 12 percent return, but compounded daily. What will be their deflated rate of return on this investment, if the other data remain the same?

11.33. You have the opportunity to invest \$10,000 in a certificate of deposit for 2 years. The nominal interest rate per year on this certificate of deposit is 10 percent, compounded quarterly. You expect that the inflation rate will be 9 percent per year. What is your effective rate of return on this investment, after inflation has been taken into account?

11.34. In Example 11.13, page 280, the inflation rate has been increased to 25 percent, and the constant-dollar opportunity cost of capital to 8 percent.

What is the present worth of the cash flow in terms of year 0 constant dollars? All other information remains unchanged.

11.35. McDonald Engineering Company plans to open an office in Argentina. The following net cash flow, after all taxes and expenses are paid, is expected for the first 4 years of the venture:

Year	Current pesos (000,000,000)
0	−6
1	−2
2	+4
3	+40
4	+125

Inflation is expected to remain at 100 percent during the 4 years of the study. A 50 percent rate of return per annum is the minimum that McDonald Engineering will accept, on constant money, for investments in *any* South American country.

Should McDonald Engineering make this investment?

Answer: No

11.36. In Example 11.14, page 281, what must be the sale price of the property in current dollars, *ceteris paribus*, if the rate of return is increased to 45 percent and the expected inflation rate to 16 percent?

11.37. Regina Barletta, a photographer, is trying to decide on the investment of her life savings in a real estate property. Her accountant estimates annual after-tax profits to be $85,500 in terms of dollars of the year of purchase, that is, year 0. During the first year and each succeeding year, the profits in current dollars will rise by 7 percent annually. Inflation is expected to rise at the same rate. The property will cost Barletta $350,000. She estimates a resale value of $500,000 in current dollars after 6 years.

(*a*) What is the rate of return on constant year 0 dollars of this investment? Give the answer to the nearest whole percent.

(*b*) If Barletta chooses to invest her capital in an investment of equal risk, what is the minimum annual current-dollar rate that she should be willing to accept, given the opportunity to invest in real estate?

11.38. The Teton National Forest has received bids from three logging companies. These have been analyzed, and the following costs and benefits to Teton were determined. The discount rate is 10 percent. In using this rate, you assume that the federal government is now requiring that inflation be included in the analytical procedures and that 10 percent is to be used on deflated dollars. All discount rates used must be accurate to the nearest whole percent. The inflation rate is projected at 9 percent annually over the next 5 years. No taxes will be paid since Teton National Forest is a federal government entity. Which company, if any, should be chosen to do the job, based on dollars of constant purchasing power? A benefit/cost ratio analysis is the *only* acceptable approach to this problem.

Year	Ashland ($000)	Bertwick ($000)	Chapman ($000)
0	-30	-45	-55
1	0	+15	+100
2	0	+15	0
3	+50	+30	0
4	+50	+30	0
5	+50	+30	0

11.39. In Example 11.15, page 282, the coal port manager wishes to investigate the same cash flow over the years from 1994, when the investment of $10.25 million would have been made, to 2006, when the handling equipment would be scrapped. The inflation rate is the average annual compound rate from 1981 to 1992 inclusive, rounded to the nearest whole percent. Find the net present value in terms of 1994 dollars.

11.40. In a certain Middle Eastern country, a design engineer in the bridge department of the Ministry of Transport had before him estimates of the costs in U.S. dollars of two mutually exclusive alternatives for a highway bridge, one of steel and the other of reinforced concrete. One must be selected:

	Steel	Reinforced concrete
First cost (millions)	60	75
Annual maintenance cost (millions)	5	1
Life (years)	20	40

The above costs were estimated in constant dollars of year 0. An inflation rate of 50 percent per year was estimated for the following 40 years. If the social discount rate on current dollars is 80 percent, which is the more economical alternative?

The B/C ratio method is to be used. No other method will be credited.

11.41. Your consulting contract with the Corps of Engineers requires an analysis of certain navigational aids on the Mississippi River which will cost $10 million at the beginning of the project (time 0). Benefits are estimated at $600,000 annually and costs of operation and maintenance at $200,000 annually, both in constant dollars based on year 0. Both benefits and costs will increase at the rate of 4 percent per year. A 30-year time horizon for the analysis is used.

Removal costs at the end of a 30-year life are estimated at $800,000—a disbenefit—in current dollars.

A discount rate of 7 percent on current dollars is stipulated for the analysis. An inflation rate of 4 percent per year is projected.

All discounting, inflating, or deflating rates used in the analysis must be rounded to the nearest whole percent so that compound-interest tables can be consulted in the calculations. Since this is a government project, no income taxes are to be considered.

Should the project be built? Use the present-worth method. No other method will be accepted under your contract.

11.42. In 1988, a chemical processing plant invested in a new sulfuric acid facility. The cost that year was $3,426,000. In the subsequent 5 years, the net profit after taxes made by the facility and the appropriate price index were as shown:

Year	Net profit (current $)	Price index (1967 = 100)
1988	—	125.3
1989	1,800,000	133.1
1990	1,932,000	147.7
1991	1,740,000	161.2
1992	1,816,000	170.5
1993	1,540,000	181.5

If the minimum attractive rate of return after taxes and based on constant dollars was 5 percent, determine the NPV of the investment as of 1988 in terms of 1988 dollars. Was the investment a sound one?

11.43. The table below shows two mutually exclusive projects with their current cash flows by year. It also shows the appropriate price index for each year that the service will be performed. There are no salvage values for either project. A 20 percent constant-dollar discount rate is used.

Year	Project 1	Project 2	Price index
0	−150	−190	100
1	50	60	101
2	70	90	102
3	100	150	104
4	130	170	106
5	170	190	108

(*a*) Compute the *B*/*C* ratio for project 1 before taxes and corrected for inflation.

(*b*) Compute the incremental *B*/*C* ratio between the two projects, corrected for inflation.

(*c*) Which of the two projects would you select, *ceteris paribus*?

(*d*) If you did not correct for inflation, what would be the incremental *B*/*C* ratio?

11.44. In Example 11.16, page 283, the estimates of future inflation are changed to 6 percent for the coming year, 7 percent for the following year, and 9 percent for the last year. What is the rate of return in constant dollars, before taxes, for this investment, to the nearest whole percent? All other data remain the same.

11.45. Horatio Hercules, a professional wrestler, is trying to decide whether to invest his life savings in an apartment house. It will cost $3.3 million. His accountant's estimates of rental revenues and all costs will result in an annual profit of $264,000, after taxes, based on the time of purchase, that is, year 0. During the first year and each succeeding year, net profits will rise by 5 percent annually, as will inflation. A real estate appraiser reports that the property can be sold for $4.21 million in current dollars in 5 years. What is the rate of return on constant year 0 dollars of this investment? Give your answer to the nearest whole percent.

11.46. A small, temporary shingle factory will cost $150,000. It is projected to gain revenues of $80,000 annually and will cost $55,000 annually—both in constant year 0 dollars. At the end of the 5 years that the factory will last, the property on which the factory will be built will be sold for $30,000 in constant dollars.

Over 5 years, costs are expected to rise at the rate of 4 percent per year and revenues at the rate of 8 percent per year based on year 0 dollars. The consumer price index, which the analyst will use to measure the buying power of her gains, if any, will rise at about 6 percent per year, the same rate at which the property is expected to appreciate.

Before the effect of taxes is included, what will be the net present value of this investment, if the investor expects a constant-dollar deflated rate of return on her investment of 9.4 percent? Discount rates may be rounded to the nearest whole percent.

11.47. In Example 11.17, page 284, the Uruguayan engineer revises his estimate on the first cost of corrugated steel pipe (CSP) to $225,000. The other data remain the same as before. Which pipe should be chosen?

11.48. Repeat Prob. 11.47, using an inflation rate of 5 percent and a deflated rate of return of 10 percent.

11.49. Prices on a steel highway bridge to be built by Yakima County are presently as shown:

Piers and abutments	$2,300,000
Steel and construction in place	4,100,000
Concrete wearing surface	695,000
	$7,095,000

If construction is delayed for 2 years, the following inflation rates will be experienced per year:

Piers and abutments	3%
Steel structure in place	4%
Concrete wearing surface	2%

(*a*) What will be the cost of the bridge 2 years hence, to the nearest $1,000?

(*b*) What will be the dollar result, in present-worth terms, of delaying the construction of the bridge for 2 years, when only the first cost of the bridge is considered? Yakima County's opportunity cost of capital on current dollars is 10 percent. Is it better to delay construction for 2 years? Why?

11.50. Three types of storm sewer pipe are being considered for installation in a small city in Virginia. The facts about the installed pipe are given below:

Type	Cost ($)	Life (years)
Reinforced-concrete pipe (RCP)	800,000	75
Corrugated steel pipe (CSP) A12	450,000	25
Asphalt-coated galvanized steel CSP/Zn	400,000	15

Inflation in pipe costs is estimated at 5 percent per year over the next 75 years. Removal costs are included in the first costs. Maintenance cost will be nondifferential among the pipe types. A discount rate recommended by the Water Resources Council is 8 percent. This rate refers to current dollars. Because this is a municipal project, no income taxes will be paid. Using the NPV method, select the most economical pipe.

11.51. Four types of solar home designs have been suggested to replace a house presently heated by gas. They are summarized below:

	Additional first cost compared to gas-heated house	Annual gas saving at the end of first year
Two-level	11,000	400
Superinsulated	8,000	290
Thermal storage wall	9,500	310
Thermal storage tank	10,300	305

Gas costs are expected to rise by 10 percent per year for the next 20 years. One house design must be chosen as optimal. All designs are assumed to appreciate in value at about 10 percent per year, based on year 0, over the next 20 years, the study period.

The rate of return expected in real estate investments of similar risk, based on current dollars, is 32 percent. All money amounts are in constant dollars as of time 0. Tax effects may be taken as equal for all designs.

Which of the four designs is optimal?

All interest rates may be approximated to the nearest whole percent.

11.52. A large eastern seaboard city in the United States is considering the decision to rebuild an elevated railway. The job will be completed while the railway, carrying 50,000 passengers daily, is in operation. The rebuilding will take 3 years. Costs in millions of *constant* dollars of year 0 are as shown:

Year	Labor	Materials	Design
0	—	—	4
1	10	6	—
2	12	8	—
3	11	5	—

Labor costs will increase by 5 percent per year, and materials costs by 4 percent. The opportunity cost of capital for this city is 12 percent in *current* dollars.

A municipal bond issue will cover the cost. What should be the cost recoverable from the issue at time 0 in order to finance all the above costs?

11.53. In Example 11.18, page 286, the Lancer aircraft is now estimated to cost $40,000. MACRS GDS will be used with the half-year convention and a 7-year recovery period. The tax rate has been revised to 34 percent. All other data remain the same as before. Should the aircraft be purchased?

11.54. Spartacos, Inc., is planning to invest in equipment that promises a uniform annual saving in production costs of $5,000 in terms of each year's dollars. This piece of machinery has a first cost of $15,000 and is expected to have an economic life of 5 years. There will be no salvage value. The depreciation allowance will be based on the pre-1981 straight-line method. Spartacos is in the 40 percent tax bracket. Assume an annual inflation rate of 4 percent over the next 5 years. If Spartacos' minimum attractive rate of return on current dollars after taxes is 9.2 percent, should it buy this piece of equipment?

If necessary, round discount rates to the nearest whole percent.

11.55. Alpha International, Inc., is planning to invest in equipment that promises a uniform annual saving in production costs of $300,000 a year in current dollars for a period of 5 years. This machinery will cost $1 million. It will have a zero salvage value at the end of the fifth year. The depreciation allowance will be based on the pre-1981 straight-line method required for this type of overseas property. Alpha International is in the 50 percent tax bracket. Assume an inflation rate of 10 percent annually over the next 5 years. The opportunity cost of capital for Alpha International is 14 percent after taxes and in current dollars. Should Alpha International buy this equipment? Take into account the effect of taxes and inflation.

11.56. Omicron International, a mining company, is studying an investment which will return as follows:

Year	Cash flow (000,000)
0	−1,500
1	+500
2	+500
3	+500
4	+300
5	+300
6	+300

All figures are in constant dollars based on year 0.

The country where the investment will be made will allow depreciation of the entire first cost on a U.S. pre-1981 straight-line basis over 3 years. No salvage value is anticipated. The tax rate will be 32 percent of taxable income.

The opportunity cost of capital for Omicron International in current after-tax dollars is 12 percent. Inflation for the coming 6 years is projected to increase at 6 percent per year.

Should Omicron International accept the investment?

11.57. In Example 11.19, page 287, the inflation index for 1993 has been recalculated and is now 1,279.0. How will this affect the results, if no other changes in data take place?

11.58. In Example 11.20, page 289, your tax bracket is changed to 25 percent. Will this affect your decision on the loan, if no other effects are present?

11.59. In Example 11.21, page 291, the time is 1980; the place, the United States. Inflation has reached 13.5 percent. A commercial bank will pay 10 percent interest for $1,000 that you have to invest. You know of no other place to put the money.

(*a*) What is your deflated rate of return on your investment?

(*b*) How much money in terms of beginning-of-year buying power will you have at the end?

11.60. In a developing country, inflation averages 30 percent per year. For lack of any other acceptable opportunity, you invest in a bank certificate of deposit which will result in the following cash flow:

Year	Cash flow
0	−1,000
1	+170
2	+170
3	+1,170

(*a*) What is your rate of return in terms of the buying power of year 0 money?

(*b*) If this investment is the only one available to you, should you not put the $1,000 cash into a safe deposit box for the next 3 years? Why or why not?

11.61. Salvador Landeros Limon, a Mexican farmer who works 17 hectares of land, has just sold his corn crop for 6 million pesos (about $2,000). He and his family must live for a year on this amount as well as buy fertilizer for next year's crop. In the meantime, he must do something with the cash.

Inflation for that year, 1991, was 22.7 percent. The local bank in the nearest small town will pay 15 percent interest. Should he bury the money in a pot or put it in the bank? What is his real interest rate on these sole choices open to him?

11.62. In Example 11.22, page 292, the inflation rate has gone up to 16 percent, and the deflated opportunity cost of capital is 25 percent. Will these changes accentuate the difference between considering and not considering inflation, if all other data remain the same?

11.63. "The interest rate, or opportunity cost of money, is also important in selecting alternatives. Since public agencies don't compete for funds in the private sector, it is difficult to identify a reasonable rate for the purpose of life-cycle analysis. One acceptable approach would be to assume an interest rate equal to the prime bank interest rate minus the rate of inflation."[4]

(*a*) Identify the *prime bank interest rate* and *rate of inflation* of the quotation in terms of i, f, or u.

(*b*) Is the "interest rate," as defined in the last sentence of the quotation, to be used with constant- or current-dollar cash flows? Assume that the method is correct for the moment.

(*c*) The prime bank interest rate is 8 percent. The inflation rate is 6 percent. The interest rate, according to the method described in the quotation, is therefore 2 percent. Is this calculation correct? If not, what is the correct interest rate, to two decimal places?

(*d*) In a country, not the United States, the prime bank interest rate is 80 percent, and the inflation rate is 100 percent. What is the correct discount rate—the "interest rate" mentioned in the quotation?

(*e*) Is it true, in times of $200 billion federal deficits, that "public agencies don't compete for funds in the private sector"? Why, very briefly?

---

[4]James K. Cable, "The View from Iowa," *Civil Engineering*, November 1985, p. 63.

CHAPTER 12

RISK

Up until now, in this text we have assumed a world without risk. Along with income tax and inflation, the riskiness of investments was ignored in order that the foundations of economic analysis might be laid. In Chaps. 10 and 11, income tax and inflation were introduced as important factors in the analysis of investments. Now, in this chapter, the risk involved in capital investments will be discussed. Because a riskless world does not and cannot exist, some method of accounting for risk must be used. Even if, as in the case of a U.S. savings bond, the risk of default is almost infinitesimal, still, it does exist. Although we may ignore such a small risk, it is unwise to neglect the large risk involved in wildcat oil drilling or land speculation. With such investments, risk is an aspect that must be included in our thinking about them and, if possible, in our calculations. Decision under risk is the subject of this chapter.

DEFINITIONS

Certainty means that events cannot be other than we have imagined them. If we have estimated a fifth-year income of $10,512.39 from an investment in real property, this number is treated as though it was bound to occur. We may doubt our estimate, but because we have assumed away risk, we treat it as a certainty.

If we have the possibility of more than one outcome in the future and the probability of each future outcome is known, or can be estimated, then any decision with regard to these outcomes will be made under risk. (This is not the definition of risk itself. A quantitative definition of risk will be made later in this chapter.) For example, we estimate a fifth-year income from an investment in real property at a 20 percent chance of being $9,432, a 50 percent chance of being

$10,316, and a 30 percent chance of being $11,558. The investment will be made under risk, because more than one possible outcome exists, the chances related to each outcome are known, and the chances sum to 100 percent.

Now imagine that the possibility of more than one outcome in a set of mutually exclusive outcomes exists, the set includes all possible outcomes, and the probability of each outcome's occurring is unknown and cannot be estimated. A decision made under these conditions is made under uncertainty. For example, suppose that the previously used estimates of $9,432, $10,316, and $11,558 do not have probabilities associated with them. Any one of three may occur, but the chances of each one's occurring are unknown.

PROBABILITY MEASUREMENT

The *probability* of an event $p(E)$ is the limit approached by the ratio of the number of times the event actually occurs to the number of trials in which the event can occur, as the number of trials is increased without limit. In symbols,

$$p(E) = \lim\left(\frac{\text{events}}{\text{trials}}\right) \tag{12.1}$$

The probability of an event $p(E)$ is always a fraction between 0 and 1 where more than one trial is made.

$$0 \leq p(E) \leq 1$$

If the probability equals 1, the event is certain to happen; if it equals 0, the event will never happen. A simple example will show how Eq. (12.1) is used.

■ **Example 12.1.** A pair of dice is believed to be unfair, that is, "loaded." To determine the probability of each number from 2 to 12 appearing on a given throw, the dice are thrown a great many times. Each throw is a *trial* in the definition illustrated in Eq. (12.1). The *event* is the appearance of 2 through 12. For any single event—the appearance of a 2, say—tabulation of the results of 7,200 throws shows 2 coming up 210 times. The probability as a result of these trials is

$$p(E) = \frac{210}{7{,}200} = .02917$$

The number of trials is increased to 14,400, and 2 appears 422 times. The resulting probability is

$$p(E) = \frac{422}{14{,}400} = .02931$$

As the number of trials is increased, the probability of a 2 appearing continues to approach a value whose accuracy increases with the number of trials. ■

In many cases, it is not necessary to determine the probability of events by trials. The probabilities may be measurable by what might be called the *mechanics* of the system. If an event can happen in a ways and cannot happen in b ways, and if all these ways are equally likely to occur, then the probability of the event's happening is

$$p(E) = \frac{a}{a+b} \tag{12.2}$$

■ **Example 12.2.** A gambler wishes to determine the probability of drawing the queen of spades from a standard deck of 52 cards. This event can happen in only one way and fails to happen in 51 ways. The probability of drawing the queen of spades is therefore, according to Eq. (12.2),

$$p(E) = \frac{1}{1+51} = \frac{1}{52} = .01923$$

Each drawing of a card is an *independent event* if the card drawn is replaced each time in the deck. The probability of .01923 will hold for this independent event. It in no way depends on the result of a previous draw. But now imagine that the experiment of Example 12.2 is repeated. The queen of spades is not drawn. Instead a two of clubs appears.

What is the probability of drawing the queen of spades on the next draw, if the two of clubs is not replaced in the deck? It is

$$p(E) = \frac{1}{1+50} = .01961$$

The probability has increased because the second drawing was *dependent* on the first. ■

EXPECTED VALUE

On the basis of the previous discussion of probability, the useful concept of *expected value* can now be developed. But first we must define a *random* or *stochastic* variable.

Variables, each one of which, depending on chance, can assume a given value with a definite probability, are called *random variables*. A random variable is defined (1) if the set of its possible values is given and (2) if the probability of each value's appearance is also given. The toss of a single die can result in the appearance of a number from 1 through 6, each with a probability of $\frac{1}{6}$, or .1667. Each appearance of a number, or point, on the die is a random variable. A six of hearts drawn from a deck of 52 cards is the result of the draw, which is a random variable with a probability of $\frac{1}{52}$ of appearing on any single independent draw.

Now suppose that a random variable X possesses n values:

$$X_1, X_2, \ldots, X_n$$

and that $$p_1, p_2, \ldots, p_n$$

are the probabilities associated with each X. The *expected value* of X is

$$\mu = E(X) = p_1X_1 + p_2X_2 + \cdots + p_nX_n$$

$$\mu = \sum_1^n p_iX_i \tag{12.3}$$

■ **Example 12.3.** The dice of Example 12.1 are discarded, and a single, perfectly "fair" die is substituted for them. A die is *fair* if each of its faces has an equal chance of appearing on a throw. What is the expected value of an unlimited number of tosses of the die?

Solution. Substituting in Eq. (12.2)

$$p(E) = \frac{a}{a+b} \tag{12.2}$$

we find the probability of any face of the die showing up is

$$\frac{1}{1+5} = \frac{1}{6}$$

From Eq. (12.3), the expected value of an unlimited number of tosses is

$$E(X) = \frac{1}{6}(1) + \frac{1}{6}(2) + \frac{1}{6}(3) + \frac{1}{6}(4) + \frac{1}{6}(5) + \frac{1}{6}(6)$$
$$= \frac{1}{6}(21) = 3.5$$

■

To further discuss expectation, certain rules of probability must be established.

1. The probability of an event may never be negative:

$$p(E) \geq 0 \tag{12.4}$$

2. The sum of the probabilities corresponding to each outcome in a set of mutually exclusive and exhaustive outcomes must be unity, denoted by

$$p(R) = 1 \tag{12.5}$$

where R is the set of probabilities corresponding to the outcome set S.

3. For two mutually exclusive events E_1 and E_2, the probability that *either* E_1 or E_2 will transpire is the sum of the probabilities of the two separate events:

$$p(E_1 \cup E_2) = p(E_1) + p(E_2) \tag{12.6}$$

4. For two independent events, the probability of both occurring simultaneously is the product of their separate probabilities:

$$p(E_1, E_2) = p(E_1)p(E_2) \tag{12.7}$$

■ **Example 12.4.** If we substitute fair dice for the loaded dice of Example 12.1, we will not have to make a large number of trials to determine the expected value of unlimited tosses of the dice. Using the foregoing rules will enable us to do so because the dice are a mechanically perfect system—not because such a perfect pair is possible, but because we have assumed them so.

Let us take variable X to be the total number of points on the two dice on a toss. For X to be 2, a 1 must show up on each die. The probability of a 1 on a single die is $1/6$, as we have seen in Example 12.3. The probability of a 1 showing on both dice is, by Eq. (12.7),

$$p = \left(\frac{1}{6}\right)\left(\frac{1}{6}\right) = \frac{1}{36} = .02777$$

This is a different result from that obtained by 14,400 trials in Example 12.1. Nor will 100,000 trials in Example 12.1 get us any closer to the value .02777, because the dice of that example are loaded. On the other hand, if we were to toss our fair dice 7,200, 14,400, and 100,000 times, we would find that the probability corresponding to a 2 would approach closer and closer to .02777.

The probability of obtaining a value of 3 must be ascertained by enumerating first the number of ways a 3 may show up. It may show up as a 1 on the first die and a 2 on the second die, or vice versa. The probability of a 1 on the first die is $1/6$, and a 2 on the second die is also $1/6$. The probability of both these events occurring simultaneously is, by Eq. (12.7),

$$\left(\frac{1}{6}\right)\left(\frac{1}{6}\right) = \frac{1}{36}$$

The probability of a 2 on the first die and a 1 on the second die is also $1/36$.

The probability of either of those two ways of obtaining a 3 on a roll of two dice is, by Eq. (12.6),

$$\frac{1}{36} + \frac{1}{36} = \frac{2}{36}$$

All the probabilities of achieving points on a pair of fair dice may be determined similarly. For obtaining a 7, for example, six ways exist, each with probability $1/36$. The probability of a 7 appearing is

$$\frac{1}{36} + \frac{1}{36} + \frac{1}{36} + \frac{1}{36} + \frac{1}{36} + \frac{1}{36} = \frac{6}{36}$$

The events in set S and the probabilities in set R correspond as follows:

S	R	S	R	S	R	S	R
2	$\frac{1}{36}$	5	$\frac{4}{36}$	8	$\frac{5}{36}$	11	$\frac{2}{36}$
3	$\frac{2}{36}$	6	$\frac{5}{36}$	9	$\frac{4}{36}$	12	$\frac{1}{36}$
4	$\frac{3}{36}$	7	$\frac{6}{36}$	10	$\frac{3}{36}$		

The most likely result of a toss is a 7; the least likely is a 2 or a 12. The expected value is, by Eq. (12.3),

$$\mu = \sum_{i=1}^{n} p_i X_i$$

$$= \frac{1}{36}(2) + \frac{2}{36}(3) + \frac{3}{36}(4) + \frac{4}{36}(5) + \frac{5}{36}(6) + \frac{6}{36}(7) + \frac{5}{36}(8)$$

$$+ \frac{4}{36}(9) + \frac{3}{36}(10) + \frac{2}{36}(11) + \frac{1}{36}(12)$$

$$= \frac{252}{36} = 7$$ ■

■ **Example 12.5.** A major U.S. airline made the following observations on its daily Boeing 747 service between New York and London during 1994:

Flights	Passengers	Flights	Passengers
79	350 (full)	23	290–299
85	340–349	18	280–289
40	330–339	15	270–279
34	320–329	12	260–269
28	310–319	7	250–259
24	300–309	365	

Calculate the expected value of future load factors (the number of passengers divided by the capacity). To calculate the load factor (LF), use as the midpoint of the given range a number whose last digit is 5.

Solution. Table 12.1 shows the solution of the problem. The first column shows the number of flights as given in the problem. The second column shows the calculated load factor. For 79 out of 365 flights, the airplane was full and the load factor was therefore 1.0. For 85 flights the number of passengers was between 340 and 349. We use the midpoint as 345. The load factor was therefore:

$$\frac{345}{350} = 0.99$$

and so on.

The third column, p_i, was calculated by dividing the number of flights by 365. The probability of achieving a load factor of 1.00, that is, a full airplane, was

$$\frac{79}{365} = .22$$

The rest of the probabilities were calculated similarly.

The expected value of the load factor E(LF) was calculated by multiplying each load factor by the probability of its appearance and summing the results. The answer is

$$\mu = E\,(\text{LF}) = \sum_{i=1}^{11} p_i X_i = 0.92$$

Expected value can be used to introduce risk into problems where inputs are considered as certainties rather than, as is often the case, variables able to assume a wide range of values. The following two examples illustrate the procedure with such inputs.

TABLE 12.1
Boeing 747 load factors

Flights	Load factor X_i	p_i	E(LF)
79	1.00	.22	0.22
85	0.99	.23	0.23
40	0.96	.11	0.11
34	0.93	.09	0.08
28	0.90	.08	0.07
24	0.87	.07	0.06
23	0.84	.06	0.05
18	0.81	.05	0.04
15	0.79	.04	0.03
12	0.76	.03	0.02
7	0.73	.02	0.01
365		1.00	0.92

■

■ **Example 12.6.** A telephone company, in a western state, with many miles of rural lines made a study of telephone pole mortality. The results are condensed in the following table:

Age	Percentage retired
5	3
10	10
15	17
20	22
25	20
30	16
35	7
40	4
45	1
	100%

A telephone pole cost $423.10, installed, according to the accounting department. The cost of capital, before taxes, to the telephone company was 15 percent. What annual cost, before taxes, including principal and interest, should be used for a pole in economic studies by this company?

Solution. The expected value of the annual cost of a telephone pole is, by Eq. (12.3),

$$E(A) = \sum_{i=1}^{n} p_i X_i$$

$$= 423.10 \sum_{i=1}^{N} p_i \left(A/P, 15, N \right)$$

$$= 423.10\left[.03\left(A/P, 15, 5\right) + .10\left(A/P, 15, 10\right) + \cdots + .01\left(A/P, 15, 45\right)\right]$$

$$= \$70.45$$

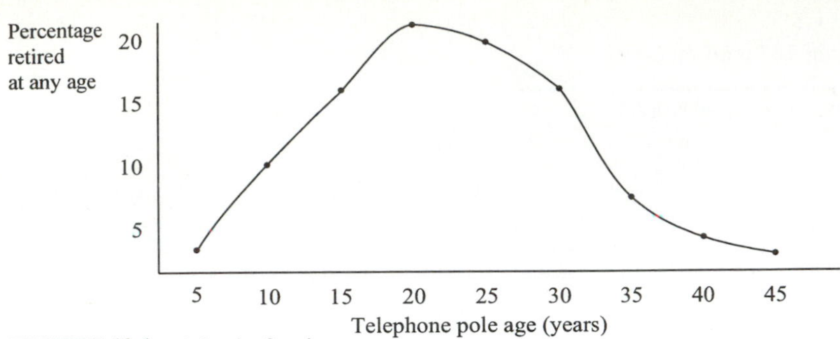

FIGURE 12.1 A density function

If we plot the age of the pole on the abscissa and the percentage retired as the ordinate, the resulting graph is called a *density function*, a plot of p versus X. See Fig. 12.1. ■

The following example shows how expected value is considered on an annual basis.

■ **Example 12.7.** A large publicly owned metropolitan bus system has accumulated data on two models of buses it has used over the past 5 years. Both models have exhibited frame defects that have required repair. Frame repairs cost an average of $1,076.62 per failure. Cost accounting data did not permit any more accurate breakdown as to model repaired, year repaired, or frequency per bus. The following data on frame repairs are available:

	Model 1	Model 2
Fleet size	267	102
Year		
1	16	5
2	35	11
3	64	22
4	76	40
5	87	67

A new fleet of 1,000 buses is to be purchased, and model 1 or 2 will be chosen. The following data are available for each model per bus.

	Model 1	Model 2
First cost	$107,000	$111,000
Salvage value (end of year 5)	$10,000	$12,000
Maintenance cost other than frame repair (annual)	$600	$600

A 5-year study period is chosen. The opportunity cost of capital is 10 percent. No taxes or inflation are to be considered. Benefits are the same no matter which model is chosen. Expected cost is the criterion of choice. The *present-worth method* is required.

Solution. Evidently, it must be assumed that the available data on frame repairs are sufficient for the purpose at hand. It is also assumed that each bus breaks down only once a year.

The probability that a frame repair will occur during year 1 is

$$\frac{16}{267} = .060$$

for model 1 and

$$\frac{5}{102} = .049$$

for model 2.

The expected value of the frame repairs during year 1 for the new fleet of 1,000 buses is

$$\frac{16}{267}(1{,}000)(1{,}077) = 64{,}539$$

for model 1 and

$$\frac{5}{102}(1{,}000)(1{,}077) = 52{,}794$$

for model 2.

If similar calculations are performed for the rest of the years, the following table of expected value of frame repairs for each model results.

	***E*(frame repairs) ($)**	
Year	**Model 1**	**Model 2**
1	64,539	52,794
2	141,180	116,147
3	258,157	232,294
4	306,562	422,353
5	350,933	707,441

The differential costs and benefits between the two models, if we neglect the $600 maintenance cost other than frame repairs, are:

	Model		**Increment**
Year	**1**	**2**	**2 - 1**
0	-107,000,000	111,000,000	4,000,000
1	-64,539	-52,794	+11,745
2	-141,180	-116,147	+25,033
3	-258,157	-232,294	+25,863
4	-306,562	-422,353	-115,791
5	-350,933	-707,441	-365,508
5	+10,000,000	+12,000,000	+2,000,000

The expected NPV of the incremental cash flow is, at 10 percent, approximately -$3 million. Therefore model 1 should be selected and model 2 rejected. ■

OBJECTIVE AND SUBJECTIVE PROBABILITIES

Objective probabilities of a set of events can be estimated from the attributes of the system in operation. For example, a die with six sides has six points on the die. The probability of the appearance of any number on a single toss is $\frac{1}{6}$. (See

Example 12.3.) Objective probabilities can also be determined from observed data. For example, many tosses of loaded dice in Example 12.1 resulted in an estimation of the probability of a 2 showing up on any one toss. In Example 12.5, observations of the number of passengers per flight resulted in objective probabilities of load factors. But what do we do about events that will occur but once and that have not happened before? We can generate subjective (intuitive) probabilities.

Considerable controversy surrounds intuitive probabilities. On the pro side, invention of such probabilities introduces the idea that outcomes are not certain. This, in itself, is an advantage, because it then forces prospective investors to consider what the outcomes of a course of action may be. The next step relates outcomes to probabilities. Even if only rough guesses can be made as to what the probabilities are, the discussion aroused by the attempt will emphasize again the riskiness of the investment.

On the con side, subjective probabilities are condemned as mere guesses that lend an undeserved air of exactness to an analysis. Such criticisms point out that no matter how sophisticated the methods used to arrive at the subjective probabilities, they remain a weak framework on which to base a decision.

The following examples use subjective probabilities.

■ **Example 12.8.** A construction company engaged in building a new roadbed for the Union Pacific railroad immediately adjacent to the existing lower-level track, soon to be engulfed by the lake behind McNary Dam on the Columbia River, is considering the risks associated with blasting a rock cut of 180 feet to make way for the new roadbed. If the existing railroad is blocked, a penalty clause in the contract prescribes a fine of $5,000 per hour per blockage, regardless of whether a train is actually halted. Any possible blockage from the blast can be cleared in 1 hour or less.

The rock cut can be drilled and blasted in three ways with the following associated probabilities of blockage and costs.

Alternative	Probability of blockage	Cost of blast ($)
A One 180-foot blast	.50	30,000
B Two 90-foot blasts	.30	32,000
C Three 60-foot blasts	.10	36,000

The cost of the blast is composed of costs for labor, powder, equipment, surveying, and moving the equipment in and out. The blockage probability is the total probability of blocking the tracks associated with all the blasts in the alternative.

Which method of rock cut should be chosen?

Solution. The example reflects the unpredictability of explosives and the results of their use. The costs of the alternatives are

$$A \quad 30{,}000 + (0.50)(5{,}000) = \$32{,}500$$
$$B \quad 32{,}000 + (0.30)(5{,}000) = \$33{,}500$$
$$C \quad 36{,}000 + (0.10)(5{,}000) = \$36{,}500$$

The cheapest alternative is *A*. This is the one selected. ■

■ **Example 12.9.** Bethesda Aviation plans to introduce a new aircraft which will compete directly with the Cessna 150. The following are three cash flows, with their associated probabilities, considered by the firm's management to be possible during the first 5 years of production:

Year	Cash flows ($000,000)		
	1 ($p = .5$)	2 ($p = .3$)	3 ($p = .2$)
0	−10	−10	−10
1	+2	+3	+4
2	+4	+4	+4
3	+6	+5	+4
4	+8	+6	+4
5	+10	+7	+4

The sums are in constant, after-tax, year 0 dollars. If the constant-dollar, after-tax opportunity cost of capital for Bethesda Aviation is 6 percent, what is the expected value of the NPV of the venture?

Solution. The NPV of the first cash flow is

$$\begin{aligned}\text{NPV}_1 &= -10 + 2(P/F, 6, 1) + 4(P/F, 6, 2) \\ &\quad + 6(P/F, 6, 3) + 8(P/F, 6, 4) + 10(P/F, 6, 5) \\ &= +14.29\end{aligned}$$

The net present values of the remaining flows, calculated similarly, are

$$\text{NPV}_2 = +10.57 \quad \text{and} \quad \text{NPV}_3 = +6.85$$

The expected value of the three flows is

$$\begin{aligned}\mu = E(\text{NPV}) &= (14.29)(.5) + (10.57)(.3) + (6.85)(.2) \\ &= +11.69\end{aligned}$$

■

RISK MEASUREMENT

Throughout this text, the phrase *alternatives of equal risk* has been used. What is meant by *equal risk*? To answer this question, variance must be defined. *Variance* is a measure of how far an outcome may be expected to "vary" from an expected value. It is, of course, affected by the probability of an outcome's occurrence. Variance is the measure of risk we have been seeking. It is defined as

$$\sigma^2 = E(X_i - \mu)^2 \tag{12.8}$$

or

$$\sigma = \left[\sum_{i=1}^{n} p_i (X_i - \mu)^2\right]^{0.5} \tag{12.9}$$

where σ is called the *standard deviation.*

In a set of mutually exclusive alternatives, the one with the largest σ is the riskiest. If two alternatives have the same σ, they are said to be of *equal risk.*

TABLE 12.2
Calculating the standard deviation

Alternative i (1)	NPV X_i (2)	$E(X_i)$ (3)	$X_i - \mu$ (4)	p_i (5)	$(X_i - \mu)^2$ (6)	$p_i (X_i - \mu)^2$ (7)
1	14.29	11.69	+2.60	.5	6.76	3.38
2	10.57	11.69	−1.12	.3	1.25	0.38
3	6.85	11.69	−4.84	.2	23.43	4.69

$$\sigma^2 = \sum_{i=1}^{3} p_i (X_i - \mu)^2 = 8.45$$

$$\sigma = \left[\sum_{i=1}^{3} p_i (X_i - \mu)^2\right]^{0.5} = 2.91$$

Let us first calculate σ for an example, to see just how it fulfills its duty of measuring the deviation of a set of outcomes from their expected value.

■ **Example 12.10.** In Example 12.9 we calculated the expected value of the net present value (NPV) of a competitor to the Cessna 150. The three alternative cash flows are shown in present-worth form in column 2 of Table 12.2. The expected value of these NPVs is shown in column 3. The deviation from the expected value is $X_i - \mu$ and is shown in the next column. The probability of such a deviation occurring is shown in column 5. It is the same as the probability of the alternative occurring at all. The deviation is squared, thus removing the effect of negative signs. The probability multiplied by the squared deviation appears in the rightmost column. The summation of this column is the variance σ^2 and its positive square root is σ, the standard deviation. It may be thought of as the expected value of how much the NPV will be different from the mean μ. ■

The expected value of an outcome and its standard deviation both enter into risky decisions. Four cases may be distinguished. These are shown in Table 12.3.

Case 1 is the situation we have dealt with until this chapter—the alternatives are equally risky ($\sigma_2 = \sigma_1$), but one outcome is better than another. See Fig. 12.2.

Case 2 involves unequal risk and equal expected values. The choice between the two alternatives will reflect the risk preference of the decision maker. If the preference is for high risk, alternative 1 will be chosen. If risk is to be avoided as much as possible, then alternative 2 will be selected. In a gambling casino, some

TABLE 12.3
Comparing expected value and standard deviation in decisions

Case			
1	$\mu_2 > \mu_1$	$\sigma_2 = \sigma_1$	Choose alternative 2
2	$\mu_2 = \mu_1$	$\sigma_1 > \sigma_2$	Depends on risk preference of decision maker
3	$\mu_2 > \mu_1$	$\sigma_1 > \sigma_2$	Choose alternative 2
4	$\mu_1 > \mu_2$	$\sigma_2 < \sigma_1$	Depends on tradeoff between μ and σ

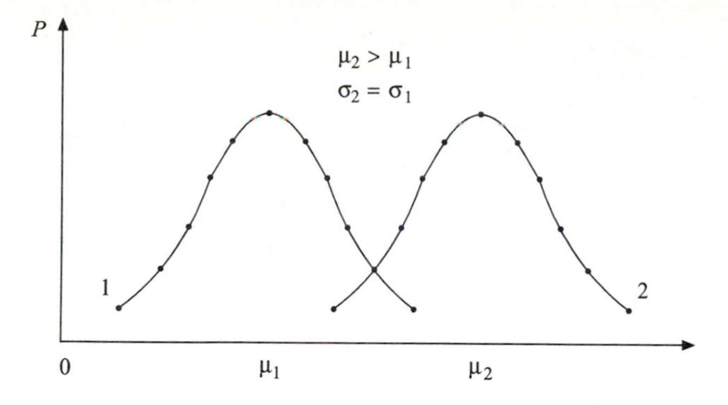

FIGURE 12.2 Equal-risk alternatives

players will choose higher-risk games than other players, even when the expected values of the payoffs of all games are the same. See Fig. 12.3.

Case 3 is easy to decide. Alternative 2 has a higher expected value and is also less risky than alternative 1. Most deciders will choose alternative 2. See Fig. 12.4.

Case 4 presents us with a choice that involves a tradeoff between risk and expected value. If we know exactly how much a dollar of expected value will equal in standard deviation—for example, an increase in an expected value of $1 must be balanced by at most a 25¢ increase in standard deviation—we can make our decision. See Fig. 12.5.

■ **Example 12.11.** In Example 12.9 the expected value of the NPV for the Bethesda Aviation competitor to the Cessna 150 was $11.69 million with a standard deviation of $2.91 million. If another proposal is presented, for a competitor to the Piper Comanche, with an expected NPV of $12.72 million and a standard deviation of $6.72 million, which will we choose, assuming that we own Bethesda Aviation?

Solution. This is case 4 in Table 12.3. If we decide that a tradeoff between NPV and risk is the one previously mentioned—an increase of $1 of NPV must be balanced by at most a 25¢ increase in standard deviation—then

$$\Delta \text{NPV} = 12.72 - 11.69 = 1.03$$

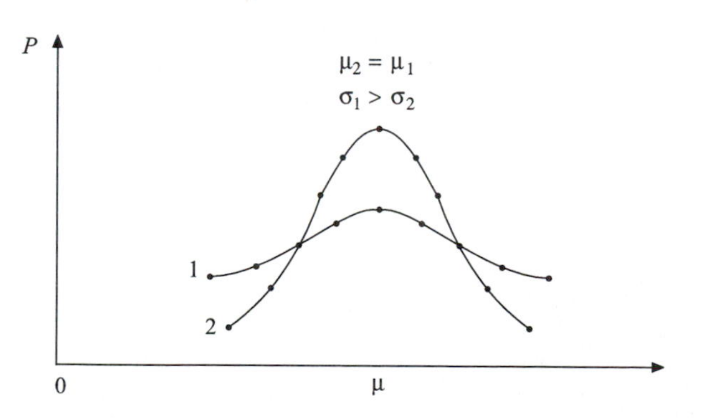

FIGURE 12.3 Different-risk alternatives

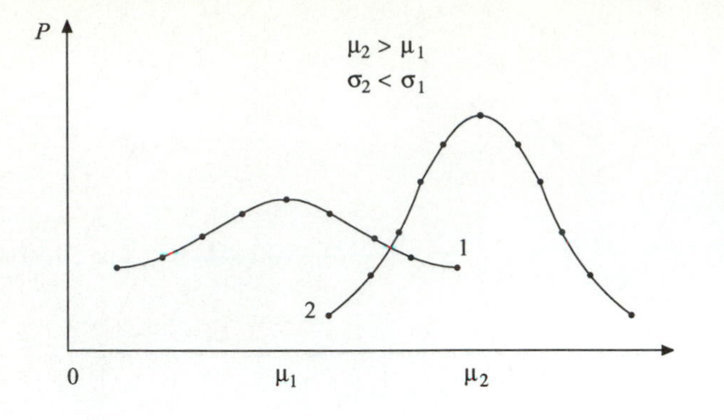

FIGURE 12.4 Different risks and expected values

$$\Delta\sigma = 6.72 - 2.91 = 3.81$$

$$\frac{\Delta\sigma}{\Delta \text{NPV}} = \frac{3.81}{1.03}$$

$$= 3.70 > 0.25$$

Therefore, accept the Cessna 150 competitor.

To make the solution clearer, we may ask, What is the maximum standard deviation we can accept in order to build the Piper Comanche competitor?

$$\Delta\sigma_{\text{max}} = (1.03)(0.25) = 0.26$$

$$\sigma = 2.91 + 0.26 = 3.17$$

This is less than 6.72. Therefore choose the Cessna 150 competitor. ■

The tradeoff between return and risk must be determined by the individual decision maker. He may look to past decisions under risk in his organization for guidance. He may consult the owners of a business or its managers. He may look to other enterprises in similar circumstances and observe what they have done. He may look to the securities market.

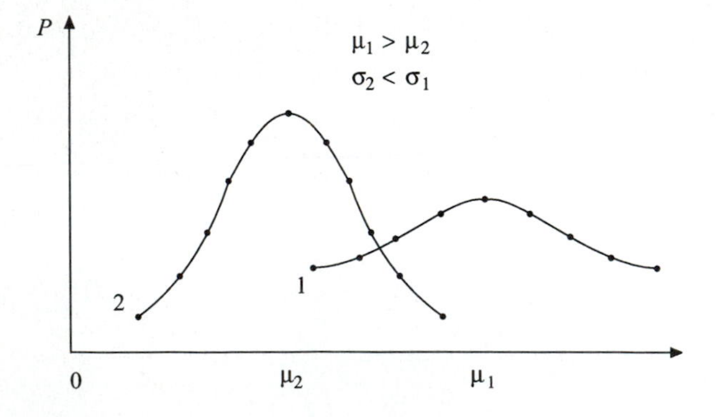

FIGURE 12.5 Different risks and expected values

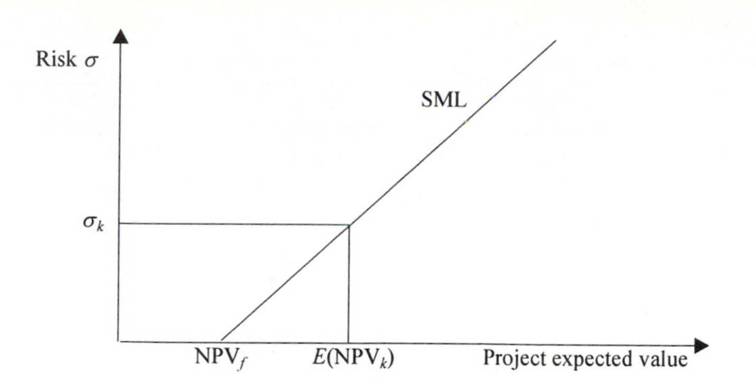

FIGURE 12.6 A possible tradeoff between risk and expected value

Some evidence exists to suggest that most businesses consider the relation between risk and return as linear, as follows:

$$\frac{\sigma_k}{E(\text{NPV}_k) - \text{NPV}_f} = \text{constant} \tag{12.10}$$

where σ_k is the standard deviation of project k, $E(\text{NPV}_k)$ is the expected net present value of project k, and NPV_f is the riskfree NPV, perhaps that of government securities. The line is called the *security market line* (*SML*), indicating its source in the financial world. Fig. 12.6 shows this relationship. Any project above the SML line will be rejected as having too much risk. Any project below the line will be accepted. The projects defining the SML, such as project k, are drawn from past experience, expressed as probability distributions of NPVs from completed projects. The projects themselves may be investments in securities. This means that the SML must be adjusted to reflect current returns on investments. The NPV_f will change over time, for example, as the rate of return on government securities changes. An increase will cause the SML to shift to the right; a decrease, to the left.

The SML has been shown as a straight line for simplicity's sake. Its true shape depends on the entity whose past projects and future possible projects it reflects.

It follows from all that has been said in this chapter that expected value has its limitations as a guide to decisions. Without the standard deviation to inform it, a decision based on the expected value alone can be misleading, as the following example shows.

■ **Example 12.12.** A lead mining company in Africa attempted to evaluate its prospecting operations. Table 12.4 shows in column 1 the size of the mine under five categories, beginning with a fruitless search and ending with the largest discovery believed possible. Column 2 shows the probability associated with each outcome. The outcomes are exhaustive, as shown by their probability summation to 1.0. Column 3 shows the estimated NPV of each outcome, beginning with –\$500,000—the cost of the operation. Column 4 shows the calculation of the expected NPV of the effort: +\$33,000.

TABLE 12.4
African lead prospecting

Mine size (1)	Probability of discovery (2)	NPV (000,000) (3)	*E*(NPV) (000,000) (4)
No discovery	.934	−0.5	−0.467
1	.050	5	0.250
2	.010	10	0.100
3	.005	20	0.100
4	.001	50	0.050
	1.000		0.033

The search appears to be worthwhile.

What about standard deviation? How risky is the investment compared to the certainty of obtaining the opportunity-cost-of-capital return on investment with an NPV of zero?

Calculation of the standard deviation will be based on a variation of Eq. (12.8):

$$\sigma^2 = E(X_i - \mu)^2$$

$$\sigma^2 = E(X_i)^2 - \mu^2 \tag{12.11}$$

$$\sigma = \left[E(X_i)^2 - \mu^2\right]^{0.5} \tag{12.12}$$

$$\begin{aligned} E(X_i)^2 &= \sum_{i=0}^{4} (X_i)^2 p \\ &= (-0.5)^2(.934) + (5)^2(.050) + (10)^2(.010) + (20)^2(.005) + (50)^2(.001) \\ &= 0.2335 + 1.2500 + 1.0000 + 2.0000 + 2.5000 \\ &= 6.983 \end{aligned}$$

So $$\sigma = \left[6.983 - (0.033)^2\right]^{0.5} = \$2.642 \text{ million}$$

The standard deviation from the mean is high, \$2.6 million and 80 times the expected NPV. If the investment being compared is at the opportunity cost of capital with no risk, then μ and σ are both are zero. The decision will depend on the tradeoffs between μ and σ—case 4 of Table 12.3. ■

SUMMARY

With the conclusion of this chapter we have made the journey back to the real world where income tax, inflation, and risk are everyday facts of life.

Certainty means that only one set of outcomes in the future is considered possible. Risk implies the existence of a function that relates a set of outcomes to a set of probabilities such that each outcome refers to a single probability. *Uncertainty* is the absence of certainty and of the risk function as well.

The *probability* of an event is the limit approached by the ratio of the number of times the event actually occurs to the number of times it can occur, as the number of trials is increased without limit.

Events may be *independent* of or *dependent* on each other. Variables, each one of which, depending on chance, can be attained with a definite probability, are called *random variables*.

The *expected value* of a set of outcomes is

$$E(X) = \sum_{i=1}^{n} p_i X_i$$

Probabilities can be *subjective* or *objective*. The latter can be determined from the attributes of a system in operation. The former are intuitive.

Risk is quantitatively measured by the *variance* σ^2 or by the standard deviation σ.

$$\sigma^2 = E(X_i - \mu)^2$$

$$\sigma = \left[E(X_i - \mu)^2\right]^{0.5}$$

The *standard deviation* is the expected value of how far a random variable may deviate from its mean.

Both the standard deviation and the expected value of a random variable must enter into a decision about that variable. The SML expresses the tradeoff between σ and μ as a linear function.

PROBLEMS

12.1. In Example 12.1, page 310, the probability of a 2 appearing is shown as .02931 after 14,400 tries. Another way of expressing this fact is to say 1 in how many, approximately? Can you predict how often, that is, 1 in how many, a throw of fair dice would come up a 2?

12.2. The queen of spades of Example 12.2, page 311, has not yet appeared after 13 draws, with none of the cards drawn being replaced in the deck.

(*a*) What is the probability of the queen of spades appearing on the 14th draw?

(*b*) If someone is willing to wager $40 against your $1 that the queen of spades will not appear on the 14th draw, should you accept the bet? You are a gambler who plays strictly by the odds.

12.3. What is the probability that two cards drawn from a deck will be the same suit, if the first card is not restored to the deck? See Example 12.3, page 312.

12.4. In Example 12.4, page 313, enumerate the six ways a total of 7 can appear on a toss of fair dice.

12.5. A year later, the airline of Example 12.5, page 314, discovered that 86 flights were made during the preceding year with a full load. Between 280 and 289 passengers were carried on 15 flights, and between 250 and 259 on 3 flights. *Ceteris paribus*, calculate the expected value of the new load factor.

12.6. Gold, silver, and platinum prices have the following probabilities of reaching certain levels at the end of 1 year:

Metal	Current price ($)	Price 1 year hence ($)	Probability
Gold	650	1,000	.10
		850	.40
		750	.50
Silver	19	40	.20
		30	.40
		25	.40
Platinum	671	1,200	.20
		900	.50
		750	.30

If you have on hand $100,000, in which metal should you invest to maximize your expected return at the end of 1 year?

Answer: Silver

12.7. In Example 12.6, page 315, the cost of an installed pole is now $768.56. In addition, improved weather protection of the poles has increased the age in each category of the percentage retired by 5 years. For example, the shortest pole life is now 10 years instead of 5, and the longest is 50 years instead of 45. What annual cost should now be used for a pole, providing all other data and conditions remain the same?

12.8. Tests have demonstrated that a well-known make of home can opener—Can-Easy—has 4 times the chance of failure in any one year as an equally well-known but considerably better can opener, Open-Up. The Open-Up can opener has a chance of failure of 1 in 10 in any single year. The Can-Easy costs $23.75; and the Open-Up, $41.95. Both have a life of 3 years. All other aspects of the openers—energy use, parts availability, etc.—are the same. Failure means that the can opener must be discarded. It also may be assumed that the can opener will be replaced with one of the same make if it fails and at the end of its life.

Your MARR for this type of investment is 12 percent. What can opener should you buy? Use the expected annual-cost method.

Answer: Choose Can-Easy

12.9. Repeat Prob. 12.8, using the individual present-worth method. *Draw the cash flows.*

12.10. In a developing country in Asia, agricultural fertilization by a machine spreader was being considered by the Ministry of Agriculture. Corn was the crop to be studied on the agricultural experiment station, and a parcel of 200 hectares was used.

The spreader cost $4,000 delivered, had a life of 10 years, and had no salvage value. (The monetary unit of the country is indicated by the dollar sign.) An opportunity cost of capital that included an estimated risk factor was 25 percent for this type of operation.

Fertilizer cost $45 per metric ton. The parcel was to be fertilized annually with 10 metric tons, then 20 metric tons, and finally 30 metric tons, and the results were observed for each. Other cultivation costs were $9,000 for the entire parcel.

Corn prices per hectoliter over the next 10 years were expected to vary according to the following probabilities:

Price ($)	*p*
1.00	.20
1.20	.35
1.50	.45

A preliminary calculation was run to permit some estimate of the accuracy of past data. This indicated the following yields (hectoliters per hectare) for each quantity of fertilizer:

Yield (metric tons)	Hectoliters per hectare	Probability
10*	30	.25
	35	.30
	38	.45
20	40	.20
	43	.40
	46	.40
30	50	.40
	52	.30
	54	.30

*The fertilizer was spread over the entire parcel.

According to the above facts and estimates, what amount of fertilizer per hectare should be spread by the new machine?

Answer: 30 metric tons

12.11. A drought that affected the corn crop of a Mexican area had a probability of annual occurrence and resulting cost of lost crops as follows:

p	Crop loss ($000,000)
.01	3,000
.05	1,500
.10	500
.20	200

Irrigation works have been proposed which will reduce the drought damage incurred, but at the first costs and annual operations costs shown below:

First cost ($000,000)	1,000
Annual cost ($000,000)	10
Life (years)	50
Salvage value	0

If this plan is adopted, the probability of damage will be reduced to the following:

p	Crop loss ($000,000)
.005	3,000
.02	1,500
.03	500
.04	200

If the public sector MARR is 12 percent and no inflation effects are considered, should the dam be built?

12.12. In a Bureau of Indian Affairs project, the question arose as to how long a certain bridge should be. The longer the bridge over this particular streambed, the larger the maximum flow that could be accommodated without flooding. It was estimated that lands and buildings upstream of the bridge would incur damage that would vary inversely with the opening provided for the flooding stream. An extensive study revealed the following facts:

Bridge length (feet)	First cost ($000)	Maintenance cost per year ($000)	Flood capacity* (year)	Flood damage ($000)
200	650	10	100	0
175	500	9	50	600
150	400	8	40	750
125	325	7	30	900
100	275	6	25	1,200

*The numbers in this column refer to the frequency at which a flood may be expected. For example, a 100-year flood means a flood that is likely to occur once in 100 years. The probability of its occurring in any one year is therefore .01.

All bridge lives were 40 years with no salvage value. The opportunity cost of capital was 10 percent. Which bridge should be chosen?

12.13. In Example 12.7, page 316, an error is discovered in the data on model 1. For the fifth year, 107 frame repairs had to be made. The first cost of model 1 is now listed at $115,000 instead of $107,000. All other data being the same, which model should be chosen?

12.14. A Maryland stream passing through farmlands caused the following damages in crop losses when the stream overflowed its banks. The probabilities associated with each damage estimate were determined from stream flow data collected during the past 150 years and were the probabilities of flood peaks being equaled or exceeded during any year.

Probability	.005	.01	.04	.15	.433
Damage ($000)	1,000	800	400	300	100

Calculate the expected annual flood damage. A flood control project under consideration, design *A*, will reduce the flood damage to the following:

Probability	.005	.01	.04	.15	.433
Damage ($000)	500	250	80	0	0

Design *A* will cost $1,700,000, and have an estimated life of 40 years, with no salvage value. It will cost $5,000 per year to maintain.

Design *B* will cost $1,200,000, have an estimated life of 30 years, and cost the same amount as design *A* to maintain. It will have no salvage value. It will affect the flood damage probability distribution as follows:

Probability	.005	.01	.04	.15	.433
Damage ($000)	700	350	175	110	0

This agency uses a 7 percent discount rate. No inflation effects will be considered. Should the flood control project be built? If it is built, should design *A* or *B* be used?

12.15. A road in Texas is at grade with a railroad. The crossing is being studied with a view to building a highway overpass to eliminate the grade crossing. The agency doing the study is the Texas Department of Highways and Transportation. Thus, the point of view of the society is being taken. Costs of accidents at the crossing include material damages such as to automobiles and trains, loss of life, and injuries. The frequency of accidents has been recorded over 50 years, and the following table composed. It reflects the best estimate of what the accident rates and costs will be for the immediate future.

A highway overpass that would eliminate all accidents will cost $2,500,000, have a life of 20 years, and have no salvage value. The MARR for this Texas department is 10 percent.

Grade-crossing accidents: Yearly cost

Total cost of accident(s) ($000)	Probability of accident's annual occurrence
100–199	.41
200–299	.27
300–399	.19
400–499	.12
500–999	.01
	1.00

Should the overpass be built at this time? Use the approximate midpoint of the cost of accident ranges, for example, $150,000 for the range from $100,000 to $199,000.

12.16. A large real estate development corporation is considering the design of a multistory building in San Francisco. An earthquake of some magnitude is considered a certainty within the building's life of 50 years. The force of the earthquake is measured on the Richter scale. The building can be designed to survive a given force with minor damage; but if the force is exceeded, major damage equal to the cost of the building is believed certain. The designs under consideration, the magnitude of earthquake that each design will resist, the probabilities of such a Richter-scale force occurring and of being exceeded during any one year, and the building cost of each design are as shown:

Design	Earthquake force on Richter scale that design will resist	Probability of force occurring	Probability of force being exceeded	Design cost ($000,000)
	< 7	.7	.3	
1	7	.2	.10	26
2	8	.05	.05	30
3	9	.03	.02	40
4	10	.02	0	60
	> 10	0		
		1.00		

The opportunity cost of capital for this company before taxes is 20 percent. Inflation is nondifferential among the alternatives because the building designs differ only in earthquake resistance. Thus revenues will be the same whatever building is constructed. The salvage value is the same for all alternative designs.

Which design should be built?

12.17. In Example 12.8, page 318, Steeples, an explosives expert, believes that the probabilities associated with each alternative are incorrect and corrects them thus:

Alternative	Probability
A	.65
B	.25
C	.10

If all other data are the same as before, which alternative should be chosen?

12.18. An investor in an IRA account paying 11 percent annually plans to deposit $1,500 per year for the next 30 years. He will then withdraw in equal amounts over the following 10 years, that is, from year 31 through year 40, an amount such that zero balance will remain in the account at the end of the 40th year.

According to Internal Revenue Service (IRS) rules, $1,500 may be subtracted from the taxable income yearly. On withdrawing any money from the account, he must pay regular income tax on it. The investor is in the 32 percent income tax bracket, where he expects to remain for the next 40 years.

Subjectively, the investor believes that there is a 20 percent chance that over the 40 years inflation will be 5 percent, a 60 percent chance that it will average 8 percent, and a 20 percent chance that it will be 11 percent.

After taxes and inflation have been accounted for, what is the true, deflated, after-tax rate of return on this investment? Assume that the end-of-year convention is used for all payments and rates. Sufficient accuracy will be attained by the use of whole percentage rates only.

12.19. In a residential and office development being planned by Atlas Company, the following table shows the vacancy rates—the percentage of the time that the units will not be rented—and the effect on annual company revenue, with the associated probabilities of occurrence.

Vacancy	Residential		Office	
rate (percent)	*p*	Revenue ($)	*p*	Revenue ($)
5	.30	−2,000,000	.20	−6,000,000
7	.45	−3,500,000	.30	−10,000,000
10	.25	−5,000,000	.50	−13,000,000
	1.00		1.00	

(*Note*: As vacancy rates increase, rents must be lowered to attract renters.) Calculate the expected value of the loss in revenue (*a*) in the residential part, (*b*) in the offices, and (*c*) in the two combined, that is, in the development as a whole.

12.20. Home mortgage interest rates are declining. A real estate lender believes that 3 months from now the chances of lower rates will be as shown:

Nominal rate (percent)	*p*
8	.20
8½	.20
9	.30
9½	.20
10	.10
	1.00

At the moment, the rate at which you can finance a $180,000 loan for 30 years is a nominal 10 percent.

(*a*) What is the monthly payment you will make if you take out the loan now? Carry out the calculations to the nearest dollar, using the proper formula.

(*b*) What is the nominal rate you can expect to pay if you wait 3 months?

(*c*) How much money will you expect to pay monthly if you wait 3 months? Carry out the calculations to the nearest dollar, using the proper formula.

12.21. In Example 12.9, page 318, the third cash flow is now considered to be impossible for Bethesda Aviation to achieve, and thus it disappears from the problem. In addition, cash flow 1 now has a probability of .4. Also the opportunity cost of capital has been raised to 10 percent. What is the expected value of the venture? All other data remain the same.

12.22. A production process which will continue to be needed for the next 3 years has annual costs of $310,000 which are expected to remain constant. A novel redesign of the process will cost $150,000 to install. It is believed that the new process has a 50 percent chance of cutting costs to $210,000 per year. However, increasing labor costs may (with a probability that is thought to be .25) result in cost increases from $210,000 in the first year to $230,000 in the second year and $250,000 in the third year. There is also a 25 percent probability that costs will increase from $210,000 in the first year to $285,000 in the second year and $360,000 in the third year. If a 12 percent rate of return is required before income taxes and no salvage values are foreseen, should the new design be installed?

12.23. Following the procedure in Example 12.10, page 320, calculate the standard deviation of the data on dice outcomes in Example 12.4, page 313.

12.24. In Example 12.5, page 314, a major U.S. airline made the following observations on its daily Boeing 747 service between New York and London during 1994:

Flights	Passengers
79	350 (full)
85	340–349
40	330–339
34	320–329
28	310–319
24	300–309
23	290–299
18	280–289
15	270–279
12	260–269
7	250–259
365	

(*a*) Calculate the standard deviation of the load factor.
(*b*) What does the standard deviation mean in this case?

12.25. Investment in vacant land is usually risky. In an area of large growth near Austin, Texas, a lot was for sale at $35,000. The payoffs and their probabilities were estimated as follows:

Net gains after taxes ($)	p
−10,000	.02
−5,000	.03
0	.10
+5,000	.50
+10,000	.20
+50,000	.10
+100,000	.05
	1.00

(*a*) What is the expected value of the net gains after taxes?

(*b*) What is the standard deviation? Comment on the riskiness of the venture.

12.26. In Example 12.11, page 321, a third proposal is presented—to reproduce the Stearman PT-17. It will result in an expected NPV of $13.56 million with a standard deviation of $3.00 million. *Ceteris paribus*, what should be our decision?

12.27. Certain authors advocate the use of the *coefficient of variation*, defined as

$$\text{Coefficient of variation} = \frac{\text{standard deviation}}{\text{expected cash flow}}$$

Applied to mutually exclusive alternatives, it will decide among them by comparing the risk (σ) per dollar. The smaller the coefficient of variation, the less risk associated with the alternative.

For example,[1] compare two mutually exclusive projects:

Project	First cost ($)	Annual return ($)	p
Oil drilling	60,000	0	.25
		10,000	.50
		20,000	.25
Real estate	60,000,000	9,980,000	.25
		10,000,000	.50
		10,000,000	.25

Assume that the projects will each be held for 5 years and that they will be sold for a price equal to their first cost. If the rate of return on projects is a minimum of 10 percent for such projects in constant-dollar, after-tax conditions, which project should be selected on the basis of the coefficient of variation? Is this a correct method?

12.28. In Example 12.12, page 323, compute the standard deviation of the project by using Eq. (12.9), page 319.

12.29. An investor is attempting to choose among three courses of action over a time horizon of 1 year. The first option is to place $5,000 in a Mexican

[1]This problem was suggested by L. D. Schall and C. W. Haley, *Introduction to Financial Management*, 2d ed. New York: McGraw-Hill, 1980, pp. 265–269.

bank investment account, in dollars, at 16 percent interest annually, after Mexican taxes, on a 3-month minimum period of deposit. This means that the total amount of her investment, plus interest, is available any time after 3 months. Under Mexican law, it is not withdrawable before the 3-month minimum period is concluded. The second option is to convert her dollars to pesos and deposit them in a similar Mexican bank investment account, but at 21.5 percent annually, after Mexican taxes, on the same 3-month minimum. The third option is to buy gold, which has doubled in value during the past year.

The investor estimates the risk of each course of action as follows: With option 1, associated with holding her money in a dollar account in a Mexican bank is the risk of a major revolution, such as occurred in Mexico in 1910, that might result in confiscation of all foreign investments. This possibility is so remote, however, that the investor discards it completely. The second option, a savings account in Mexican pesos, involves a special kind of risk—devaluation. Since the investor expects to continue to live in the United States, she will convert her peso earnings to dollars. The present rate of exchange for bank drafts is 22.69 pesos per dollar, less a bank charge of 66 pesos per $1,000 exchanged. The conversion rate for pesos to dollars is $0.0437 per peso (22.89 pesos per dollar), with no accompanying bank charge. However, a risk of devaluation of the peso during the coming year exists. The investor estimates this risk as follows:

Change	Probability
Constant at $0.0437 per peso (22.89 pesos/$)	.6
Change to $0.0310 per peso (32.30 pesos/$)	.3
Change to $0.0250 per peso (40.00 pesos/$)	.1

The third option, gold investment, involves risk also. The investor estimates the chances of gold's doubling in value in the coming year, as it did last year, at 10 percent, of rising to 1.5 times its present price at 50 percent, and of holding its present price at 40 percent.

In addition, certain income tax advantages accrue to the Mexican investments. Mexican income taxes are deducted directly by the bank from the returns, before these are placed in the investor's account. Before Mexican taxes are deducted, the dollar investment account pays 20.2532 percent. Similarly, the peso investment account pays 24.02 percent.

Mexican income tax paid qualifies as a tax credit against U.S. income tax. Thus it can reduce U.S. income tax by the amount paid to the Mexican government. Notice that this is not a deduction from taxable income; it is a tax *credit*. Although the IRS limits the tax credit that may be taken as a result of foreign taxes paid, it is expected that the investor's tax credit will not reach this maximum.

The return on the investment, if it is held for more than 1 year, as this investor assumes will happen, will be taxed as capital gains in the United States, providing these gains are realized. Forty percent of the capital gain will be added to ordinary income and taxed at the investor's income tax bracket rate, which is expected to be about 37 percent.

The $5,000 is presently in the investor's checking account in the United States at 0 percent interest, where it serves as emergency cash. The investor believes, however, that 3 months will be enough warning to withdraw the money, should she need it.

(*a*) Which course of action is advisable, given the information presented?

Answer: Invest in gold

(*b*) Is it necessary to consider inflation rates in both Mexico and the United States?

12.30. The local telephone company in central Florida is reviewing options for the design and construction of microwave transmission towers. One option is certain to be chosen. The towers must be designed to withstand hurricane wind velocities. The construction costs for three mutually exclusive alternatives for the tower design are shown in the table below, along with the results of historical data on wind velocities in the region and their probabilities of occurrence.

Tower design	Construction cost ($)	Wind velocity range (miles per hour)	p
		0–99	.7
A	560,000	100–119	.12
B	600,000	120–149	.1
C	650,000	150–180	.08

Each tower option will withstand the range of wind velocities opposite it in the table. If the range of velocities is exceeded, however, the tower will fall. If a tower is blown over, it must be replaced immediately at its original cost. Thus the towers will remain in service, as originals or replacements, for the foreseeable future.

It may be assumed that only one storm of sufficient magnitude to cause wind velocities in excess of 100 miles per hour will occur per hurricane season, that is, per year. No wind velocity has ever been recorded at higher than 180 miles per hour. The probabilities shown in the table measure the likelihood of hurricane winds reaching the range of velocities *in any one year*.

The towers have an estimated life of 25 years. Yearly maintenance costs will be the same for each tower option. The telephone company uses a 10 percent opportunity cost of capital. Tax and inflation effects will be the same for any option and may be ignored.

Which option should be chosen? Use the *annual-cost method* for your solution.

CHAPTER 13

LOANS

Borrowing money from someone else for any purpose means that a benefit is received by the borrower at the start of the loan period (time 0) and costs are incurred for the repayment period. The amount of money given by the lender to the borrower is called the *principal* of the loan. It is the sum that constitutes the base of the interest percentage. The internal rate of return (IROR) of the cash flow comprising payments—the costs—and the amount of the loan—the benefit—is the *interest rate* of the loan. The term of the loan is the time during which the loan will be *amortized*, i.e., repaid.

■ **Example 13.1.** Figure 13.1*a* illustrates a loan of $10,000 to be repaid in three equal installments of $4,380 each. The principal of the loan is $10,000. The term is 3 years. The interest rate is calculated by finding the internal rate of return of

$$0 = +10{,}000 - 4{,}380(P/A, i^*, 3)$$

$$(P/A, i^*, 3) = \frac{10{,}000}{4{,}380}$$

$$= 2.283$$

$$i = 15\%$$

The cash flow diagram, where the point of view is that of the borrower, is the inversion of the investment cash flow we have dealt with so far.

Figure 13.1*b* profiles the present worth of the loan, as drawn from Table 13.1, which shows the net present value (NPV) associated with selected discount rates.

Notice the horizontal line at NPV = 10,000. As i is increased without limit, the NPV will approach this line. At i = 1,000, for example, the NPV is +9,562. ■

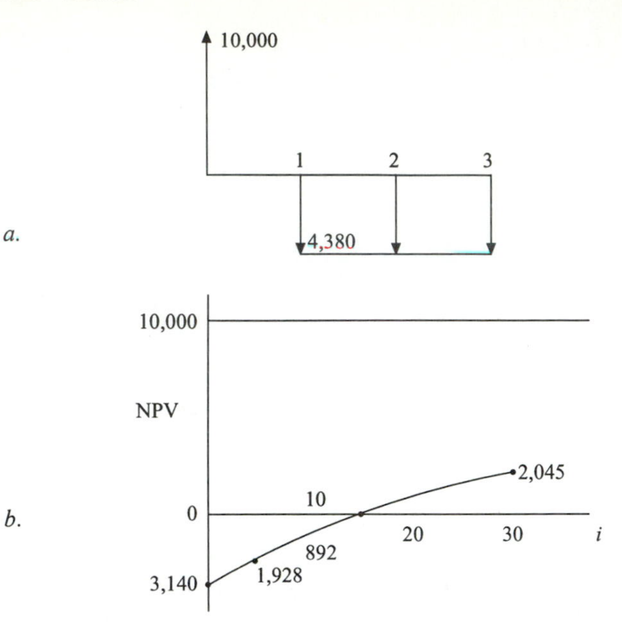

FIGURE 13.1 An automobile loan at 15 percent

All this seems simple enough, and indeed it is, for this simple case. However, the choice among loans is as difficult as the choice among investments. A well-known savings and loan association in the Washington, D.C., area lists the following types of mortgages it makes available to its clients:

Adjustable-rate mortgage
Graduated-payment adjustable-rate mortgage
Balloon mortgage
30-Year fixed-rate mortgage

We shall see, via some examples, the complicated financing plans even so ubiquitous a creature as the homeowner encounters. How is one to choose among the variety of mortgage loans that have arisen because of "creative financing"?

It may seem that the choice of mortgage can be made on the basis of interest rate alone, especially because the Truth-in-Lending Law prevails in the United States.[1]

Let us leave aside the question of loan rates quoted outside the borders of the United States and the protection of U.S. laws. We shall see that, even in the United States, true interest rates are no guide for choosing among loan plans. This statement, surprising as it will seem to many, will be amply proved in the discussion to come in this chapter.

---

[1] Although many persons believe that this law requires that the effective annual interest rate be stated for all loans, such is not the case. Only the nominal interest rate need be stated, or failing that, only the total amount of interest paid regardless of the timing of the payments.

TABLE 13.1
A loan profile

i	NPV
0	-3,140
5	-1,928
10	-892
15	0
20	+774
25	+1,450
30	+2,045
1,000	+9,562

LOAN CRITERIA

As a first step in showing that loans should not be judged by the criterion of interest rate, i.e., their internal rates of return, let us investigate how loans are used. Loans are used by the borrower for two purposes only: investment or consumption. If the loan is made for investment, then the funds to repay the loan must come from the enterprise's available cash when the payment comes due. These monies are not available for investment. Thus their opportunity cost is the opportunity cost of capital. If the purpose of the loan is consumption, then its cost is the opportunity lost by consuming rather than investing. Once again it is the *opportunity cost of capital.* This fundamental rule of economics is inescapable.

But the skeptical reader may reply, "I used $100,000 to take care of my mother's hospitalization, and this sum was offered to me at 15 percent. My opportunity for investment is at only 10 percent. Do you mean to say that I am only paying 10 percent rather than 15 on the loan?"

The answer to that is no; the loan is at 15 percent, not 10 percent. But the standard for judging between loans is your individual cost of capital for the repayment period—that is, 10 percent. We shall see how these matters are handled in the examples that follow.

Need it be added that the lender is, of course, making 15 percent on his or her investment in you?

■ **Example 13.2.** Henry Higgins Company is trying to decide between two loan proposals for financing a new molding shop. The loan amount is $1 million, and the term of both loans is 15 years.

The first loan proposed requires equal annual payments of principal and interest of $146,820. This loan is at 12 percent interest rate.

The second loan proposal requires a single payment of $7,137,900 at the end of 15 years. Its rate of interest is 14 percent.

Both proposals provide $1 million now. Which should be accepted?

The opportunity cost of capital of the company is 25 percent. Ignore tax and inflation effects.

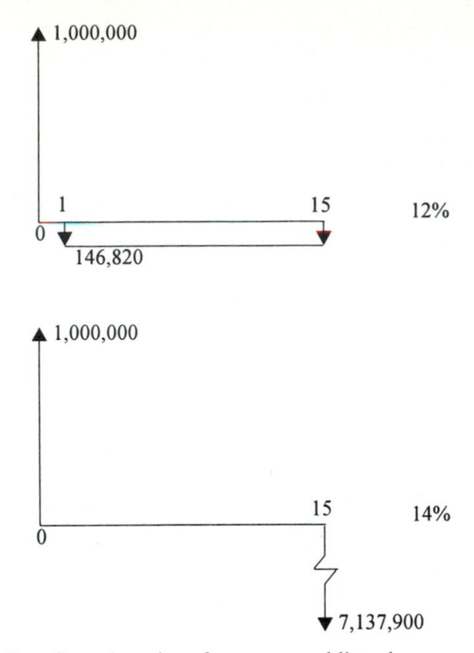

FIGURE 13.2 Two financing plans for a new molding shop

Solution. The financial vice president performs the following calculations. See Fig. 13.2.

$$\begin{aligned}\text{NPV}_1 &= +1{,}000{,}000 - 146{,}820(P/A, 25, 15)\\ &= +1{,}000{,}000 - 146{,}820\,(3.859)\\ &= +1{,}000{,}000 - 566{,}578\\ &= +433{,}422 \approx +433{,}400\end{aligned}$$

$$\begin{aligned}\text{NPV}_2 &= +1{,}000{,}000 - 7{,}137{,}900(P/F, 25, 15)\\ &= +1{,}000{,}000 - 251{,}254\\ &= +748{,}746 \approx +748{,}700\end{aligned}$$

Take plan 2.

$$\begin{aligned}\text{AC}_1 &= 146{,}820\\ \text{AC}_2 &= 7{,}137{,}900(A/F, 25, 15)\\ &= 7{,}137{,}900(0.00912)\\ &= 65{,}098\end{aligned}$$

Take plan 2.

It is clear that the second proposal at 14 percent should be accepted, even though its interest rate is 2 percent higher than the other mutually exclusive alternative. The fact is that funds are to be drawn to pay off the loan from uses where they would yield 25 percent. Calculations based on this rate rather than the interest rates of the loans are thus appropriate for deciding between the loans. ■

■ **Example 13.3**

(*a*) Your wife's new fur coat will cost \$8,000. You decide to borrow this sum. Your bank offers you a 10 percent loan for a term of 36 months. The monthly payment will be \$258.14.

Your savings and loan association offers the same amount of loan at 11 percent for 3 years. The monthly payment will be \$261.91.

Your opportunity cost of capital is a nominal 12 percent. Which loan should you choose, inflation and tax considerations aside?

(*b*) Now a friend offers to lend you \$8,000 if, at the end of 3 years, you pay him \$10,000. What should you do, if the other conditions are the same as in part (*a*)?

Solution

(*a*) Clearly one cash flow *dominates* the other. This means that it will always be chosen. See Fig. 13.3. All other things being equal, the amount of monthly payment is the criterion of choice. The 10 percent loan should be selected.

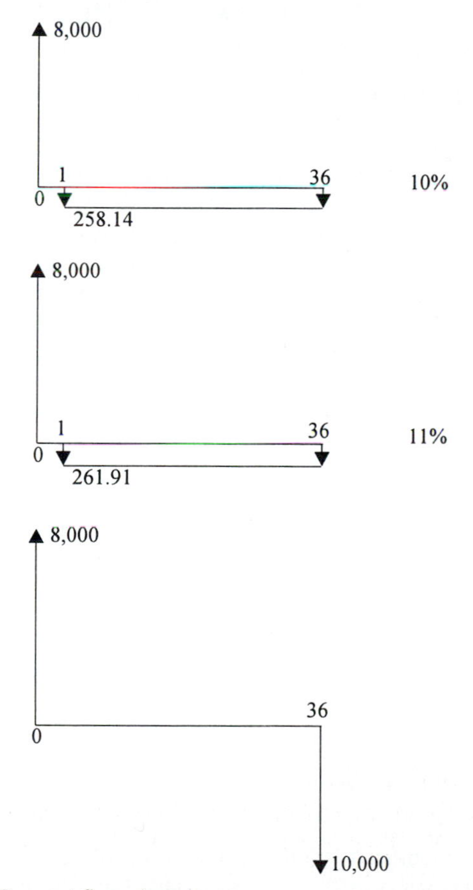

FIGURE 13.3 Fur coat financing plans

(*b*) The opportunity cost of capital of 12 percent must be used to decide among the plans. The easiest way to do this is to convert the $10,000 final payment to a monthly payment. Put another way, the question is, How much will I need to put away at 12 percent, in equal monthly payments, to have $10,000 at the end of 3 years? The effective rate is

$$i = \frac{r}{M} = \frac{12\%}{12} = 1\% \text{ per month}$$

$$A = F(A/F, 1, 36)$$

$$(A/F, i, N) = \frac{i}{(1+i)^N - 1}$$

$$= \frac{0.01}{(1.01)^{36} - 1}$$

$$= 0.023214$$

$$A = 10{,}000(0.023214)$$

$$= \$232.14$$

The friend's offer should be accepted because it requires the lowest monthly repayment cost.

In this example, the periodic cost of the loan is the decision criterion.

Could another method have been used to make the decision? The net present-worth method requires that each of the three cash flows be discounted at the nominal 12 percent opportunity cost of capital. The periodic-worth method, which we have called the *annual-worth method*, was the one used. The benefit/cost ratio method is not appropriate for a loan decision because it is usually restricted to investment decisions by governments. But from a technical standpoint, there is no reason why it could not have been used, discounting at 12 percent nominal. The IROR method also may be used, but the analyst must remember that the alternatives are mutually exclusive, and therefore incremental analysis is the only correct treatment, as is the case with the benefit/cost ratio method. ■

A loan is like an investment except that in a loan the benefits are received first and the costs come later. However, loans are not the only situation where a choice among cash flows involving early benefits, or even benefits alone, must be made. The following example illustrates such a case.

■ **Example 13.4.** A prospective tenant for your residential rental house in an African country proposes two plans for the rental payments on an 11-month lease:

1. She will pay you $600 on the first day of each month for 11 months, starting on the date she occupies the house.
2. Or she will pay you $6,000 the day she occupies the house.

If your rate of return on funds is a nominal 9 percent compounded monthly, which plan should you take? Ignore income tax considerations and inflation.

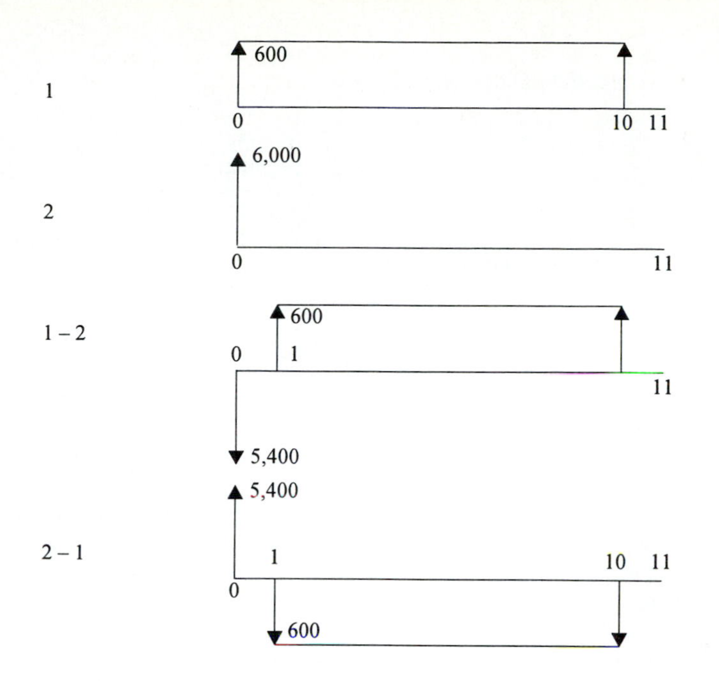

FIGURE 13.4 Renting an African house

Solution. Figure 13.4 shows the two plans. The net present-value method for each plan reveals the following:

$$\begin{aligned}\text{NPV}_1 &= 600 + 600\,(P/A, 9/12, 10)\\ &= 600 + 600\,(9.600)\\ &= 6{,}360\end{aligned}$$

and

$$\text{NPV}_2 = 6{,}000$$

The NPV solution indicates plan 1 is preferable.

Let us now try the incremental NPV solution. If no attention is paid to how the increment is taken, 1 − 2 or 2 − 1, the solution is still unequivocal. See Fig. 13.4.

$$\text{NPV}_{1\text{-}2} = -5{,}400 + 600\big(P/A, 3/4, 10\big) = +360$$

Therefore accept plan 1.

$$\text{NPV}_{2\text{-}1} = +5{,}400 - 600\big(P/A, 3/4, 10\big) = -360$$

Again, plan 1 is accepted.

What if the IROR method is used, again without regard to how the increment is taken? (Bergmann's rule applies only to investment alternatives of equal first costs—not the case in point.)

For 1 − 2:

$$-5{,}400 + 600(P/A, i^*, 10) = 0$$
$$(P/A, i^*, 10) = 9.000$$
$$i^* = 2\% \text{ per month}$$

For 2 − 1:

$$+5{,}400 - 600(P/A, i^*, 10) = 0$$
$$(P/A, i^*, 10) = 9.000$$
$$i^* = 2\% \text{ per month}$$

The IROR solution is equivocal. It shows either plan 1 or plan 2 as preferable. In both, i^* (2 percent) is greater than i (¾ percent).

It was stated at the beginning of this chapter that borrowing cannot be optimized by comparing each plan's interest rate and choosing the lowest. Although the circumstances of this example are similar to those encountered in borrowing, the lack of cash outflow in both plans is unusual and demands careful analysis.

The difficulty is resolved if we remember that funds received are destined for investment or consumption. Under either situation the opportunity cost of capital and the investment—made or forgone—conditions apply. Plan 1 minus 2, the investment cash flow showing the first payment as an outflow, is the one to be used, therefore. It indeed supplies the correct answer. ■

UNEQUAL-TERM LOANS

Consider mutually exclusive loan proposals whose terms differ. This is actually the more common case for the large group of those who borrow to finance a home, as opposed to the one where equal-term loans are offered. Frequently these borrowers must choose among a 30-, 25-, or 15-year term amid other complications. Governments and businesses issuing bonds also must consider the effect of differing periods of amortization. The question arises, How should the borrower judge loans of unequal terms, income tax and inflation considerations aside?

Let us also exclude the factor of ability to pay, even though it is important in real situations. We shall treat the U.S. government with its enormous financial resources and the poorest citizen as though both could consider dispassionately a 1-year loan versus a 30-year loan without thinking about whether the shorter-term loan would put them under severe financial stress.

■ **Example 13.5.** Two loans are offered to a prospective borrower. Both are for \$100,000, and both are at 10 percent interest rate. However, the first loan proposal requires repayment at the end of 1 year. The second requires repayment at the end of 10 years. Leave aside considerations of income tax and inflation effects. Do not take into account financial stress in regard to one loan plan or the other.

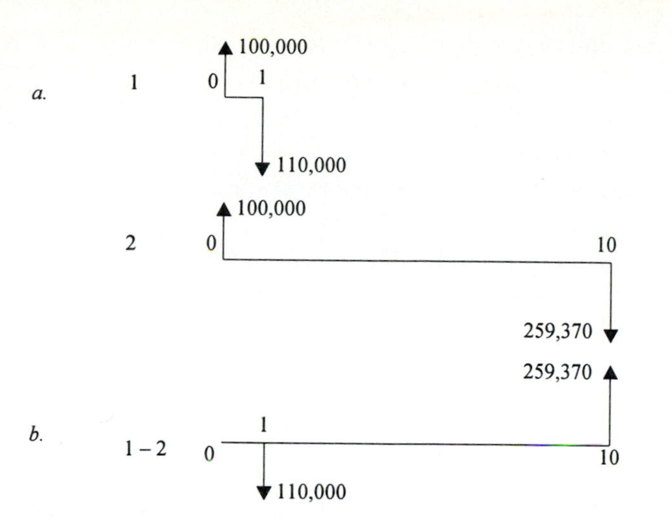

FIGURE 13.5 Unequal-term loans

(*a*) Which should the borrower choose if her opportunity cost of capital is also 10 percent?
(*b*) If her opportunity cost of capital is 12 percent?
(*c*) If her opportunity cost of capital is 8 percent?

Solution

(*a*) Figure 13.5*a* shows the two plans. Plan 2 will be repaid by a single payment, at the end of 10 years, of

$$\begin{aligned} F &= P(F/P, 10, 10) \\ &= 100{,}000(2.5937) \\ &= \$259{,}370 \end{aligned}$$

The incremental cash flow diagram of Fig. 13.5*b* reveals the elements of choice between the two plans. Choosing plan 1 over plan 2 means paying out \$110,000 at the end of year 1 and thus avoiding the payment of \$259,370 at the end of year 10. At a 10 percent opportunity cost of capital, the net present value of plan 1 over plan 2 is

$$\begin{aligned} \text{NPV}_{1\text{-}2} &= -110{,}000(P/F, 10, 1) + 259{,}370(P/F, 10, 10) \\ &= -110{,}000(0.9091) + 259{,}370(0.3855) \\ &= -100{,}001 + 99{,}987 \\ &= 0 \end{aligned}$$

This result, within tabular accuracy, was to be expected. It means that the borrower will be indifferent between the two loan plans because the opportunity cost of capital is the same as the interest rate of both loans.

(*b*) At $i = 12$ percent, the cash flow 1 – 2 has a net present value of

$$\begin{aligned} \text{NPV}_{1\text{-}2} &= -110{,}000(P/F, 12, 1) + 259{,}370(P/F, 12, 10) \\ &= -110{,}000(0.8929) + 259{,}370(0.3220) \\ &= -98{,}219 + 83{,}517 \\ &= -14{,}702 \approx -14{,}700 \end{aligned}$$

The borrower should reject loan 1, therefore, and accept loan 2.

(*c*) At $i = 8$ percent,

$$\begin{aligned} \text{NPV}_{1\text{-}2} &= -110{,}000(P/F, 8, 1) + 259{,}370(P/F, 8, 10) \\ &= -110{,}000(0.9259) + 259{,}370(0.4632) \\ &= -101{,}849 + 120{,}140 \\ &= +18{,}290 \end{aligned}$$

The borrower should accept loan 1 rather than loan 2. ■

The point of all this: *Equal service lives do not work for loans*. The borrower receives a benefit at time 0. This is invested or spent. Whatever happens to it, it disappears from the loan problem. (It can be invested perpetually, e.g., in an annuity.) There are only two relevant questions: What cash flow faces the borrower as a result of the loan? and What is the opportunity cost of the borrower in respect to the loans?

When two or more loans are being compared, the sum borrowed—once again, and even more clearly—disappears from the problem. The principal of the loan is a nondifferential benefit to the borrower. That this is a fact appears in the incremental cash flow. The principal of the loan disappears, leaving only the payments.

Do the payment periods need to be equalized, as in investments? They do not. Equal service lives do indeed demand to be recognized when one is considering culvert pipe, or long-lived versus short-lived automobiles. But in loans, the borrower receives the benefit of the principal and is free to spend it on a perpetual annuity lasting into many generations in the future or something as ephemeral as a European vacation. With loans, as with all mutually exclusive alternatives, only the differences in cash flows are relevant to a decision among loan plans, when the principal amounts are the same.

■ **Example 13.6.** You are offered a loan of \$100,000 at 14 percent for 30 years. You are also offered a loan of the same amount at 13 percent for 15 years. Both loans will be paid off by level annual payments of principal and interest. If your opportunity cost of capital is 20 percent, which loan should you accept? Do not consider inflation or tax effects. Solve by the net present-value method.

Solution. Figure 13.6 illustrates the two loans. The loan payments are

$$\begin{aligned} \text{AC}_1 &= 100{,}000(A/P, 14, 30) \\ &= \$14{,}280 \text{ annually} \end{aligned}$$

and

$$\begin{aligned} \text{AC}_2 &= 100{,}000(A/P, 13, 15) \\ &= \$15{,}474 \text{ annually} \end{aligned}$$

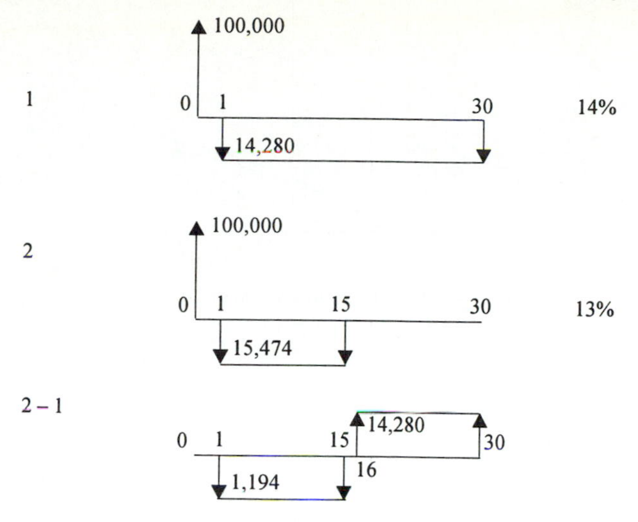

FIGURE 13.6 Loans for 15 and 30 years

The net present value of each loan at 20 percent opportunity cost of capital is

$$\text{NPV}_1 = +100{,}000 - 14{,}280(P/A, 20, 30) = +28{,}900$$

and

$$\text{NPV}_2 = +100{,}000 - 15{,}474(P/A, 20, 15) = +27{,}660$$

We choose loan 1 rather than loan 2 at this MARR.

We are led to another question, brought on by the increment between the two loans, as shown in the bottom cash flow of Fig. 13.6. Evidently, because both down and up cash flows exist, an opportunity cost of capital also exists at which the decision will change to favor the loan at 14 percent rather than the one at 13 percent. This point of sensitivity is found by use of the following equation:

$$\text{NPV}_1 = \text{NPV}_2$$

$$+100{,}000 - 14{,}280(P/A, i^*, 30) = +100{,}000 - 15{,}474(P/A, i^*, 15)$$

It is also the IROR of the incremental cash flow, as in

$$0 = -1{,}194(P/A, i^*, 15) + 14{,}280(P/A, i^*, 15)(P/F, i^*, 15)$$

$$i^* = 17.99\%$$

A plot of the two loan cash flows and their incremental cash flow is shown in Fig. 13.7 as drawn from Table 13.2.

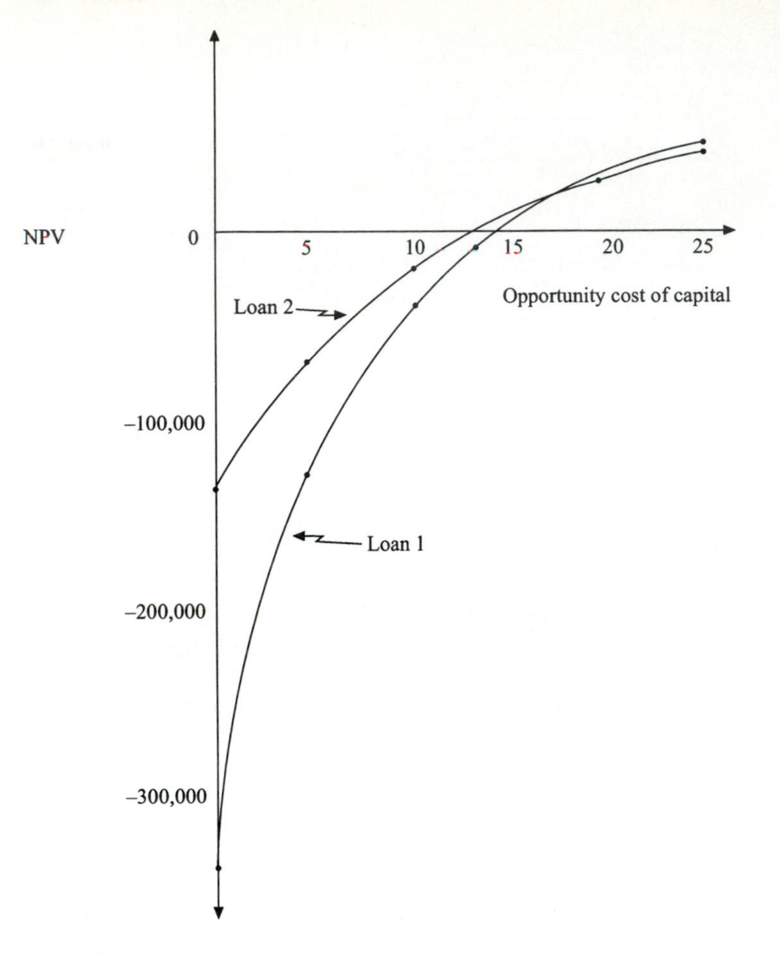

FIGURE 13.7 Two loan profiles

TABLE 13.2
Two loans of $100,000

	NPV		
i	**Loan 1 at 14%**	**Loan 2 at 13%**	**2 − 1**
0	−328,400	−132,100	+196,300
5	−119,500	−60,600	+58,900
10	−34,620	−17,700	+16,920
13	−7,040	0	+7,040
14	0	+4,960	+4,960
18	+21,200	+21,200	0
20	+28,900	+27,660	−1,240
25	+42,950	+40,280	−2,670

■

TABLE 13.3
Tax effects of an automobile loan

Year	Payment	Interest portion at 15%	Principal portion	Remaining balance	Tax relief	After-tax payments
0	—	—	—	10,000	—	—
1	4,380	1,500	2,880	7,120	555	3,825
2	4,380	1,068	3,312	3,808	395	3,985
3	4,380	572	3,808	0	212	4,168

TAX EFFECTS

The reader will recall that the subject of tax effects was first mentioned in Chap. 10 under the section entitled "Deductions." There it was stated, and illustrated in Example 10.1, that interest paid on business and residence loans is deductible from taxable income. We investigated this effect, as well as inflation, in Example 11.20. The point will be emphasized again here.

■ **Example 13.7.** In Example 13.1, page 337, imagine now that the loan will be used to purchase an automobile for use in your business. Thus it will have income tax effects. Recall that the automobile loan was to be repaid in three annual installments of $4,380 each. Figure 13.1*a* shows the cash flow.

The first step in the analysis is to separate the payments into principal and interest. (The reader may refer to the original explanation of how this is done in Chap. 4 in the section entitled "Separation of Interest and Principal.") Table 13.3 shows the result. Each of the "interest portion" quantities is deductible from taxable income. If the borrower is in the 37 percent tax bracket, the tax relief shown in that column of Table 13.3 will result. The tax relief was computed by multiplying the interest portion for each year by 0.37. The rationale is that the taxpayer will not have to pay taxes at 37 percent on any deductible amount. The after-tax payments are shown in the rightmost column: they are computed by subtracting the tax relief from the payments for each year. ■

INFLATION EFFECTS

It is often said that inflation helps debtors and hurts lenders. A borrower in an inflationary economy receives money with a certain buying power and pays back money with a lesser buying power. At even low rates of inflation the effect on loan choice is marked. The following example shows this.

■ **Example 13.8.** Adolph Galland Company wishes to borrow $10,000 for 10 years. The bank offers two loan repayment schedules: Plan 10, at 10 percent interest, requires a level payment of $1,628 per year. Plan 11, at 11 percent, will be paid back by a single payment of $28,394 at the end of 10 years. The company requires a return on constant dollars of 25 percent, and inflation is projected at 4 percent per year.

(*a*) Is the bank quoting the correct payments on the loan?

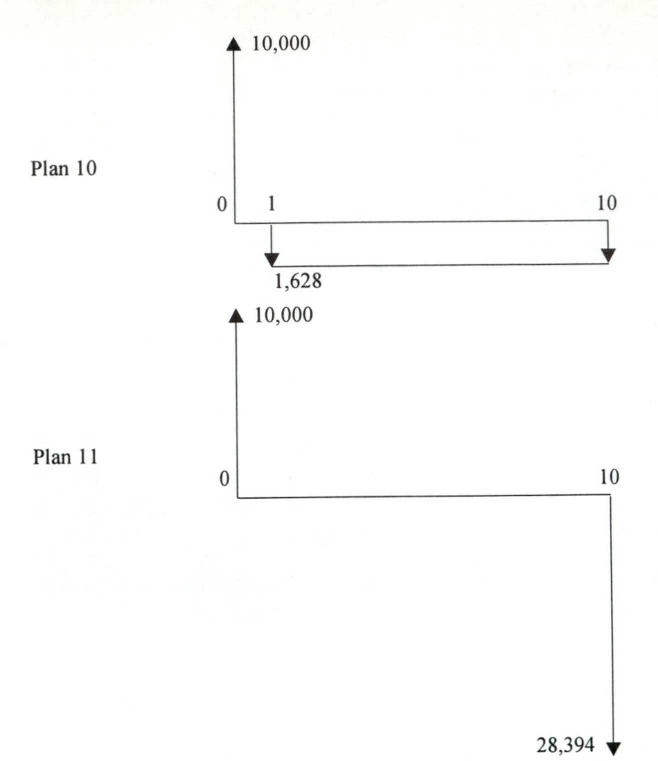

FIGURE 13.8 The Adolph Galland Company loans

(*b*) If the loans are equal in risk of default and all other things are equal, which loan should be chosen? Use the individual NPV method to make your choice.

Neglect income tax effects and possible financial strain.

Solution

(*a*) These loan plans are shown in Fig. 13.8. To discover whether the bank is quoting the correct payments on the loans, we must find the internal rate of return of each cash flow.

$$+10{,}000 - 1{,}628(P/A, i^*, 10) = 0$$

$$(P/A, i^*, 10) = \frac{10{,}000}{1{,}628}$$

$$= 6.143$$

This result is within tabular accuracy for 10 percent. We conclude that the payments are correct for plan 10. For plan 11:

$$+10{,}000 - 28{,}394(P/F, i^*, 10) = 0$$

$$(P/F, i^*, 10) = \frac{10{,}000}{28{,}394}$$

$$= 0.3522$$

$$i^* = 11\%$$

The plan 11 payment is also correct for the quoted interest rate.

(*b*) The plans will be judged by the criterion of the company's opportunity cost of capital on current dollars. (The loan will be paid in current dollars, of course.) This opportunity cost of capital is

$$\begin{aligned} u &= (1+i)(1+f)-1 \\ &= (1.25)(1.04)-1 \\ &= 0.30 = 30\% \\ \text{NPV}_{10} &= +10{,}000 - 1{,}628(P/A, 30, 10) \\ &= +10{,}000 - 1{,}628(3.092) \\ &= \$4{,}966 \\ \text{NPV}_{11} &= +10{,}000 - 28{,}394(P/F, 30, 10) \\ &= +10{,}000 - 28{,}394(0.0725) \\ &= \$7{,}941 \end{aligned}$$

Plan 11 should be chosen in spite of its higher interest rate than that of plan 10. The combined effect of the opportunity cost of capital of the company and an inflation rate caused a deferred-payment plan to be more attractive than one requiring annual payments. ■

The next example gives some flavor of the difficulties involved in a decision among mortgage plans, even with plans having the same term. The analyst must often guess at the rate of return on 90-day Treasury bills, at the opportunity cost of capital 30 years hence, at the inflation rate over 30 years, at an income tax bracket, which may well change over 30 years, and so forth. Yet it is better to make such guesses and do such an analysis than to trust to luck.

■ Example 13.9

(*a*) The following plans for repayment of a mortgage are offered by a lender. The mortgage is for $100,000. The lender maintains that the plans are equivalent at 12 percent. Is this so?

Plan 1	11,502	for 30 years
Plan 2	8,883	for years 1 through 5
	13,439	for years 6 through 30
Plan 3	8,883	for years 1 through 5
	12,470	for years 6 through 10
	14,201	for years 11 through 30

(*b*) The above plans are examples of "creative financing" in real estate sales. They allow borrowers a choice that will depend on their estimate of their future financial situation. In this example, the percentage rates on each plan have been converted by the lender to yearly payments. Normally, the borrower must perform this conversion.

If the borrower believes that the opportunity cost of capital will approximate 9 percent over the next 30 years, which plan should the borrower choose? Do not consider taxes or inflation.

(*c*) In part (*b*), 9 percent is the borrower's estimate of the returns on a certificate of deposit at a bank. If the borrower assumes that inflation will average 5 percent annually over the next 30 years, will it change the decision? If so, how? Ignore tax effects.

Solution

(*a*) Equivalence here simply means that each loan carries an interest rate of 12 percent. Because all the loans provide \$100,000 at time 0, the following relations will hold if they are indeed equivalent payment schedules:

$$\text{NPV}_1 = \text{NPV}_2 = \text{NPV}_3 = \$100{,}000$$

$$\begin{aligned}\text{NPV}_1 &= 11{,}502(P/A, 12, 30)\\ &= 11{,}502(8.055)\\ &= 92{,}649\end{aligned}$$

$$\begin{aligned}\text{NPV}_2 &= 8{,}883(P/A, 12, 5) + 13{,}439(P/A, 12, 25)(P/F, 12, 5)\\ &= 8{,}883(3.605) + 13{,}439(7.843)(0.5674)\\ &= 91{,}828\end{aligned}$$

$$\begin{aligned}\text{NPV}_3 &= 8{,}883(P/A, 12, 5) + 12{,}470(P/A, 12, 5)(P/F, 12, 5)\\ &\quad + 14{,}201(P/A, 12, 20)(P/F, 12, 10)\\ &= 8{,}883(3.605) + 12{,}470(3.605)(0.5674) + 14{,}201(7.469)(0.3220)\\ &= 91{,}684\end{aligned}$$

Evidently an error has been made. The loans do not have an IROR of 12 percent. They are not equivalent.

(*b*) The plans must be judged at the borrower's opportunity cost of capital. Therefore the net present worths of each loan will be determined and the highest NPV chosen.

$$\begin{aligned}\text{NPV}_1 &= +100{,}000 - 11{,}502(P/A, 9, 30)\\ &= +100{,}000 - 11{,}502(10.274)\\ &= -18{,}172\end{aligned}$$

$$\begin{aligned}\text{NPV}_2 &= +100{,}000 - 8{,}883(P/A, 9, 5) - 13{,}439(P/A, 9, 25)(P/F, 9, 5)\\ &= +100{,}000 - 8{,}883(3.890) - 13{,}439(9.823)(0.6499)\\ &= -20{,}349\end{aligned}$$

$$\begin{aligned}\text{NPV}_3 &= +100{,}000 - 8{,}883(P/A, 9, 5) - 12{,}470(P/F, 9, 5)(P/A, 9, 5)\\ &\quad - 14{,}201(P/A, 9, 20)(P/F, 9, 10)\\ &= +100{,}000 - 8{,}883(3.890) - 12{,}470(3.890)(0.6499)\\ &\quad - 14{,}201(9.129)(0.4224)\\ &= -20{,}841\end{aligned}$$

The first plan is least expensive.

(*c*) $f = 5\%$ and $u = 9\%$

The constant-dollar cost of capital is:

$$\begin{aligned} i &= \frac{u-f}{1+f} \\ &= \frac{0.09-0.05}{1+0.05} \\ &= 0.038 \end{aligned}$$

The deflated rate of 3.8 percent applied to the loan plans in constant dollars will result in the same net present values and the same decision. ■

Loans are usually compared on the basis of periodic cost. We have seen that this is a valid measure when comparing loans of the same initial principal and term. When inflation must be considered, however, the only relevant measure of periodic cost is the cost after inflation has been taken into account. Particularly clear thinking is required to arrive at a correct solution in such cases, as the following example reveals.

■ **Example 13.10.** At 26 percent interest compounded annually, $100,000 is borrowed for 2 years. During these 2 years, the inflation rate is projected to be 20 percent per year.

(*a*) What is the annual cost of the loan? That is, what is the amount that must be paid back in current dollars at the end of each of 2 years to amortize the loan?

(*b*) In addition, calculate the equivalent uniform annual cost of the loan in year 0 (constant) dollars.

Solution

(*a*) Through the use of the capital recovery factor, the annual cost of the loan in current dollars is computed as

$$A = P(A/P, 26, 2) = 100{,}000(0.70248) = \$70{,}248$$

(*b*) The constant-dollar interest rate on the loan is

$$i = \frac{u-f}{1+f} = \frac{0.26-0.20}{1.20} = 0.05 = 5\%$$

Using the capital recovery factor at the 5 percent deflated rate, we see that the early payment is

$$A = P(A/P, 5, 2) = 100{,}000(0.53780) = \$53{,}780$$

However, if we deflate the payments for year 1 in current dollars by dividing by 1 plus the inflation rate for the first year and by 1 plus the inflation rate squared for the second year, then we find that the yearly payments in constant dollars are

$$\text{Year 1} \qquad \frac{70{,}248}{1.20} = 58{,}540$$

$$\text{Year 2} \qquad \frac{70{,}248}{1.44} = 48{,}753$$

Now, it might seem that it is possible to equate the two methods of calculating annual cost of the loan in constant dollars by finding the present worth of the two payments just calculated and equating the result to the present worth of the two equal payments. And this is indeed the case, because at 5 percent the following is true:

$$58{,}540(P/F, 5, 1) + 48{,}783(P/F, 5, 2) = 53{,}780(P/A, 5, 2)$$
$$55{,}752 + 44{,}248 = 100{,}000$$
$$100{,}000 = 100{,}000$$

The possibility of error enters in performing this calculation if we decide that the opportunity cost of capital to the borrower is not 5 percent but some other percentage, say, 9 percent. It will be discovered, if we attempt to perform the previous operation at 9 percent, that an inequality occurs. This means that a borrower with an opportunity cost of capital of 9 percent will prefer one schedule of payments to the other. Equivalence occurs at and only at the 5 percent constant-dollar rate between the two methods of payment. This is exactly in line with the concept of equivalence developed in Chap. 3.

Thus the only correct method of answering the question posed above, when inflation is taken into account, is to use the capital recovery factor with the deflated interest rate. After this cost in constant dollars is obtained, an opportunity cost of capital percentage may be applied to it to discover, e.g., the present worth of the loan to a particular borrower. ■

Finally, let us combine, in a single example, the effect of different loan amortization periods, income taxes, and inflation.

■ **Example 13.11.** Two loan plans are offered for a $100,000 loan:

Plan 1 13 percent for 30 years
Plan 2 12 percent for 15 years

Your constant-dollar opportunity cost of capital is 10 percent. Your tax bracket is projected at 43 percent. Inflation is expected to average 20 percent. All these projections are for the 5 years you expect to keep the house that acts as collateral for the loan.

Which loan is preferable over the 5-year time horizon?

Solution. The annual loan payments in current dollars are

$$AC_1 = \$13{,}341$$
$$AC_2 = \$14{,}682$$

The before-tax cash flow for each loan appears in Table 13.4. Both loans provide you with $100,000. It is not possible to choose the lower-payment loan as the better, because the effects of income tax relief and inflation will cause tradeoffs to occur. For example, the higher interest payments of the 13 percent loan result in more tax relief than with the 12 percent loan. More principal will be paid off during each year under loan 2 at 12 percent. This will result in a lower payoff at the end of year 5. Inflation will have an effect on all these payments.

TABLE 13.4
Two loan plans for $100,000 compared after taxes and inflation

Plan	Year	BTCF (1)	Interest (2)	Principal (3)	Balance (4)	Tax relief at 0.43 (5)	Current $ ATCF (6)	(P/F, 32, N) (7)	Constant $ PW (8)
	0	+100,000			100,000		+100,000		+100,000
	1	−13,341	13,000	341	99,659	5,590	−7,751	0.7576	−5,872
	2	−13,341	12,956	385	99,274	5,571	−7,770	0.5739	−4,459
1 @ 13%	3	−13,341	12,906	435	98,839	5,550	−7,791	0.4348	−3,388
	4	−13,341	12,849	492	98,347	5,525	−7,816	0.3294	−2,575
	5	−13,341	12,785	556	97,791	5,498	−7,848	0.2495	−1,957
	5	−97,791					−97,791	0.2495	−24,399
									+57,350
	0	+100,000			100,000		+100,000		+100,000
	1	−14,682	12,000	2,682	97,318	5,160	−9,522	0.7576	−7,214
	2	−14,682	11,678	3,004	94,314	5,022	−9,660	0.5739	−5,544
2 @ 12%	3	−14,682	11,318	3,364	90,950	4,867	−9,814	0.4348	−4,267
	4	−14,682	10,914	3,768	87,182	4,693	−9,989	0.3294	−3,290
	5	−14,682	10,462	4,220	82,962	4,499	−10,183	0.2495	−2,541
	5	−82,962					−82,962	0.2495	−20,699
									+56,445

Let us first calculate the current-dollar, inflated cost of capital.

$$i = 10\% \qquad f = 20\%$$

$$\begin{aligned} u &= i + f + if \\ &= 0.10 + 0.20 + (0.10)(0.20) \\ &= 0.32 = 32\% \end{aligned}$$

This is the discount rate that must be applied to the current-dollar after-tax cash flow in order to arrive at the constant-dollar present worths of the cash flows. The higher positive amount will govern the choice.

Column 1 of Table 13.4 shows the *before-tax cash flow* (*BTCF*) with the last amount in the column being the loan payoff. The annual payment is divided into principal and interest in columns 2 and 3. The remaining balance after principal payments are subtracted is shown in column 4. The final amount in this column is the payoff on the mortgage. Column 5 is the result of multiplying column 2 by 0.43, and it shows the tax relief. Column 6 is the difference between columns 1 and 5. It shows the cash flow after tax effects have been calculated. Discounting (column 6 times column 7) the *after-tax cash flow* (*ATCF*) at 32 percent gives column 8. Summing column 8 for plan 1 and plan 2 shows the decision criterion: $57,350 for plan 1 and $56,445 for plan 2. Plan 1 is chosen, but by the surprisingly small margin of $905. ■

COLLATERAL

In Example 13.11, the word *collateral* was used to describe the security for a loan—a mortgage—that the house represents. This word and its interpretation in specific circumstances have a direct bearing on the conduct of an economic analysis. Is there ever an occasion when the loan payments should be included in the costs of investment? The answer to this frequently heard question is yes, when the asset being acquired acts as collateral for the loan. The loan in such a case is restricted to investment in the asset that acts as collateral and may not be used for any other purpose.

Compare this situation to the case where money is lent to a company for any use it cares to make of it—operating expenses, investment, repayment of existing debt. In all these uses, the funds have an opportunity cost determined by the cutoff line of the capital budget, as will be seen in a later chapter.

A common example of a loan secured by collateral is the home mortgage. The money loaned may not be used for any other purpose than the purchase of a specific house. In much larger real estate loans, as in the residential mortgage, the collateral is the real estate to be acquired. The money loaned may not be used for any other purpose. If the loan payments are not met, the collateral is taken over by the lender. Home owners behind in their payments face foreclosure of their mortgages.

■ **Example 13.12.** A small investor, Dorsey Pender, wishes to acquire a residential rental property, hold it for 5 years, and then sell it. The property's selling price is $130,000 with 10 percent down payment required. His net profit each year, exclusive of mortgage payments, he estimates at \$9,000. At the end of 5 years, he believes he can net \$160,000 on the sale, before paying off the mortgage. Excluding income tax and inflation considerations, if his opportunity cost of capital is 10 percent, should he make the investment?

His bank will lend him \$117,000 at 13 percent for 30 years with the property to be bought acting as collateral.

Solution. His mortgage payments, on an annual basis, will be

$$\begin{aligned} AC &= 117{,}000(A/P, 13, 30) \\ &= 117{,}000(0.13341) \\ &= \$15{,}609 \text{ per year} \end{aligned}$$

His cash flow, which must include the mortgage payments, is shown in Table 13.5.

The remaining balance on the loan at the end of year 5, after the fifth mortgage payment is made, is

$$R_Y = A(P/A, i, N - Y)$$

in which

$$\begin{aligned} Y &= 5 \qquad\qquad A = 15{,}609 \\ i &= 13\% \qquad\quad N = 30 \\ R_5 &= 15{,}609(P/A, 13, 30-5) \\ &= 15{,}609(7.330) \\ &= 114{,}414 \end{aligned}$$

The reader will remember from Chap. 4 that the remaining balance on a loan is simply the present worth of the remaining payments. The amount is shown on an auxiliary line 5 of Table 13.5.

The net present worth of the cash flow shown in the last column of the table is

$$\begin{aligned} \text{NPW} &= -13{,}000 - 6{,}609(P/A, 10, 5) + (160{,}000 - 114{,}414)(P/F, 10, 5) \\ &= -13{,}000 - 6{,}609(3.791) + 45{,}586(0.6209) \\ &= -9{,}750 \end{aligned}$$

The investment should not be made, but the deal should be further analyzed with income tax and inflation effects included.

TABLE 13.5
Mortgage payments included in the cash flow

Year	Profit	Mortgage payments	Cash flow
0	—	—	-13,000
1	9,000	15,609	-6,609
2	9,000	15,609	-6,609
3	9,000	15,609	-6,609
4	9,000	15,609	-6,609
5	9,000	15,609	-6,609
5			+160,000
5			-114,414

No double counting was involved. Only actual cash amounts were shown in the cash flow. Because the asset was collateral for the loan, mortgage payments were included in the cash flow. ■

The reader may now ask for a clearer explanation than has been previously given for the inclusion, or not, of financing costs. It is evident that financing costs must be included where the asset under consideration is collateral for the loan that finances it, as in the previous example. But could not financing costs be included where the asset is not financed by a loan for which it is the collateral? This situation occurred in many cases we have considered in this book. Opportunity cost did exist.

Financing costs could have been included for the examples considered where funding came from within the entity doing the investing. They were not, because such costs were not differential among the alternatives being considered. The same percentage of first cost would have been included for financing all alternatives, an unnecessary exercise. The following example is germane.

■ **Example 13.13.** A dragline is to be purchased by two construction companies, one small (Tiny Construction Company), the other large (Gargantua Construction Company). Each is considering exactly the same make and model of dragline. Tiny is forced to finance its purchase by a loan at 15 percent for 8 years to be paid back in equal annual payments with the dragline itself acting as collateral for the loan. The Gargantua dragline will be purchased out of general funds from many sources assembled in one capital budget.

In the case of Tiny Construction Company, financing costs are included because they are differential; i.e., they are directly connected to the purchase of this machine. If the machine is not purchased, the financial costs will not be incurred; and if it is purchased, they will be incurred.

In the case of Gargantua, the money used to buy the dragline comes from a general investment fund within the company whose sources are varied: retained earnings, common stock sales, loans, and other sources. Thus, financing cost is not differential in the case of equipment purchase. The purchase or rejection of the dragline will have no effect, e.g., on whether dividends are paid.

The dragline will cost \$120,000. Over its life of 8 years, its maintenance and operation costs will start at \$3,000 for the first year and rise to \$5,100 for the eighth year at a constant increase of \$300 per year. No salvage value will be considered. Ignore tax and inflation effects. Both companies have a MARR of 20 percent.

Solution. Tiny Construction Company will repay the loan in eight equal payments, computed as follows:

$$\begin{aligned} A &= P(A/P, i, N) \\ &= 120{,}000(A/P, 15, 8) \\ &= \$26{,}742 \end{aligned}$$

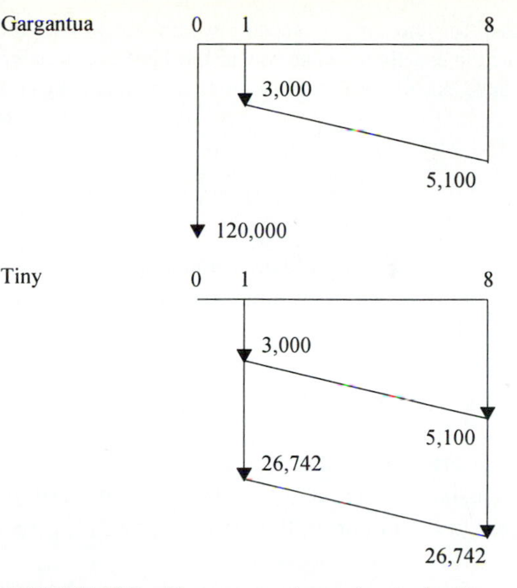

FIGURE 13.9 Two companies analyze a dragline

Both companies' cash flows are shown in Fig. 13.9. Tiny Construction Company's cash flow must be discounted at 20 percent, even though the borrowing took place at 15 percent.

$$\begin{aligned} PW &= 26{,}742(P/A, 20, 8) + 3{,}000(P/A, 20, 8) + 300(P/G, 20, 8) \\ &= 26{,}742(3.837) + 3{,}000(3.837) + 300(9.8831) \\ &= 114{,}120 + 2{,}965 \\ &= 117{,}085 \end{aligned}$$

Gargantua Construction Company will discount its cash flow at 20 percent also, that being its cutoff capital budget rate—its opportunity cost of capital.

$$\begin{aligned} PW &= 120{,}000 + 3{,}000(P/A, 20, 8) + 300(P/G, 20, 8) \\ &= 120{,}000 + 3{,}000(3.837) + 300(9.8831) \\ &= 120{,}000 + 11{,}511 + 2{,}965 \\ &= 134{,}476 \end{aligned}$$

Why the difference? Is Tiny Construction Company actually able to buy equipment more cheaply, with operating and maintenance costs being the same? No, because the financing costs of Gargantua will be paid as dividends, interest on bonds, interest on short-term loans, and so forth, but need not appear in the decision to buy this machine or that one because of nondifferentiality. The analyst could introduce a standard financing cost, if desired, but it would be nondifferential among alternatives. (For bidding on a job, the financing cost would appear as part of an overhead cost applied to machine use per hour.)

Introducing a financing cost for the large company would mean that the Gargantua dragline would be treated in exactly the same way as the Tiny Construction Company dragline: i.e., it would have exactly the same cash flow, if the financing cost were at 15 percent. ■

SUMMARY

Borrowing money from someone else constitutes a loan. Choosing among loan plans is an exercise in economic analysis and decision making. The choice cannot be made on the basis of the interest rates charged.

Loans are used for two purposes only—investment and consumption. Both uses require that the decision be based on the opportunity cost of capital of the borrower.

In equal-initial-principal, equal-term loans, the periodic payment may be used to decide among mutually exclusive alternative loans. The present worth of the alternative loan plans, calculated at the opportunity cost of capital, may also be used.

In cases where the incremental cash flow between loan plans, or benefit streams, is used, the increment should be taken so that the first payment of the increment is an outflow.

In unequal-term loans, when one is deciding among mutually exclusive loan proposals, the net present-worth method may be used, either incrementally or individually. Periodic payments are unreliable as a basis for judgment.

Usually interest paid on business loans is deductible from taxable income. This means that tax effects must be calculated in order to choose among loan plans. The after-tax cash flow decides the issue.

Generally speaking, inflation helps debtors and hurts lenders. It also has an effect on the choice among possible loans, accentuated as inflation rates rise. The only relevant measure of periodic cost is the cost after inflation has been taken into account.

Differing loan periods, taxes, and inflation must all be considered when one is deciding among loan plans.

Collateral is the security for a loan. When the loan is secured by collateral and may not be used for any other purpose than the purchase of that collateral, loan payments must be included in the investment cash flow.

PROBLEMS

13.1. In Example 13.1, page 337, the payments are increased to $5,000 each. Draw the cash flow diagram for the loan. What interest rate is being paid, to the nearest one-tenth of a percent? All other data remain the same.

13.2. Laura Slocum notices a sign, posted in the window of an appliance store, "Color TV Sale—Cash Price $528, EZ Credit, 12% Financing." After selecting her TV, Slocum signs an installment purchase agreement for $600, to be paid at $50 per month for 12 months. She objects that the TV price is only $528, but the salesman explains that the store's policy is to deduct the interest on its loans all at once at the beginning of the contract. Twelve percent of $600 is $72, and $528 plus $72 equals $600. He states that this saves the store the added expense of bookkeeping for all its monthly accounts. It can then pass the savings on in the form of low interest rates.

What effective annual interest rate is Slocum actually paying?

13.3. "Our clients' problems started in February 1981, when they went to get a second mortgage for $12,000. On signing, they received only $7,000, the lender keeping $5,000 in fees and commissions for arranging the loan. In addition, our clients had to pay $1,800 in interest (also at closing). Worst of all, the loan had to be repaid in 1 year...."[2]

Calculate the interest rate on this loan to the nearest whole percentage.

Answer: 131%

13.4. The All-Africa Construction Company, operating out of Lagos, requests a loan from a local lender. The lender offers U.S. $3 million at 11 percent per year over 5 years. Payments will be made at $930,000 per year, computed as $330,000 interest (11 percent of $3 million) plus $600,000 per year principal payment.

(*a*) Is the loan actually at 11 percent?

(*b*) If not, what is the correct interest rate on the loan, to the nearest tenth of a percent?

13.5. A home purchaser agrees to assume a seller's $75,000 existing 7 percent mortgage. The seller provides a second mortgage of $45,000 at an interest-only rate of 13 percent for 5 years. The sale price is set at $150,000. The down payment is $30,000. A truth-in-lending statement is given to the purchaser.

(*a*) What are the total finance charges on the second mortgage?

Answer: $29,250

(*b*) How much would the buyer owe the seller each year on the second mortgage?

Answer: $5,850

Compute the amounts asked for as though the payments on the mortgages were made on an annual basis, rather than a monthly one.

13.6. You are offered a second mortgage loan of $80,000 for 15 years at 11½ percent annually with one point (1 percent of the loan amount) to be charged at closing (the time when the loan is made). Two points will also

---

[2]National Law Center, *The Advocate,* vol. 16, no. 5, Washington, D.C.: George Washington University, November 25, 1985.

be charged as closing costs (loan fee) at closing. An additional appraisal fee of $185 will be charged.

(*a*) If payments are made yearly, what will be the level payment sufficient to amortize the loan of $80,000?

(*b*) How much money will you actually receive at closing?

13.7. In Example 13.2, page 339, the opportunity cost of capital for Henry Higgins Company is now 11 percent. Will this change the decision on which loan to select? *Ceteris paribus* applies.

13.8. Two possible repayment schemes of a loan of $2 million are offered to the Spring Valley Water District as follows:

Year	Loan 1	Loan 2
1	$188,780 annually	260,000
2	for 20 years	252,000
3		244,000
4		236,000
⋮		⋮
20		108,000

The second scheme requires a payment that decreases at a constant $8,000 per year.

Which of these schemes should the water district choose and why? The MARR is 10 percent, and no inflation effects are to be considered.

13.9. An oil exploration company was offered financing for its operations in an African country for the following two plans, both fulfilling the requirement of $15 million for each of 5 years during which exploration would take place.

Year	Plan 1 ($000,000)	Plan 2 ($000,000)
0	+15	+15
1	+15	+15
2	+15	+15
3	+15	+15
4	+15	+15
5	-25.797	0
6	-25.797	0
7	-25.797	0
8	-25.797	-50.270
9	-25.797	-50.270
10	-25.797	-50.270
11	-25.797	-50.270
12	-25.797	-50.270
13	-25.797	-50.270
14	-25.797	-50.270
15	-25.797	-50.270

The first payment to the company will be made at the signing of the agreement, i.e., time 0.

Which of the two plans should the exploration company accept if its MARR is 30 percent?

Taxes have been included, and inflation has been considered in the cash flow.

13.10. A Middle Eastern country has received the following payment schedules from a private German industrial complex. The loan of $300 million will cover financing, design studies, construction, and 5 beginning years of operation of a large river basin development. The country must choose a repayment schedule if it decides to go ahead with the project.

The development will take 4 years to construct, and no loan payment will be required during those years. Plan *A* delays payment for the first 2 years. Plans *B* and *C* require payment from year 5 through year 15.

	Payment ($000,000)		
Year	**Plan *A***	**Plan *B***	**Plan *C***
1	0	0	0
2	0	0	0
3	0	0	0
4	0	0	0
5	0	50.526	24.000
6	0	50.526	24.000
7	45.158	50.526	24.000
8	45.158	50.256	24.000
9	45.158	50.256	24.000
10	45.158	50.256	24.000
11	45.158	50.256	24.000
12	45.158	50.256	24.000
13	45.158	50.256	24.000
14	45.158	50.256	24.000
15	345.158	50.256	324.000

If the country estimates its opportunity cost of capital at 15 percent, which plan should be chosen?

13.11. In Example 13.3, page 341, the bank offers you a 10 percent loan with a term of 60 months. The savings and loan company agrees to the same term, but at 11 percent. Your friend now requires $15,000 at the end of 5 years. All other data being unchanged, which loan should you take?

13.12. In Example 13.4, page 342, your African house can be rented to a tenant who is willing to lease it for 2 years. In addition, he offers either $13,000 at the beginning of the lease, on the day he occupies the house, or $650 per month for 24 months, with the first payment made on the day he moves in, and each succeeding payment on the first day of each month following. Everything else remaining the same, which alternative should you accept?

13.13. In Example 13.5, page 344, will a change in the loan amount make any difference in the decision in the face of changing opportunity costs of capital? Try it, *ceteris paribus*, for a loan of $350,000.

13.14. You may select one of the following loans: (1) $100,000 to be paid annually for 10 years at 11 percent or (2) $100,000 to be paid annually for 15 years at 12 percent.

(*a*) Assume your opportunity cost of capital is 8 percent. Which loan would you choose?

(*b*) Assume your opportunity cost of capital is 16 percent. Which loan would you choose?

13.15. Beneficial Savings and Loan Company suggests the following schedules as acceptable alternatives for repayment of a $200,000 mortgage on a house you plan to buy.

(*a*) A 30-year mortgage at 14 percent compounded annually.

(*b*) A balloon mortgage with payments to be made based on a 12 percent interest rate compounded annually for 25 years, but with the balance to be paid at the end of 10 years.

(*c*) A 15-year fixed-rate mortgage at 15 percent compounded annually.

Your opportunity cost of capital on current dollars before taxes is 10 percent. Without considering tax or inflation effects, which plan should you select? Simplify the calculation by considering all repayment on an annual basis.

13.16. In Example 13.6, page 346, your opportunity cost of capital has decreased to 8 percent. All other data remaining the same, which loan should you accept?

13.17. The Damrongs are about to purchase a house for $150,000. They have the following two options for financing:

(*a*) Borrow $150,000 at 10 percent interest per year for 25 years.

(*b*) Borrow $50,000 from the homeowner at 11 percent interest for 10 years and, in addition, assume the remaining balance of $100,000 on the owner's mortgage at 9 percent for 20 years.

The Damrongs have an opportunity cost of capital of 8 percent per year. No taxes or inflation will be considered at this time. Although payments will be made monthly, annual payments will be sufficiently accurate for decision purposes.

Which loan plan should they choose?

13.18. Estelle Skalavoyski wishes to buy a house costing $100,000. She has three loan alternatives on the full $100,000 amount:

(*a*) Pay $35,000 every 2 years, starting at year 2 and ending at year 10, for a total of five payments.

(*b*) Pay the loan monthly at a 14 percent nominal rate for a term of 30 years.

(*c*) Pay $10,000 starting at year 1 and increasing each year by $3,500 for the next 8 years. Thus the payment at year 2 will be $13,500, at year 3, $17,000, and so on.

Which of the loan plans should Skalavoyski choose and why? Her opportunity cost of capital is 10 percent. Do not consider inflation or taxes.

13.19. You are considering lending $100,000 to R&H Construction Company to help it finance a commercial building program. R&H has proposed quarterly loan payments based on a 16 percent nominal rate compounded quarterly and a 5-year term.

(*a*) If the first payment is made 3 months after the date you lend R&H the money, what will be the amount of each payment to the nearest dollar?

(*b*) While you are considering R&H's first proposal, the company withdraws it and asks you to consider canceling the first 8 payments and allowing the company to pay off the loan in 12 payments, starting with what would have been the ninth payment under the previous arrangement. What will be the amount of each of the remaining 12 payments to the nearest $10?

(*c*) Instead of accepting either of the first two proposals, you stipulate that you will cancel the first 8 payments and accept the 9th and subsequent payments through the 20th payment in the amount of the answer to part (*a*) above. What will still be owed you, request in a balloon payment at the same date as the original 20th payment. What will be the amount of that balloon payment?

13.20. In Example 13.7, page 349, assume now that the loan was actually for $16,000, at 15 percent interest rate. All other data remain the same. Compute the tax relief and after-tax payments.

13.21. In Example 13.8, page 349, Adolph Galland Company now wishes to borrow $150,000. Compute the payments necessary under each plan. Which loan should be chosen?

13.22. Captain Taylor wishes to buy a house costing $120,000. She is financing the house by using U.S. Army–backed financing. This allows a home buyer to finance 100 percent of the cost of the house. She has two alternatives: (1) $120,000 to be paid annually for 15 years at 10 percent interest or (2) $120,000 to be paid annually for 30 years at 12 percent interest. Inflation is estimated to be averaging 6 percent annually over the term of the loan. Captain Taylor's constant-dollar opportunity cost of capital is 3 percent before taxes. Which plan should she select? Round all calculated percentages to the nearest whole percentage. Neglect income tax relief.

13.23. HWM Company specializes in hazardous waste management. It is contemplating capital equipment purchases that will require funding of $500,000.

A loan of $500,000 can be negotiated at an interest rate of 10 percent, for a term of 25 years. During this period, inflation is expected to average 5 percent per year. No income tax effects will be considered at this time.

(*a*) What will be the annual payment on the loan if it is held to term? Assume that payments may be made on an annual basis.

(*b*) What will be the deflated interest rate on the loan?

(*c*) Assume that HWM Company's opportunity cost of capital on current dollars is 16 percent. Is it better to hold the loan for 15 years and pay off the balance at that time, or to hold the loan for the full 25 years?

13.24. In Example 13.9*a*, page 351, what rates are being charged for plans 1, 2, and 3?

13.25. An international lender, HCE Company, offers your firm, headquartered in the Bahamas, the following repayment plan for a loan of $50 million: interest at 12 percent, to be charged on signing the loan agreement, computed as 12 percent of $50 million times 5 years, the time period of the loan. "Reduced principal payments," as HCE calls them, are to be paid in five payments of $9.828 million at the end of each of 5 years. HCE Company describes this arrangement as a "better than 12 percent" loan.

Another lender, James Joyce Ltd., offers $50 million with deferred payments of $61.68 million at the end of the third, fourth, and fifth years. The James Joyce Ltd. proposal seems outrageous in comparison to HCE's, your boss remarks.

If your opportunity cost of capital is 50 percent, which loan should you accept?

13.26. What would be the decision in Prob. 13.15 if an inflation rate of 8 percent were projected over the next 30 years?

13.27. C.F. Moreno Engineering Company, newly established by close friends of yours, is in need of working capital. If you will lend them $26,785, they will pay you $1,740 per year at the end of each 5 years from the time you make the loan. In addition, they will pay, at the end of the fifth year, a lump sum which will represent your original loan appreciated at 15 percent, compounded annually, for 5 years.

You anticipate an inflation rate of 4 percent per year over the next 5 years.

In terms of constant year-0 dollars, what will be your rate of return, to the nearest whole percentage, if you make the loan under the above conditions? Ignore income tax.

13.28. In Example 13.9*c*, page 352, what will an estimate of 4 percent inflation annually over the next 30 years do to the decision? *Ceteris paribus* rules.

13.29. The car buyer now wishes to compute the annual cost of the Mercedes-Benz 300D of Probs. 6.7 and 10.4 after taxes and inflation. Over the next 5 years, he believes inflation will average 12 percent annually. The interest rate on the loan is 15 percent, compounded annually. The term of the loan is 4 years. A 10 percent down payment is required. The opportunity cost of capital to the buyer in current dollars is 9 percent.

Compute the *uniform* annual cost of the Mercedes-Benz 300D, in 1995 dollars, i.e., *after* inflation has been taken into account but *before* taxes. As previously stated, all nonfinancial costs and the resale value have been deflated to 1995 dollars. These are $555 for fuel and maintenance, $900 for insurance, annually, and $7,300 resale value. Consider each yearly payment as being made once per year on December 31.

13.30. Now consider all effects of purchasing the Mercedes-Benz 300D after taxes and after inflation. What will be its present cost to the buyer over the 5-year time horizon? Use the information given in Probs. 6.7, 10.4, and 13.29. Use MACRS GDS for depreciation.

13.31. In Example 13.10, page 353, suppose that the opportunity cost of capital to the borrower is actually 9 percent in deflated dollars, mentioned as a possibility in the example.

(*a*) What will be the amount of the loan payments?

(*b*) What will be the constant-dollar interest rate on the loan?

(*c*) What will be the opportunity cost to the borrower, per year, of the loan?

13.32. In Prob. 13.8, page 362, the MARR in constant dollars is 10 percent. Inflation is estimated to be 4.5 percent per year over the next 20 years. Inflation must now be included in the analysis.

Repeat Prob.13.8. All calculated percentages may be rounded to the nearest whole percentage so that the tables may be used.

13.33. In Example 13.11, page 354, your constant-dollar opportunity cost of capital is revised downward to 8 percent, while the inflation rate is estimated at 25 percent. What effect will this have on your decision? All other inputs remain unchanged.

13.34. Perpetual Home Funding offers 30-year fixed-rate mortgages at 10 percent with 2¾ points (percentage points, based on the amount of the mortgage, to be paid at time 0). It also offers 15-year fixed-rate mortgages at 9 percent with 2¼ points.

A mortgage of $200,000 is envisaged, to be paid off annually.

(*a*) Do not consider inflation. Which plan will a home buyer prefer if her opportunity cost of capital before taxes is 8 percent and no tax effects are considered?

(*b*) If inflation averaging 4 percent annually over the next 30 years is estimated, her constant-dollar opportunity cost of capital before taxes is 8 percent, and no tax effects are taken into account, which plan should the home buyer choose?

All rates may be rounded to the nearest whole percentage for calculations.

13.35. The following information is advertised for condominiums in Puerto Vallarta, Mexico. Prices are given in U.S. dollars.

Beginning of month	Payment
1	3,517.00
2–6	1,200.00
7–186	1,083.72

Interest is stated as a fixed 13 percent and is included in the payments. The total cash price is stated as U.S. $95,170. The buyer estimates an annual inflation rate of 10 percent over the next 15 years in Mexico. This inflation rate will apply because even though prices are quoted in dollars, actual payments will be made in pesos. The buyer assumes that the exchange rate between pesos and dollars will match inflation in both countries and thus may be ignored.

(*a*) Is the interest rate stated correct? If not, what is it?

(*b*) What is the deflated interest rate after inflation is considered?

13.36. Your consulting engineering company, headquartered in the United States, wishes to borrow $900,000 for operating expenses over the next 3 years to cover a temporary cash flow difficulty.

First National Bank offers a loan to be repaid in a lump sum of $1,478,700 at the end of 3 years. The interest rate on this loan is 18 percent.

Second National Bank offers a 3-year loan at 16 percent to be repaid in three equal payments of $400,734 at the end of each of 3 years.

The company, which operates internationally in many countries with widely differing inflation rates, requires a deflated minimum attractive rate of 6 percent. It uses 45 percent as its average tax bracket for all local, state, federal, and country-of-operation taxes. It projects an inflation rate of 5 percent for the United States over the term of the loan. After taxes and inflation have been considered, which loan should your company take?

13.37. In Example 13.12, page 356, the seller reduces his price to $120,000. The 10 percent down payment is $12,000. Pender's bank will lend him $108,000 on the same terms as before. All other data being the same, should he now make the investment?

13.38. Your engineering management consulting company is considering several second-mortgage loan possibilities to finance future operations. Collateral for the loans will be an office building owned by your firm. The principal of the loan is $2 million. The loan alternatives are (1) a 15-year, fixed-rate loan at 13 percent and (2) a 10-year variable-rate loan with the first 5 years at 9 percent and the last 5 years at 15 percent. This means that the annual payments during the first 5 years will be at 9 percent based on a 10-year term. The remaining balance at the end of the first 5 years will be paid off at 15 percent based on a 5-year term.

Your company's opportunity cost of capital is 20 percent on before-tax, current dollars.

Payments on the loan selected will be made *annually*.

(*a*) As financial vice president, which loan do you find more attractive?

(*b*) If inflation is estimated at 5 percent per year over the next 15 years, will your decision change? Why or why not?

13.39. In Example 13.13, page 358, explain why the financing cost is not included in Gargantua's cash flow. Why must it be included in Tiny's cash flow?

13.40. An oil refinery receives a budget from its company to finance changes in equipment. These changes are suggested by its maintenance engineers and designed in the refinery engineering office.

The chief engineer decides, on the basis of an economic analysis, which of the suggested projects will be constructed. She is aware that the budget sum allocated to the refinery by the company comes from a general fund supplied by retained earnings, sale of stock, and bonded indebtedness. The average interest paid on this capital supply is 10 percent. The chief engineer insists that all project analyses include this financial cost.

As head project engineer, how would you react to her demand?

CHAPTER 14

CAPITAL BUDGETING

In Chap. 1, pages 3 and 4, mutually exclusive, independent, and interdependent investments were defined, and examples were given. The reader should review those pages before proceeding.

Engineering projects are normally indivisible or, as the economist characterizes them, "lumpy." You cannot build 65 percent of a road between points *A* and *B*; the road fulfills its function only if you build all of it. A dam cannot be built to 20 percent of completion; it does its job only if it is completely finished. Projects such as roads, dams, and buildings are either built or not built. There is no workable point between building and not building. As we will see, this characteristic means we can use only integer programming and that the integers must be either 0 or 1, that is, binary.

Contrast this situation with the usual one in linear programming. In deciding among a number of products, or the *mix*, it is perfectly possible to choose to produce 1,020 of *A*, 240 of *B*, 3,100 of *C*, and so on, where *A* can be produced in quantities from 0 to 10,000, *B* from 0 to 2,000, *C* from 0 to 5,000, and so on. To require binary solutions—all or none of a product—would be to restrict the outcome unnecessarily, and thus practically guarantee an erroneous solution.

This chapter will not deal with ordinary production problems, usually covered in a course on linear programming, but only with the special category of linear programming problems encountered in engineering works where a capital budget is required.

Up to this point, only mutually exclusive investments have been considered. The methods used in their choice have been based on equal service lives, incremental analysis, and elementary algebraic solutions. Now independent investments—an equally important category—will be taken up. The methods of treatment will be opposite to those used for mutually exclusive alternatives. If economic lives are unequal, they need not, and must not, be equalized. Incremental analysis must not be used. Where a budget must be adhered to, mathematical programming provides the only correct approach to a solution if we disregard the usually unworkable complete-enumeration method. Evidently, then, the fundamental and first question an analyst must ask, when confronted with an investment problem, is, Are these mutually exclusive or independent alternatives? The methods and treatment of the problem depend on the answer to that question. The methods and treatment will be completely different, depending on the answer.

Toward the end of the chapter, we will deal with problems where mutually exclusive and independent alternatives are combined. Incremental analysis disappears, absorbed into mathematical programming, for we will not always choose the best of a set of mutually exclusive alternatives.

In this chapter, the discount rate will be exogenous, i.e., will come from an outside source, such as a board of directors or a central office. The discount rate determined arbitrarily, outside the capital budgeting process, is the most usual condition. However, as will be explained further in Chap. 15, the discount rate itself can be a function of the capital budget selected. For this chapter we will consider only the exogenous rate, such as 10 percent mandated by the U.S. Department of Transportation.

Heretofore, we have been concerned with the selection of projects where no budget was involved. It was possible to build any of the projects considered. We had only to choose one. Now we will consider the case where a set of projects must be selected. Where the budget is unlimited, we can make up the set by choosing all projects that qualify, i.e., whose discounted benefits exceed or equal their discounted costs. But what if there is not enough money to build all the projects? Then we must resort to capital budgeting.

Capital budgeting rests on the assumption that a budget must be made up. Unlimited funds are not available for all the investments the enterprise would like to make. Capital budgeting without the necessity of a budget is a contradiction in terms. *Capital budgeting* is the problem of selecting from a number of available projects those that will optimize the return on the capital invested and whose cost will still remain within the amount of money available for investment. Expressed mathematically, the *objective function* is

$$\text{Maximixe } Z = \sum_{i}^{m} b_i x_i \tag{14.1}$$

where m is the number of projects being considered, b_i is a figure of merit, say, the net present value of project i, and x_i is the decision variable that indicates how much of project i should be undertaken. In indivisible projects, such as highways, x_i is either 0 or 1—0 if the project should not be undertaken within a particular budget and 1 if it should.

The preceding objective function is subject to the following constraints:

$$\sum_{i}^{m} c_i x_i \leq M \tag{14.2}$$

where c_i is the cost of project i, x_i is the decision variable as explained above, and M is the budget. Also

$$x_i \geq 0$$

That is, we cannot have a negative project. And

$$x_i \leq 1$$

because we can do a project only once. Therefore,

$$0 \leq x_i \leq 1 \tag{14.3}$$

If the projects are indivisible, as they will be in our engineering examples, then

$$x_i = 0, 1 \qquad \text{for } i = 1, 2, \ldots, m \tag{14.4}$$

Taken together, the above equations comprise a 0-1 integer program.

The following example illustrates the method.

■ **Example 14.1.** A road agency in a small African country presents the projects shown in the accompanying table to the World Bank for possible funding.

(*a*) If the World Bank were to fund all the projects with a *B/C* ratio greater than 1.0, which would be funded and in what priority?

(*b*) The World Bank decides that the country can afford no more than a $50 million addition to its external debt at this time. Under this budget, which projects should be funded?

Project	Present worth of benefits ($000,000)	First cost ($000,000)
1	150	30
2	45	30
3	40	10
4	24	20
5	17	9
6	10	5
7	8	1

Solution

(*a*) The benefit/cost ratios are as follows:

Project	*B/C*	Priority
1	5.00	2
2	1.50	6
3	4.00	3
4	1.20	7
5	1.89	5
6	2.00	4
7	8.00	1

Without a constraint on cost, the bank would fund projects 7, 1, 3, 6, 5, 2, and 4 with

$$\Sigma \text{NPV} = \$189 \text{ million} \qquad \Sigma C = \$105 \text{ million}$$

All projects have B/C ratios greater than 1.0. Therefore, all would be funded.

Notice that this is not a capital budgeting problem, because there is no budget. It is introduced here to emphasize the difference between unconstrained cost and budgeted cost problems.

(*b*) Define b_i, used in Eq. (14.1), as the NPV for project *i*. Then, subtracting first cost from the present worth of the benefits, we have

Project	NPV
1	120
2	15
3	30
4	4
5	8
6	5
7	7

Substituting in Eq. (14.1) gives

$$\text{Max } Z = \sum_{i}^{m} b_i x_i$$
$$= 120x_i + 15x_2 + 30x_3 + 4x_4 + 8x_5 + 5x_6 + 7x_7$$

subject to, from Eqs. (14.2) to (14.4),

$$\sum_{i}^{m} c_i x_i \leq M$$
$$30x_1 + 30x_2 + 10x_3 + 20x_4 + 9x_5 + 5x_6 + 1x_7 \leq 50$$
$$0 \leq x_i \leq 1$$
$$x_i = 0,\ 1 \qquad \text{for } i = 1, 2, \ldots, 7$$

At this point a computer program must be used. (See App. A, page 467.) Hand methods are possible but prohibitively laborious. The computer provides the solution:

7, 1, 3, 5

The figure of merit of this solution is

$$\Sigma \text{NPV} = 7 + 120 + 30 + 8 = \$165 \text{ million}$$

at a cost of

$$\Sigma C = 1 + 30 + 10 + 9 = \$50 \text{ million}$$

which is exactly equal to *M*, the budget.

If the computer program has been written correctly, this is the correct answer. The program may be checked by putting through it a problem to which the answer is known and obvious. In the program used in this example, this has been done. Such a check does not absolutely guarantee that the program is correctly written. Only careful inspection of the algorithm can do that. ■

PROGRAMMING VERSUS EFFICIENCY OF CAPITAL CRITERIA

In Example 14.1, the net present value was used as the measure of worth in the program that resulted in the optimal allocation of resources to the projects. The NPV could not have been used by itself to perform this allocation. Comparing the solution to the associated NPVs reveals the impossibility of ranking the projects from best to worst by starting with the highest NPV and proceeding to the lowest. It is impossible because there is no relation in the NPV that indicates the return per dollar spent. To overcome this difficulty, the *efficiency of capital criteria* were developed: the benefit/cost ratio and the internal rate of return (IROR). For many years, it was believed by students of capital budgeting that a correct solution of the problem could be achieved by simply ordering the projects from best IROR or B/C to worst and cutting off at the point in the list where the budget was exhausted. These authors failed to recognize that they were dealing with a problem in mathematical programming that could only be solved by methods appropriate to that branch of mathematics. This point, well known to those engaged in operations research, escaped the attention of many authors, particularly in the area of engineering economy. The following example, an extension of Example 14.1, illustrates the point.

■ **Example 14.2.** An independent consultant suggests to the country of Example 14.1 that the projects to be funded by the World Bank under a $50 million budget ought to be determined by their B/C ratios. He bases his recommendation on the following list:

Project	B/C
1	150/30 = 5.00
2	45/30 = 1.50
3	40/10 = 4.00
4	24/20 = 1.20
5	17/9 = 1.89
6	10/5 = 2.00
7	8/1 = 8.00

Accordingly, the projects to be funded are

7, 1, 3, 6

This differs from the World Bank's recommendation. Which list is correct and why?

Solution. Under $M = 50$, the B/C ratio solution—7, 1, 3, 6—possesses the following attributes:

$$\Sigma\text{NPV} = 7 + 120 + 30 + 5 = \$162 \text{ million}$$

and

$$\Sigma C = 1 + 30 + 10 + 5 = \$46 \text{ million}$$

By programming, the solution is 7, 1, 3, 5:

$$\Sigma \text{NPV} = 7 + 120 + 30 + 8 = \$165 \text{ million}$$

and

$$\Sigma C = 1 + 30 + 10 + 9 = \$50 \text{ million}$$

Clearly the programming solution is correct. The *B*/*C* ratio solution is not. The programming solution shows a total NPV of $165 million, compared to the *B*/*C* ratio solution of $162 million. Both solutions stay within the budget of $50 million.

Incorrect solutions of the same type may occur regardless of the method used—e.g., internal rate of return.

Some readers may point out that the *B*/*C* ratio solution uses less funds, by $4 million, than the programming solution. What will happen to these excess funds? Will they not affect the solution?

The excess funds will be invested at the opportunity cost of capital, whatever that may be. The net present value of any amount invested at the opportunity cost of capital is zero. For example, let us say that the opportunity cost of capital is 20 percent. The $4 million invested at this rate for any length of time will have a net present value of

$$\text{NPV} = \text{PW}_B - \text{PW}_C$$

or the present worth of the benefits minus the present worth of the costs. In our problem

$$\text{NPV} = 4 - 4 = 0$$

The stumbling block in the understanding of this relationship is usually the present worth of the benefits. Readers should test their understanding by confirming that $4 million invested at 20 percent for, say, 15 years will indeed have a present worth of benefits of $4 million. In fact, any amount, invested at any rate that is the opportunity cost of capital, for any period will have an NPV equal to zero. Thus the $4 million has no effect on the solution. ■

■ **Example 14.3.** Using the data of Example 14.1, we now change the measure of worth b_i from the net present value to the present worth of the benefit of project *i*. This will require that we treat leftover funds explicitly.

Solution. Test each solution:
Programming: 7, 1, 3, 5

$$\Sigma C = 1 + 30 + 10 + 9 = 50$$
$$\Sigma B = 8 + 150 + 40 + 17 = 215$$

B/*C* ratios: 7, 1, 3, 6

$$\Sigma C = 1 + 30 + 10 + 5 + 4 = 50$$
$$\Sigma B = 8 + 150 + 40 + 10 + 4 = 212$$

The $4 million at the end of the benefits summation in this last equation is the leftover funds that must be invested at the opportunity cost of capital. The present worth of the benefits of the leftover funds is $4 million.

Once again the programming solution gives the correct choice of projects. The *B*/*C* ratio solution does not.

This solution, using as a measure of worth the present worth of the benefits, can be compared to the solution using the measure of worth as the net present value by the aforementioned equation:

$$\text{NPV} = \text{PW}_B - \text{PW}_C$$

In the programming solution,

$$\text{NPV} = 215 - 50 = 165$$

In the *B*/*C* ratio solution,

$$\text{NPV} = 212 - (46 + 4) = 162$$

Both solutions check the answers of Examples 14.1 and 14.2. ■

CAPITAL RATIONING—INDEPENDENT PROJECTS ONLY

Let us examine the remaining two differences between mutually exclusive and independent projects, now that we have disposed of the differences in mathematical optimization between them. These differences lie in the treatment of economic lives and incremental analysis.

When selecting an optimal list of projects that will fit into a capital budget, we must not equalize the lives of the projects. The situation is completely opposite to that confronting the analyst who must provide alternatives that are equal in service and in service lives. Recall the example of the reinforced-concrete sewer pipe versus the corrugated-steel sewer pipe. To make a fair comparison between them, the life of the shorter-lived alternative, the corrugated steel pipe, had to be made equal to that of the longer-lived alternative by repeated cycles of investment. This need disappears when we are dealing with independent projects. In mutually exclusive alternatives, the analyst must select one project *or* the other. In selecting a list of investment projects constrained by a budget, the engineer is concerned with one project *and* another. In this latter case, there is no relation between the projects that demands an equalization of lives. In government budgets, consider the lack of relation between a school in north Texas and a farm-to-market road in south Texas. Their only relation is that they are in the same state and thus appear in the same budget.

For this same reason, incremental analysis must not be used to compare budget projects. Incremental analysis requires that mutually exclusive alternative projects—like different, and thus mutually exclusive, levels of production in the same plant—be related. If you select one, all others are excluded. If the mutually exclusive relationship is absent, the economic rule of "extra benefit must at least balance extra cost" is thus no longer applicable.

TABLE 14.1
Six independent projects

Project	Life	Present worth of benefits	First cost	NPV at 12%	*B/C*	Priority
1	5	90	30	60	3.00	2
2	10	25	10	15	2.50	3
3	8	40	20	20	2.00	4
4	6	17	9	8	1.89	5
5	12	10	5	5	2.00	4
6	2	8	1	7	8.00	1

It is well to make clear that the order of the projects on the list is unimportant to the choice of projects. What we want is an optimal selection of projects, however they may be listed. Alphabetical listing by name makes just as much sense as ordering by *B/C* value.

The following example involves projects with differing lives.

■ **Example 14.4.** Table 14.1 shows six independent projects. All the projects have different lives. None of the lives have been equalized to any other life, as would be the case if the alternatives were mutually exclusive. By using Eqs. (14.1) through (14.4), the programming solution follows from

$$\text{Max } Z = 60x_1 + 15x_2 + 20x_3 + 8x_4 + 5x_5 + 7x_6$$

s.t.:

$$30x_1 + 10x_2 + 20x_3 + 9x_4 + 5x_5 + 1x_6 \leq 50$$

$$0 \leq x_i \leq 1$$

$$x_i = 1, 0 \qquad \text{for } i = 1, \ldots, 6$$

where the NPV is the measure of worth and 50 is the budget.

The programming solution reveals that the optimal capital budget is composed of projects 6, 1, 2, and 4 with

$$\Sigma\text{NPV} = 7 + 60 + 15 + 8 = 90$$

and

$$\Sigma\text{C} = 1 + 30 + 10 + 9 = 50$$

This is the correct solution.

Notice that this solution does not agree with the *B/C* ratio solution, which is 6, 1, 2, and 5 with

$$\text{NPV} = 7 + 60 + 15 + 5 = 87$$

and

$$C = 1 + 30 + 10 + 5 = 46 < 50$$

The leftover funds of 4 will have an NPV equal to 0. Two projects have a *B/C* ratio of 2.00. The one selected is the one that fits within the budget, project 5.

Recall that there is a method we have seen that effectively equalizes the lives of mutually exclusive alternatives. If it is used with independent alternatives, its effect will be the same; i.e., it will equalize the lives of the independent alternatives under consideration. This method is *annual worth.* We have seen it called *annual profit* and, where benefits of all alternatives were equal, *annual cost*. This method must not be used in capital budgeting problems. The reader may emphasize the point by doing Prob. 14.16. ■

CAPITAL RATIONING—INDEPENDENT AND MUTUALLY EXCLUSIVE PROJECTS

Now let us suppose that some projects on the list of those being considered for inclusion in the capital budget may be built, using any one of a number of mutually exclusive alternatives. The arrangement of the projects is shown with independent ones forming rows of a matrix and mutually exclusive ones forming the columns. Clearly the choice among such projects is an exercise in integer programming. Now row i and column j subscripts must be included. The objective function is

$$\text{Max } Z = \sum_{i}^{m} \sum_{j}^{n} b_{ij} x_{ij} \tag{14.5}$$

subject to

$$\sum_{i}^{m} \sum_{j}^{n} c_{ij} x_{ij} \leq M \tag{14.6}$$

$$x_{ij} = 0,\ 1 \qquad \text{for } i = 1, 2, \ldots, m; j = 1, 2, \ldots, n$$

where

b_{ij} = measure of worth or figure of merit
x_{ij} = decision variable, 0 or 1
c_{ij} = first cost of project
M = budget
m = number of rows (independent projects)
n = number of columns (mutually exclusive projects)

By definition, no more than one mutually exclusive project may be chosen per row. Therefore, another constraint must be added.

$$\sum_{j}^{n} x_i \leq 1 \tag{14.7}$$

To see how this equation works, consider four mutually exclusive alternatives for project 2. To select only one of them, Eq. (14.7) becomes

$$x_{21} + x_{22} + x_{23} + x_{24} \leq 1$$

TABLE 14.2
Real estate projects

Project	Description	First cost	NPV	Benefit
1	Multistory building			
	15 floors	10	5	15
	18 floors	12	6	18
	21 floors	15	7	22
2	Shopping center			
	200,000 ft^2	30	8	38
	250,000 ft^2	36	10	46
	300,000 ft^2	44	12	56
3	Residential tract			
	Plan *A*	21	8	29
	Plan *B*	25	15	40

If any one of these four is selected, the inequality prevents any other from being chosen because the right-hand side can never be more than 1.

The measure of worth is usually net present value. With proper adjustments to Eq. (14.5), it can also be the present worth of the benefits.

■ **Example 14.5.** A real estate investment company is considering its capital budget for the coming year, shown in Table 14.2. First cost is present cost at time 0 for those projects whose construction will require more than 1 year. Benefits are in present-value terms. All monies not used up in the budget may be invested at the opportunity cost of capital, which is furnished by the company.

The budget consists of the three projects, each with a number of mutually exclusive alternatives, presented in Table 14.2.

(*a*) Write the integer programming equations for an optimal solution.
(*b*) With an unconstrained budget, which projects would be chosen?
(*c*) Perform some experimental reductions of the budget. What are the results of your explorations?

Solution

(*a*) By choosing the NPV as the measure of worth, Table 14.3 appears in matrix form. The equations are

$$\text{Max } Z = 5x_{11} + 6x_{12} + 7x_{13} + 8x_{21} + 10x_{22} + 12x_{23} + 8x_{31} + 15x_{32}$$

subject to

$$10x_{11} + 12x_{12} + 15x_{13} + 30x_{21} + 36x_{22} + 44x_{23} + 21x_{31} + 25x_{32} \leq M$$
$$x_{11} + x_{12} + x_{13} \leq 1$$
$$x_{21} + x_{22} + x_{23} \leq 1$$
$$x_{31} + x_{32} \leq 1$$

and $\quad x_{ij} = 0, 1 \quad$ for $i = 1, 2, 3;\ j = 1, 2, 3$

(*b*) In the absence of capital budgeting, alternatives 1, 3 (21 floors); 2, 3 (300,000 ft^2); and 3, 2 (plan *B*) would be chosen with

$$Z = \Sigma\text{NPV} = 7 + 12 + 15 = \$34 \text{ million}$$

and $\quad \Sigma C = 15 + 44 + 25 = \$84$ million

TABLE 14.3
Mutually exclusive and independent projects

	Net present-value matrix				
	1	or	2	or	3
1	5		6		7
and					
2	8		10		12
and					
3	8		15		—
	Initial-cost matrix				
	1	or	2	or	3
1	10		12		15
and					
2	30		36		44
and					
3	21		25		—

(*c*) If the budget is reduced to \$80 million, the solution becomes 1, 1 (15 floors); 2, 3 (300,000 ft^2); and 3, 2 (plan *B*) with the following characteristics:

$$Z = \Sigma\text{NPV} = 5 + 12 + 15 = \$32 \text{ million}$$

and $$\Sigma C = 10 + 44 + 25 = \$79 \text{ million} < \$80 \text{ million}$$

The extra \$1 million is invested at the opportunity cost of capital with an NPV equal to zero.

A second solution as favorable as the first is 1, 3 (21 floors); 2, 2 (250,000 ft^2); and 3, 2 (plan *B*) with

$$Z = \Sigma\text{NPV} = 7 + 10 + 15 = \$32 \text{ million}$$

and $$\Sigma C = 15 + 36 + 25 = \$76 \text{ million} < \$80 \text{ million}$$

Here the extra \$4 million will be invested at the opportunity cost of capital.

Reducing the budget to \$70 million causes the optimal selection to become 1, 3 (21 floors); 2, 1 (200,000 ft^2); and 3, 2 (plan *B*). The characteristics of this budget are

$$Z = \Sigma\text{NPV} = 7 + 8 + 15 = \$30 \text{ million}$$

and $$\Sigma C = 15 + 30 + 25 = \$70 \text{ million}$$

with no excess funds.

A budget of \$50 million causes the optimal choice to be 1, 3 (21 floors) and 3, 2 (plan *B*) with project 2 completely disappearing.

$$Z = \Sigma\text{NPV} = 7 + 15 = \$22 \text{ million}$$

and $$\Sigma C = 15 + 25 = \$40 \text{ million} < \$50 \text{ million}$$ ■

CAPITAL RATIONING—INTERDEPENDENT PROJECTS

Interdependent projects are those that cannot exist by themselves. The construction of one project is predicated upon the completion of another project. For example, a refinery and a pipeline to it are related. The pipeline to the refinery will have no reason for existence unless the refinery is built. The refinery, however, can exist without the pipeline because it can also be supplied with crude petroleum from an existing port.[1] If the pipeline is designated project 1 and the refinery is project 2, the relationship can be expressed as a constraint

$$-x_2 + x_1 \leq 0$$

where x_1 and x_2 are the decision variables.

If the program selects the refinery by itself and rejects the pipeline, the equation is

$$-1 + 0 \leq 0$$

which is true. The refinery thus passes the test of the constraint.

If the program selects the pipeline alone, the equation becomes

$$-0 + 1 \leq 0$$

which is false. The pipeline by itself thus is barred by the constraint. It cannot appear in the selection.

If both refinery and pipeline qualify for inclusion in the optimal capital budget, the constraint is

$$-1 + 1 \leq 0$$

which is true. The test of the constraint has been passed. Both projects may appear, therefore.

The following example combines independent, mutually exclusive, and interdependent projects.

■ **Example 14.6.** The road projects of Table 14.4 are being considered by a state highway department for inclusion in the next year's budget, along with the first cost and the net present value of each project at a 10 percent opportunity cost of capital. Projects designated by a number and the same letter represent different locations of the same road and thus are mutually exclusive. Project E cannot be built without project Ds also being built. However, project D may be constructed alone. Eight hundred million dollars is believed to be the maximum that the state legislature will approve for the first cost of new road construction.

(*a*) Write out the equations necessary for a solution.

(*b*) Select the projects to be included in an unrestricted budget.

(*c*) Select the projects that will be included in the budget request of no more than \$800 million and the amount of that request.

(*d*) Is this problem solvable, with absolute confidence in the answer, at all budget levels below \$1,992 million by an incremental benefit/cost ratio, incremental internal rate of return, net present-value, or any other approach that does not include integer programming?

---

[1] Suggested by an example from James C. T. Mao, *Quantitative Analysis of Financial Decisions*, New York: MacMillan, 1969, pp. 253–257.

TABLE 14.4
Road projects

Project		First cost ($000,000)	Net present value ($000,000)
1.	*A*1	15	11
	*A*2	20	14
	*A*3	36	23
	*A*4	41	29
2.	*B*1	60	49
	*B*2	72	56
	*B*3	80	63
3.	*C*1	223	75
	*C*2	251	80
	*C*3	285	93
	*C*4	301	101
4.	*D*	150	18
5.	*E*	209	72
6.	*F*1	321	106
	*F*2	360	131
7.	*G*1	283	65
	*G*2	290	71
	*G*3	297	86
8.	*H*	195	20
9.	*I*	359	32

Solution

(*a*)

$$\begin{aligned}\text{Max } Z = {} & 11x_{11} + 14x_{12} + 23x_{13} + 29x_{14} \\ & + 49x_{21} + 56x_{22} + 63x_{23} \\ & + 75x_{31} + 80x_{32} + 93x_{33} + 101x_{34} \\ & + 18x_{41} \\ & + 72x_{51} \\ & + 106x_{61} + 131x_{62} \\ & + 65x_{71} + 71x_{72} + 86x_{73} \\ & + 20x_{81} \\ & + 32x_{91}\end{aligned}$$

subject to

$$\begin{aligned} & 15x_{11} + 20x_{12} + 36x_{13} + 41x_{14} \\ & + 60x_{21} + 72x_{22} + 80x_{23} \\ & + 223x_{31} + 251x_{32} + 285x_{33} + 301x_{34} \\ & + 150x_{41} + 209x_{51} + 321x_{61} + 360x_{62} \\ & + 283x_{71} + 290x_{72} + 297x_{73} \\ & + 195x_{81} + 359x_{91} \le 800 \\ & x_{11} + x_{12} + x_{13} + x_{14} \le 1 \\ & x_{21} + x_{22} + x_{23} \le 1 \\ & x_{31} + x_{32} + x_{33} + x_{34} \le 1 \\ & x_{61} + x_{62} \le 1 \\ & x_{71} + x_{72} + x_{73} \le 1 \\ & -x_4 + x_5 \le 0 \\ & x_{ij} = 0, 1 \quad \text{for } i, j = 1, 2, \ldots, 9\end{aligned}$$

TABLE 14.5
Example 14.6

Project		B	$\frac{\Delta B}{\Delta C}$	First cost	Part (*b*) ΣC	Part (*c*) ΣC	NPV	ΣNPV
x_{11}	*A*1	26	1.73					
x_{12}	*A*2	34	1.60					
x_{13}	*A*3	59	1.56					
x_{14}	*A*4	70	2.20	41	41	41	29	
x_{21}	*B*1	109	1.82					
x_{22}	*B*2	128	1.58					
x_{23}	*B*3	143	1.88	80	121	121	63	
x_{31}	*C*1	218	1.34					
x_{32}	*C*2	331	1.18					
x_{33}	*C*3	378	1.38					
x_{34}	*C*4	402	1.50	301	422	422	101	
x_4	*D*	168	1.12	150	572			
x_5	*E*	281	1.34	209	781			
x_{61}	*F*1	427	1.33					
x_{62}	*F*2	491	1.64	360	1,141	782	131	324
x_{71}	*G*1	348	1.23					
x_{72}	*G*2	361	1.86					
x_{73}	*G*3	383	3.14	297	1,438			
x_8	*H*	215	1.10	195	1,633			
x_9	*I*	391	1.09	359	1,992			

The objective function, cost constraint, mutual exclusivity constraints, interdependency constraint, and integer constraint are shown in order above.

(*b*) Table 14.5 shows the results of calculations necessary to build all projects with a *B*/*C* ratio greater than 1. The incremental *B*/*C* ratio ($\Delta B/\Delta C$) was used to determine this. An unrestricted budget includes projects *A*4, *B*3, *C*4, *D*, *E*, *F*2, *G*3, *H*, and *I*. Its cost is $1,992 million.

(*c*) The projects selected under an $800 million budget are *A*4, *B*3, *C*4, and *F*2. This programming solution results in

$$\Sigma\text{NPV} = \$324 \text{ million} \quad \text{and} \quad \Sigma C = \$782 \text{ million}$$

with $18 million left over at the opportunity cost of capital. The amount of the budget request will be $782 million.

(*d*) No. ■

SUMMARY

The main point to be recognized in capital budgeting is that it is inherently a different kind of problem from a choice among mutually exclusive alternatives. Among independent alternatives, incremental analysis has no place. Lives of alternatives must not be equalized. The methods we have found useful thus far no longer serve. A new mathematical approach provides the only guaranteed success—0-1 integer programming.

The so-called efficiency of capital criteria—the *B*/*C* ratio, internal rate of return, and like measures—cannot be used with confidence to solve capital budgeting problems.

Three general types of capital rationing problems exist:

1. Independent projects—these may be listed.
2. Independent projects combined with mutually exclusive projects—these may be shown in matrix form.
3. Interdependent projects—these may be handled by special constraint equations.

All the above may be found in the same capital budget.

Annual-worth or annual-cost methods may not be used in capital rationing because of the inherent equalization of project lives that characterize them.

PROBLEMS[2]

14.1. Mexican bank stocks were evaluated in an analysis by a prominent investment broker, as shown in the following table.

Mexican bank analysis

Bank	Stock price April 10, 1994 (pesos)	Earnings per share past 12 months (pesos)	Estimated 1994 earnings (pesos)	Multiple for past year's earnings	Multiple based on projected earnings
Banamex	187	28.65	44.68	6.5	4.2
Bancomer	42.75	7.53	9.32	5.7	4.6
BCH	16	3.34	4.15	4.8	3.9
Comermex	41.50	8.27	9.75	5.0	4.3
Confia	169	40.10	57.82	4.2	2.9
Cremi	13.5	2.87	3.19	4.7	4.2
Nafinsa	100	22.27	30.00	4.5	3.3
Serfin	395	35.47	57.10	11.1	6.9

(*a*) Based on the price/earnings ratio over the past 12 months, rank the stocks for an unlimited capital budget. The number of shares bought of each stock will depend on availability.

Answer: Confia, Nafinsa, Cremi, BCH, Comermex, Bancomer, Banamex, Serfin

(*b*) Based on percentage annual rate of return for estimated 1994 earnings, rank the stocks for an unlimited capital budget.

Answer: Confia, Nafinsa, BCH, Banamex, Cremi, Comermex, Bancomer, Serfin

(*c*) List other factors that should be taken into account prior to investment in these Mexican bank stocks.

14.2. In Example 14.1, page 373, the World Bank raises the capital budget to $60 million.

(*a*) Which projects will be selected under this constraint?

[2] Appendix A, page 467, describes a solution method, using a personal computer, for capital budgeting problems.

(*b*) What will be the total NPV?
(*c*) What will be the total cost?
Ceteris paribus rules.

14.3. A small manufacturing company has before it projects for improving the manufacturing process, for new products, and for increasing sales. They are completely independent of each other. The projects are as follows:

Project	Total investment required ($000)	Net annual cash flow ($000)	Life (years)
1	250	40	8
2	275	45	6
3	320	55	15
4	380	70	9
5	600	95	10
6	750	130	12
7	1,100	190	11
8	1,300	250	7

No salvage values are considered. The company's minimum attractive rate of return is 10 percent. All figures are in current dollars. Income tax effects have been included in the estimates. Use the *net present-value* criterion.

(*a*) Which projects will be selected if unlimited funding is available? What is the budget necessary to fund them?
(*b*) Write the equations for selecting projects at a time when limited funding is expected.

14.4. A river basin development in the northwestern region of the United States requires that a number of dams be built. The time horizon for all dams is 100 years. The discount rate imposed by the federal agency planning the project is 5 percent. If all dams are independent of one another, or assumed so, which dams would qualify for construction if the budget were set at no more than $800 million in construction costs, all adjusted to time 0? All unused funds must be returned to the federal government.

The table covers the 10 projects with the net present worth of each shown.

Project	First cost	Annual benefits	(*P*/*A*, 5, 100) ($00,000)	Present worth	Net present worth
1	1,500	84.7	19.847	1,681	181
2	2,500	131.5	19.847	2,610	110
3	3,000	160.2	19.847	3,179	179
4	1,000	69.3	19.847	1,375	375
5	2,000	132.0	19.847	2,620	620
6	3,000	169.3	19.847	3,360	360
7	1,500	77.3	19.847	1,535	35
8	1,000	67.5	19.847	1,340	340
9	1,500	84.4	19.847	1,675	175
10	2,000	126.7	19.847	2,515	515
	19,000				

14.5. A national park in the United States has before it a number of independent possibilities for investment in facilities. First costs and annual costs are shown in the following table:

Item	First cost ($)	Operation and maintenance cost per year	Life (years)
1. New trails	1,100,000	35,000	Infinite
2. Repave portions of existing roads	600,000	25,000	20
3. New visitors' center	900,000	100,000	30
4. Tour buses	350,000	150,000	6

The opportunity cost of capital is 10 percent.

(*a*) What must be the capital budget to fund all proposed facilities?

(*b*) What is the total annual cost of the proposed facilities, both capital and operation and maintenance?

(*c*) The capital budget is reduced to $2,500,000. Show the equations necessary to optimize project selection, and explain as briefly as possible what purpose each equation serves.

14.6. J. H. Lorie and L. J. Savage called attention to a hitherto unnoticed problem in rationing capital.[3] Later, H. Martin Weingartner solved the problem by the use of integer programming.[4] The table below shows the Lorie-Savage capital budgeting problem. All sums in the table are in terms of present-worth dollars.

Only $50 in *terms of present value* is available to the company in the first year and only $20 in the second year. Which projects should be selected if the objective is to maximize net present value? Show the necessary equations and solve them, if possible.

Answer: 1, 3, 4, 6, 9

Lorie-Savage capital budgeting problem

Investment	Cost in year 1	Cost in year 2	NPV
1	12	3	14
2	54	7	17
3	6	6	17
4	6	2	15
5	30	35	40
6	6	6	12
7	48	4	14
8	36	3	10
9	18	3	12

[3] J. H. Lorie and L. J. Savage, "Three Problems in Rationing Capital," *Journal of Business*, vol. 28, October 1955, p. 231.

[4] H. Martin Weingartner, *Mathematical Programming and the Analysis of Capital Budgeting Problems,* Englewood Cliffs, N.J.: Prentice-Hall, 1963, pp. 44–47.

14.7. In Example 14.2, page 375, the budget is increased to $60 million as in Prob. 14.2.

(*a*) If the incorrect method proposed by the consultant is used, which projects will be selected? What are their total NPV and total cost?

(*b*) Correct the results of part (*a*) by using the mathematical programming solution.

No change occurs in the other data in the example.

14.8. Seven independent projects of major importance are being considered by the Tennessee plant of a computer manufacturing corporation. Their particulars are as shown:

Project	Present worth of benefits ($000,000)	First cost ($000,000)
1	90	30
2	60	30
3	22	10
4	40	20
5	17	9
6	10.5	5
7	8	1

(*a*) Which projects would be chosen for an unconstrained budget?

Answer: All projects

(*b*) Which projects would be chosen by 0-1 programming if the budget were reduced to $50 million? This question must be answered by trial and error, in the absence of a computer.

Answer: 7, 1, 3, 5

(*c*) Which projects would be chosen by the use of the benefit/cost ratio if the budget were reduced to $50 million?

Answer: 7, 1, 3, 6

(*d*) What are the implications of the answer to part (*b*) compared to the answer to part (*c*)?

14.9. Management of the Shenandoah Power Company has decided it can invest in any of the following independent projects:

Year	Project 1	Project 2	Project 3
0	–3,000	–4,500	–2,000
1	600	2,000	350
2	1,000	2,000	650
3	1,400	2,000	950
4	1,800	2,000	1,250
5	2,200	2,000	1,550

The budget is set at $5,000. The minimum attractive rate of return is 15 percent.

(*a*) An analyst at Shenandoah insists that use of the internal rate of return to rank the projects is the only appropriate method. Using their IRORs, rank the projects and select a capital budget.

(*b*) Solve the problem by inspection, using a 0-1 linear programming model.

(*c*) Discuss very briefly the implications of differences, if any, in the solutions obtained through the two different methods. Which is the correct approach and why?

14.10. In Example 14.3, page 376, with the budget increased to $60 million, use the present worth of the benefits as the figure of merit. Solve the problem, using (*a*) mathematical programming and (*b*) *B*/*C* ratios. Comment on the solutions. All other data in the problem remain the same.

14.11. In Example 14.4, page 378, the budget is reduced to 45. Compare the programming and the *B*/*C* ratio solutions. All other elements of the example stay the same.

14.12. The capital budget for the Nevada State Transportation Department shows the following projects and alternatives:

Project	First cost ($000,000)	Uniform annual benefit ($000,000)	*B*/*C*
*A*1	30	14.50	2.43
*A*2	40	19.00	2.26
*B*1	50	18.50	1.86
*B*2	60	21.50	1.51
*C*1	5	1.80	1.81
*C*2	10	3.90	2.11
*D*1	10	4.00	2.01

Each project has a 10-year life with no salvage value. The MARR is 15 percent. No taxes or inflation is to be considered. The budget is $100 million.

The Nevada State Transportation Department analyst performs the calculations whose results are shown in the above table. She then selects, on the basis of her training and a number of prominent engineering economy textbooks she has purchased, projects *A*1, *B*1, *C*2, and *D*1, using the incremental benefit/cost ratio approach guaranteed to be correct by said textbooks.

On a hunch, she decides to submit the budget and projects to an operations research professor at the University of Nevada, Reno. This analyst selects, via an integer program present on the university's computer, projects *A*2, *B*1, and *D*1.

Check all calculations and determine which procedure is correct.

Answer: *A*2, *B*1, *D*1

14.13. An analyst, when faced with the following capital budgeting problem, selected projects 5 and 1, using the exogenously mandated opportunity cost of capital of the organization at 15 percent and a budget of \$50,000. All projects listed are independent. An integer programming algorithm was used.

(*a*) Verify the solution of the analyst.

(*b*) Is the above selection the optimal one, considering the specific capital budget? Explain your answer. Any funds not used in the budget may be invested at 15 percent per year over the next 5 years.

Year \ Project	1	2	3	4	5
0	–20,000	–20,000	–20,000	–20,000	–20,000
1	0	8,210	9,010	13,980	9,828
2	5,510	8,210	9,010	10,485	9,828
3	11,020	8,210	9,010	6,990	9,828
4	16,530	8,210	9,010	3,495	9,828
5	22,040	8,210	9,010	0	9,828

14.14. In Example 14.5, page 380, consider the projects with none mutually exclusive, i.e., as though all eight projects were independent ones.

Write out the integer programming equations.

Which projects would be chosen under the following conditions, if all other data remained unchanged?

(*a*) M unlimited

(*b*) $M = \$80$ million

(*c*) $M = \$70$ million

(*d*) $M = \$50$ million

Comment on the results.

14.15. The Long Beach and Western Railroad is considering its capital budget. The engineering division wishes to present the board of directors with the following proposals for improving operations:

Project	First cost	Annual net benefit	Economic life
Signaling and train control			
1	2,000,000	600,000	10
2	2,800,000	800,000	10
3	3,400,000	850,000	10
Track maintenance			
1	10,000,000	2,000,000	15
2	12,000,000	2,700,000	15
3	13,500,000	2,900,000	15
4	15,000,000	3,200,000	15
Right-of-way relocation			
1	22,000,000	4,400,000	Infinite
2	28,000,000	6,000,000	Infinite
3	34,000,000	7,100,000	Infinite

The projects under each section are mutually exclusive, e.g., the four projects under track maintenance. Thus, only one project from each section may be built. All projects have economic lives as shown.

The board has allocated the engineering division a budget of $40 million for the coming year.

Given an 18 percent cost of capital, what will the engineering division recommend, providing all tax and inflation effects have been included?

(*a*) Set up the programming equations that will answer the question.

(*b*) The computer recommends track maintenance 2 and right-of-way 2. What are the values of the objective function and the constraint with this answer?

14.16. In an oil refinery, the following proposals are received by the engineer in charge from the refining division for inclusion in the capital expenditure budget for the facility. The opportunity cost of capital is 35 percent. All installations have a zero salvage value. No taxes or inflation is involved.

(*a*) With an unlimited capital budget, which projects would be chosen? What is their total first cost? What is their total annual cost?

(*b*) Write the integer programming equations for optimization with reduced budgets.

	First cost	Annual maintenance cost	Economic life
1. Install high-speed pumps (four alternative makes)			
A	450,000	70,000	5
B	560,000	60,000	7
C	590,000	45,000	9
D	640,000	40,000	10
2. Design and install mixer for distilled water evaporator			
A	130,000	1,500	8
B	150,000	1,500	16
3. Replace valves on no. 1 crude unit (three alternative manufacturers)			
A	120,000	1,000	4
B	135,000	800	5
C	140,000	1,200	7

14.17. A ports study of an African country considered the following projects with their associated discounted costs and discounted benefits.

Project		($000,000)						
Port Al Khobar								
		35ft	or	50 ft				
1. Dredging	Cost	100		200				
	Benefits	100		250				
2. Conveyors	Costs	15						
	Benefits	20						
Port Ras Tanura								
3. New berths		2	or	3	or	4		
	Costs	60		75		90		
	Benefits	80		90		100		
4. Warehouse	Costs	20						
	Benefits	25						
Port Manama								
5. Warehouse	Costs	10						
	Benefits	14						
6. Access road improvement		*A*	or	*B*	or	*C*	or	*D*
	Costs	5		10		14		15
	Benefits	8		13		18		19

Mutually exclusive alternative projects are listed horizontally in rows. Six independent projects are listed vertically. However, in Port Ras Tanura, the warehouses project will not be undertaken unless new berths are constructed.

(*a*) With unlimited funding, which project will be undertaken? What will be the total cost and net present value?

(*b*) With funding limited to *M*, write the equations for a capital budget.

14.18. The Ministry of Agriculture in a small Latin American country has before it projects for rice production in the provinces of Aconcagua, Cotopaxi, and Popocatepetl. Each project has a number of mutually exclusive alternatives for its implementation, all of them technically possible.

Project	First cost (000,000)	NPV (000,000)
1. Province of Aconcagua		
A	20	6
B	30	10
C	50	20
2. Province of Cotopaxi		
A	10	4
B	15	5
C	21	8
D	23	9
3. Province of Popocatepetl		
A	6	2
B	8	3

The budget, however, will not permit going forward on the best alternative in each project.

(*a*) Write the programming equations for a budget of $70 million.

(*b*) One of your assistants reports that the computer has selected 1, 3; 2, 2; 3, 1. What is your response?

14.19. Seven projects are proposed in a certain chemical plant, as follows:

Project	First cost	NPV
1	30	60
2	30	30
3	10	10
4	20	20
5	20	15
6	5	5
7	1	7

The maximum budget available is 50. Which projects will be chosen?

Write out the mathematical programming equations and solve. A solution using a computer program will be necessary.

14.20. Repeat Prob. 14.19, but now assume that projects 1, 2, and 3 are mutually exclusive alternatives, as are projects 4 and 5, and 6 and 7. Answer the same questions, with all other data remaining unchanged.

14.21. Repeat Prob. 14.19, but use the present worth of the benefits as the figure of merit. Will the *B*/*C* ratio solution give the same answers? All other data are unchanged.

14.22. In Example 14.6, page 382, what selection would be made if the interdependency constraint were removed? Answer the same questions as in the example.

14.23. The manager for research and development at Omega International Company is planning to replace outdated machines *A*, *B*, and *C* of his testing facilities. For replacing machine *A* he has two options to choose from, *A*1 and *A*2. For replacing machine *B* he has another two options to choose from, *B*1 and *B*2; and for replacing machine *C* he has three options: *C*1, *C*2, and *C*3. The first costs and annual savings associated with each option appear in the following table.

Alternative options	First cost ($)	Annual savings ($)	Life (years)
*A*1	60,000	12,500	10
*A*2	35,000	7,200	10
*B*1	12,000	3,100	10
*B*2	8,000	2,000	10
*C*1	14,000	3,100	10
*C*2	10,200	2,850	10
*C*3	5,000	1,700	10

Further, he knows that machine *C*1 is compatible with machine *B*2. In other words, he cannot use machine *C*1 unless he buys machine *B*2. But machine *B*2 may be used without machine *C*1. The opportunity cost of capital for this company is 15 percent. Taxes and inflation should not be considered.

(*a*) Formulate the integer programming equations for a budget limiting his first cost to $76,000. Use the NPV as the figure of merit.

(*b*) Select the machines he would buy in case there were an unlimited budget.

14.24. A Calistoga County road supervisor has a budget of $30 million. The following projects are available for implementation:

1. Rebuild stretch between Jonesville and Dalton with two mutually exclusive alternatives.
   a. Replace wearing surface: cost, $10 million; NPV, $6 million.
   b. Patch surface and rebuild worst sections: cost, $5 million; NPV, $3 million.

2. Repair bridges: cost, $4 million; NPV, $2 million. Project 2 will not be carried out unless project 1 is also. However, project 1 can be performed without project 2.

3. Construct a new bypass at Iron Rock: cost, $16 million; NPV, $10 million.

4. Repair or replace damaged culverts along the Little River–Lambeth road.
   a. Replace: cost, $5 million; NPV, $3 million.
   b. Make emergency repairs in worst culverts: cost, $2 million; NPV, $1 million.

5. Construct new maintenance center: cost, $3 million; NPV, $1 million.

(*a*) With an unrestricted budget, which projects will be undertaken? Use the *B*/*C* ratio method to decide.

(*b*) Write the mathematical programming equations for a budget of $30 million.

(*c*) If 1*a*, 2, 3, and 4*a* are selected, is this a feasible solution? Show your work.

(*d*) If 1*b*, 3, 4*b*, and 5 are selected?

(*e*) If 1*a*, 1*b*, 2, and 4*a* are selected?

(*f*) If 2, 3, 4*a*, and 5 are selected?

(*g*) What is the correct solution?

14.25. Consider the following projects:

Project	First cost	Annual cost for years 1 to 4, each year	Benefit
A	50	10	100
B	60	5	100
C	40	15	80
D	50	20	120
E	70	2	90

All values are present-value dollars. All entries in the table are in millions of dollars. Answer the following, showing all work.

(*a*) If all projects are independent, but for nonbudgetary reasons only two can be chosen, which two should be selected if you use a B/C analysis?

(*b*) If the performance of any project will exclude all the other projects, which project should be selected if you use a B/C analysis?

(*c*) Given that projects A, D, and E are mutually exclusive, project C cannot be performed unless project B is completed, the first-cost maximum budget is \$120 million, and the maximum budget for each of the next 4 years is \$18 million, identify all the equations necessary to solve this problem on a computer and all the feasible combinations.

(*d*) Select the optimum combination of projects.

14.26. In Prob. 14.6, investments 1 and 3 are interdependent; i.e., investment 1 cannot be undertaken without 3, but 3 can be undertaken without 1.

In addition, investments 2, 4, and 5 are mutually exclusive, as are investments 6 and 9.

(*a*) Show the necessary equations for solution.

(*b*) A friend suggests that the optimal solution is 1, 4, and 9. What is the value of the objective function if you use this selection? Is this solution feasible? Why or why not?

(*c*) Another friend suggests 2, 3, and 6. What is the value of the objective function? Is the solution feasible? Why or why not?

(*d*) A third friend suggests 4, 6, and 9. What is the value of the objective function? Is this a feasible solution? Why or why not?

(*e*) Make a guess as to the optimal solution.

CHAPTER 15

OPPORTUNITY AND FINANCIAL COSTS OF CAPITAL

This text has often used *capital*, *investment*, *opportunity cost of capital in the public sector*, *discount rate*, and associated terms. It is time to define these accurately and to say as much as can be said at this writing about the controversial subject of the discount rate in the public and private sectors.

The reason that this chapter immediately follows the one on capital budgeting is that there is a direct link between the capital budget and opportunity cost of capital—the discount rate. We will now investigate exactly what that connection is.

DEFINITIONS

Capital is usually thought of as a certain amount of money available for investment, or an amount of money tied up in a business. "My capital in the laundry is so-and-so." We have a more or less clear idea of what the speaker means. But capital, to an economist, is not a sum of money; instead, it is real goods or assets. A building is a capital asset—or simply *capital*. So are inventories of finished goods or half-finished goods. So is a bulldozer, or a yard full of tubing. So were Robinson Crusoe's fishhooks, discussed in Chap. 1.

Capital is any asset used in the production of other goods and services.

Investment is the use of resources to create such a capital asset. For example, the building of a bridge by the Ministry of Public Works of Ecuador is an investment in a capital asset. It takes steel, concrete, labor, and time to build the bridge. This taking of time is characteristic of investment in a capital good. Resources are taken away from production for a time and are put to the creation of a capital good.

The *opportunity cost of capital* means the benefit forgone for a time by choosing to invest rather than consume. This investment may be in either the public or the private sector. The opportunity cost of capital is measured by the discount rate in percent.

PERSONAL COST OF CAPITAL

The opportunity cost of capital is also the marginal rate of time preference. The *marginal rate of time preference* means the amount an individual would have to be paid in order to restrict consumption for a given period, usually 1 year. It is the person's opportunity cost of capital. For example, if an individual demands at least $10 as payment for investing, i.e., forgoing the pleasure of spending, $100 for 1 year, then that individual's marginal rate of time preference is 10 percent per year. (You may wish to review the "Money Has a Double Value" section of Chap. 4, page 41.) If an individual is indifferent between spending $1 now or $(1 + i)^N$ dollars at some period N in the future, then i is that person's marginal rate of time preference, or opportunity cost of capital.

It is an easy matter to find out one's personal cost of capital if one's opportunities for investment are limited, because there are so few possibilities to investigate. A telephone call to the local bank will usually suffice to discover the cost of spending $1,000 rather than saving it. A recent listing at a local bank in Washington, D.C., appears in Table 15.1. For some of the accounts, interest rates fluctuate. For others, interest rates are fixed. A number of compounding rules apply as well, depending on the choice made. Risk is minimal for such investments, although it is not zero. A 2-year investment being considered, at the same risk as at the bank, should be discounted at 8.75 percent, the rate payable on a certificate of deposit for that period. The rate quoted is a nominal one, compounded quarterly. The effective annual rate is higher. (The reader may review the calculation necessary to find the higher rate by referring to the section entitled "Rates of Interest: Nominal and Effective," Chap. 4, on page 55.)

Personal discount rates are dependent on the opportunities known and available to the investor. For example, bank rates may vary by region. A more favorable rate in San Francisco may not be known to the investor in Austin.

TABLE 15.1
Bank investment opportunities

Account	Annual rate (%)	Special information
Money market account	7.20	$1,000 minimum
91-Day certificate of deposit	7.45	$500 minimum
6-Month money market certificate	7.60	$500 minimum
1-Year certificate of deposit	8.10	$500 minimum
2-Year certificate of deposit	8.75	$500 minimum
3-Year money market certificate	9.00	$500 minimum
4-Year certificate of deposit	9.10	$500 minimum

Savings and loan companies usually pay higher rates, but at greater risk. Personal discount rates depend on the risk preference of the investor.

Finally, personal discount rates depend on the length of time an investor is willing to relinquish the money. Usually the longer the time, the higher the rate of interest paid.

LINK BETWEEN THE OPPORTUNITY COST OF CAPITAL AND THE CAPITAL BUDGET

As we have seen, the private sector uses an opportunity cost of capital, also called the *minimum attractive rate of return* (*MARR*), or *discount rate*, or *marginal cost of capital*, or *marginal interest rate*, before and after taxes. If taxes are a nondifferential cost among alternatives, they can be excluded, and the opportunity cost of capital before taxes can be used. We have seen what an important part income tax plays in an analysis.

The public sector of the economy also uses an opportunity cost of capital. But because public sector enterprises pay no income taxes, they employ a single discount rate for evaluating projects.

How is the opportunity cost of capital determined? Its importance cannot be overrated. As the discount rate, its magnitude directly influences choice among mutually exclusive alternatives. The opportunity cost of capital is generated along with the capital budget of an enterprise whether in the private or public sector. *It is the rate of return on the least attractive project included in that budget.*

■ **Example 15.1.** In Example 14.6, page 382, the projects selected under an $800 million budget were *A*4, *B*3, *C*4, and *F*2. A NPV of $324 million was achieved at a total first cost of $782 million. See Table 15.2. What is the opportunity cost of capital under these conditions?

Solution. Note that to compute the quantities of Table 15.2, highway life was assumed at 20 years, the salvage value of the highway was zero, and most important of all, the discount rate, mandated by the state legislature, was 10 percent. This discount rate, you will recall, was used to calculate the NPVs. These, in turn, were used as coefficients of the objective function. The result was the capital budget.

TABLE 15.2

IRORs of Example 14.6 projects

Project	Discounted first cost ($000,000)	Annual benefit ($000,000)	Benefit ($000,000)	IROR (percent)
*A*4	41	70	8.22	19.5
*B*3	80	143	16.80	20.5
*C*4	301	402	47.22	14.7
*F*2	360	491	57.67	15.0
	782			

We rearrange the projects so that their individual rates of return—not incremental, notice—run from highest to lowest:

Project	IROR (percent)
*B*3	20.5
*A*4	19.5
*F*2	15.0
*C*4	14.7

The lowest IROR, which indicates the least attractive of the four projects, is 14.7 percent. It is the opportunity cost of capital for the state highway department in its present investment situation. The reasoning goes like this: If the $301 million for project *C*4 were used for any other purpose, then the state would give up a 14.7 percent return on investment. The benefit given up by selecting one option rather than another is the definition of opportunity cost.

Remember that the discount rate, 10 percent, has been imposed from outside the particular agency: i.e., it has been imposed on the state highway department by the state legislature. Were the state highway department free to set its own discount rate, it would now set it at 14.7 percent. Because it is not free to do so, it will continue to discount its projects at 10 percent. In the private sector, where no such outside (exogenous) restrictions exist, the enterprise is free to change its discount rate to that indicated by the capital budget.

And now it is evident that the budget and the discount rate are interdependent. For if we discount the projects available for funding of Example 15.1 at the new 14.7 percent rate, the NPVs will change and an entirely new project selection will appear for the capital budget. Out of the selection a new lower IROR and thus discount rate may appear. Evidently, we are dealing with an iterative process, whose inclusion in a basic text is beyond its scope.[1] Suffice it to say that using 14.7 percent would reduce the NPVs of all the projects. And they would be reduced in some similar fashion because the cash flow patterns of the projects are the same, in this particular case. The ordering of the projects from highest to lowest would not change. ■

How long may we continue to use a budget-determined discount rate? There is no particular reason why the internal rate of return of the project just above the cutoff line must be the same for every period. (Imagine the cutoff line is just below project *C*4 in the above list of Example 15.1.) In the case of Example 15.1, the period is 1 year, because the budget of the highway department is an annual one. Next year it will use the same discount rate, 10 percent, because it is fixed by law. But for a private company, e.g., a dynamic and progressive one, the rate of return of the project just above the cutoff line will probably rise. The object of the game, after all, is to create profitable projects. However, in a private company, experience over a number of years will dictate policy with respect to the discount rate, particularly as actual rates of return on completed projects become available.

---

[1]For a detailed description of how this works, see Peter Lusztig and Bernhard Schwab, "A Note on the Application of Linear Programming to Capital Budgeting," *Journal of Financial and Quantitative Analysis*, December 1968, pp. 427–43. For a further critique, see R. H. Bernhard, "Some Problems in the Use of a Discount Rate for Constrained Capital Budgeting," *AIIE Transactions*, September 1971, pp. 180–84.

■ **Example 15.2.** Imagine that the situation in Example 14.4, page 378, describes a division of a private sector company. What is the opportunity cost of capital implied by the mathematical programming solution?

Solution. The mathematical programming solution is projects 6, 1, 2, and 4. However, no internal rates of return are calculated. Would it be correct to assume that the internal rate of return of project 4 is the cutoff rate, i.e., the opportunity cost of capital? Let us calculate the IRORs and see.

For project 1:

$$\begin{aligned} \text{NPV} &= 60 = -30 + A(P/A, 12, 5) \\ &= -30 + A(3.605) \\ A &= \frac{60+30}{3.605} \\ &= 25 \end{aligned}$$

The equation necessary to compute the IROR is

$$\begin{aligned} 0 &= -30 + 25(P/A, i^*, 5) \\ i^* &= 78.8\% \end{aligned}$$

After the remaining projects are treated similarly, the following table appears:

Project	IROR (percent)	Priority
1	78.8	2
2	43.0	3
3	37.0	5
4	39.9	4
5	30.9	6
6	457.8	1

The priority column is added to see if it matches the results obtained by mathematical programming or by the *B*/*C* ratio method.

It does not match the *B*/*C* ratio solution of 6, 1, 2, and 5; but it does match the programming one of 6, 1, 2, and 4. Any match, however, would be accidental, for if we calculated the NPVs of the six projects at discount rates other than 12 percent, different discounted benefits, discounted costs, and NPVs would appear. The result would be different *B*/*C* ratios and objective functions. The different objective functions would result, very probably, in a different choice of projects. As matters stand, however, given a discount rate of 12 percent, mandated exogenously, the new opportunity cost of capital for this division of the company is now 39.9 percent, which it will not be able to use as a discount rate.

The division will continue to be subject to the opportunity cost of capital mandated by higher management. However, the results of the above table can be used by the division head to argue for a larger budget than the 50 currently allocated. The same results can also be employed to review the policy that has set the 12 percent discount rate, with a view toward raising it. ■

Paying off loans is always a possible project and thus always one of the capital projects considered in the formation of the budget. This is the reason why the loan interest rate is the floor to any opportunity cost of capital generated within the enterprise by the capital budget. Of course, the higher-interest-rate loans are paid off first.

FINANCIAL COST OF CAPITAL IN THE PRIVATE SECTOR

Is the capital budget actually used to set the opportunity cost of capital for a private company? Much more frequently it is the financial cost of capital that is employed. The correctness of this use will be discussed later in this section.

An individual's financial cost of capital is easily determined, as we have seen. Usually it is the rate at which funds can be borrowed from a local lender—a bank, a savings and loan company, a mortgage loan company. For a firm, it is not so simple. If it is publicly owned, i.e., if its stock is sold on the open market, then its financial cost of capital is determined by the weighted average of the interest rate being paid on all its sources of capital. These may be bonds. They may be preferred and common stocks. (The interest rate used on stock is usually the ratio of the dividends to the market value of the stock. The market value is used because it reflects opportunity cost, not book value, which is based on historical—sunk—cost and is therefore irrelevant for decision making.) They may be retained earnings or any other source of investment funds. They may be a combination of all these sources and more. The following example illustrates the technique.

■ **Example 15.3.** Consider the Godfrey Corporation, which has a 1995 capital structure shown in column 1.

Item	(1) Amount ($000,000)	(2) Average dividend or interest rate (percent)	(3) Total ($000,000) (1) × (2)
8% bonds, compounded annually, 20 years	10	8	0.8
Preferred stock	84	7	5.9
Common stock	35	6	2.1
Retained earnings	5	6	0.3
Short-term bank loans	5	10	0.5
	139		9.6

What is the financial cost of capital for Godfrey?

Solution. The average financial cost if we exclude income tax effects, as shown at the foot of column 3, is \$9.6 million per year on a total capitalization of \$139 million. Thus the weighted average cost of capital is

$$\frac{9.6}{139} \times 100 = 6.9\%$$

Any project having an internal rate of return less than 6.9 percent would not be funded, no matter what the capital budget, because it makes no sense to use borrowed money at a lower rate of return than its interest rate cost.

But this observation gives rise to a new idea. Should we not use as a floor to the IROR of the capital budget the marginal—rather than the average—rate? In the table, the highest rate shown is 10 percent on the short-term bank loans. This is the maximum rate that can be avoided. If paying off the loan were considered a project in the capital budget, then its rate of interest would become the discount rate determined by the capital budget, provided that the budget was large enough that a loan payoff was always included as the last project funded by the budget. This last proviso is the assumption behind the use of the marginal financial cost of capital as the discount rate. It is, of course, an unrealistic one. A company cannot always exercise such control over its budget, as to always include a loan payoff as the last project. Many more projects, paying off at much higher rates, will probably intervene. But the U.S. government uses the previous argument to justify formulas based on its borrowing rates—e.g., the U.S. savings bond 30-year rate—to set its project discount rates. ■

Such questions as the one proposed in the forgoing paragraph surround the whole subject of the financial cost of capital. For example, if we value common and preferred stock at market value, what date shall be chosen for setting a price? If the stock price is fluctuating, what period shall we use for the sampling price? Choosing a period in which prices were high gives one answer; choosing a period in which prices were low gives another.

In view of the opportunity to buy back our own stock and thus avoid paying dividends on it, we project its value over some future period, say, 5 years. For many companies, such a projection is impossible—or almost so—because external influences—unpredictable and uncontrollable—play a decisive role in stock price behavior on the market. A U.S. oil company is a good example. Its stock price depends on the price of oil on the world market, which depends on supply and demand, cartels, weather, and wars—all difficult to predict.

Another grave difficulty in determining the cost of capital in relation to stock prices is that the projects we choose affect the price of our stock. This in turn affects the floor of our capital budget, which contains the projects we choose, and so forth. This interdependence of stock prices and choice of project is the subject of many an article in the literature of finance.

Many theories have been presented on the determination of the financial cost of capital. It is outside the scope of this text to review them here.

More may be said on the difficulty of obtaining a figure for the financial cost of capital, but suffice it to repeat that its use as an opportunity cost, except in the extreme case where loans are paid off, is wrong.

PUBLIC SECTOR OPPORTUNITY COST OF CAPITAL

Imagine an analyst whose viewpoint is that of the entire society. Imagine, too, that she is concerned with all projects in the society, both public and private. The question facing her is what discount rate to use for all projects—from something as mammoth as the water resources development of the Columbia River basin to something as tiny as the replacement of one production-line machine with another.

If all funds for public sector investment come from the private sector through taxes and government borrowing, then she looks to the opportunities lost to the private sector because of taxes paid or money lent to the government. She then observes the rate of return that could have been obtained with this money, had it not been taxed away. This rate of return is the rate of return on the least attractive (marginal) projects undertaken in the private sector. It does not matter whether the funds come from taxes or borrowing. The marginal rate of return lost to the private sector ought to be the discount rate used for evaluating public projects.

It may be that the government is heavily engaged in state-owned enterprises, as in the socialist countries and some developing ones. If this is the case, the analyst looks to the same area—the rate of return on marginal projects undertaken—to determine the discount rate. These marginal projects may be located in either the public or the private sector.

Income taxes paid from the private to the public sector are a transfer payment. (See Chap. 17.) Thus the analyst considers only the marginal rate of return on projects before income taxes. This means that before taxes of any kind—property, income, capital gains, sales—are paid, a project in the private sector returns as though it were a public project. Or put another way, if the government were to undertake the same project, the before-tax rate of return would be the same as the social rate of return. It is assumed that corporate income taxes are actually paid and not avoided, so that the rate of corporate income tax may be taken at whatever its highest rate is for the particular year, 50 percent, for example. In a riskless world, it seems that the proper rate of discount for public projects to be used by the society's analyst should be the marginal rate in use by the private sector. If this rate is i_p and the public sector opportunity cost of capital i_s, then

$$i_p = i_s$$

All this is highly theoretical and has little relation to what happens in the real world. In reality, politics in every country plays the major role in setting the discount rate. For example, a large dam is proposed for a western state. It has a life of 100 years, which is to say that its benefit stream—and the future stream is overwhelmingly benefits—will stretch out that far. For long future streams to have any effect on the benefit/cost ratio, the discount rate has to be low. Figure 15.1

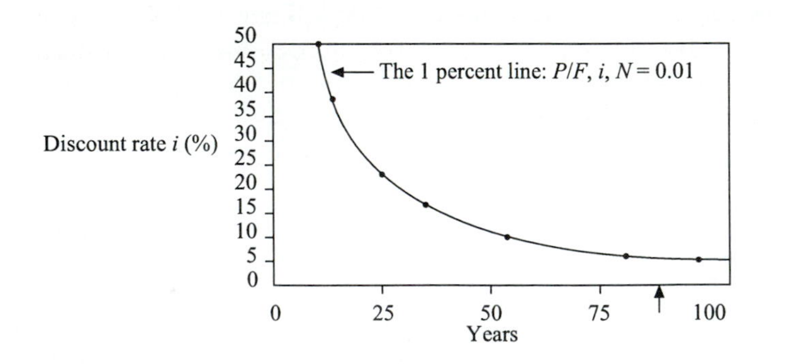

FIGURE 15.1 Where the discount factor equals 1 percent

shows that at a discount rate of 5 percent, benefits 95 years out in time will be included at only 1 percent of their value in the present-value computation of benefits. Small wonder that agencies concerned with large dam construction advocate low discount rates. The Corps of Engineers and the Bureau of Declaration have never supported high discount rates; nor, by and large, have senators from western states.

For years the Department of Transportation has used a 10 percent discount rate. Figure 15.1 shows that, at that rate, the single-payment present-worth factor reaches 0.01 at 50 years. If the Corps of Engineers used this rate, the benefits of half of the dam life of 100 years would have no effect on the *B*/*C* ratio.

Now suppose that a private sector discount rate equal to 20 percent were to be used. Figure 15.1 shows that no effect more than 25 years in the future would be considered at more than 1 percent of its undiscounted value. It should be evident now why higher discount rates exclude effects on future generations. No doubt it is presumptuous of us to imagine we can decide anything beyond 25 years in the future. Developments—scientific, engineering, political—seem to come faster and faster, and the world changes its aspect more and more rapidly. Some modesty in regard to our predictive powers seems to be called for.

In practice in the United States, discount rates for federal projects are set by two agencies: the Office of Management and Budget (OMB) and the President's Water Resources Council.

The OMB, in a 1972 circular,[2] ordered a discount rate of 10 percent by all agencies of the U.S. government except the U.S. Postal Service. Water resource projects, the government of the District of Columbia, and nonfederal recipients of federal loans or grants were specifically exempted. "The prescribed discount rate of 10 percent represents an estimate of the average rate of return on private investment, before taxes and after inflation" (p. 4). With this statement the OMB supported the position that the social opportunity cost of capital should be based on the rate of return the funds would have earned in the private sector, and not on the financial cost to the government in the form of the interest rate paid on government bonds. OMB's 1972 directive is still in force at the time of this writing, 1995.

In 1962, the President's Water Resources Council stated that the discount rate for planning and evaluation shall be based on "marketable securities" of the United States.[3] A formula for doing this was set forth. The result was that water resources projects were based on financial cost to the government rather than opportunity cost in the private sector. The former has always been lower than the latter.

---

[2]U.S. Office of Management and Budget, "Discount Rates to Be Used in Evaluating Time-Distributed Costs and Benefits," Circular A-94, rev., Washington, March 27, 1972.

[3]President's Water Resources Council, "Policies, Standards, and Procedures in the Formulation, Evaluation and Review of Plans for Use and Development of Water and Related Land Resources," Senate Document no. 97, 87th Cong., 2d Sess., Washington, May 29, 1962, p. 12, as quoted in Robert L. Banks and Arnold Kotz, "The Program Budget and the Interest Rate for Public Investment," *Public Administration Review*, vol. 27, no. 4, December 1966, pp. 283–292.

For the same source of funds—the taxpayers—the federal government uses at least two different discount rates based on two opposed economic concepts. The different discount rates result in favored treatment being given to one class of projects—water resources—at the expense of all the other classes—education, transport, judicial, and so forth. This is the result of political pressures in the conduct of government.

Local and state governments also use discounting techniques and therefore a discount rate. The rate is set by the authority involved, usually with an eye toward federal government practice.

In developing countries, the situation is much the same as in the United States, except that generally higher discount rates are used in order to reflect the higher costs of capital. Setting rates by fiat is the normal procedure.

The reader may well wonder at this point, Is there no way to develop a rational policy in reference to the discount rate? Imagine a grand design where all the projects available to the society were listed with their costs and benefits. Both public and private sector projects will be shown. Now the internal rate of return of all projects is calculated. With a given budget and the internal rate of return of each project as the figure of merit, an optimal capital budget is determined by using programming methods explained in Chap. 14, but with some more advanced techniques and concepts. The internal rate of return of the last project selected is the social opportunity cost of capital—the social discount rate.

The data, the means of collecting the data, the impetus behind such a project, which requires a general agreement on the correctness of the determination of the discount rate by means of the capital budgeting method—all these and many more requisites of such a project are lacking. Until they are available, the fundamentally crucial discount rate for public projects must be left to the political process. Even if determination of the discount rate by the process described here were possible, it would only assist the political decision.

RISK AND THE COST OF CAPITAL

In the private sector, risk is often accounted for by setting a higher discount rate for riskier projects. The extra percentage points added to the firm's ordinary minimum attractive rate of return in order to account for risk is called the *risk premium*. The method is a gross one. It is better to account for risk by establishing, as far as possible, the frequency distributions of the input variables and using them to calculate a figure of merit—e.g., the NPV—with its expected value and standard deviation. This would then be employed as we have seen in the chapter on risk.

In the public sector, the recommended method is described in the OMB circular[4] (p. 4):

---

[4]U.S. Office of Management and Budget, "Discount Rates to Be Used in Evaluating Time-Distributed Costs and Benefits," Circular A-94, rev., Washington, March 27, 1972.

> *Treatment of uncertainty*. Actual costs and benefits in future years are likely to differ from those expected at the time of decision. For those cases for which there is a reasonable basis to estimate the variability of future costs and benefits, the sensitivity of proposed programs and projects to this variability should be evaluated.
>
> The expected annual costs and benefits (or effects) should be supplemented with estimates of minimum and maximum values. Present value cost and benefits should be calculated for each of these estimates. The probability that each of the possible cost and benefit estimates may be realized should also be discussed, even when there is no basis for a precise quantitative estimate. Uncertainty of the cost and benefit (or effects) estimates should be treated explicitly, as described above. The prescribed discount rate should be used to evaluate all alternatives. Specifically, the evaluations should *not* use different discount rates to reflect the relative uncertainty of the alternatives.

How often the rules of the above quote are actually observed in practice is a matter so far undetermined.

Now an interesting question arises. How is the social rate of discount to be determined, theoretically, if we accept the recommendation of economists that it be the marginal rate of return from private sector projects, and if we recognize that the private sector has included risk premiums? For example, say the marginal discount rate in use in the private sector is 20 percent. The risk premium is estimated at 5 percent. Should the social discount rate be 15 percent?

The marginal private discount rate at which projects are being rejected is 20 percent. This is the crucial point. It does not matter that projects not earning at least 20 percent are being rejected because of risk or inflation or patriotism or prejudice or any other reason. If they are being rejected, the acceptance of public sector projects at less than the cutoff rate in the private sector, *no matter how that rate is determined*, is open to question.

SUMMARY

The terms relating to capital are defined once again. The personal cost of capital is easy to determine for most people. The rate of interest paid by a bank or a savings and loan company is normally what must be forgone if we use our funds for another purpose such as investment.

The opportunity cost of capital and the capital budget of an organization are closely related. The rate of return of the last project accepted in the capital budget is the opportunity cost of capital for that entity at that time.

The financial cost of capital is the weighted-average interest rate paid by the company for its funding. The funding may come from equity, stocks, bonds, short-term loans, and so forth. The financial cost of capital should not be used as the discount rate.

The public sector opportunity cost of capital should theoretically be the rate of return of the least attractive project forgone by the private sector. After all, money for government projects comes from private sector taxes or borrowing.

The discount rate used has a powerful effect on a project's acceptance. Therefore the discount rate used in public projects is a political issue. Government organizations engaged in long-lived projects support low discount rates.

In practice, the discount rate for government projects is set by political fiat.

A more rational determination of the social discount rate must await general acceptance of certain economic principles that are now controversial.

Project risk in the private sector is often accounted for by raising the discount rate. In the public sector, it is sometimes allowed for by involving risk in the input variables and determining the figure of merit as a distribution with an expected value and a standard deviation.

PROBLEMS

15.1. Determine your personal opportunity cost of capital by investigating the rate of return at your bank under a checking account, savings account, and certificates of deposit. For the CDs, note how the interest rate will change for increasing periods of time. Note also the required minimum amounts to obtain the rate of return. Are the compounding periods the same for all the options?

15.2. Comment on the (*a*) risk, (*b*) inflationary aspects, and (*c*) income tax aspects of each option in Prob. 15.1.

15.3. What is the effective annual rate of return of each option in Prob. 15.1? Is this the same as what the bank calls the *yield*? Comment on how these effective rates determine your opportunity cost of capital for an investment in a municipal bond, AAA-rated.[5] Choose a bond from the financial pages of your local paper or from *The Wall Street Journal*.

15.4. Investigate the possibilities for investment in a stock. Choose one from the financial pages of your local newspaper or from *The Wall Street Journal*. You will have to investigate the stock's performance over the last few years to determine its historical rate of return in the previously cited references.

Comment on how your investigation may help you to determine your opportunity cost of capital.

15.5. In Table 15.2, the capital budget has been reduced to $481 million. What is the opportunity cost of capital to the state under this new budget? Will the state highway department use this opportunity cost as a discount rate, given the facts of the case as before?

---

[5]Standard and Poor's Corporation and Moody's Investor Services rate bonds by a number of classifications. Almost any library in the United States and many overseas will have a collection of these publications on stocks and bonds.

15.6. (*a*) In Example 15.1, page 399, why is the state highway agency unable to use the new opportunity cost of capital (discount rate) obtained from its optimization of its capital budget?

(*b*) You are not so limited. Try out 14.7 percent as the discount rate, and substitute the resulting NPVs in the objective function. Use the same equations and data in the remainder of Example 14.6, page 382. Comment on the result.

15.7. In Example 15.2, page 401, verify the IRORs of projects 2, 3, 4, 5, and 6. Why does the selection determined by this method differ from that determined by *B*/*C* ratios?

15.8. In Example 15.2, page 401, compute a new capital budget, using the cutoff rate of 39.9 percent, with the data of Example 14.4, page 378. Comment on the results. (You are assuming that this rate will be approved by top management.)

15.9. The firm of Hornblower and Associates, Engineers and Architects, has a capital structure as follows:

Item	Amount ($000,000)	Average dividend or interest (percent)
7 Bonds, compounded annually, 15 years	4	7
Preferred stock	15	10
Common stock	30	8
Retained earnings	3	8
Short-term bank loans	2	11

What is the financial cost of capital for this publicly owned company?

15.10. Hightower Investment Company has the financial structure shown in the following table.

Item	Amount ($000,000)	Average interest or divident rate (percent)
Common stock	25	8
Preferred stock	20	10
7% Bonds, compounded annually, 15 years	5	7
Bank loans	7	12
Retained earnings	3	8

What is the financial cost of capital for this company?

15.11. Should an allowance for future generations be included in the public sector project discount rate by lowering the rate? Why or why not?

15.12. Should a risk premium be included in the public sector discount rate by lowering the rate? Why or why not?

15.13. For many years, the Army Corps of Engineers and Bureau of Reclamation water projects have been judged at a much lower discount rate than Department of Transportation projects. The former organizations have enjoyed a cutoff rate on their capital budget in recent years of 7 percent or near. The Department of Transportation has been required to discount its projects at 10 percent.

(*a*) Discuss the implications of this difference in its effect on the allocation of resources in the United States.

(*b*) Would you be in favor of continuing this policy? In as objective a manner as possible, outline the pros and cons of your position on the matter.

CHAPTER 16

REPLACEMENT

The replacement problem in engineering economy, like other areas in our study, has been the subject of much debate and controversy. Replacement of assets—everything from a bridge to a pocket calculator—allows us to use a number of the techniques we have acquired so far in this text. By *replacement* is meant that the asset is either removed from service entirely—scrapped—or put into some secondary use such a standby. Called *abandonment* in business use, replacement can come about for many reasons: The physical life of the asset has come to an end. It is worn out, like a burned-out lightbulb. Or the economic life of the asset is over. (Economic life will be explained later in this chapter.) Or newer and better machines have been developed that do what it does, but faster and more cheaply. The handheld electronic calculator pushed a multitude of cumbersome, noisy, slow, mechanical calculators off the desks of engineers and onto the scrap heap in what seemed the wink of an eye.

In this chapter, replacement is considered under two headings:

1. Replacement by an identical asset, before and after taxes
2. Replacement by an entirely different asset, after taxes

In all cases, we emphasize the following:

1. It is a decision *now*, at this moment, that we are attempting to make. What we will do a year from now will depend on a new analysis at that time.
2. There is no question of doing away with the service, that is, the benefit, that the asset provides, for example, a milling machine. Thus a replacement *must* be obtained that will perform the same job. We are not dealing with, say, the advisability of disposing of an apartment house.
3. Only annual costs will enter our investigations. Benefits will appear *only* as salvage or resale values.

4. We will use in our analysis the market value of the asset to be replaced—*not* the replacement cost, book value, trade-in value, etc. The reader may profit from reviewing the concepts of sunk cost and opportunity cost as presented in Chap. 2.

DEFENDERS AND CHALLENGERS

Two terms have been used for many years in discussions of replacement. George Terborgh of the Machinery and Allied Products Institute is given credit for these descriptive terms.

A *defender* is an asset, presently in service, whose replacement is being considered. A *challenger* is the possible replacement.

A defender can be replaced by an essentially identical challenger. A construction company's favorite make and model of pickup truck is an example. An electric lightbulb is another. Or a defender can be replaced by a different kind of challenger. A personal computer of one make challenged by a new generation of personal computers of another make with better capacity, better computational abilities—a larger memory—is an example of this kind of problem. In both these types of replacement problems, the question of economic life arises.

ECONOMIC LIFE

Up to this chapter in our study of engineering economy, the service life has always been a given. Now it will be determined.

Economic life in a replacement situation is the point in time when the equivalent uniform annual cost (EUAC) of an asset is a minimum. The physical life of an asset must always exceed or equal its economic life.

REPLACEMENT

In general, a replacement situation occurs when an existing asset is compared to another asset that may replace it. The implication in the matter of equal service lives of this distinction is important and will be discussed later in this chapter.

Identical Replacement

Here, a defender will be replaced by an identical challenger. The reader may object, "But surely prices will have risen in any country in the world these days. How can the replacement be identical with the asset in service when its price has risen in comparison?" One answer is that if all prices and costs rise proportionately, then the analysis can be made in constant dollars. A standard pickup truck, or a hammer, or an electric motor, does not change much over the years, given the same make and model. What does change is the price, the operations and maintenance costs, and the salvage value. The assumption of constant dollars and proportionate rises takes care of the objection.

TABLE 16.1
A construction company car

Year j	First cost (1)	Operations cost (2)	Maintenance cost (3)	Salvage value (4)
0	–12,000			
1		–2,400	–1,000	8,000
2		–2,400	–2,200	6,000
3		–2,400	–3,400	4,000
4		–2,400	–4,600	2,000
5		–2,400	–5,800	0

■ **Example 16.1.** A small U.S. construction company has used a standard, two-door, business coupe as its company car for the past 20 years. Discussion has arisen in the estimating group as to the economy of the particular make and model. The estimating group decides to ascertain, before taxes, when the company car should be replaced with an identical make and model. When this analysis is concluded, an after-tax analysis will be made.

The company is in the 40 percent tax bracket and uses a 20 percent before-tax and a 10 percent after-tax opportunity cost of capital. The analysis will be performed based on constant, year-0 dollars.

Table 16.1 shows the original price of the car and the cost of operation and maintenance over 5 years in the first three columns. The rightmost column shows the salvage or resale value of the car at the end of each year. See Fig. 16.1.

Solution. A before-tax decision will be made first. In Table 16.2, the *equivalent uniform annual cost* (*EUAC*) of the first cost is computed for retirement after each of 5 years. The operations cost of column 2 is invariant no matter how long the car is kept in service up to 5 years. The EUAC of annual maintenance varies for holding periods from 1 to 5 years as the maintenance cost increases from \$1,000 in the first year by jumps of \$1,200 yearly. Column 4 shows the *equivalent uniform annual benefit* (*EUAB*) realized if the car is kept in service for each of 5 years. The rightmost column shows the total EUAC, including the effects of capital recovery, operations, maintenance, and salvage. The total EUAC is at a minimum (\$9,052) at the end of 3 years. This is the economic life of the car before tax effects are considered. See Fig. 16.2. The car should be replaced every 3 years.

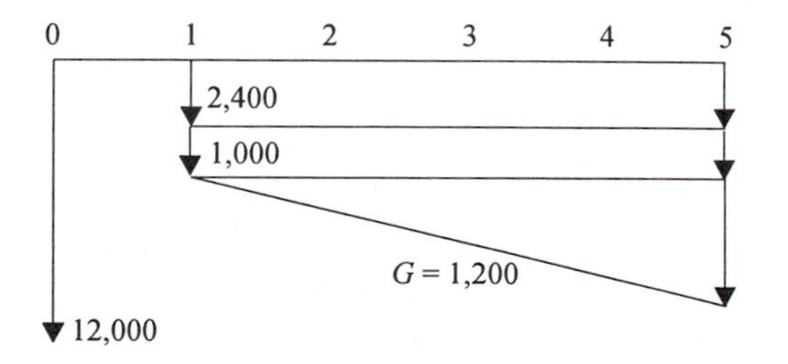

FIGURE 16.1 Cash flow for a construction company car.

TABLE 16.2

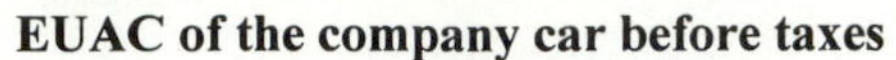

EUAC of the company car before taxes

Year of retire-ment j	EUAC of capital recovery 12,000 $(A/P, 20, j)$ (1)	Operations (2)	EUAC of maintenance 1,000 + 1,200 $(A/G, 20, j)$ (3)	EUAB of salvage $S_j\,(A/F, 20, j)$ (4)	Total annual cost (1 + 2 + 3 − 4)
1	14,400	2,400	1,000	8,000	9,800
2	7,854	2,400	1,546	2,727	9,073
3	5,696	2,400	2,055	1,099	9,052
4	4,636	2,400	2,529	373	9,192
5	4,013	2,400	2,969	0	9,382

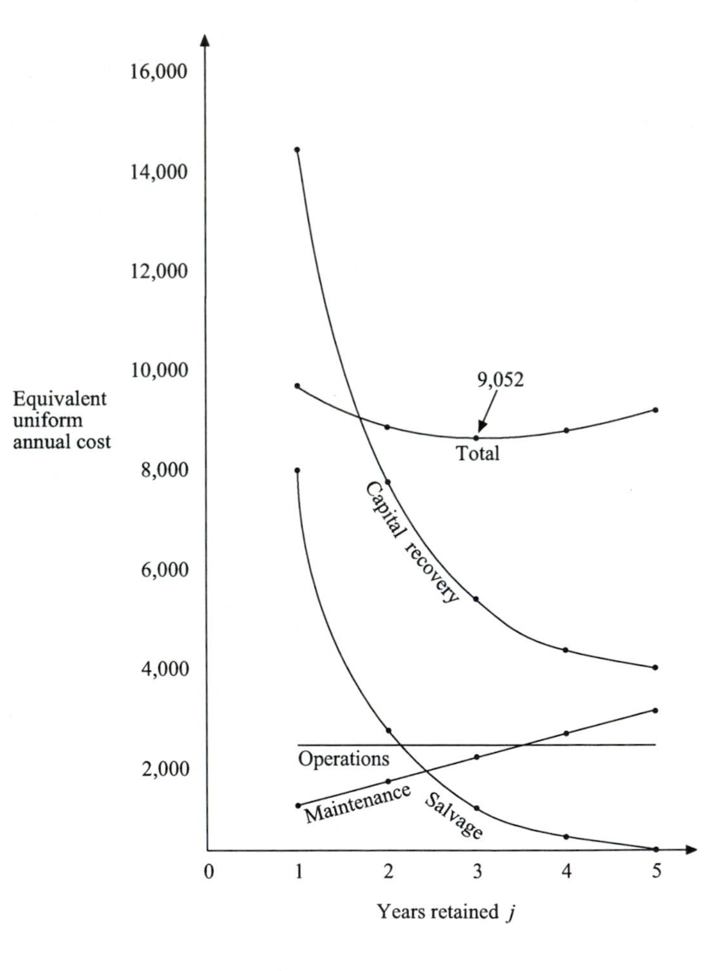

FIGURE 16.2 A construction company car's EUAC.

TABLE 16.3
After-tax cash flow of a company car if held for 4 years

Year j	Before-tax cash flow	Depreciation	Taxable income	Tax @ −0.40	After-tax cash flow
0	−12,000				−12,000
1	−3,400	−4,000	−7,400	+2,960	440
2	−4,600	−2,000	−6,600	+2,640	1,960
3	−5,800	−2,000	−7,800	+3,120	2,680
4	−7,000	−2,000	−9,000	+3,600	3,400
4	+2,000				+2,000

To compute the after-tax EUAC for service years j = 1 to 5, the following conditions apply:

1. The resale value at the end of 5 years is zero.
2. The Internal Revenue Service has made a special ruling for this company's fleet, allowing 33.33 percent first-year depreciation providing a switch to pre-1981 straight-line depreciation is made thereafter. The result is a first-year depreciation of

$$\text{First-year depreciation} = 12{,}000(0.3333) = \$4{,}000$$

3. A switch is made to straight-line depreciation for the second year. Straight-line depreciation will be used after that.

$$\text{Year 2 and subsequent depreciation} = \frac{(\text{book value at end of year 1}) - S_5}{\text{remaining life}}$$
$$= \frac{8{,}000 - 0}{4}$$
$$= \$2{,}000$$

4. The salvage value will always equal the book value of the car. Thus, no capital gains or losses will be incurred. Deductible items will provide the only income tax effects, therefore.

To illustrate the method used, the after-tax EUAC for 4 years is calculated. Table 16.3 shows the results. The before-tax cash flow is taken directly from Table 16.1. The depreciation is shown in the next column, and the remaining columns are calculated in the way already familiar to you. Figure 16.3 shows the after-tax cash flow for 4 years of company car operation.

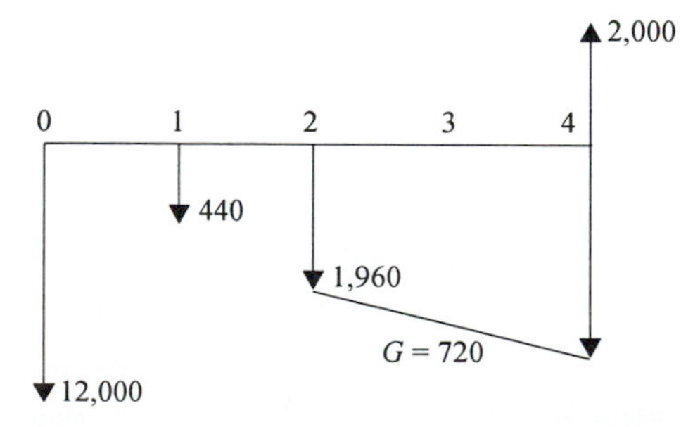

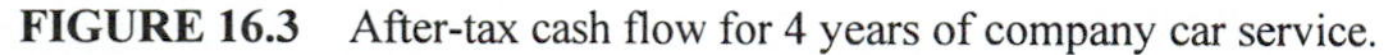
FIGURE 16.3 After-tax cash flow for 4 years of company car service.

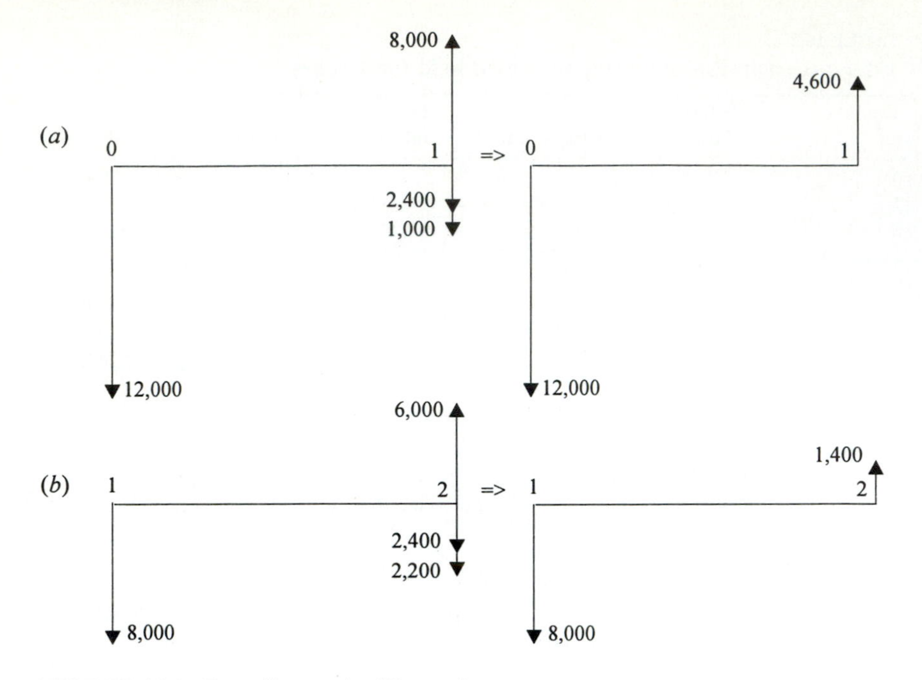

FIGURE 16.4 Extending service life year by year.

The equation for the EUAC is

$$\begin{aligned}\text{EUAC}_4 = &-12{,}000(A/P, 10, 4) - 440(P/F, 10, 1)(A/P, 10, 4)\\ &+[-1{,}960\,(P/A, 10, 3) - 720(P/G, 10, 3)]\,(P/F, 10, 1)\,(A/P, 10, 4)\\ &+2{,}000(A/F, 10, 4)\\ = &\ \$5{,}360\end{aligned}$$

Table 16.4 shows the results for all 5 years of service. After taxes, the economic life is either 2 or 3 years. The extra satisfaction of driving newer cars moves the estimating group to recommend that company cars be replaced every 2 years. The after-tax EUAC calculation has proved to be well worth the extra effort, in addition to being more realistic. ■

Up to this point, we have imagined ourselves standing at time 0, *now*, and asking ourselves, How long should we hold onto this asset in order to achieve the cheapest EUAC? It may occur to you that what we might actually do in the future is to examine the cost of extending the asset's service life by 1 year at a time, moving our position forward from year to year. We then, puzzled, will ask ourselves, What relation has this to our original analysis? Example 16.2 will illustrate and answer the question. We will use the situation and data of Example 16.1.

■ **Example 16.2.** Standing at year 0, the annual cost (AC) before taxes for the first year is the same as in Example 16.1. See Fig. 16.4*a*, right side, netted out.

TABLE 16.4

EUAC of company car after taxes

Year of retirement j	EUAC of capital recovery 12,000 $(A/P, 10, j)$ (1)	EUAC of operations and maintenance $[440 + 1960\,(P/A, 10, j-1) + 720\,(P/G, 10, j-1)] \times (P/F, 10, 1)\,(A/P, 10, j)$ (2)	EUAB salvage $S_j\,(A/F, 10, j)$ (3)	Total EUAC, $(1 + 2 - 3)$ (4)
1	13,200	440	8,000	5,640
2	6,914	[440 + 1,960 (0.909)](0.9091) (0.5762) = 1,164	2,857	5,221
3	4,825	[440 + 1,960 (1.736) + 720 (0.826)](0.9091) (0.4021) = 1,622	1,208	5,239
4	3,786	[440 + 1,960 (2.487) + 720 (2.329)](0.9091) (0.3155) = 2,005	431	5,360
5	3,166	[440 + 1,960 (3.170) + 720 (4.378)](0.9091) (0.2638) = 2,352	0	5,518

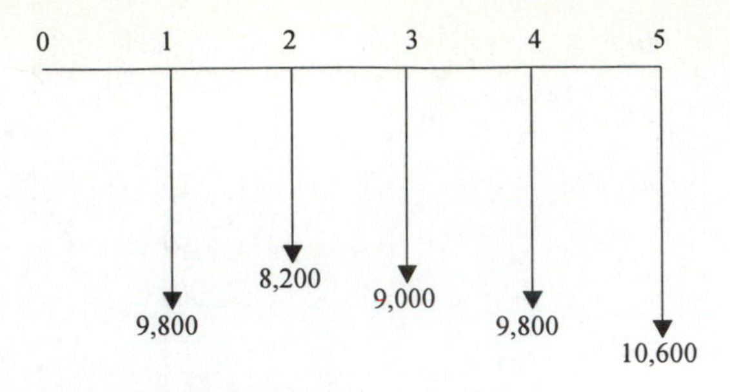

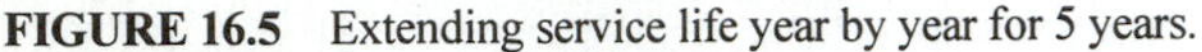

FIGURE 16.5 Extending service life year by year for 5 years.

$$\begin{aligned} AC_1 &= 12{,}000(A/P, 20, 1) - 4{,}600(A/F, 20, 1) \\ &= 12{,}000(1.20000) - 4{,}600(1.0000) \\ &= 14{,}400 - 4{,}600 \\ &= 9{,}800 \end{aligned}$$

For the second year, now standing in time at the end of the first, the $12,000 original cost is now sunk, and the $8,000 salvage value is now an opportunity cost. The salvage value at the end of year 2 is $6,000, and the corresponding operations ($2,400) and maintenance ($2,200) costs appear. See Fig. 16.4*b*, netted out.

$$\begin{aligned} AC_2 &= 8{,}000(A/P, 20, 1) - 1{,}400(A/F, 20, 1) \\ &= 8{,}000(1.20000) - 1{,}400(1.00000) \\ &= 9{,}600 - 1{,}400 = 8{,}200 \end{aligned}$$

The result is not the same as that shown in Table 16.2 for the second year (9,073).

Similarly, for each of the remaining years of car life:

$$\begin{aligned} AC_3 &= 6{,}000(A/P, 20, 1) + (2{,}400 + 3{,}400 - 4{,}000)\,(A/F, 20, 1) \\ &= 6{,}000(1.20000) + 1{,}800(1.00000) \\ &= 7{,}200 + 1{,}800 \\ &= 9{,}000 \\ AC_4 &= 4{,}000(A/P, 20, 1) + 5{,}000(A/F, 20, 1) \\ &= 4{,}000(1.20000) + 5{,}000(1.00000) \\ &= 4{,}800 + 5{,}000 \\ &= 9{,}800 \\ AC_5 &= 2{,}000(A/P, 20, 1) + 8{,}200(A/F, 20, 1) \\ &= 2{,}400 + 8{,}200 \\ &= 10{,}600 \end{aligned}$$

These individual-year costs of extending service life year by year provide no basis for a decision as they stand. (See Fig. 16.5.) But we may logically expect that the EUAC of the annual costs for any service life j will match the results of Table 16.2 and thus lead to the same decision. Like this:

$$\begin{aligned}
\text{EUAC}_1 &= [9{,}800(P/F, 20, 1)]\,(A/P, 20, 1)\\
&= [9{,}800(0.8333)](1.20000)\\
&= 9{,}800\\
\text{EUAC}_2 &= [9{,}800(P/F, 20, 1) + 8{,}200(P/F, 20, 2)](A/P, 20, 2)\\
&= [9{,}800(0.8333) + 8{,}200(0.6944)](0.65455)\\
&= (8{,}166 + 5{,}694)(0.65455)\\
&= (13{,}860)(0.65455)\\
&= 9{,}072\\
\text{EUAC}_3 &= [9{,}800(P/F, 20, 1) + 8{,}200(P/F, 20, 2) + 9{,}000(P/F, 20, 3)](A/P, 20, 3)\\
&= [9{,}800(0.8333) + 8{,}200(0.6944) + 9{,}000(0.5787)](0.47473)\\
&= (8{,}166 + 5{,}694 + 5{,}208)(0.47473)\\
&= 9{,}052
\end{aligned}$$

Similarly,

$$\text{EUAC}_4 = 9{,}192 \qquad \text{and} \qquad \text{EUAC}_5 = 9{,}381$$

The results match the total EUACs of Table 16.2. The decision is the same.

Thus we recognize that the situation viewed from time 0 or year by individual year leads us to the same conclusion. ■

Different Replacement

What if the challenger is not the same as the defender? This is at least as reasonable a supposition as that of the previous section, where the challenger and defender were exactly alike. For fast-moving improvement in technology such as is occurring in the computer industry, a challenger better than a defender is to be expected.

■ **Example 16.3.** Imagine that the company car of Example 16.1 is now a challenger only, and not the defender as well. In Fig. 16.6 are shown the after-tax EUACs for the challenger, taken from Table 16.4. The defender has the annual cost characteristics after taxes shown as the dashed line.

The analyst calculates that the annual cost of retaining the defender for 1 year is $6,120 after taxes. This is more than the $5,221 EUAC for a 2-year life of the challenger, its economic life. Therefore the defender should not be retained for 1 year, but might be retained for longer.

Should the defender be retained for 2 years? The annual cost after taxes for the defender is calculated as $4,940 for a 2-year remaining life. This is less than the annual cost of the challenger at its economic life of $5,221. Therefore the defender should continue in service for 2 years more.

Notice that we need not proceed further. It may be that the defender's third-year annual cost will be greater than the challenger's EUAC at its economic life, or it may

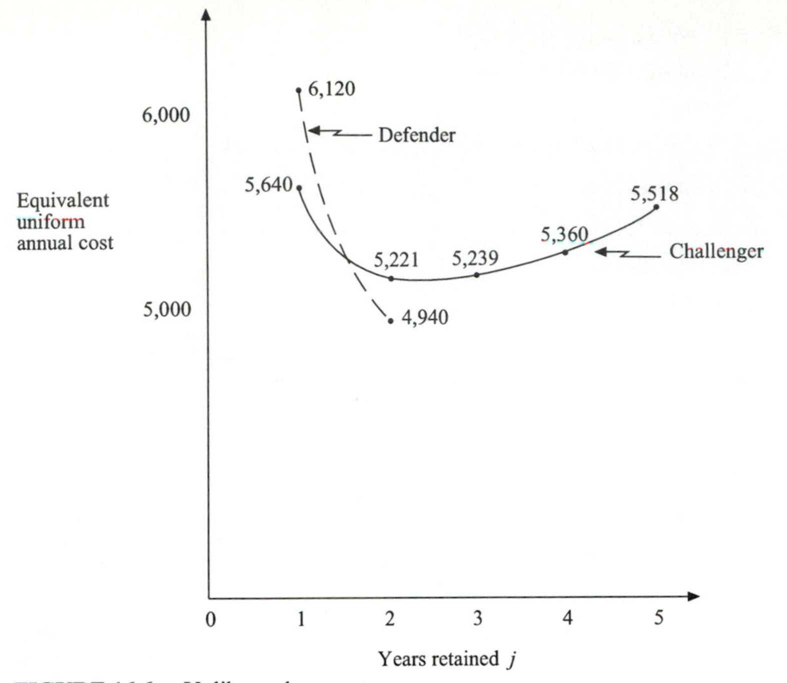

FIGURE 16.6. Unlike replacement.

be less. It does not matter. Our decision has been made for the next 2 years—retain the defender. Notice also that we do not compare the EUAC of the economic life of the defender to that at the economic life of the challenger. In fact, it is unnecessary to determine the economic life of the defender. We only seek a decision on retention or replacement *now*. And we will investigate only to the point where we will retain the defender. If that point never comes, if the EUAC at the economic life of the challenger is always less than the defender's EUAC at any life, then we will replace the defender with the challenger.

Said another way, the defender will finally be replaced by a challenger. Meanwhile, the defender is cheaper per year than the challenger and thus worth retaining.

To summarize, with more than one challenger:

1. Choose the best challenger by selecting the one whose EUAC at its economic life ($\text{EUAC}_{C\,\text{min}}$) is least.
2. The defender's EUAC_D is always compared to $\text{EUAC}_{C\,\text{min}}$.
3. Calculate EUAC_D for $j = 1$. If EUAC_D is less than $\text{EUAC}_{C\,\text{min}}$, retain the defender. If EUAC_D is greater than $\text{EUAC}_{C\,\text{min}}$, recalculate EUAC_D for $j = 2$ and so on until EUAC_D becomes less than $\text{EUAC}_{C\,\text{min}}$, in which case the defender is retained. If not, continue testing at successive j values until you are sure that EUAC_D will never be less than $\text{EUAC}_{C\,\text{min}}$ or until the physical life of the defender is reached. If EUAC_D is never less than $\text{EUAC}_{C\,\text{min}}$, then replace the defender by the challenger immediately and exit. See Fig. 16.7.

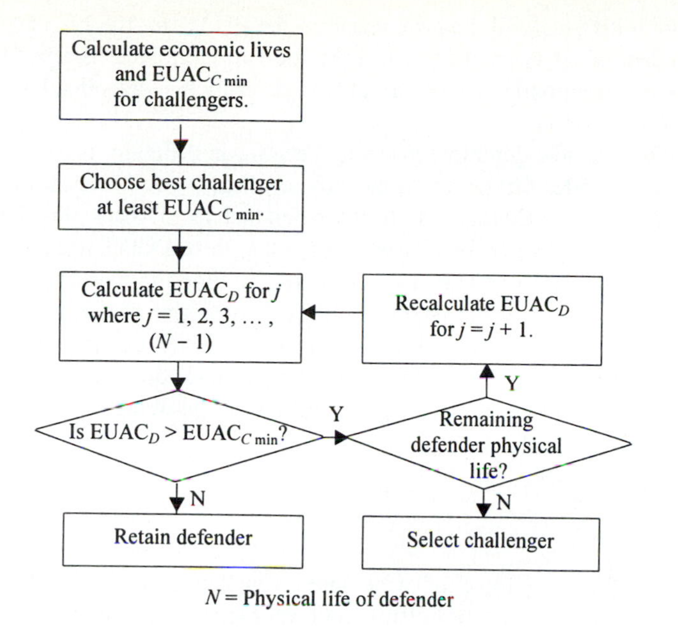

FIGURE 16.7 Replacement decision flowchart.

Once again, notice that it is not necessary to determine the economic life of the defender. In fact, doing so may lead to confusion and error.

4. Reinvestigate when new challengers arrive on the scene or when the analyst suspects the defender's EUAC may have become greater than the EUAC at the economic life of the challenger. Replacement takes place when no defender life results in an EUAC less than that of the challenger at its economic life. ■

Now a doubt may arise in the reader's mind. What about the fundamental rule in economic analysis that the lives of mutually exclusive alternatives must be equalized? In Example 16.1 and 16.2, they were equalized, and no question arises. But it is quite possible, under the decision rules shown in Example 16.3, to choose to retain the defender for a different remaining life than the economic life of the challenger. What then?

Recall that the annual-cost method used in these examples, by its very nature, equalizes the lives of the alternatives. One cycle is followed by another in each of the alternatives until the lives are equal. And each cycle, for each alternative, has the same EUAC as the previous one. This idea presents no difficulties in regard to the challenger, but what of the defender? The defender is already partway through its life. Its next cycle then requires, not a new identical defender, but a partially used defender in exactly the same condition as the present one. The idea is absurd. Notice that a new identical defender will have an EUAC different from that of the remaining life of the defender. For example, the EUAC of the remaining 5 years of life for a defender will be different from the EUAC for the life of a new identical defendant of 10 years. This will make the two combined have a life equal to 15 years, the economic life of the challenger; but the defender has two different EUACs.

The practical commonsense way out of this difficulty is to remind ourselves that the defender for 5 years against the challenger for 15 years are really not the alternatives. The alternatives are the defendant for 5 years plus the life of some new, better challenger for 10 years versus the present challenger. If we choose the defendant, we are committed for now. At the time the new challenger arrives, it will be compared to the defender. If we choose the challenger, it becomes the defender and will be tested against an even newer challenger. The process is continuous and thus does not violate the equal-lives dictum of economic analysis. It is also subject to most uncertain projections of the future.

SUMMARY

Replacement is a topic of wide interest which makes use of the techniques explained in this text. Problems concern replacement by an identical asset or replacement with a different asset. Income and capital gains taxation may be taken into account as well as risk and inflation.

As we have often seen, tax effects can be all-important in making economic decisions. Replacement decisions are no exception.

Defenders may be challenged by like or unlike challengers.

The economic life of an asset is reached when its annual cost (EUAC) becomes a minimum.

For identical replacement of an asset, the rule is: *Replace at the end of the economic life*. The rule holds in both the before- and after-tax cases.

For unlike replacement, the rule is: *Test the annual cost of each year of the defender's remaining life successively against the annual cost at the economic life of the challenger. Choose the year of defender replacement at that point where the defender's EUAC becomes greater than the EUAC at the challenger's economic life; but continue the analysis, if necessary. If the defender's EUAC never becomes less than that of the challenger's at the challenger's economic life, replace immediately with the challenger.*

The question of unequal lives in a replacement situation may be resolved by remembering that the comparison is really of the defender plus some new and better challenger of unknown characteristics versus the known challenger.

PROBLEMS

16.1. What is meant by *economic life*? Explain with an example.

16.2. In Example 16.1, compute the EUAC of holding the car for 3 years, *before taxes*. All data remain the same, except that the salvage value has been revised to $3,000.

16.3. Two production machines are under consideration. Machine A, the defender, bought 1 year ago, exhibits the following characteristics:

First cost	$15,000
Physical life	5 years
Maintenance cost	$2,400 per year
Salvage value at any time	0

Similar information on machine B, the challenger, is given:

First cost	$20,000
Physical life	4 years
Maintenance cost	$2,400 per year
Salvage value at any time	$2,000

The opportunity cost of capital for the company considering the machines is 12 percent. No taxes or inflation will be considered for this preliminary analysis. Use EUAC to make your decision.

(*a*) What is the economic life of the challenger?

(*b*) Should the challenger be purchased at this time? Say why, indicating which costs and lives should be compared.

16.4. Metro, the subway and bus system in the Washington, D.C., area, is considering replacement of several of its escalators in certain subway stations. The following table presents the situation for each station.

	Replacement	Original
First cost ($)	1,350,000	420,000
Physical life (years)	25	25
Remaining life (years)	25	5
Maintenance and operation cost per year ($)	50,000	100,000
Salvage value ($)	0	0

The existing escalators will last 5 more years. The replacement will be performed by the manufacturer and installer of the original escalators.

Metro, as a public entity, pays no income taxes. Do not consider inflation. The opportunity cost of capital is 10 percent.

Should Metro replace the original escalators?

16.5. The Georgetown office copying machine was purchased 3 years ago by your electrical engineering firm. It may be sold at present for $225 on the open market. Its resale value at any time in the future is estimated as $125. Operating costs for material, labor, and maintenance will remain at $4,100 per year for the next 4 years, the study period.

Office Leasing, a reliable company, will furnish a similar machine for $1,000 per year with a guaranteed operating expense of $3,200 annually over the next 4 years.

Should you replace the present machine with the leased one if your MARR is 15 percent and tax and inflation effects are disregarded?

16.6. An electrical contractor has in his service a fleet of 10 vans of varying ages. With a view to standardizing his fleet with a single make, he wishes to know what the annual cost, at the economic life, will be for a van with the following characteristics:

Life	First cost	Operations cost	Maintenance cost	Salvage value
0	13,200			
1		2,160	880	9,080
2		2,160	1,980	7,340
3		2,160	3,080	5,600
4		2,160	4,180	3,860
5		2,160	5,280	2,120
6		2,160	6,380	380

All costs are in constant dollars. His opportunity cost of capital before taxes is 20 percent. A before-tax analysis is required.

16.7. The Bureau of Reclamation is studying a stretch of the leaky Coachella Canal in California with a view to lining the presently unlined canal.

The new plastic lining will cost $170 million. Yearly maintenance will cost $1 million for the first year of operation and will increase by $1.2 million per year for each succeeding year. Thus the cost for year 2 will be $2.2 million, for year 3, $3.4 million, and so on. The life of the lining is 25 years, with no salvage value.

The original canal cost $800 million 30 years ago. It will last in its present unlined condition for another 20 years at a minimum, but with an increasing water loss each year. Water loss is estimated at $20 million for next year. For each succeeding year, the water loss will increase by $5 million per year. The cost of water loss for year 2 will thus be $25 million, for year 3, $30 million, and so on.

The opportunity cost of capital in constant dollars is 8 percent; all estimates are made in constant dollars. Assume that, for the purposes of a preliminary calculation, the new lining can be installed at time 0, that is, immediately. For a public enterprise, no income taxes need be considered.

(*a*) What is the economic life of the challenger?

(*b*) At what year from now, time 0, should the new lining be installed, or should it not be installed at all?

16.8. Your financial advisers have assured you that they can always get at least a 5 percent return on your funds. But you find yourself with a transportation problem. Your present car seems to be costing a great deal. You would like a newer and better one.

Your present car is worth $6,000. Its useful remaining life and costs are as shown:

Years retained	Operations and maintenance	Resale value
1	2,500	5,000
2	2,750	4,000
3	3,000	3,000
4	3,250	2,000
5	3,500	1,000

A better car will cost you $20,000. Its useful life and costs are these:

Years retained	Operations and maintenance	Resale value
1	500	14,000
2	1,100	12,000
3	1,700	10,000
4	2,300	8,000
5	2,900	6,000

Should you buy the better car now? Because the car is not used in business, no income tax effects will appear. All sums are in current dollars.

16.9. AAA company, a contractor, specializes in handling emergency service calls. The company currently owns five Ford panel trucks to cover the Annapolis area. The Ford truck has the following characteristics:

Life	First cost	Operation cost	Maintenance cost	Salvage value
0	$15,000			
1		2,000	800	11,000
2		2,000	1,850	9,000
3		2,000	2,900	7,000
4		2,000	3,950	5,500
5		2,000	5,000	3,800
6		2,000	6,050	2,500

All costs are in constant dollars. The contractor wishes to know when the trucks should be replaced with an identical model. The company opportunity cost of capital before taxes is 20 percent. No taxes are to be considered.

16.10. The metropolitan police department of a major U.S. city believes that significant savings in operator time and computer maintenance would be achieved by the acquisition of new computers. The new Fayram computers will cost $1.5 million and will last 5 years. Their operations and maintenance costs, with salvage value, appear in the following table, along with data on the existing Reynolds computers. All sums are in constant dollars of 1994. The Reynolds computer may be sold at present (December 31, 1994) for $300,000. It will last 4 more years.

Year	Operations cost ($000,000)	Maintenance cost	Salvage value
Reynolds (defender)			
1995	836	54	300
1996	836	59	300
1997	836	64	300
1998	836	69	300
Fayram (challenger)			
1995	402	23	400
1996	402	25	400
1997	402	27	400
1998	402	29	400
1999	402	31	400

The discount rate used by all city services is 10 percent on constant dollars. No taxes are paid because the police department is a public agency. Inflation will not be considered at the moment.

Should the existing computers be replaced at this time?

16.11. A public, nonprofit hospital is considering the replacement of an artificial kidney machine that 4 years ago cost $35,000. It will last, physically, 4 more years. If the machine is kept 1 more year, its operating and maintenance costs are expected to be $25,000. Operating and maintenance costs in the second, third, and fourth years are expected to be $27,000, $29,000, and $31,000, respectively.

A new machine is calculated to have an economic life of 5 years. The EUAC at its economic life is computed at $30,450. The company selling the new machine will allow $9,000 on the old machine for a trade-in. If the hospital delays its purchase for 1, 2, or 3 years, the trade-in value on the existing machine is expected to decrease to $7,000, $5,000, and $3,000, respectively, with no value thereafter.

At a discount rate of 15 percent, should the existing machine be replaced now? If your answer is negative, when would you replace it, if at all? If your answer is affirmative, how long should you retain the new machine, assuming that nothing changes?

No income tax need be considered because the hospital is a nonprofit enterprise.

16.12. You have accepted a consulting contract with Iraqi Airlines. They are considering replacing Boeing 707 passenger aircraft, now 30 years old, with Russian Ilyushins.

The challenger—the Ilyushin—will cost $55 million each. The airplane will require a major overhaul, which will cost $2 million every 10 years, excluding year 30. Maintenance costs each year will be $300,000 for each Ilyushin. At the end of its 30-year life, it will be sold for $1 million. During its life, its resale value will diminish by $1.8 million annually.

Because Iraqi Airlines is government-owned, it will pay no taxes.

The costs given above are in constant, year-0 dollars. The constant-dollar discount rate to be used is 5 percent.

What is the economic life of the challenger, and what is its annual cost at that life?

16.13. A steel bridge on a Louisiana state highway near the Gulf of Mexico is costing $450,000 yearly in maintenance—largely chipping, priming, and painting. It originally cost $1,600,000 when it was built 15 years ago. The Louisiana bridge engineers estimate that its remaining life is 10 years; then it will need to be replaced because of increased traffic. Its salvage value at any point in time is zero, because the cost of demolition will most likely equal its value as scrap steel.

A concrete bridge is considered to be the best challenger. It will cost $3,000,000 to build and $100,000 annually in maintenance costs. Its estimated life is 50 years. Its resale value may be counted as zero at any time during its life.

No taxes of any kind will be considered for this government project. All costs are in constant dollars of year 0. Inflation may be ignored. Assume that annual benefits for either structure are exactly the same. A discount rate of 10 percent is to be used in the analysis.

(*a*) What is the economic life of the challenger?

(*b*) Should the steel bridge be replaced now?

16.14. In Example 16.2, check the computation of the $EUAC_4$.

16.15. Choose one of the following answers for each of the six figures shown on the following page. All defenders have a physical life of 7 years. Challenger physical lives are as shown:

1. Select *defender* now and assume retention for at least ____ year(s).
2. Select *challenger* now and assume retention for ____ year(s).
3. Cannot decide on the basis of information shown. Investigate more years of retention for the defender.

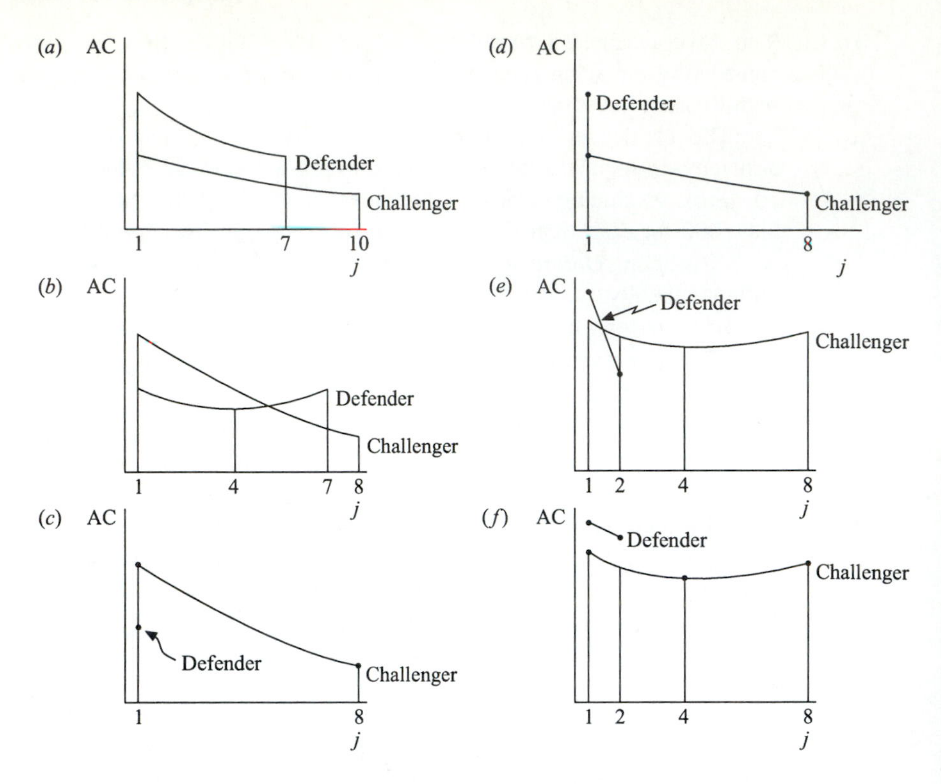

16.16. In Example 16.1, calculate the *after-tax* EUAC of holding the car for 3 years. All data, including depreciation, remain the same, except that the salvage value has been revised to $3,000. Notice that the salvage value is no longer equal to the book value. Capital gains or losses may be treated as ordinary revenue or cost for the year involved.

16.17. A milling machine cost $39,000. Three years after its purchase, its disposal for $25,000 is being considered. The owner is attempting to decide whether to hold the machine for another year, making a total of 4 years in her possession, or to sell it now for $25,000. What will be the *capital recovery cost,* i.e., the annual capital cost, of holding the asset for another year at the end of which its resale value will be $22,000?

The machine has been depreciated by ADS, the half-year convention, and a recovery period of 14 years. Compute an after-tax figure based on a MARR after taxes of 10 percent.

One hundred percent of capital gains are taxed as ordinary income. All capital losses may be deducted from taxable income. The company is in the 45 percent tax bracket for all taxes.

16.18. In Example 16.1, calculate the annual cost, after taxes, for the second year of ownership of the company car, if it is held for 2 years, assuming that a 5-year MACRS GDS is used with the half-year convention. Also assume that the full second-year depreciation can be taken. Capital gains will be taxed as ordinary income at 45 percent of the gain. Capital losses are fully deductible from ordinary income in the year of sale.

16.19. A submersible pump and motor for your Peruvian agricultural company were purchased 3 years ago for $120,000. They were depreciated by ADS, half-year convention, and a recovery period of 10 years, a method required by the IRS for overseas property. One member of your value engineering group presents the following analysis of the EUAC of the pump for the remaining 17 years of its life.

First cost	$120,000
Less depreciation	30,000
Equals book value	90,000
Maintenance and operation costs per year	5,000
Salvage value in 17 years	10,000
MARR = 25 percent	
Tax bracket = 40 percent	

Capital gain will be treated as income for the year of disposal. She arrives at a figure of $29,432 per year. Would you approve of this analysis? Why or why not?

16.20. An office building in Washington, D.C., is served by three elevators. These were installed 6 years ago when the building was constructed and have required frequent maintenance of the cabs. New cabs are being considered as replacements.

The existing cabs cost $15,000 for all three 6 years ago and will last another 14 years. Maintenance costs for all three average $10,000 annually. The resale value of the cabs at any time will be assumed to be zero.

The three new cabs will cost $25,000 total. Total maintenance costs will be $3,000 per year. The cabs will have a physical life of 7 years. Like the existing cabs, their resale value at any time will be assumed to be zero.

The building owners are in the 45 percent tax bracket. Their after-tax cost of capital is 9 percent. No inflation effects will be considered in this analysis at this time. Capital gain or loss will be counted as income or loss for the year of disposition.

ADS depreciation, the half-year convention, and a 12-year recovery period are used for this type of asset.

(*a*) What is the economic life of the challenger, after taxes?
(*b*) Should the defender be replaced?

16.21. A Boyington motor grader is owned by a private contractor. Replacement by the Kamakazi grader is being considered. Data on the machines are these:

	Boyington	Kamakazi
First cost ($)	45,000	55,000
Age (years)	11	0
Estimated remaining physical life (years)	8	8
Resale value now ($)	30,000	N/A
Salvage value at end of life ($)	0	0
Maintenance and operating costs annually ($)	3,000	2,000

The estimated remaining life can be achieved only by a major overhaul of the Boyington grader at $18,000 now and at the end of year 4. The overhauls will be expensed, not capitalized, i.e., not depreciated. Its resale value will be zero at any time in the future.

The contractor used ADS depreciation, half-year convention, and a recovery period of 6 years for the Boyington machine. It has been completely depreciated. In case of an extended life through overhaul, he will not be able to take more depreciation on this machine. Capital gains will be taxed as ordinary income.

The Kamakazi machine will be depreciated by the same method. Its resale value will approximate its book value.

If the contractor is in the 40 percent bracket for all taxes, should he buy the new machine or continue with the old one, and for how long? His marginal attractive rate of return is 12 percent after taxes. Neglect inflation.

16.22. An existing sheet-metal press in a metals specialty company was purchased 5 years ago for $15,000. Its estimated life was 18 years at the time of purchase. ADS depreciation, half-year convention, and a 12-year recovery period were used for tax purposes. Its market value at any time is about equal to its book value. Operation and maintenance costs are $6,000 per year.

An improved version of the press is now available for $10,000. Its life is estimated at 15 years. The same depreciation conditions will be used for accounting purposes. Its book value at any point in its life is thought to be a probable market value for the press. Operation and maintenance costs are $1,000 per year.

The company is in the 48 percent income tax bracket. Its MARR after taxes is 15 percent. Any loss or gain on disposal of fixed assets will be treated as loss or income for the year in which the disposal occurs.

Based on an economic analysis, should the new machine be bought? When? The old machine will be sold as soon as the new machine is acquired.

16.23. The HV make of small desktop computers purchased in large numbers by the South American Aviation Company is 1 year old. A new type made by RS Company is now on the market. It is being considered as a replacement for the HV model. The HV may be sold for $600 at present. Facts about the computers are shown:

	Make	
	HV	**RS**
First cost ($)	1,500	1,750
Original life (years)	2	2
Remaining life (years)	1	2
Maintenance cost annually ($)	200	120

The company is in the 48 percent tax bracket. Gain or loss on the disposal of assets is taxed at 40 percent of the gain or all the loss applied to the current taxable income. It may be assumed that the salvage value during the life of either computer may be approximated by its book value at the end of that year, except for the aforementioned $600 for the HV now.

Depreciation of the HV was computed by ADS, half-year convention, and a 5-year recovery period. The same method will be used for the RS.

If the MARR after taxes for this company is 15 percent, should the HV type be replaced by the RS?

16.24. A new type of car battery has just come on the market. After extensive tests on its probable physical life, a large automobile leasing company estimates a physical life of 4 years. The new battery will cost $65 if bought in fleet-size quantities now. At any point in its life, its market value may be assumed to equal its book value. It will be depreciated by MACRS GDS, half-year convention, and a 5-year recovery period.

Maintenance and operation costs of the battery are zero.

This company is in the 40 percent tax bracket. Its after-tax minimum attractive rate of return is 18 percent. Gains or losses on disposal of batteries are treated as ordinary income or expense. No inflation effects will be considered.

What is the economic life of the battery?

16.25. The following data describe the characteristics of two tractor units used in long-distance household moving.

	1	**2**
First cost ($)	50,000	116,000
Original physical life (years)	4	6
Remaining life (years)	2	6
Net annual revenues ($)	20,000	40,000

The moving company is in the 45 percent tax bracket for all taxes. The MARR of this company is 20 percent after taxes have been accounted for and 35 percent before taxes. Inflation effects will be ignored.

The company presently owns the make 1 equipment, bought 2 years ago. At any point in its life, the resale value of either make of equipment will approximate its book value.

The company uses ADS depreciation, half-year convention, and a recovery period of 4 years. Loss or gain on disposal is treated as ordinary expense or revenue.

Determine whether the defender should be sold and the challenger bought. Because the company intends to continue in the long-distance moving business, an alternative is certain to be chosen. An after-tax solution is required.

16.26. Franklin Computer Associates is considering the acquisition of a new SBY model 61 computer to replace its existing BMC 21 for its New Jersey operation.

The BMC 21 was purchased 3 years ago for $2 million. It is now worth $1 million in disposal value. It will last 3 years more before it will no longer do the job for Franklin Computer Associates. It will have a salvage value during its remaining life equal to its book value at any time. It has been and will continue to be depreciated for income tax purposes by the ADS. Its labor and maintenance charges total $180,000 per year.

The SBY 61 will cost $1.8 million. During its 6-year life it will also be depreciated by the ADS with the half-year convention. Its labor and maintenance charges will average only $10,000 per year, and its disposal value at any time during and at the end of its life will be equal to its book value.

Assume for both machines that full depreciation will be allowed for the year of disposal and that disposal takes place at the last instant of any year. The disposal gain or loss will be treated as revenue or cost for the year involved.

The income tax rate for this company is 49 percent. The opportunity cost of capital after taxes in constant dollars is 18 percent. The company's financial cost of capital is 10 percent. All figures are in current dollars.

Should the SBY model 61 be purchased now on the basis of the above information?

CHAPTER 17

SECTOR ANALYSIS AND VIEWPOINT

Sector is an economist's word. It means an economic area of action. In the *private sector*, resources are owned by individuals—persons or associations of persons called *corporations*. The firms that manage these resources and draw production of goods and services from them are in private hands. In the *public sector*, the ownership and management of resources in order to produce goods and services are in government hands—local, state, and national.

Viewpoint is the institutional position that the economic analyst takes in relation to the project under consideration. The word *institutional* is used here in its social sense in which elements of the public and private sectors of society are considered to be institutions. For example, if an analyst is working for a city government, she may be expected to judge the consequences of the project she is considering in relation to their effect on that government. Costs are therefore costs to the city, and benefits are benefits to the city. In effect, she draws a line around the city and says, "Any adverse consequences that cross the line in an outward direction, I will count as costs; and any favorable consequences that come in across the line, I will count as benefits of this project." For the very same project, another analyst may be imagined as working for a higher unit of government, say, the national government. Here the institutional line he draws will include the entire society of the country. Notice that he is not taking the viewpoint of the government, an entirely different entity from the society itself. In fact, the government and the society may be opposed to each other. It is possible for a society to be in bondage to its government, with the will of the people entirely ignored. However, the government analyst, who is taking the viewpoint of society, will make exactly the same statements as the city analyst, but the line he draws will enclose the whole society of the nation. Thus, the costs and benefits noticed by the city analyst may change and even disappear when viewed from the national position.

HOW VIEWPOINT AFFECTS CASH FLOWS—TRANSFER PAYMENTS

Transfer payments in a society are simple exchanges of money that involve no use of resources or changes in production. Money is a medium of exchange, and that alone. It is not the resources themselves—land, labor, and capital—but simply a measure of them. Most people find it difficult to understand how this can be, so used are they to thinking in terms of money as a good in itself. It is what money will buy that is the good or service.

The transfer payments in our example will be neither costs nor benefits. Now some of the costs and benefits of the proposed project will not cross the extended line encompassing the whole society, but will remain entirely within it. Other consequences will cross both lines and will remain costs and benefits from either viewpoint. A glance at Fig. 17.1 clarifies this important notion. Arrow 1 is a mass transit subsidy by the national government to the city government. From the city government's point of view, it is a benefit; but from the national government's viewpoint, it is a transfer of funds within the society and does not enter into the analysis. Arrow 2 represents an increase in the welfare of the citizens of the city because the pollution level will be reduced by this project. It is not just an exchange of funds from the national society to the local society; it is a true increase in the joy that the people of the community take in life. Thus it is an increase in the total well-being of the national society—if the well-being of society is thought of as a vast sum to which all the satisfactions of each individual member of the populace are added. Arrow 3 represents increased taxes that will be paid by the citizens of the city to the national government as a result of the project, and they are thus transfer payments in the national context. Arrow 4 represents the labor of an electrician who will be paid by the city and whose labor is thus lost to any other use in the society. It is a true cost in social resources.

Evidently, favorable and adverse consequences of courses of action appear and disappear depending on whose viewpoint the analyst takes. It is therefore important that analysts have a clear idea of exactly where they stand in the economy, for their position determines the costs and benefits they will consider—not only the amount, but even whether these costs and benefits exist. For example, an agricultural dam is first observed from the government's point of view and then from the point of view of a private company. The result of the switch in viewpoint is that taxes are included as items to be considered under the private viewpoint, items completely absent from the government's one. The result of including depreciation and capital loss may be an entirely different answer to the project question.

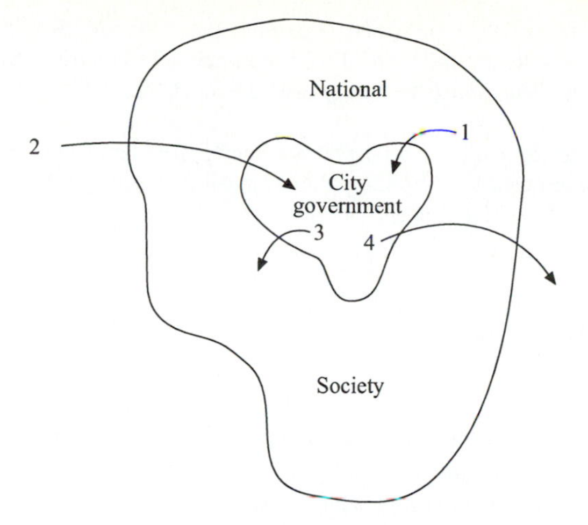

FIGURE 17.1 Effect of viewpoint.

It can be argued that analysts who take the viewpoint of their own organization are being selfish, if not actually antisocial. The analyst should, it is said, take consequences to everyone into account as much as possible. Unfortunately, this attitude would probably be considered overly altruistic by those for whom the analyst is working, for "charity begins at home" in most institutions in both the private and public sectors.

The importance of the effect of viewpoint cannot be overestimated. It can change the acceptance of a project to rejection, or vice versa. A proper appreciation of viewpoint can give a firm, logical foundation to the project analysis. What follows from this is a defensible position that the most skeptical critic will find difficult to attack.

■ **Example 17.1.** In a large metropolitan area, a major bridge was still subject to tolls. The board of directors of the Metropolitan Transportation Authority of the state was considering whether to remove the toll.

Private cars had to pay \$0.75 to cross the bridge east to west and nothing to return. Trucks and buses paid \$3 under the same conditions. The total revenue per year was \$40 million. Costs per year, including maintenance of the bridge at \$7 million and the cost of collecting tolls at \$3 million, were \$10 million.

The average waiting time at the toll booths was 3 minutes per vehicle. Vehicle trips were 100,000 automobiles and 10,000 trucks and buses daily.

A firm of consulting engineers and economists was contracted to investigate the situation. It determined that the wait at the toll booths resulted in costs to the bridge users, both private and business, of \$83.2 million per year.

What should the board of directors decide?

Solution. The decision depends on the viewpoint taken. If only the Metropolitan Transportation Authority's revenues and costs are considered, the tolls should be continued because a profit of 40 − 10 = \$30 million is being made annually. Presumably, it is being returned to the state's general funds.

From the point of view of society, however, the costs are 10 + 83.2 = $93.2 million per year, if tolls are collected. If tolls are removed, the cost is $7 million for bridge maintenance. The savings to society is 93.2 − 7 = $86.2 million. Tolls should be removed.

What about the $40 million in tolls? It is a transfer payment from the bridge users to the state as represented by the Metropolitan Transit Authority. It does not enter the matter if society's viewpoint is taken. ■

PARETO AND KALDOR

If there is any branch of economics on which engineering economy may be said to be based, it is welfare economics. (This has nothing to do with the so-called welfare state.) Welfare economics investigates the effect, in the sense of benefits and costs of policy recommendations which deal with an allocation of resources. Two names, among many others, are associated with this branch of economics. *Pareto* advanced his criterion for judging the worthwhileness of a policy or a project: *Any change that makes some people better off and hurts no one is justifiable*. The *Kaldor* criterion states that a project is acceptable *if the total sum of individual benefits exceeds the total sum of individual costs, providing that the government makes up the losses* of individuals whose personal balance shows the costs to exceed the benefits. Both these criteria may be applied to projects in the public or private sector.

The reader should also recall or reread the discussion in Chap. 1 on opportunity cost. *Opportunity cost* is the loss in benefits that comes about from choosing one mutually exclusive alternative rather than another. The notion of opportunity cost underlies everything said here.

PRIVATE SECTOR VIEWPOINT

The viewpoint of the private sector presents few problems. The line is drawn around the business itself. Benefits are consequences of a course of action that are favorable to the company in some way. Costs are unfavorable consequences. The owners of the business are presumed to set the criteria for judging what consequences are to be considered favorable and what unfavorable.

The last statement is open to some doubt. What if the owners of the business are thousands of scattered stockholders and the actual decision-making power is in the hands of those managing the business? Can it always be presumed that these two groups have the same view of what consequences are to be considered favorable and what unfavorable? Take the example of the decision to purchase a company airplane or a resort house for the use of the executives of the company. The management may well consider the benefits of both these conveniences to exceed the costs, whereas the stockholders may not, because the stockholders' standards of what is favorable and unfavorable are different. The management may be concerned with the prestige and image of the company, whereas the stockholder is usually concerned solely with profit.

The idea that all companies are run to make a profit for their owners is oversimplified. Other goals exist, such as expansion of the business and accumulation of power, or preservation of the salaries of the management. Many a company, especially in regulated industry, tots up a loss year after year while its executives continue to be paid large salaries.

In big companies often a smaller unit than the entire company is taken as having an institutional position. One department—the traffic department, for instance—may make decisions on investment purely on the basis of its own budget and its own well-being. It will probably come as no surprise to the reader that fiefdoms tend to be set up in large companies, and in them the rule is: *We will look after ourselves and our budget. What happens to the rest of the company—that's the worry of top management.* In such cases, interdepartmental charges become costs to the department, and interdepartmental credits become benefits instead of what they really are—a transfer of funds from one pocket of the company's trousers to another, and nothing more. Use of such figures in economic analysis causes much confusion. The study of company operations and investments from the viewpoint of the entire company instead of individual departments is one aspect of systems analysis. Among other advantages, such studies avoid the problem of interdepartmental transfer payments and draw the analyst's line around the company as a whole.

It hardly needs saying that taxes for a company are real costs and should be treated as such. The social effect of so doing will be touched on in the next section.

■ **Example 17.2.** The operations division of Intercontinental Steamship Company receives an annual budget. It is considering the acquisition of a container cargo ship, now in the possession of a private owner. The cost to Intercontinental for this ship will be $25 million. Annual operational costs will be $1,900,000. In addition, the purchase of this ship will require certain increases in the computer use of the company, at an increased cost of $450,000. This will not be charged to the operations division. Pierre Calvin, head of operations division, believes it should be ignored, therefore. Judy Wesley, the general manager, disagrees. Comment on these positions.

Solution. Calvin is taking the narrow view of his own department. The decision to purchase the ship cannot be properly evaluated unless $450,000 for increased computer use is included in the calculations.

Wesley in her disagreement should advance the system point of view. That is, the consequences to the whole company should be taken into account, not just those that the department will experience. ■

PUBLIC SECTOR VIEWPOINT

In the United States, the economic analysis of public projects was given a viewpoint by Congress in the Flood Control Act of 1936. In that act it was stated that "... if the benefits to whomsoever they may accrue are in excess of the estimated costs ...," the federal government should participate in such public works. The key phrase on viewpoint is *to whomsoever they may accrue*, which implies that the viewpoint of the whole society of the United States is to be taken in such projects. This attitude was extended to the economic analysis of many other types of projects, including highways, air pollution control, forest conservation, and so on. It is a reflection of the ideas of Pareto and Kaldor.

Large lending agencies such as the World Bank, Interamerican Development Bank, and others take the societal point of view when dealing with project analysis. It is the society as a whole that participates in economic development. The costs and benefits to the total society are what must be taken into account, therefore.

It is possible for an analyst to take the societal viewpoint but to restrict the area of society observed. For example, an analyst working for a state highway transportation department may well take the viewpoint of the state in consideration of costs and benefits. For the analyst, it is perfectly possible to include as a benefit the 90 percent subsidy that the federal government supplies on highways belonging to the interstate system. The analysis of a particular stretch of proposed interstate highway results in a high benefit/cost ratio in the documents approved by the analyst's superiors. It will show a much lower benefit/cost ratio in the documents given to the federal authorities, whose viewpoint will be that of the country as a whole and for whom the total cost of the project will be exactly that. Subsidies ideally should be treated as transfer payments, no matter what unit of government the analyst takes as a viewpoint. Practically, as in the mass transit subsidy of Fig. 17.1, it is not done, and subsidies are treated as benefits to the entity receiving them.

What are we to say of world viewpoints, if we accept the rule that subsidies are to be treated as transfer payments, no matter what unit of government the analyst takes as a viewpoint? Should an analyst for the United Nations take the world viewpoint in project analysis within individual countries? International loans and any other transfers of funds would be considered transfer payments and thus not enter the analysis. Only costs that represent actual use of world resources or real decrease in well-being for the persons affected by the project, and benefits that represent real increases in well-being should enter into the analysis. Needless to say, it would be difficult to convince an analyst whose duties demand absolute loyalty to the nation that a subsidy from an international organization should not be taken as a direct benefit of the project being analyzed.

Taxes are analogous to subsidies. Although local governments do not pay taxes as institutional entities, still the citizens that they govern do. If the viewpoint of the people of a community—and not just the agency that governs it—is taken, then taxes edge into the analysis. In the example of Fig. 17.1, arrow 3 represents increased taxes paid by the citizens of the city to the national government. Should the increase in taxes be considered a cost of the project? If the viewpoint of the city government is taken, the answer is no. If the viewpoint of people of the community is taken, the answer is yes. Whether or not taxes and subsidies are included in project analysis depends on the viewpoint taken.

A final point on taxes: Taxes for a private company are real costs and for the government agency receiving them are real benefits. From society's viewpoint they are only transfer payments. Thus a project that, from a private viewpoint, is unworthy may well be justifiable from a societal viewpoint, or vice versa. Does not the existence of taxes cause misallocation of resources from a societal viewpoint? Certainly, but because taxes are unlikely to disappear—like death, they are certain, as Benjamin Franklin noted—the private sector will continue to base decisions on them. Thus what may be good or bad for the firm will not necessarily be good or bad for the society.

COST OF UNEMPLOYED RESOURCES

An example best explains the much misunderstood point of unused resources, whether of land, labor, or capital.

■ **Example 17.3.** An oil company is operating in the Amazon jungle just east of the Andes. After the exploration has ended and the first well has been brought in, the company decides to build a road to supply the oil field. Exploration teams, equipment for survey of the site, and finally the drilling rig, casing, and pipe for the first well have been brought by air. Now many hundreds of tons of supplies will be needed in the field, and these will be trucked in across the Andes. The crude oil that the field produces will be trucked out in tankers. The company decides to do a cost study along with the engineering design of the road. The benefits are so obvious that no benefits study will be undertaken.

Among the important costs is that of local labor. If we convert the cost of the labor per hour from the currency of the country to dollars, a laborer will receive $1.00 per hour. This cost is correctly entered as the unit cost of local labor by the company's economist. It is multiplied by the number of laborers needed, by the number of hours needed to complete the job, and this total enters the cost breakdown for the entire project under the heading *local labor*. The company's economist has, of course, taken the company's viewpoint.

The project proposal is taken to the government of the country for approval. The road, even though constructed with private funds, will enter the national system of roads, bridges, and highways. It will be numbered as a national highway. When the company leaves the country, as it may eventually, the road will be maintained by the government through the public works department. An engineer in the department of public works is asked to analyze the costs of the project from the point of view of the country. In his consideration of the cost items, he encounters local labor at $1.00 per hour. He notes, however, that the laborers will be drawn from tribes of Indians in the area who previously have been unconnected with the national economy. They have been unemployed in the sense that they produced nothing but their own food, and just enough of that for subsistence. The engineer asks himself, What cost to the country should I put down for these laborers? He concludes that the cost of the labor should be the value of the food they produced as hunters.

He reasons thus: The laborers produced nothing except food which they ate when they worked as hunters. As laborers, they will produce, along with other inputs, a road.

The opportunity cost of their labor to the country is therefore $0.02 per hour, the cost of the food they produced as hunters. It costs the country nothing for them to give up their jobs as hunters except the food they produced for their own subsistence.

He writes two statements to clarify for himself the effect of shifting this labor from hunting to road building:

- Contribution to national income as hunters per hour = $0.02
- Contribution to national income as road builders per hour = $1.00.

It is now clear that the nation loses the opportunity for food when these hunters shift from hunting to road building. That is, the nation loses the opportunity to employ them as hunters once they become road builders. The cost—the opportunity cost—to the nation is therefore $0.02 per hour, and this is the cost that should be placed after the item in the cost breakdown entitled *local labor*. Notice the immense difference in labor costs that viewpoint has made for the very same workers on the very same project. Yet both costs are perfectly correct: $1.00 per hour from the private viewpoint and $0.02 per hour from the public viewpoint. ■

What has been said about workers' wages is true also of other resources. The private viewpoint is straightforward, but the public sector analyst must ask, What is this resource costing the nation in its next best use? What is the opportunity cost of this resource?

The concept presented in this example can be summarized: Where unemployment of resources exists—land, labor, or capital—the price system cannot be depended upon to provide correct social costs and benefits. Thus, under unemployment, adjustments that depend on opportunity costs must be made if a proper allocation of resources is to result.

SHADOW PRICES

Prices or costs assigned to inputs for the purpose of evaluating a project are called *shadow prices*. They are used when market prices do not represent the correct social cost of resources, particularly in the case of unemployment.

■ **Example 17.4.** Imagine a road project in a developing country which will open up to colonization and agricultural production, land that was previously inaccessible. In the economic analysis of the project, the prices of farm labor, land, and seed are assigned zero values. Labor is given a zero value because the colonizers are expected to be drawn from a pool of unemployed persons. Thus the economy is not forgoing any production—i.e., any benefit—by transferring this labor from its unproductive state to a remote farm. The same can be said of the inaccessible land, because it has no market value prior to the building of the road. The seed is worth something in the market, but its cost is negligible.

Are all these items shadow priced? The labor price is definitely a shadow price because its zero value is less than the market price of labor in the country. The land price is not a shadow price because it equals the market price of zero. The market price of the land reflects its productivity at the moment. The price of the seed is a shadow price, even though the reasons for assigning a zero cost do not rest on the opportunity cost doctrine. ■

The reader may wonder at the use of the words *price* and *cost* synonymously. Costs are associated with supply—and price with demand—in economic theory. Cost to the seller is price to the buyer. The number in monetary terms is the same.

In Chap. 1, in the sections entitled "Externalities" and "Social Benefits and Costs," both these concepts were defined and briefly discussed. The reader should review those sections before proceeding. In the economy of pure competition, shadow prices are unnecessary because market prices represent the opportunity cost to the society for all goods and services. In the idealized purely competitive economy, market prices represent social costs because there are no such things as externalities, transfer payments, and unemployment. In the real world, correct economic analysis from a societal point of view requires that opportunity costs be used, which at times may differ markedly from market price. Labor, foreign exchange that represents the costs of imported goods and services, and fuel are examples of inputs to a project that may have differing opportunity costs and market prices. Prices within a country may also be controlled—the case of foreign exchange is a good example—and thus cannot represent free-market prices. Subsidies of all kinds may exist. Taxes, in the case of private sector projects judged from a social viewpoint, may also have to be excluded.

Economic analysis from the viewpoint of society also requires that externalities be priced. Externalities often do not appear in the market except in the most indirect way and far in the future. Consider the case of air pollution adjacent to a cement plant. The cost of the air pollution is not reflected in the market price of the cement the plant produces—certainly not in any direct way. It appears, for example, in the abnormal quantity of soap that householders living near the plant must buy in order to keep their clothing clean. It shows up years later in the medical costs of those who suffer from lung disease brought on by living near the plant. How can such costs be priced?

The answer to this question is an unsatisfactory one, it must be admitted from the start. How can we assign an opportunity cost, and thus a social cost, to the abnormal soap use, or the illness of a machinist, who works across the city but lives next to the cement plant and who will come down with severe emphysema 15 years from now? How are we, as analysts, to assign an extra cost to the plant because of these and myriad other external costs? Theoretically it could be done. Studies of extra soap use around cement plants could be undertaken by researchers. The incidence of emphysema among those who live near cement plants could also be discovered. But what of all the other externalities—the ugliness of the gray layer of dust over everything, the depressed spirits of the plant's neighbors, the cost to other folks who only drive through. The list is long. The task of assigning social costs and benefits seems impossible.

Indeed, it is difficult. What actually happens is that those who undertake such studies—consulting engineers and economists—become familiar with the more important externalities and their prices through familiarity with, for example, cement plants. Or they may be specialists in steel plants, or lead mines, or ports, or railroads. Assignments of social costs and benefits are made by persons of long experience. They may be—and are—argued about. But they are certainly recognized, and they do enter the project analyses. It is only common sense to recognize externalities.

SUMMARY

Costs and benefits can change and even disappear, depending on the viewpoint that the analyst takes when evaluating the project. Both private and public projects can be judged from any institutional viewpoint.

Transfer payments in a society are simple exchanges of money that involve no use of resources or changes in production. They may be included or excluded from the costs and benefits of a project depending on the viewpoint taken.

The Pareto and Kaldor criteria form the basis of engineering economic analysis.

The private sector viewpoint is straightforward. Some complications may occur if the owners of the business are not its managers. The public sector viewpoint may be that of government agencies or of society itself. The viewpoint of society may be taken in relation to either private or public enterprise. Subsidies and taxes require that the viewpoint of the analyst be defined with care for their correct treatment.

Unemployed resources—land, labor, or capital—cause the price system to become undependable; it no longer reflects social costs. By social costs are meant opportunity costs in the use of resources. Thus adjustments must be made to costs, if a proper allocation of resources is to result.

Shadow prices are assigned to inputs in the cases where prices in use do not represent the social cost of resources, local or imported. Externalities must also be priced in order to be included in an analysis.

PROBLEMS

17.1. In Example 17.1, page 435, $40 million in revenues is being returned to the state treasury. This, the argument goes, is equivalent to its being returned to the people of the state. Should it not, therefore, be counted as a social benefit?

17.2. The transportation agency of the state of Utah is considering acquisition of right-of-way for a new highway. Only one right-of-way will be purchased from private owners, but it is certain that the purchase will be made. The three candidates would cost the agency as follows:

Location	Cost ($000,000)
1	42
2	54
3	61

The land in question is now in full tree-growing production. The present value of the forest products that would be lost to the economy because of the right-of-way is estimated as follows for each location:

Location	Cost ($000,000)
1	50
2	45
3	47

Which location should be chosen on the basis of social cost?

17.3. In a Latin American country, the major beer producers introduced the throwaway bottle, copying a marketing innovation in the United States a few years previously. (Until the disposable bottle came along, bottled drinks required a deposit which the purchaser received when the empty bottles were returned.) The cost of the new disposable bottle was passed on to the consumer. The decision was a profitable one for the beer companies.

Discuss the costs and benefits, both commensurable and irreducible, associated with this decision from

(*a*) The point of view of the beer companies
(*b*) The point of view of the society of the country

17.4. In January 1996, a medical doctor and her husband are killed in a single-car accident on icy interstate highway 35 near San Antonio, Texas.

(*a*) What costs will appear on the books of public and private firms and agencies as a result of this accident? Itemize and estimate the costs as well as you can.
(*b*) What are the social costs of the accident in addition to the ones cited in (*a*)? How might these be determined?

17.5. The state of California received a subsidy from the federal government for the construction of interstate highway 80 between Sacramento and the state line to the east. This *subsidy* amounted to 90 percent of the construction cost. One of the major costs of the highway was the *right-of-way* near Sacramento which removed valuable farmland from production. A major benefit of the highway was the *reduction in accidents* because the new highway provided much greater vehicle safety than the highway that it replaced.

(*a*) How would each of the above italicized items be treated—cost, benefit, or transfer payment—from the viewpoint of an economic analyst working for the state of California who is required by superiors to take strictly the state's point of view?
(*b*) How would each of the above items be treated from the viewpoint of an economic analyst working for the Brookings Institution and observing the society of the United States as a whole?

17.6. In Example 17.2, page 437, Calvin argues that the increase in computer use will not be solely because of the acquisition of the new ship. He wants a more accurate breakdown of the computer charges. In the next moment, he maintains that because the computer charges are borne by the entire company, they should not be charged against his new ship. Comment.

17.7. A subcontractor to Idaho Construction Company presented a bid for earth movement for $600,000. This bid was accepted as a reasonable amount by Idaho's engineers because it represented what they thought earth movement should cost, using the conventional method of moving earth with tractors and scrapers. The method actually used was hydraulic mining, and the cost was only $400,000.

(*a*) From the viewpoint of the subcontractor, what was the profit on the job?

(*b*) From the viewpoint of Idaho Construction Company, what was the cost of this particular earth movement?

(*c*) From the viewpoint of society, what was the cost of this earth movement, provided the hydraulic mining operation costs represented true social costs?

(*d*) Will the true social costs of the hydraulic mining operation ever be recorded in the accounts of the U.S. Army Corps of Engineers (the job owner), and therefore the U.S. government? Discuss the implications of your answer.

17.8. In Example 17.3, page 439, Andean Oil Company, a foreign-owned enterprise, decides to institute a training program for the more intelligent Indians in its employ. Spanish and arithmetic will be taught, at a cost of $5,000 per employee for the whole program. The company wishes to deduct the cost from its taxable income, but the absence of the deduction will not cause the program to be dropped.

How will a government economist view this request, providing she takes the societal point of view?

17.9. New York City has been plagued for some years now with graffiti. This is especially in evidence in the New York subway system, where it covers the interior and exterior of the cars as well as the stations. The problem seems insolvable. However, recently it has been suggested that the sale of spray paint cans to persons under age 18 be banned.

Comment on the possible effects of such a policy from the viewpoint of:

(*a*) The spray paint companies and their distributing network

(*b*) The government of New York City with all its boroughs

(*c*) The society of New York City and all its boroughs

(*d*) The society of the United States

Pay particular attention to the description of benefits and costs under each viewpoint. Include those costs that will appear on the account books of the entity involved, those costs which are measurable in terms of dollars but which will not appear on the books of any company or unit of government, and finally those costs and benefits which are unlikely to appear on anyone's balance sheet.

17.10. Congress is presently considering a program similar to the Civilian Conservation Corps (CCC) of the 1930s. It will enlist young, single men who are unemployed and who may be considered unemployable under present economic conditions for reasons of intelligence, education, social adjustment, and so forth.

The program will cost the government $330 million per year, of which $120 million will be spent in wages and the rest will be spent on food, lodging, clothing, medical and dental services, recreation, and transportation.

Income taxes at 15 percent, or $18 million, will be collected by the government from these laborers.

It is estimated that their work on land reclamation, forest roads in national parks and national forests, flood control, and similar works will result in benefits to society of $270 million in the form of roads built, trees planted, and so forth. A complicated allocation calculation has been performed to separate the benefits attributable to labor alone. It is thought to be accurate, however.

If the government's point of view is taken, is this program economically viable on the basis of only the costs and benefits cited above?

If society's point of view is adopted, should the program be recommended on the basis of the above benefits and costs?

Show numerical reasons to support your answers.

17.11. In Example 17.3, page 439, the opportunity cost of Indian labor was recalculated at $0.05 per hour, and they received a raise to $1.10 per hour as day laborers.

An economist calculated their social cost per hour at $1.10 less $0.05, or $1.05. Comment on this calculation.

17.12. Leroy Hardcastle is out of work. The government of the country where he resides pays him the equivalent of $300 per month to help support him and his family. To this sum he finds he must add another $200 to cover the bare necessities of existence, meaning food, rent, and a minimum of clothing.

(*a*) If the price system in this country is working perfectly, i.e., all prices represent the social costs of resources for the production of goods and services, what is the social cost to the society of the country of the Hardcastle family? What is the social benefit to the country in this situation?

(*b*) A senator of the national assembly of the country has introduced a bill which, in essence, proposes to subsidize private businesses to the extent that they hire persons who have been unemployed for over 1 year. The subsidy would be the amount that the government has been paying in support payments. This measure, it is hoped, will increase employment.

Hardcastle's employer under this bill would receive $300 per month in government subsidy. Imagine that Hardcastle finds a job which will pay him $900 per month as a house painter. Employed, Hardcastle will provide what benefits to the society? What will the Hardcastle family cost the society? Assume that Hardcastle, a frugal man, does not increase his expenditures for himself or his family.

(*c*) Taking the viewpoint of the government only, not the society, have government costs changed as a result of Hardcastle's employment?

(*d*) From the viewpoint of the painting contractor who hired Hardcastle, have the contractor's costs changed as a result of the hiring? Assume that Hardcastle received the standard wage rate for his services.

(*e*) This bill of the senator's sounds like a perfect example of the Pareto optimum. Do you think it will work in practice? Why or why not?

17.13. In Example 17.4, page 440, seed is now assigned its market value. Is this also a shadow price?

If an agricultural experiment station is established in the zone of influence of the new road, how should its personnel be costed? Base your answer on *social* cost.

17.14. The Ministry of Public Works of a developing country is considering the introduction of a labor-intensive program in agricultural irrigation. The program will cost 200 million dinars. Maintenance costs per year will be 25 million dinars.

The costs of the project are divided as follows:

	First cost	Annual maintenance cost (000,000)
Equipment	10	2
Labor	190	23

Labor is 50 percent unemployed in the area where workers will be recruited. This situation will continue in the future.

The equipment will be imported into the country, using some of the government's stock of foreign exchange in U.S. dollars. The government-controlled dollar/dinar rate is 1 $U.S. to 40 dinars. The market rate in London is actually 1 $U.S. to 240 dinars. These rates are expected to continue indefinitely. The first cost and maintenance amounts for equipment given above are computed at the government-controlled exchange rate.

The project has an infinite life, if properly maintained. The social discount rate is 18 percent.

(*a*) What is the annual cost of this project as computed by the Ministry of Public Works?

(*b*) What is the annual social cost to the country of this project?

17.15. In San Marcos, a developing country, a new rural roads project in the public sector is under consideration. The Ministry of Public Works estimates the following first costs:

	First cost (000,000)
Labor	1,000
Mechanical equipment	2,500
Land	500
	4,000

Labor will be drawn from an area where there is 75 percent unemployment. The estimated cost of 1,000 million is based on 200,000 pesos for each of 5,000 workers.

Mechanical equipment will be bought overseas, using foreign exchange at 2,000 pesos per dollar, the international market rate. The government-controlled rate within San Marcos is 1,000 pesos per dollar. It is on this controlled rate that the 2,500 million cost has been computed.

Land that is to be used by the road right-of-way is presently producing 700 million pesos per year in crops. It will be acquired by eminent domain for the estimated 500 million.

Make adjustments to the Ministry of Public Works' estimate, and calculate the total social cost of the project.

17.16. In a small Latin American country, the following costs of a Hermanos Gomez cement plant have been estimated:

	Pesos (000,000)
Materials	1,000
Labor	
Skilled	400
Unskilled	200
National property taxes	50
Project administration	100

The national government intends to subsidize the project at 300 million pesos. The subsidy occurs because the government is trying to replace imports by in-country production. The property taxes will be paid to the national government. The cement plant is totally owned by citizens of the country.

In the area where the plant is to be built, the rate of unemployment of unskilled labor stands at 100 percent. Skilled labor is fully employed throughout the country.

(*a*) From the point of view of Hermanos Gomez analysts, what is the net cost of the plant to the company?

(*b*) From the point of view of the government of the country, what is the cost to it of this project?

(*c*) From the viewpoint of a World Bank analyst, taking the point of view of the entire society of the country concerned, what is the net social cost of the new plant?

CHAPTER 18

SENSITIVITY ANALYSIS

In this chapter we shall cover payback and break-even point. Then we will go on to questions of more general interest to the decision maker: How much profit do I need to make to reach a particular rate of return? Do I really need to make a study of the probable economic life of this project, or is this a relatively unimportant question? If I change the depreciation method from MACRS GDS to ADS, how much difference will that make in the net present value of the project? What is the variable in my analysis that must be studied with the greatest care, that presents the greatest danger to the success of the project?

Questions such as these are vital. The methods used to answer them are one of the most interesting and challenging parts of engineering economy, and incidentally an area of the subject where the electronic computer can be of great use.

PAYBACK

Payback is included in a chapter on sensitivity analysis more because it provides an opportunity to criticize the method than because it belongs here. It is as widely used as it is fundamentally unsound, as will soon be seen. Its wide use demands that it be explained here. You may be forced, at sometime in your career, to use the method, but you must guard your organization against costly errors that may come about because of its use. In other words, you should employ correct methods to make sure of the result. As the name implies, it measures the worth of an investment by the time it takes to pay it back from the net profits that the investment generates. The shorter the payback period, the better the investment.

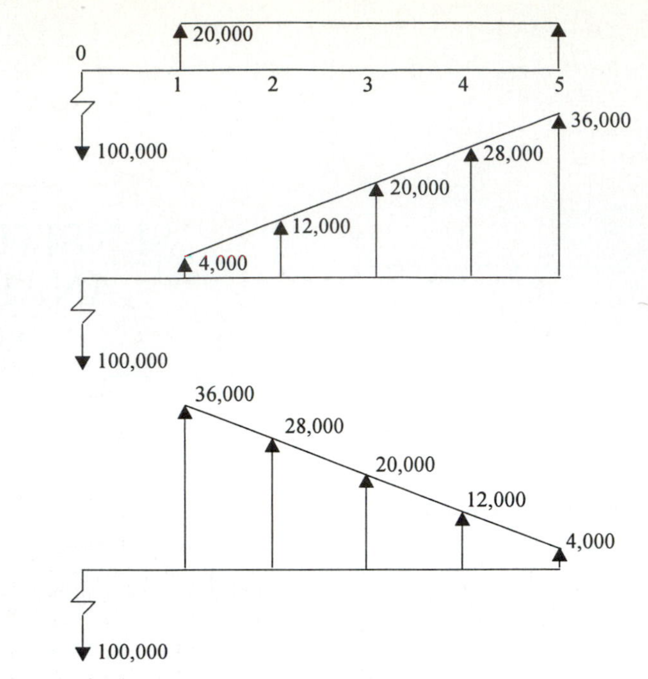

FIGURE 18.1 Payback patterns.

For example, an investment of $100,000 that generates profits of $20,000 annually has a payback of

$$\text{Payback} = \frac{100{,}000}{20{,}000} = 5 \text{ years}$$

Payback is an unreliable method because it makes no allowance for the time value of money. The payback method gives the same result—5 years—for each of the cash flows shown in Fig. 18.1. It requires only common sense to choose pattern 3 over the other two because it provides more money sooner. For those who have followed this text, it will also be evident that at any opportunity cost of capital above 0 percent, by any method seen so far, investment 3 will prove to be better than either of the other two.

BREAK-EVEN ANALYSIS

Break-even analysis is a limited form of sensitivity analysis. It restricts itself to the question, At what point in regard to such and such a variable will I just break even? That is, at what point will my revenues just equal my costs? The point will be that of zero profit and zero loss. In management science, as it applies to production, the variable is Q—the number of units of product. Figure 18.2 shows the break-even point.

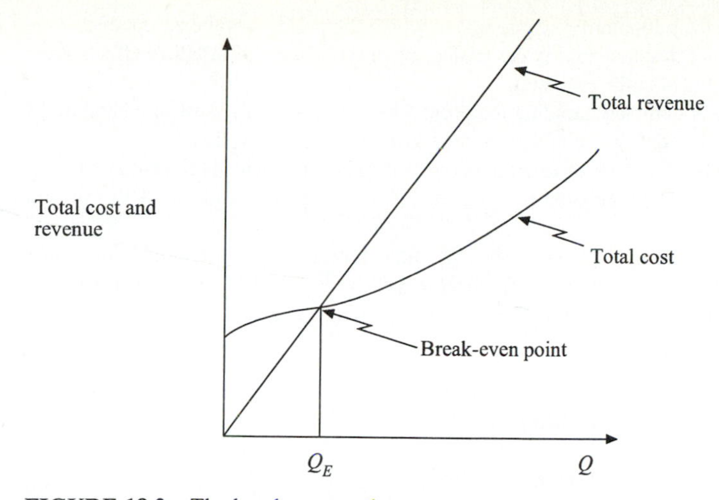

FIGURE 18.2 The break-even point.

Moreover, the analyst may use *discounted* revenues and *discounted* costs when posing the break-even question, thus refining the method somewhat.

In engineering economy texts, break-even analysis has taken on the meaning of that value of a variable at which the decision maker will be indifferent between two alternatives. For example, the variable might be the discount rate.

With the advent of relatively cheap, small, easily programmed computers, "Where will we break even?" has assumed its correct place as just one of the many questions that may be asked by an investor confronted with a venture. The multivariable sensitivity analysis section at the end of this chapter deals with the generalized problem of sensitivity analysis. In the example of that section, one of the questions asked is, At what property appreciation rate will we just break even—i.e., our rate of return on the venture will be exactly zero?

SENSITIVITY—THE NARROW DEFINITION

Under the narrow definition, the *sensitivity* of the decision to one of the input variables involved means that at some value of that variable the decision will reverse. A simple example shows what this signifies.

■ **Example 18.1.** A nuclear power plant is projected to cost $1 billion when it is completed. These facts and estimates are given:

Life	30 years
Benefits	$300 million per year
Annual operating cost	$150 million per year
Salvage value	0

(*a*) What is the maximum discount rate at which the project will be acceptable? That is, at what discount rate is the project sensitive? Tax and inflation affects have been included in the estimates.

(*b*) The power company building the project has an opportunity cost of capital of 15 percent. The data given in part (*a*) lead to a rejection of the plant.

At what amount of benefits per year is the project sensitive? That is, to what figure would the benefits per year need to be raised in order that the project qualify?

(*c*) The power company now wishes to know how much construction cost would have to be reduced in order for the project to qualify. The other original conditions remain the same.

Solution

(*a*) The maximum discount rate at which the project is acceptable is the internal rate of return of the project cash flow.

$$1 \times 10^9 = \left(0.3 \times 10^9 - 0.15 \times 10^9\right)\left(P/A, i^*, 30\right)$$

$$i^* = 14.76\%$$

(*b*) Let B stand for the yearly benefits at which the project is sensitive. Then

$$1 \times 10^9 = \left(B \times 10^9 - 0.15 \times 10^9\right)\left(P/A, 15, 30\right)$$

$$\frac{1 \times 10^9}{\left(P/A, 15, 30\right)} = \left(B - 0.15\right) \times 10^9$$

$$B = \frac{1}{\left(P/A, 15, 30\right)} + 0.15$$

$$= \frac{1}{6.566} + 0.15$$

$$= 0.302300 = \$302{,}300{,}000$$

Electricity rates would have to be raised to yield another \$2.3 million.

(*c*) Let the construction cost be designated C. Then

$$C = \left(0.3 \times 10^9 - 0.15 \times 10^9\right)\left(P/A, 15, 30\right)$$

$$= \left(0.15 \times 10^9\right)\left(6.566\right)$$

$$= \$984{,}900{,}000$$

The cost must be reduced by $1 \times 10^9 - 0.9849 \times 10^9$, or \$15.1 million. ■

SENSITIVITY—THE BROAD DEFINITION

Sensitivity may also be measured by the change that results in an output variable—e.g., the internal rate of return of the project—from changing one or more input variables. The change may or may not affect the decision.

■ **Example 18.2.** In Example 18.1, the effect on the IROR of shortening or lengthening the economic life of the plant may be of considerable interest to the nuclear power plant developers. What is the IROR associated with a life of 25 years? Of 20 years? Of 35 years?

Solution. For a life of 30 years, we already know the IROR—14.76 percent. From Example 18.1*a*,

$$(P/A, i^*, N) = \frac{1 \times 10^9}{0.15 \times 10^9}$$
$$= 6.6667$$

Substituting for N in this factor gives the following i^* values:

N	i^* (percent)
20	13.89
25	14.49
30	14.76
35	14.88

A range of 15 years in the life of the project changes the IROR by only 1 percent. The economic life is not a key variable, therefore. ■

MULTIVARIABLE SENSITIVITY ANALYSIS

For years sensitivity analysis has been recognized as one of the areas of economic analysis where the creation of mathematical equations to represent situations is a fruitful exercise. Such situations can be represented in a computer program that anyone with access to a personal computer may use. The comparative ease of creation of personal computer programs removes the necessity of a mathematical expression being written. The program itself is such an expression.

To demonstrate the enormous usefulness of sensitivity analysis in its broadest sense, an example will be used. It is a seemingly simple one, whose complications will be made apparent. It then will be used to explain how sensitivity analysis can clarify the most subtle issues in project analysis.

■ **Example 18.3.** A commercial building is being considered as a real estate venture by an investor. The venture will be analyzed under the following conditions, thought to be most likely:

Cost	\$740,000
Taxes on sale, legal fees, escrow fees	\$50,000
Mortgage	9.5 percent nominal, compounded monthly for 30 years
Down payment	10 percent
Rent	\$4,750 per month
Property taxes	\$12,000 per year
Fire insurance	\$1,240 per year
Repairs and maintenance	\$6,000 per year
Rental agent's commission at 9 percent	\$5,130 per year
Vacancy rate	5 percent
Investment period	8 years
Opportunity cost of capital	15 percent

The most likely estimate of what will happen if the property is purchased is as follows:

The property will appreciate in value at 10 percent per year based on its original cost at year 0, as a result of inflation.

Operating costs will rise, also as a result of inflation, by 5 percent per year based on the first-year estimates. Included under operating costs are

Property taxes
Fire insurance
Repairs and maintenance
Agent's commission

Income, i.e., rents, will rise by 5 percent per year, based on the first-year estimates.

A 32.5 percent tax bracket, made up of 25 percent federal tax, 5 percent state tax, and 2.5 percent local tax, will be assumed.

A real estate commission on the sale of the property of only 1 percent will be paid, because the investor expects to sell the property himself.

Fifty percent of capital gains will be taxed as ordinary income.

(*a*) What rate of return, after income tax, will be realized on the venture?

(*b*) How sensitive is the rate of return to the input variables? To which variables is the rate of return most sensitive?

(*c*) Based on parts (*a*) and (*b*), what should the investor do?

Solution

(*a*) Table 18.1 shows the computations for determining the after-tax cash flow for the most likely estimate of what will happen.

Column 1 is the estimated cost of the venture, led by, at year 0, the cash flow of

10 percent down payment	74,000
Settlement fees	50,000
	$124,000

The first-year cost is

Property taxes	12,000
Fire insurance	1,240
Repairs and maintenance	6,000
Rental agent's commission (0.09)(4,750)(12)	5,130
	$24,370

Increased by 5 percent per year, the remainder of the column appears. The auxiliary eighth year is reserved for the capital gains computation.

Column 2 is the revenue per year. The first year's revenue is

$$(4{,}750)(12)(1.00 - 0.05) = \$54{,}150$$

The 0.05 is the 5 percent vacancy rate. Year 2 is computed by

$$(54{,}150)(1.05) = \$56{,}860$$

and so on.

TABLE 18.1

Most likely estimate analysis*

	(1)	(2)	(3)	(4)	(5)	(6)	(7)	(8)	(9)	(10)
			Mortgage					Taxable income	Tax (at −0.325)	ATCF
Year	Cost	Revenue	*A*	*I*	*P*	*R*	Depreciation	(1 + 2 + 4 + 7)	(8) × −0.325	(1 + 2 + 3 + 9)
0	−124,000					666,000			—	−124,000
1	−24,370	+54,150	−70,200	−66,070	−4,130	661,870	−20,000	−56,290	+18,290	−22,130
2	−25,590	+56,860	−70,200	−65,660	−4,540	657,320	−20,000	−54,390	+17,680	−21,250
3	−26,870	+59,700	−70,200	−65,210	−4,990	652,330	−20,000	−52,380	+17,020	−20,350
4	−28,210	+62,690	−70,200	−64,710	−5,490	646,840	−20,000	−50,230	+16,330	−19,390
5	−29,620	+65,820	−70,200	−64,170	−6,030	640,810	−20,000	−47,970	+15,590	−18,410
6	−31,100	+69,110	−70,200	−63,570	−6,630	634,180	−20,000	−45,560	+14,810	−17,380
7	−32,660	+72,570	−70,200	−62,910	−7,290	626,890	−20,000	−43,000	+13,980	−16,310
8	−34,290	+76,190	−70,200	−62,190	−8,010	618,880	−20,000	−40,290	+13,100	−15,200
8		1,586,260							−152,820	+798,700

*The figures in this table will not exactly match Table 18.3 of the book diskette because of rounding differences.

The selling price after 8 years will be

$$740{,}000(F/P, 10, 8) = 740{,}000(2.1436)$$
$$= \$1{,}586{,}260$$

shown at the foot of column 2.

Columns 3, 4, 5, and 6 show the costs related to the mortgage. These costs are included because they are associated with the loan for the particular property under consideration. No other opportunity for the use of these funds exists because they may not be used for any other purpose than the purchase of a specific property. If mortgage payments are not made on time, the property is forfeit. When the property is sold, the loan must be paid off. The property is collateral for the mortgage.

Column 3 is the yearly loan payment, which will be paid monthly. The equivalent yearly payment is

$$i_y = (1 + i_M)^M - 1$$
$$= \left(1 + \frac{0.095}{12}\right)^{12} - 1$$
$$= 9.92\%$$

$$A = (740{,}000 - 74{,}000)(A/P, 9.92, 30)$$
$$= (666{,}000)(0.10541)$$
$$= \$70{,}200 \text{ annually}$$

Column 4 is the interest portion of the payment:

$$(666{,}000)(0.0992) = \$66{,}070$$

The remainder of column 4 and columns 5 and 6 are computed as explained in Chap. 4 in the "Separation of Interest and Principal" section. The bottom line of column 6 is the remaining balance of the loan at the end of the eighth year.

The depreciation, column 7, is computed on the basis of a \$140,000 value for the land and \$600,000 for the building. Only the building may be depreciated. A special ruling of the IRS assigns a life of 30 years and a zero salvage value to the building. The annual depreciation is

$$\frac{600{,}000}{30} = \$20{,}000$$

It is now possible to compute taxable income, column 8. It is the algebraic sum of columns 1, 2, 4, and 7.

Income tax at 32.5 percent is the product of column 8 and −0.325. The result is positive for 8 years, indicating tax relief.

The *after-tax cash flow* (*ATCF*) is obtained by adding columns 1, 2, 3, and 9.

Capital gains are calculated as follows:

Sale price	\$1,586,260
Less sale expense	−15,860
	\$1,570,400

From this the basis of the property must be subtracted. The basis is the acquisition price less the accumulated depreciation:

Acquisition price	740,000
Plus acquisition expenses	+50,000
	790,000
Less accumulated depreciation	−160,000
	630,000
The capital gain is	1,570,400
Less	−630,000
	\$940,400

Only 50 percent of this is taxable:

$$940{,}000 \times 0.50 = \$470{,}200$$

and it is taxed at 32.5 percent:

$$470{,}000 \times 0.325 = \$152{,}820$$

The after-tax cash flow is

Sale price	\$1,586,260
Less sale expense	−15,860
	1,570,400
Less capital gains tax	−152,820
	1,417,580
Less remaining balance	−618,880
	\$798,700

This last figure is shown at the bottom of column 10.

The internal rate of return of the after-tax cash flow is 19 percent. This is greater than the opportunity cost of capital of 15 percent, and therefore the project is acceptable under the most likely conditions.

(*b*) But what will happen to the internal rate of return, and thus the decision, if one or more of the input variables happen to change? And how much of a change will be needed to change the decision? Moreover, which input variables will cause the greatest change? Put another way, which input variables must be examined with great care, and which may be treated cursorily? We now attempt to answer these questions.

Let us begin by varying the appreciation rate from a 10 percent rise to a 5 percent. Holding all other variables constant, how will the IROR change as a result of a 5 percent rise annually over 8 years? Substitute a new sale price in the before-tax cash flow of

$$740{,}000(F/P, 5, 8) = 740{,}000(1.4775)$$
$$= 1{,}093{,}350$$

The same calculation as before is performed, summarized below:

		1,093,350
	Less	−10,930
		1,082,420
	740,000	
Plus	+50,000	
	790,000	
Less	−160,000	
		−630,000
		452,420

$$452{,}420 \times 0.50 \times 0.325 = 73{,}520$$

	1,093,350
Less	−10,930
	1,082,420
Less	−73,520
	1,008,900
Less	−618,880
	390,020

The IROR is 6 percent. This is a severe change from 19 percent, calculated under the most likely conditions. It reverses the decision. Evidently we must proceed in an orderly fashion if we wish to maximize the benefit we will receive from the sensitivity analysis we have begun.

It appears that a computer program that will allow us to substitute whatever input variables we like and observe the resulting change in the IROR would be a great help. Example 18.3 is treated in this way in App. A, page 471, and in the book diskette. You may perform sensitivity analysis yourself on a personal computer, using the book diskette.

Now we can proceed to change input variables and calculate the figure of merit, the IROR. Table 18.2 shows a number of changes in the input variables, selected at random. Figure 18.3, sometimes called a *spiderplot*, reveals the effect on the IROR of changes in the input variables. The greater the slope of the graph

TABLE 18.2
Sensitivity analysis

Analysis number	Item changed	Percentage change from most likely estimate	IROR (percent)
1	None	0	19
2	Appreciation rate = 5 percent	−50	6
3	Appreciation rate = 0 percent	−100	−21
4	Loan period = 25 years	−17	18
5	Loan period = 20 years	−33	18
6	Taxable portion of capital gains = 30 percent	−40	20
7	Taxable portion of capital gains = 40 percent	−20	19
8	Taxable portion of capital gains = 100 percent	+100	15

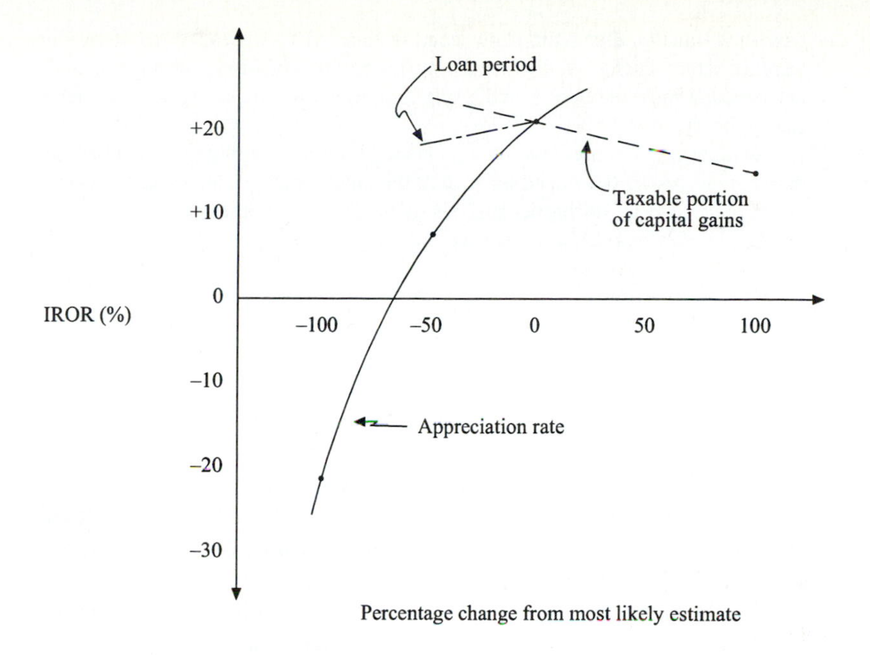

FIGURE 18.3 Sensitivity graph.

of the percentage changes in the input variable plotted against the IROR, the more sensitive is the IROR, and thus the decision, to changes in the input variable. For example, the appreciation rate graph has a slope much greater in absolute value than the taxable portion of capital gains graph. The appreciation rate is a variable that will require much more care in its prediction than the taxable portion of capital gains.

(*c*) The foregoing analysis reveals that the answer to the question as to the investor's course of action is a complicated one. If the investor accepts the most likely estimate as valid, then the investor's opportunity cost of capital at the same risk should be compared to 19 percent. If it is less than 19 percent, the investor should accept the project; if it is greater, the investor should reject the project.

But if the investor does only this, if the investor includes only the opportunity cost of capital in the decision, what is the point of the sensitivity analysis? It has shown that of the three variables considered—appreciation, loan period, and the taxable portion of capital gains—the first of these is the most important by far. Therefore a good deal more time should be spent investigating the appreciation of the property. What has been the history of property values in the same area? What is the city planning for the area? What has happened to similar properties in other cities? These questions and perhaps many others should be asked and answered before a decision is made.

And what about the remaining input variables? We still have not answered the question, How are we to choose which input variables to investigate? So far, the answer is to ask persons with a great deal of experience in the business being investigated. Another answer is, with the aid of a computer, to look into all of them individually. But what about combinations of variables? What if two or three variables change at the same time? Here, a worst-case scenario can be developed, a most likely one, and a best-case scenario. This exercise will further clarify the situation.

Moreover, if more investigation appears to be warranted, a probability distribution may be developed for each of the input variables, the expected values determined, and a stochastic analysis performed. This measure is simply a refinement of our most likely analysis. ■

SUMMARY

Simple payback is a misleading method, to be avoided. Break-even analysis is simply a small portion of general sensitivity analysis.

Under sensitivity analysis, the narrow definition states that at some value of an input variable, the decision on the worth of the project will reverse. The broad definition refers to the change that results in an output variable from changing one of the input variables without regard to whether the decision is reversed.

The heart of the matter, at once the most complicated and the most useful type of sensitivity analysis, is the multivariable problem. Here, too, is the area where the personal computer can serve a most useful function in reducing arduous and lengthy computations. The computer can be made to answer in a few minutes the most searching questions confronting an analyst.

PROBLEMS

18.1. A proposed replacement for the steam generator in a meat-cutting plant will cost \$1,530,000, will last 15 years, and will save \$200,000 per year over the existing generator. What is the payback of the project?

Answer: 7.65 years

No project will be accepted by the management of this plant that has a payback of more than 5.0 years. On the other hand, the opportunity cost of capital is set by the same management at 11 percent. Is there a contradiction in investment policy present here? Explain.

18.2. A plant producing buttons has a total-revenue (TR) curve represented by $TR = 0.12Q$ and a total-cost (TC) curve represented by $TC = 150{,}000 + 0.08Q$, where Q is the quantity produced. At what production Q will the break-even point occur?

Answer: 3.75 million buttons

18.3. The Portland Tennis Club quotes C. Le May \$330 for membership, payable every 2 months with no admittance fee. The Downtown Sports

Club, which offers the same facilities and convenience, has an admittance fee of $200 and $100 per month thereafter. After how many months will the two clubs' fees reach a break-even point? Do not consider the time value of the fees.

18.4. An analysis of a new project estimates that costs will be $60 million per year for 5 years of construction. Net benefits will amount to $20 million annually for 100 years of life.

(*a*) If money is not given a time value, in how many years *after beginning operation* will the project break even, i.e., will its benefits equal its costs?

(*b*) If resources have a social opportunity cost of 8 percent, in how many years *after beginning operation* will the project break even—i.e., will its discounted benefits equal its discounted costs?

(*c*) At the 8 percent discount rate, what will be the present value at the start of construction of the year 100 benefit?

(*d*) At the 8 percent discount rate, at what year of operational life will the present value of that year's benefits contribute 1 percent to the present value of total benefits computed at the start of operations? Is there much point in computing benefits beyond this year for this project?

18.5. A preliminary economic analysis of a proposed dam in a developing country showed that costs during 5 years of construction increased at $200,000 per year, based on a $50 million first-year cost. Net benefits are $137.2 million yearly, starting at the end of the first year of operation. The life of the dam was estimated at 100 years of operation. After how many years of operation will the discounted benefits equal the discounted costs, if a constant-dollar discount rate of 4 percent is used? During the construction and operation period, a 25 percent annual rise in the consumer price index was assumed, based on consumer prices at the beginning of construction. This question requires a break-even point as an answer.

18.6. A public housing project is expected to experience the following cash flow over 40 years:

First cost	$4,000,000
Maintenance	$500,000 per year increasing at $8,000 per year
Major overhaul every 10 years	$700,000
Salvage value at end of 40 years	$200,000

What annual income is required for this project to *break even*, based on discounted cash flow, if the opportunity cost of capital is 7 percent? Do not consider inflation. Because this is a government project, no income tax will be paid.

18.7. In Example 18.1, page 451, the nuclear plant is now estimated to cost \$1.4 billion. Answer the same questions as in the example. All other data remain the same.

18.8. In Example 11.17, page 284, a comparison was made between corrugated-steel pipe (CSP) and reinforced concrete pipe (RCP). CSP was selected. What is the sensitivity of that decision to the life of the CSP alternative?

18.9. A section of a road in Peru has benefits in terms of additions to the gross national product (GNP) of 200 million *soles* for each of the first 5 years of its existence, 300 million *soles* for each of the next 5 years, and 400 million *soles* for each of the last 10 years. Its first cost is 1 billion *soles* (1×10^9).

(*a*) If the social cost of capital is 15 percent, should the project be built? (Use the NPV method.)

(*b*) At what social cost of capital is the project sensitive? (The nearest whole percent will be sufficient to answer the question.)

(*c*) At what first cost is the project sensitive, if the social cost of capital is 15 percent?

(*d*) Is the project sensitive to a reduction in benefits of the last 10 years to 300 million annually, if the social cost of capital is raised to 18 percent?

18.10. The airline of Prob. 10.30, page 258, and Prob. 11.27, page 298, now wishes to test the sensitivity of its decision to a change in the resale value of the aircraft. If only \$1 million in current dollars may be recovered on the sale of the aircraft, is the project still sound after taxes and after inflation have been taken into account? That is, is the project sensitive to change in resale value?

Gain or loss on the sale of such assets is taxed as revenue or loss in the year of sale. Assume that all else remains the same, including the depreciation schedule of Prob. 11.27.

18.11. In Example 11.22, page 292, can you find the inflation rate at which the decision is sensitive, i.e., the rate at which the decision will be reversed?

18.12. In a wastewater treatment plant, two types of tray clarifiers are under investigation. The series-flow tray clarifier costs \$2,800,000 and will cost \$530,000 annually for operation and maintenance. The parallel-flow tray clarifier costs \$3,200,000 and will cost \$449,200 for operation and maintenance annually. Both types will last 25 years.

The city public works department studying these two alternatives uses a 9 percent discount rate, but higher rates have been suggested.

(*a*) Using the present-worth method at 9 percent discount rate, which type will be selected?
(*b*) At what discount rate will the decision on the types be sensitive? Use the present-worth method.

18.13. Refer to Prob. 16.21, page 430, in your textbook, which you did for homework.

The maintenance cost for the existing cabs has been revised to \$8,500 annually, instead of \$10,000 cited in the original problem. All other data remain unchanged. Answer the following questions.
(*a*) Is the decision sensitive to the revision?
(*b*) Regardless of your answer, according to the new maintenance estimate, how long are the cabs, whether existing or new, to be retained?

18.14. The city of Johnstown is considering building an overpass over the interstate highway. Two types of overpasses have been studied. Type *A* will cost \$2,800,000 to build and \$530,000 annually in maintenance costs. Type *B* will cost \$3,200,000 to build and \$440,000 annually in maintenance costs. Both types will last 25 years. Salvage value and dismantling costs will be ignored. The city uses a 9 percent discount rate. Because the city is a public entity, no income tax will be paid. Neglect inflation at this time.
(*a*) At the 9 percent discount rate, which type should be chosen?
(*b*) At what discount rate, to the nearest whole percent, will the decision on the type be sensitive?

18.15. In Example 13.6, page 346, a question was posed about the sensitivity of a loan decision to the opportunity cost of capital. Let us suppose that loan 2 now has a term of 25 years. What is the opportunity cost of capital at which the decision between the loans becomes sensitive? All other data remain the same.

18.16. A standard medium-range passenger aircraft is to be bought by a major U.S. airline. It will cost \$12 million per airplane. Annual after-tax constant-dollar profits over 10 years that it will be operated by the airline are given in the following table, along with the probabilities of their occurrence:

Annual profit (\$000,000)	Probability
1.5	.20
2.0	.35
2.3	.23
2.5	.22
	1.00

At the end of the 10-year operation period, the airplanes will be sold, usually to foreign operators. The after-tax constant-dollar resale value is estimated as follows:

Resale value ($000,000)	Probability
2.0	.6
3.0	.3
3.5	.1
	1.0

(*a*) If you use expected value, is this a good investment? The constant-dollar opportunity cost to the company is 15 percent after taxes. Use the net present-value method.
(*b*) At what discount rate is the decision in part (*a*) sensitive?

18.17. A college professor is offered the opportunity to buy her academic regalia on the following terms:

Gown	191.00
Hood	55.50
Cap	16.90
	263.40
Less discount at 10 percent	−26.34
	237.06
Plus sales tax at 8 percent	+18.96
	256.02
Less university subsidy	−100.00
	156.02

The professor is in the 37 percent tax bracket for this purchase. The cost to the professor ($156.02) of the regalia may be depreciated over 2 years at $78.01 per year and is tax-deductible as a business expense. The professor will be able to use the regalia for 13 years until her retirement and will then donate it to the university, but may take no tax deduction for its salvage value because the regalia will have been fully depreciated.

The ownership of the regalia will enable the professor to cease renting regalia once a year at commencement. She paid $18.67 last year (tax-deductible) and expects that the rental will increase this year and over the next 13 years at about 10 percent per year, the same as the consumer price index.

She believes that she can invest any funds in her bank at about 10 percent after taxes.

(*a*) Should she buy the regalia, taking into account tax and inflation effects?

(*b*) If she can invest at 12 percent after taxes, should she buy the regalia? In all cases, assume annual end-of-year compounding.

18.18. (*a*) In Example 18.2, page 452, will lengthening the life of the plant to 40 years change the rejection to an acceptance, given the $1 billion cost, all other data remaining the same?
(*b*) If not, at what life is the project sensitive?

18.19. In Prob. 9.15, page 212, assume that the net annual benefits are in current dollars. Inflation over the next 100 years is difficult to predict, and therefore a sensitivity analysis will be performed. With the 7 percent discount rate taken as the constant-dollar opportunity cost of capital, what effect, if any, will these different inflation rate assumptions have on the choice of designs under consideration? First, try an inflation rate of 1 percent over 100 years, then 2 percent, and finally 3 percent. Whole percentage rates may be taken as sufficiently accurate for this calculation. Round B/C ratios to one place after the decimal point.

18.20. In Example 18.3, page 453, use MACRS to compute depreciation. In accordance with 1995 law, 100 percent of capital gains is taxable as ordinary income in the year of sale. Incorporate both these changes in the example and compute the IROR. Other data are unchanged.

Assume that the building is put in service on January 1, 1995, and will be taken out of service, i.e., sold, on December 31, 2003. Thus, full depreciation may be taken for the first and final years of ownership.

18.21. A small rental house is available for purchase near a large and growing city in the United States. Information on the property is given in the following table.

Item	Range		
	Low	**Most likely**	**High**
Selling price ($)	123,000	127,500	—
Assumable mortgage (8%, 30 years, 10 years remaining)	—	86,400	—
Down payment (% of sale price)	5	10	15
Taxes annually: $1,400. Growth rate (%)	3	5	7
Mortgage interest rate (%)	15	16	17
Economic life (years)	3	5	8
Maintenance ($)	400	600	800
Agent (% of gross rent)	—	10	—
Insurance $150 per year. Growth rate (%)	3	4	5
Vacancy rate per year (%)	2	5	10
Income $800 per month. Growth rate (%)	5	8	10
Appreciation rate per year (%)	3	8	10
Owner's tax bracket (%)	35	40	45
Agent's sale commission (%)	—	6	—
Capital gains tax exclusion (%)	30	40	50

Investigate the sensitivity of the venture, using as a figure of merit the net present value after income taxes. Inflation has been included in the ranges of the items in the table.

APPENDIX A

HOW TO USE A PERSONAL COMPUTER TO SOLVE PROBLEMS IN ENGINEERING ECONOMICS

The purpose of this appendix is to introduce a faster way of solving problems through the use of the personal computer.

There are many software packages on the market to help formulate and solve economic analysis problems. These perform cash flow calculations much faster than humans can, but it should be made clear that these software programs are just another advanced and powerful tool. They cannot substitute for the analytical thinking of an engineer-economist.

If you are using a different personal computer from the IBM PC or compatible to which these instructions apply, then you must make the proper modifications. No attempt has been made to write instructions for all makes and models of personal computers.

Among the criteria used to choose a software package to be employed are ease of use and availability. Lotus 1-2-3 version 2.0 or higher fulfills these two criteria. Quattro Pro version 5.0 and higher, and Excel version 5.0 and higher will all work.

It is not the intention to teach a spreadsheet package, but rather to use it for economic analysis applications. Only the very necessary knowledge will be presented to allow the reader to go through the problems and examples of the book diskette. For information about the software, the reader is advised to consult the software manual.

To use the worksheets supplied with this text and described in this appendix, simply start the spreadsheet package as you normally would and open the file. The eight spreadsheet files on the disk and its accompanying example are listed below.

File Name	Example
AN_WORTH.WK1	6.4
B_OVER_C.WK1	7.3
INFL.WK1	11.15
IROR.WK1	8.7
MUL_ROR.WK1	8.9
NPV.WK1	5.6
SENS_AN.WK1	18.3
TAX.WK1	10.20

In addition, there are two files on the disk which can be used in conjunction with LINDO to solve problems of capital budgeting. These are discussed on page 471.

EXAMPLE 5.6

Open the file NPV.WK1.

You may move around this worksheet by using either the four arrow keys or the Page Up and Page Down keys. The last two keys will move a whole screen at a time. To move quickly to the top left cell, press the Home key.

This file contains the data for Example 5.6, the solution of this particular example, and a layout for setting up problems to be solved by using the NPV method.

Now let us examine the solution. First, enter the cash flow data for each alternative in a single column. This has to be done in order to use the @NPV function of Lotus 1-2-3. Notice that the payments for each year have been netted out and that the lives of the alternatives have been equalized, as the NPV method requires. The last step is to calculate the NPV for the two alternatives and compare them. Under FERRY we entered −200 + @NPV (.15,B29..B58). For an explanation of this function, see page 473. Here −200 is the first payment, .15 is the opportunity cost of capital, and B29..B58 is the range of cells containing all the FERRY cash flow payments but the initial one.

After you calculate the NPV for the BRIDGE in the same way, compare the two NPVs and choose the better one.

To the right of Example 5.6 the screen shows instructions and a layout for solving NPV problems. The area under row 27 and to the right of column G is unprotected. Save the file with a new filename, for example, NPV2.WK1.

EXAMPLE 6.4

This example will present procedures for solving problems by using the annual-worth method.

Open the file AN_WORTH.WK1. On your screen Example 6.4 appears. The solution consists of two parts. Part A shows the cash flows, and part B shows the calculations. Notice that the annual-worth method does not require the lives of alternatives to be equalized. To calculate the annual worth with Lotus 1-2-3, two steps are followed. The first step calculates the present worth of the cash flow by using the now familiar function NPV. The second step calculates the annual worth of the NPV calculated in the first step. The function used at step 2 is PMT, which calculates the annual worth of an amount. The arguments of this function are

The amount (in this case, the NPV of the alternative)
The discount rate
The life of the alternative

For the Mercedes we entered –@PMT(–G51,.25,10) where G51 is the address of the cell containing the NPV of Mercedes.

Notice that the discount rate in the two functions used for calculating annual worth must be the same.

You may go now in the area to the right of column H and formulate and solve your own problem. If you wish to keep any new material that you added to the worksheet, remember to save your file.

EXAMPLE 7.3

Open the file that contains the *B*/*C* method example. Its name is B_OVER_C.WK1. Look at the cash flow part of the solution. As you see, for each year a benefit and a cost figure appear, unlike the single figure (benefits – means costs) used in previous examples. The reason is that this is a convenient way for calculating, if necessary, incremental benefits and costs discounted to year 0. Notice also that we do not observe the rule of delta. As you know, this affects not the final selection but the value of the ratio of benefits to costs *B*/*C*.

In our calculations we start with comparing alternative 1 against the null. After finding that alternative 1 qualifies, we proceed to the incremental analysis of alternative 2 – 1. The function used is the Lotus 1-2-3 NPV function, which calculates the present worth of benefits or costs. By moving the cursor over the numerical value of each NPV, we see at the top left side of the screen the function and its arguments.

EXAMPLE 8.7

The internal rate-of-return method (IROR) will be used. We will be able to appreciate how powerful, compared to a pencil-and-paper solution, a PC solution can be.

Open the appropriate file IROR.WK1. Under "B. Calculations" a single Lotus 1-2-3 function calculates the IROR. This is @IRR(guess, range), and it requires a guess at the IROR. The range comprises the cells that contain the cash flow. For more information turn to page 473.

Now calculate the IROR for the first alternative. If you find that alternative 1 qualifies, then proceed, using incremental analysis. Notice in this example the use of Bergmann's rule. Lotus 1-2-3 eliminates the need for trial-and-error calculations of the NPV for various rates of return and provides us with a precise IROR. At the left column H, you may apply the @IRR function with a different guess and compare the resulting IROR; or you may solve a problem.

EXAMPLE 8.9

The file containing this example is called MUL_ROR.WK1. After you open it, notice that the incremental cash flow changes sign twice. Thus we have two internal rates of return. By moving the cursor to cell F48, you can see the @IRR formula which contains as "guess" the value of cell C48. Instead of editing the @IRR formula every time we want to make another guess, we can just enter the new guess in cell C48 and watch the IROR. Cell C48 is unprotected; thus you may enter another guess. For more information see page 474. To the right of column I, you may set up a problem of your own.

EXAMPLE 10.12

Open file TAX.WK1. It contains an example involving depreciation and taxes. Under "Solution" a table similar to Table 10.14 appears. The only difference is that in the Lotus 1-2-3 worksheet there is only one row for the third year containing the algebraic sum of the two rows appearing in Table 10.14. To find the internal rate of return before and after taxes, use the @IRR function. When you move the cursor over the cell containing the IROR, the function and its arguments appear in the left upper part of the screen. To the right of column I, you may set up a problem.

EXAMPLE 11.15

Open file INFL.WK1. Again Lotus 1-2-3 is useful in doing all the calculations. Moving to cell D23 reveals the simple formula that calculates constant 1967 dollars. Cell C43 shows the function that calculates the NPV of the constant 1967 dollar cash flow. Move the cursor to cell F49. Here appears the NPV of the constant-dollar cash flow. You can set up a problem in the area to the right of column H.

EXAMPLE 18.3

The purpose of the problems discussed so far was to acquaint us with the mechanics, or purely mathematical details, of engineering economic analysis. That is, we simplified real-world problems and based our analysis on data that are difficult to estimate accurately. However, the final decision is only as sound as the data. Assuming that the mathematical details are worked out correctly, the only thing that will improve the decision is the investigation of the data and the factors that affect them. In other words, we should study the effect of each figure on the decision. After identifying the most important variables, we should investigate them further and then make the decision. Of course, this kind of analysis involves many variables and calculations. Computers offer great help in this work. Example 18.3 will illustrate this.

The file that contains the sensitivity analysis problem is called SENS_AN.WK1. Open it. At the beginning appear much more initial data than we have been used to so far. A table appears similar to Table 18.1. By moving around this table, examine the way it is set up. The various formulas in each column of the table are similar to the ones described in Example 18.3 in the book. The difference is that we have chosen to investigate certain variables that appear under the appropriate heading in cells D57, D58, and D60. These variables are (1) the loan period, (2) appreciation rate, and (3) taxable portion of capital gains.

Move the cursor to any of those cells and change its value. Automatically, Lotus 1-2-3 calculates the IROR which appears in cell G58. Check it against the results in the Item Changed table. Try another combination of these variables, and see the resulting IROR.

CAPITAL BUDGETING

To solve problems of capital budgeting, we are going to use integer programming with software called LINDO. Again we are not going to present all the available commands in LINDO—only those necessary for solving Example 14.6. To call up LINDO, follow these steps. We assume that you exited Lotus 1-2-3 and that you are at the DOS prompt A:\>.

1. Install LINDO according to the instructions in the LINDO manual. Put the book diskette in drive A.
2. Start LINDO. After a few seconds the colon prompt appears.
3. Type TAKE A:CAP_BUDG.LND, which is the name of the file containing Example 14.6, and press Return.

4. Press the Ctrl and PrtSc keys simultaneously. This will activate your printer.
5. Type LOOK ALL and press Return. The screen will display the formulation of the problem.
6. Type GO and press Return. The problem will be solved, and the solution will be printed on your printer.
7. Type NO for no sensitivity analysis, if the question is asked.

If you wish more information about LINDO commands, you may enter COM and press Return. If you wish to exit LINDO, enter QUIT and press Return.

Repeat step 4. This will deactivate your printer.

In the printout with the solution, the value of the objective function and the values of the variables appear. When the value of a variable is 1, the associated alternative is selected. For this budget, select alternatives X14, X23, X34, and X62.

To solve similar problems, see the LINDO manual.

LOTUS 1-2-3 FUNCTIONS

Net Present Value @NPV

This function computes the present value of a series of future cash flows. The arguments, i.e., the variables, are the interest rate and the range of cells containing cash flows. Unfortunately the way this function works is not similar to the one we are accustomed to; thus we illustrate with an example. Let us say that we have the following cash flow with 10 percent interest:

Row number	Column letter	
	X	Y
	YEAR	CASH FLOW
11	0	100
12	1	200
13	2	250
14	3	400

The correct way to get the NPV at year 0 is by entering in a cell the formula +100 + @NPV(0.1,Y12..Y14) as opposed to the wrong formula @NPV(0.1,Y11..Y14). Therefore, to compute the net present value of the cash flow, add the @NPV function to the cash flow of year 0, using a range that does not include the cash flow at year 0.

Annual Worth

To compute the annual worth, we use two Lotus 1-2-3 functions: @NPV and @PMT. Unfortunately Lotus 1-2-3 does not provide a single function to compute the annual worth. There are two steps in our solution. The first step calculates the net present value of the cash flow by using the function @NPV. The second step computes the equal annual cash flows, given the net present value we calculated in the first step. The function used for this step is @PMT. Its arguments are

The amount we want to spread over the periods, which is the NPV calculated in the first step.[1]

The interest rate, which must be the same as that used to compute the net present value

The number of periods in the cash flow

Internal Rate of Return @IRR

To compute the IROR, Lotus 1-2-3 provides us with the function @IROR. The arguments of this function are guess and range: @IRR(guess, range). In Example 8.7, guess is a number that represents our estimate of the IROR of the cash flow, while range shows the cells that contain the cash flow. The guess need not be a close one. In general, this function works with iterations. If the iterations cannot approximate the IROR by 1E10 (0.0000001) after 20 tries, it will display ERR. If this happens, make another guess and proceed.

There are cash flows that have more than one IROR. Lotus 1-2-3 does not give a special message in such cases. Therefore, first find out how many IRORs exist, and then calculate them. To find the number of IRORs, see Chap. 8, p. 166.

[1]In Lotus 1-2-3 version 2.0, this amount is required to be positive. In subsequent versions this has been changed, and negative amounts are accepted as well. However, for consistency, we use a positive amount throughout the examples by putting the negative sign in front of the @PMT function. Thus in Example 6.4 because cell G51 contains a negative amount, we enter −G51, which is positive, and put another minus sign in front of the @PMT function to retain the negative sign of the payment. That is why the @PMT function in Example 6.4 is −@PMT(−G51, .25, 10).

APPENDIX B

NOTATION

A	End-of-period cash flows in a uniform series continuing for a specific number of periods
AC	Annual cost
ACRS	Accelerated cost recovery system
AW	Annual worth
B	Benefits
B/C	Benefit/cost ratio
C	Cost
CR	Capital recovery
DDB	Double-declining balance
e	Base of the naperian logarithm system, approximately equal to 2.71828
EUAC	Equivalent uniform annual cost
F	Future sum of money
F_D	Constant (deflated) dollars in year N
F_I	Current (inflated) dollars in year N
f	General rate of inflation or deflation
G	Uniform period-by-period increase or decrease in cash flows; arithmetic gradient
GDS	General depreciation system
I_Y	Annual interest payment
i	Effective interest rate per interest period; discount rate; constant-dollar rate
i_Y	Effective interest rate per year
i_M	Effective interest rate per subperiod
i^*	Internal rate of return
IROR	Internal rate of return

j	Index, generally used to denote interest period
L	Salvage value
M	Number of compounding periods per year; the capital budget
MACRS	Modified accelerated cost recovery system
MARR	Minimum attractive rate of return
N	Number of compounding periods; life of investment
P	Present sum of money; initial investment; loan principal
P_Y	The principal payment for any year Y
PW	Present worth, or net present value, of present and future cash flows
R_Y	The remaining balance for any year Y
r	Nominal interest rate per period, usually the nominal interest rate per year
S	Net salvage value of capital investment
SYD	Sum-of-the-years' digits
u	Current-dollar rate
u_M	Current-dollar rate per subperiod
u_Y	Current-dollar rate per year
Δ	Incremental
μ	Mean, or expected value
σ	Standard deviation; σ^2 = variance

APPENDIX C

COMPOUND INTEREST TABLES

FORMULAS FOR CALCULATING COMPOUND INTEREST FACTORS

Single-payment compound-amount factor $(F/P, i, N)$ $\quad (1+i)^N$

Single-payment present-worth factor $(P/F, i, N)$ $\quad \dfrac{1}{(1+i)^N}$

Sinking fund factor $(A/F, i, N)$ $\quad \dfrac{i}{(1+i)^N - 1}$

Capital recovery factor $(A/P, i, N)$ $\quad \dfrac{i(1+i)^N}{(1+i)^N - 1}$

Uniform series compound-amount factor $(F/A, i, N)$ $\quad \dfrac{(1+i)^N - 1}{i}$

Uniform series present-worth factor $(P/A, i, N)$ $\quad \dfrac{(1+i)^N - 1}{i(1+i)^N}$

Uniform-gradient conversion factor $(A/G, i, N)$ $\quad \dfrac{1}{i} - \dfrac{N}{(1+i)^N - 1}$

Uniform-gradient present-worth factor $(P/G, i, N)$

$$\frac{1}{i(1+i)^N}\left[\frac{(1+i)^N - 1}{i}\right] - \frac{N}{i(1+i)^N}$$

The following tables contain factors for these percentages: 0.25, 0.50, 0.75, 1, 1.25, 1.5, 1.75, 2, 2.5, 3, 3.5, 4, 4.5, 5, 6, 7, 8, 9, 10, 11, 12, 13, 14, 15, 16, 18, 20, 25, 30, 35, 40, 45, 50, and 60.

When $N = \infty$:

$$(F/P, i, \infty) = \infty \qquad (F/A, i, \infty) = \infty$$
$$(P/F, i, \infty) = 0 \qquad (P/A, i, \infty) = 1/i$$
$$(A/F, i, \infty) = 0 \qquad (A/G, i, \infty) = 1/i$$
$$(A/P, i, \infty) = i \qquad (P/G, i, \infty) = 1/i^2$$

0.25% Compound Interest Factors

	Single Payment		Uniform Series				Arithmetic Gradient		
N	Compound Amount Factor *F/P*	Present Worth Factor *P/F*	Sinking Fund Factor *A/F*	Capital Recovery Factor *A/P*	Compound Amount Factor *F/A*	Present Worth Factor *P/A*	Uniform Series *A/G*	Present Worth *P/G*	*N*
1	1.0025	0.9975	1.0000	1.0025	1.0000	0.9975	0.0000	0.0000	1
2	1.0050	0.9950	0.4994	0.5019	2.0025	1.9925	0.4994	0.9950	2
3	1.0075	0.9925	0.3325	0.3350	3.0075	2.9851	0.9983	2.9801	3
4	1.0100	0.9901	0.2491	0.2516	4.0150	3.9751	1.4969	5.9503	4
5	1.0126	0.9876	0.1990	0.2015	5.0251	4.9627	1.9950	9.9007	5
6	1.0151	0.9851	0.1656	0.1681	6.0376	5.9478	2.4927	14.8263	6
7	1.0176	0.9827	0.1418	0.1443	7.0527	6.9305	2.9900	20.7223	7
8	1.0202	0.9802	0.1239	0.1264	8.0704	7.9107	3.4869	27.5839	8
9	1.0227	0.9778	0.1100	0.1125	9.0905	8.8885	3.9834	35.4061	9
10	1.0253	0.9753	0.0989	0.1014	10.1133	9.8639	4.4794	44.1842	10
11	1.0278	0.9729	0.0898	0.0923	11.1385	10.8368	4.9750	53.9133	11
12	1.0304	0.9705	0.0822	0.0847	12.1664	11.8073	5.4702	64.5886	12
13	1.0330	0.9681	0.0758	0.0783	13.1968	12.7753	5.9650	76.2053	13
14	1.0356	0.9656	0.0703	0.0728	14.2298	13.7410	6.4594	88.7587	14
15	1.0382	0.9632	0.0655	0.0680	15.2654	14.7042	6.9534	102.2441	15
16	1.0408	0.9608	0.0613	0.0638	16.3035	15.6650	7.4469	116.6567	16
17	1.0434	0.9584	0.0577	0.0602	17.3443	16.6235	7.9401	131.9917	17
18	1.0460	0.9561	0.0544	0.0569	18.3876	17.5795	8.4328	148.2446	18
19	1.0486	0.9537	0.0515	0.0540	19.4336	18.5332	8.9251	165.4106	19
20	1.0512	0.9513	0.0488	0.0513	20.4822	19.4845	9.4170	183.4851	20
21	1.0538	0.9489	0.0464	0.0489	21.5334	20.4334	9.9085	202.4634	21
22	1.0565	0.9466	0.0443	0.0468	22.5872	21.3800	10.3995	222.3410	22
23	1.0591	0.9442	0.0423	0.0448	23.6437	22.3241	10.8901	243.1131	23
24	1.0618	0.9418	0.0405	0.0430	24.7028	23.2660	11.3804	264.7753	24
25	1.0644	0.9395	0.0388	0.0413	25.7646	24.2055	11.8702	287.3230	25
26	1.0671	0.9371	0.0373	0.0398	26.8290	25.1426	12.3596	310.7516	26
27	1.0697	0.9348	0.0358	0.0383	27.8961	26.0774	12.8485	335.0566	27
28	1.0724	0.9325	0.0345	0.0370	28.9658	27.0099	13.3371	360.2334	28
29	1.0751	0.9301	0.0333	0.0358	30.0382	27.9400	13.8252	386.2776	29
30	1.0778	0.9278	0.0321	0.0346	31.1133	28.8679	14.3130	413.1847	30
31	1.0805	0.9255	0.0311	0.0336	32.1911	29.7934	14.8003	440.9502	31
32	1.0832	0.9232	0.0301	0.0326	33.2716	30.7166	15.2872	469.5696	32
33	1.0859	0.9209	0.0291	0.0316	34.3547	31.6375	15.7736	499.0386	33
34	1.0886	0.9186	0.0282	0.0307	35.4406	32.5561	16.2597	529.3528	34
35	1.0913	0.9163	0.0274	0.0299	36.5292	33.4724	16.7454	560.5076	35
36	1.0941	0.9140	0.0266	0.0291	37.6206	34.3865	17.2306	592.4988	36
40	1.1050	0.9050	0.0238	0.0263	42.0132	38.0199	19.1673	728.7399	40
48	1.1273	0.8871	0.0196	0.0221	50.9312	45.1787	23.0209	1040.0552	48
50	1.1330	0.8826	0.0188	0.0213	53.1887	46.9462	23.9802	1125.7767	50
52	1.1386	0.8782	0.0180	0.0205	55.4575	48.7048	24.9377	1214.5885	52
60	1.1616	0.8609	0.0155	0.0180	64.6467	55.6524	28.7514	1600.0845	60
70	1.1910	0.8396	0.0131	0.0156	76.3944	64.1439	33.4812	2147.6111	70
72	1.1969	0.8355	0.0127	0.0152	78.7794	65.8169	34.4221	2265.5569	72
80	1.2211	0.8189	0.0113	0.0138	88.4392	72.4260	38.1694	2764.4568	80
84	1.2334	0.8108	0.0107	0.0132	93.3419	75.6813	40.0331	3029.7592	84
90	1.2520	0.7987	0.0099	0.0124	100.7885	80.5038	42.8162	3446.8700	90
96	1.2709	0.7869	0.0092	0.0117	108.3474	85.2546	45.5844	3886.2832	96
100	1.2836	0.7790	0.0088	0.0113	113.4500	88.3825	47.4216	4191.2417	100
104	1.2965	0.7713	0.0084	0.0109	118.6037	91.4793	49.2522	4505.5569	104
120	1.3494	0.7411	0.0072	0.0097	139.7414	103.5618	56.5084	5852.1116	120
240	1.8208	0.5492	0.0030	0.0055	328.3020	180.3109	107.5863	19398.9852	240
360	2.4568	0.4070	0.0017	0.0042	582.7369	237.1894	152.8902	36263.9299	360
480	3.3151	0.3016	0.0011	0.0036	926.0595	279.3418	192.6699	53820.7525	480

0.50% Compound Interest Factors

	Single Payment		Uniform Series				Arithmetic Gradient		
	Compound Amount Factor	Present Worth Factor	Sinking Fund Factor	Capital Recovery Factor	Compound Amount Factor	Present Worth Factor	Uniform Series	Present Worth	
N	*F/P*	*P/F*	*A/F*	*A/P*	*F/A*	*P/A*	*A/G*	*P/G*	*N*
1	1.0050	0.9950	1.0000	1.0050	1.0000	0.9950	0.0000	0.0000	1
2	1.0100	0.9901	0.4988	0.5038	2.0050	1.9851	0.4988	0.9901	2
3	1.0151	0.9851	0.3317	0.3367	3.0150	2.9702	0.9967	2.9604	3
4	1.0202	0.9802	0.2481	0.2531	4.0301	3.9505	1.4938	5.9011	4
5	1.0253	0.9754	0.1980	0.2030	5.0503	4.9259	1.9900	9.8026	5
6	1.0304	0.9705	0.1646	0.1696	6.0755	5.8964	2.4855	14.6552	6
7	1.0355	0.9657	0.1407	0.1457	7.1059	6.8621	2.9801	20.4493	7
8	1.0407	0.9609	0.1228	0.1278	8.1414	7.8230	3.4738	27.1755	8
9	1.0459	0.9561	0.1089	0.1139	9.1821	8.7791	3.9668	34.8244	9
10	1.0511	0.9513	0.0978	0.1028	10.2280	9.7304	4.4589	43.3865	10
11	1.0564	0.9466	0.0887	0.0937	11.2792	10.6770	4.9501	52.8526	11
12	1.0617	0.9419	0.0811	0.0861	12.3356	11.6189	5.4406	63.2136	12
13	1.0670	0.9372	0.0746	0.0796	13.3972	12.5562	5.9302	74.4602	13
14	1.0723	0.9326	0.0691	0.0741	14.4642	13.4887	6.4190	86.5835	14
15	1.0777	0.9279	0.0644	0.0694	15.5365	14.4166	6.9069	99.5743	15
16	1.0831	0.9233	0.0602	0.0652	16.6142	15.3399	7.3940	113.4238	16
17	1.0885	0.9187	0.0565	0.0615	17.6973	16.2586	7.8803	128.1231	17
18	1.0939	0.9141	0.0532	0.0582	18.7858	17.1728	8.3658	143.6634	18
19	1.0994	0.9096	0.0503	0.0553	19.8797	18.0824	8.8504	160.0360	19
20	1.1049	0.9051	0.0477	0.0527	20.9791	18.9874	9.3342	177.2322	20
21	1.1104	0.9006	0.0453	0.0503	22.0840	19.8880	9.8172	195.2434	21
22	1.1160	0.8961	0.0431	0.0481	23.1944	20.7841	10.2993	214.0611	22
23	1.1216	0.8916	0.0411	0.0461	24.3104	21.6757	10.7806	233.6768	23
24	1.1272	0.8872	0.0393	0.0443	25.4320	22.5629	11.2611	254.0820	24
25	1.1328	0.8828	0.0377	0.0427	26.5591	23.4456	11.7407	275.2686	25
26	1.1385	0.8784	0.0361	0.0411	27.6919	24.3240	12.2195	297.2281	26
27	1.1442	0.8740	0.0347	0.0397	28.8304	25.1980	12.6975	319.9523	27
28	1.1499	0.8697	0.0334	0.0384	29.9745	26.0677	13.1747	343.4332	28
29	1.1556	0.8653	0.0321	0.0371	31.1244	26.9330	13.6510	367.6625	29
30	1.1614	0.8610	0.0310	0.0360	32.2800	27.7941	14.1265	392.6324	30
31	1.1672	0.8567	0.0299	0.0349	33.4414	28.6508	14.6012	418.3348	31
32	1.1730	0.8525	0.0289	0.0339	34.6086	29.5033	15.0750	444.7618	32
33	1.1789	0.8482	0.0279	0.0329	35.7817	30.3515	15.5480	471.9055	33
34	1.1848	0.8440	0.0271	0.0321	36.9606	31.1955	16.0202	499.7583	34
35	1.1907	0.8398	0.0262	0.0312	38.1454	32.0354	16.4915	528.3123	35
36	1.1967	0.8356	0.0254	0.0304	39.3361	32.8710	16.9621	557.5598	36
40	1.2208	0.8191	0.0226	0.0276	44.1588	36.1722	18.8359	681.3347	40
48	1.2705	0.7871	0.0185	0.0235	54.0978	42.5803	22.5437	959.9188	48
50	1.2832	0.7793	0.0177	0.0227	56.6452	44.1428	23.4624	1035.6966	50
52	1.2961	0.7716	0.0169	0.0219	59.2180	45.6897	24.3778	1113.8162	52
60	1.3489	0.7414	0.0143	0.0193	69.7700	51.7256	28.0064	1448.6458	60
70	1.4178	0.7053	0.0120	0.0170	83.5661	58.9394	32.4680	1913.6427	70
72	1.4320	0.6983	0.0116	0.0166	86.4089	60.3395	33.3504	2012.3478	72
80	1.4903	0.6710	0.0102	0.0152	98.0677	65.8023	36.8474	2424.6455	80
84	1.5204	0.6577	0.0096	0.0146	104.0739	68.4530	38.5763	2640.6641	84
90	1.5666	0.6383	0.0088	0.0138	113.3109	72.3313	41.1451	2976.0769	90
96	1.6141	0.6195	0.0081	0.0131	122.8285	76.0952	43.6845	3324.1846	96
100	1.6467	0.6073	0.0077	0.0127	129.3337	78.5426	45.3613	3562.7934	100
104	1.6798	0.5953	0.0074	0.0124	135.9699	80.9417	47.0250	3806.2855	104
120	1.8194	0.5496	0.0061	0.0111	163.8793	90.0735	53.5508	4823.5051	120
240	3.3102	0.3021	0.0022	0.0072	462.0409	139.5808	96.1131	13415.5395	240
360	6.0226	0.1660	0.0010	0.0060	1004.5150	166.7916	128.3236	21403.3041	360
480	10.9575	0.0913	0.0005	0.0055	1991.4907	181.7476	151.7949	27588.3573	480

0.75% Compound Interest Factors

	Single Payment		Uniform Series				Arithmetic Gradient		
	Compound Amount Factor	Present Worth Factor	Sinking Fund Factor	Capital Recovery Factor	Compound Amount Factor	Present Worth Factor	Uniform Series	Present Worth	
N	*F/P*	*P/F*	*A/F*	*A/P*	*F/A*	*P/A*	*A/G*	*P/G*	*N*
1	1.0075	0.9926	1.0000	1.0075	1.0000	0.9926	0.0000	0.0000	1
2	1.0151	0.9852	0.4981	0.5056	2.0075	1.9777	0.4981	0.9852	2
3	1.0227	0.9778	0.3308	0.3383	3.0226	2.9556	0.9950	2.9408	3
4	1.0303	0.9706	0.2472	0.2547	4.0452	3.9261	1.4907	5.8525	4
5	1.0381	0.9633	0.1970	0.2045	5.0756	4.8894	1.9851	9.7058	5
6	1.0459	0.9562	0.1636	0.1711	6.1136	5.8456	2.4782	14.4866	6
7	1.0537	0.9490	0.1397	0.1472	7.1595	6.7946	2.9701	20.1808	7
8	1.0616	0.9420	0.1218	0.1293	8.2132	7.7366	3.4608	26.7747	8
9	1.0696	0.9350	0.1078	0.1153	9.2748	8.6716	3.9502	34.2544	9
10	1.0776	0.9280	0.0967	0.1042	10.3443	9.5996	4.4384	42.6064	10
11	1.0857	0.9211	0.0876	0.0951	11.4219	10.5207	4.9253	51.8174	11
12	1.0938	0.9142	0.0800	0.0875	12.5076	11.4349	5.4110	61.8740	12
13	1.1020	0.9074	0.0735	0.0810	13.6014	12.3423	5.8954	72.7632	13
14	1.1103	0.9007	0.0680	0.0755	14.7034	13.2430	6.3786	84.4720	14
15	1.1186	0.8940	0.0632	0.0707	15.8137	14.1370	6.8606	96.9876	15
16	1.1270	0.8873	0.0591	0.0666	16.9323	15.0243	7.3413	110.2973	16
17	1.1354	0.8807	0.0554	0.0629	18.0593	15.9050	7.8207	124.3887	17
18	1.1440	0.8742	0.0521	0.0596	19.1947	16.7792	8.2989	139.2494	18
19	1.1525	0.8676	0.0492	0.0567	20.3387	17.6468	8.7759	154.8671	19
20	1.1612	0.8612	0.0465	0.0540	21.4912	18.5080	9.2516	171.2297	20
21	1.1699	0.8548	0.0441	0.0516	22.6524	19.3628	9.7261	188.3253	21
22	1.1787	0.8484	0.0420	0.0495	23.8223	20.2112	10.1994	206.1420	22
23	1.1875	0.8421	0.0400	0.0475	25.0010	21.0533	10.6714	224.6682	23
24	1.1964	0.8358	0.0382	0.0457	26.1885	21.8891	11.1422	243.8923	24
25	1.2054	0.8296	0.0365	0.0440	27.3849	22.7188	11.6117	263.8029	25
26	1.2144	0.8234	0.0350	0.0425	28.5903	23.5422	12.0800	284.3888	26
27	1.2235	0.8173	0.0336	0.0411	29.8047	24.3595	12.5470	305.6387	27
28	1.2327	0.8112	0.0322	0.0397	31.0282	25.1707	13.0128	327.5416	28
29	1.2420	0.8052	0.0310	0.0385	32.2609	25.9759	13.4774	350.0867	29
30	1.2513	0.7992	0.0298	0.0373	33.5029	26.7751	13.9407	373.2631	30
31	1.2607	0.7932	0.0288	0.0363	34.7542	27.5683	14.4028	397.0602	31
32	1.2701	0.7873	0.0278	0.0353	36.0148	28.3557	14.8636	421.4675	32
33	1.2796	0.7815	0.0268	0.0343	37.2849	29.1371	15.3232	446.4746	33
34	1.2892	0.7757	0.0259	0.0334	38.5646	29.9128	15.7816	472.0712	34
35	1.2989	0.7699	0.0251	0.0326	39.8538	30.6827	16.2387	498.2471	35
36	1.3086	0.7641	0.0243	0.0318	41.1527	31.4468	16.6946	524.9924	36
40	1.3483	0.7416	0.0215	0.0290	46.4465	34.4469	18.5058	637.4693	40
48	1.4314	0.6986	0.0174	0.0249	57.5207	40.1848	22.0691	886.8404	48
50	1.4530	0.6883	0.0166	0.0241	60.3943	41.5664	22.9476	953.8486	50
52	1.4748	0.6780	0.0158	0.0233	63.3111	42.9276	23.8211	1022.5852	52
60	1.5657	0.6387	0.0133	0.0208	75.4241	48.1734	27.2665	1313.5189	60
70	1.6872	0.5927	0.0109	0.0184	91.6201	54.3046	31.4634	1708.6065	70
72	1.7126	0.5839	0.0105	0.0180	95.0070	55.4768	32.2882	1791.2463	72
80	1.8180	0.5500	0.0092	0.0167	109.0725	59.9944	35.5391	2132.1472	80
84	1.8732	0.5338	0.0086	0.0161	116.4269	62.1540	37.1357	2308.1283	84
90	1.9591	0.5104	0.0078	0.0153	127.8790	65.2746	39.4946	2577.9961	90
96	2.0489	0.4881	0.0072	0.0147	139.8562	68.2584	41.8107	2853.9352	96
100	2.1111	0.4737	0.0068	0.0143	148.1445	70.1746	43.3311	3040.7453	100
104	2.1751	0.4597	0.0064	0.0139	156.6843	72.0344	44.8327	3229.4936	104
120	2.4514	0.4079	0.0052	0.0127	193.5143	78.9417	50.6521	3998.5621	120
240	6.0092	0.1664	0.0015	0.0090	667.8869	111.1450	85.4210	9494.1162	240
360	14.7306	0.0679	0.0005	0.0080	1830.7435	124.2819	107.1145	13312.3871	360
480	36.1099	0.0277	0.0002	0.0077	4681.3203	129.6409	119.6620	15513.0866	480

1% Compound Interest Factors

	Single Payment		Uniform Series				Arithmetic Gradient		
	Compound Amount Factor	Present Worth Factor	Sinking Fund Factor	Capital Recovery Factor	Compound Amount Factor	Present Worth Factor	Uniform Series	Present Worth	
N	*F/P*	*P/F*	*A/F*	*A/P*	*F/A*	*P/A*	*A/G*	*P/G*	*N*
1	1.0100	0.9901	1.0000	1.0100	1.0000	0.9901	0.0000	0.0000	1
2	1.0201	0.9803	0.4975	0.5075	2.0100	1.9704	0.4975	0.9803	2
3	1.0303	0.9706	0.3300	0.3400	3.0301	2.9410	0.9934	2.9215	3
4	1.0406	0.9610	0.2463	0.2563	4.0604	3.9020	1.4876	5.8044	4
5	1.0510	0.9515	0.1960	0.2060	5.1010	4.8534	1.9801	9.6103	5
6	1.0615	0.9420	0.1625	0.1725	6.1520	5.7955	2.4710	14.3205	6
7	1.0721	0.9327	0.1386	0.1486	7.2135	6.7282	2.9602	19.9168	7
8	1.0829	0.9235	0.1207	0.1307	8.2857	7.6517	3.4478	26.3812	8
9	1.0937	0.9143	0.1067	0.1167	9.3685	8.5660	3.9337	33.6959	9
10	1.1046	0.9053	0.0956	0.1056	10.4622	9.4713	4.4179	41.8435	10
11	1.1157	0.8963	0.0865	0.0965	11.5668	10.3676	4.9005	50.8067	11
12	1.1268	0.8874	0.0788	0.0888	12.6825	11.2551	5.3815	60.5687	12
13	1.1381	0.8787	0.0724	0.0824	13.8093	12.1337	5.8607	71.1126	13
14	1.1495	0.8700	0.0669	0.0769	14.9474	13.0037	6.3384	82.4221	14
15	1.1610	0.8613	0.0621	0.0721	16.0969	13.8651	6.8143	94.4810	15
16	1.1726	0.8528	0.0579	0.0679	17.2579	14.7179	7.2886	107.2734	16
17	1.1843	0.8444	0.0543	0.0643	18.4304	15.5623	7.7613	120.7834	17
18	1.1961	0.8360	0.0510	0.0610	19.6147	16.3983	8.2323	134.9957	18
19	1.2081	0.8277	0.0481	0.0581	20.8109	17.2260	8.7017	149.8950	19
20	1.2202	0.8195	0.0454	0.0554	22.0190	18.0456	9.1694	165.4664	20
21	1.2324	0.8114	0.0430	0.0530	23.2392	18.8570	9.6354	181.6950	21
22	1.2447	0.8034	0.0409	0.0509	24.4716	19.6604	10.0998	198.5663	22
23	1.2572	0.7954	0.0389	0.0489	25.7163	20.4558	10.5626	216.0660	23
24	1.2697	0.7876	0.0371	0.0471	26.9735	21.2434	11.0237	234.1800	24
25	1.2824	0.7798	0.0354	0.0454	28.2432	22.0232	11.4831	252.8945	25
26	1.2953	0.7720	0.0339	0.0439	29.5256	22.7952	11.9409	272.1957	26
27	1.3082	0.7644	0.0324	0.0424	30.8209	23.5596	12.3971	292.0702	27
28	1.3213	0.7568	0.0311	0.0411	32.1291	24.3164	12.8516	312.5047	28
29	1.3345	0.7493	0.0299	0.0399	33.4504	25.0658	13.3044	333.4863	29
30	1.3478	0.7419	0.0287	0.0387	34.7849	25.8077	13.7557	355.0021	30
31	1.3613	0.7346	0.0277	0.0377	36.1327	26.5423	14.2052	377.0394	31
32	1.3749	0.7273	0.0267	0.0367	37.4941	27.2696	14.6532	399.5858	32
33	1.3887	0.7201	0.0257	0.0357	38.8690	27.9897	15.0995	422.6291	33
34	1.4026	0.7130	0.0248	0.0348	40.2577	28.7027	15.5441	446.1572	34
35	1.4166	0.7059	0.0240	0.0340	41.6603	29.4086	15.9871	470.1583	35
36	1.4308	0.6989	0.0232	0.0332	43.0769	30.1075	16.4285	494.6207	36
40	1.4889	0.6717	0.0205	0.0305	48.8864	32.8347	18.1776	596.8561	40
48	1.6122	0.6203	0.0163	0.0263	61.2226	37.9740	21.5976	820.1460	48
50	1.6446	0.6080	0.0155	0.0255	64.4632	39.1961	22.4363	879.4176	50
52	1.6777	0.5961	0.0148	0.0248	67.7689	40.3942	23.2686	939.9175	52
60	1.8167	0.5504	0.0122	0.0222	81.6697	44.9550	26.5333	1192.8061	60
70	2.0068	0.4983	0.0099	0.0199	100.6763	50.1685	30.4703	1528.6474	70
72	2.0471	0.4885	0.0096	0.0196	104.7099	51.1504	31.2386	1597.8673	72
80	2.2167	0.4511	0.0082	0.0182	121.6715	54.8882	34.2492	1879.8771	80
84	2.3067	0.4335	0.0077	0.0177	130.6723	56.6485	35.7170	2023.3153	84
90	2.4486	0.4084	0.0069	0.0169	144.8633	59.1609	37.8724	2240.5675	90
96	2.5993	0.3847	0.0063	0.0163	159.9273	61.5277	39.9727	2459.4298	96
100	2.7048	0.3697	0.0059	0.0159	170.4814	63.0289	41.3426	2605.7758	100
104	2.8146	0.3553	0.0055	0.0155	181.4640	64.4715	42.6884	2752.1818	104
120	3.3004	0.3030	0.0043	0.0143	230.0387	69.7005	47.8349	3334.1148	120
240	10.8926	0.0918	0.0010	0.0110	989.2554	90.8194	75.7393	6878.6016	240
360	35.9496	0.0278	0.0003	0.0103	3494.9641	97.2183	89.6995	8720.4323	360
480	118.6477	0.0084	0.0001	0.0101	11764.7725	99.1572	95.9200	9511.1579	480

1.25% Compound Interest Factors

	Single Payment		Uniform Series				Arithmetic Gradient		
	Compound Amount Factor	Present Worth Factor	Sinking Fund Factor	Capital Recovery Factor	Compound Amount Factor	Present Worth Factor	Uniform Series	Present Worth	
N	*F/P*	*P/F*	*A/F*	*A/P*	*F/A*	*P/A*	*A/G*	*P/G*	*N*
1	1.0125	0.9877	1.0000	1.0125	1.0000	0.9877	0.0000	0.0000	1
2	1.0252	0.9755	0.4969	0.5094	2.0125	1.9631	0.4969	0.9755	2
3	1.0380	0.9634	0.3292	0.3417	3.0377	2.9265	0.9917	2.9023	3
4	1.0509	0.9515	0.2454	0.2579	4.0756	3.8781	1.4845	5.7569	4
5	1.0641	0.9398	0.1951	0.2076	5.1266	4.8178	1.9752	9.5160	5
6	1.0774	0.9282	0.1615	0.1740	6.1907	5.7460	2.4638	14.1569	6
7	1.0909	0.9167	0.1376	0.1501	7.2680	6.6627	2.9503	19.6571	7
8	1.1045	0.9054	0.1196	0.1321	8.3589	7.5681	3.4348	25.9949	8
9	1.1183	0.8942	0.1057	0.1182	9.4634	8.4623	3.9172	33.1487	9
10	1.1323	0.8832	0.0945	0.1070	10.5817	9.3455	4.3975	41.0973	10
11	1.1464	0.8723	0.0854	0.0979	11.7139	10.2178	4.8758	49.8201	11
12	1.1608	0.8615	0.0778	0.0903	12.8604	11.0793	5.3520	59.2967	12
13	1.1753	0.8509	0.0713	0.0838	14.0211	11.9302	5.8262	69.5072	13
14	1.1900	0.8404	0.0658	0.0783	15.1964	12.7706	6.2982	80.4320	14
15	1.2048	0.8300	0.0610	0.0735	16.3863	13.6005	6.7682	92.0519	15
16	1.2199	0.8197	0.0568	0.0693	17.5912	14.4203	7.2362	104.3481	16
17	1.2351	0.8096	0.0532	0.0657	18.8111	15.2299	7.7021	117.3021	17
18	1.2506	0.7996	0.0499	0.0624	20.0462	16.0295	8.1659	130.8958	18
19	1.2662	0.7898	0.0470	0.0595	21.2968	16.8193	8.6277	145.1115	19
20	1.2820	0.7800	0.0443	0.0568	22.5630	17.5993	9.0874	159.9316	20
21	1.2981	0.7704	0.0419	0.0544	23.8450	18.3697	9.5450	175.3392	21
22	1.3143	0.7609	0.0398	0.0523	25.1431	19.1306	10.0006	191.3174	22
23	1.3307	0.7515	0.0378	0.0503	26.4574	19.8820	10.4542	207.8499	23
24	1.3474	0.7422	0.0360	0.0485	27.7881	20.6242	10.9056	224.9204	24
25	1.3642	0.7330	0.0343	0.0468	29.1354	21.3573	11.3551	242.5132	25
26	1.3812	0.7240	0.0328	0.0453	30.4996	22.0813	11.8024	260.6128	26
27	1.3985	0.7150	0.0314	0.0439	31.8809	22.7963	12.2478	279.2040	27
28	1.4160	0.7062	0.0300	0.0425	33.2794	23.5025	12.6911	298.2719	28
29	1.4337	0.6975	0.0288	0.0413	34.6954	24.2000	13.1323	317.8019	29
30	1.4516	0.6889	0.0277	0.0402	36.1291	24.8889	13.5715	337.7797	30
31	1.4698	0.6804	0.0266	0.0391	37.5807	25.5693	14.0086	358.1912	31
32	1.4881	0.6720	0.0256	0.0381	39.0504	26.2413	14.4438	379.0227	32
33	1.5067	0.6637	0.0247	0.0372	40.5386	26.9050	14.8768	400.2607	33
34	1.5256	0.6555	0.0238	0.0363	42.0453	27.5605	15.3079	421.8920	34
35	1.5446	0.6474	0.0230	0.0355	43.5709	28.2079	15.7369	443.9037	35
36	1.5639	0.6394	0.0222	0.0347	45.1155	28.8473	16.1639	466.2830	36
40	1.6436	0.6084	0.0194	0.0319	51.4896	31.3269	17.8515	559.2320	40
48	1.8154	0.5509	0.0153	0.0278	65.2284	35.9315	21.1299	759.2296	48
50	1.8610	0.5373	0.0145	0.0270	68.8818	37.0129	21.9295	811.6738	50
52	1.9078	0.5242	0.0138	0.0263	72.6271	38.0677	22.7211	864.9409	52
60	2.1072	0.4746	0.0113	0.0238	88.5745	42.0346	25.8083	1084.8429	60
70	2.3859	0.4191	0.0090	0.0215	110.8720	46.4697	29.4913	1370.4513	70
72	2.4459	0.4088	0.0086	0.0211	115.6736	47.2925	30.2047	1428.4561	72
80	2.7015	0.3702	0.0073	0.0198	136.1188	50.3867	32.9822	1661.8651	80
84	2.8391	0.3522	0.0068	0.0193	147.1290	51.8222	34.3258	1778.8384	84
90	3.0588	0.3269	0.0061	0.0186	164.7050	53.8461	36.2855	1953.8303	90
96	3.2955	0.3034	0.0054	0.0179	183.6411	55.7246	38.1793	2127.5244	96
100	3.4634	0.2887	0.0051	0.0176	197.0723	56.9013	39.4058	2242.2411	100
104	3.6398	0.2747	0.0047	0.0172	211.1879	58.0211	40.6038	2355.8757	104
120	4.4402	0.2252	0.0036	0.0161	275.2171	61.9828	45.1184	2796.5694	120
240	19.7155	0.0507	0.0007	0.0132	1497.2395	75.9423	67.1764	5101.5288	240
360	87.5410	0.0114	0.0001	0.0126	6923.2796	79.0861	75.8401	5997.9027	360
480	388.7007	0.0026	0.0000	0.0125	31016.0548	79.7942	78.7619	6284.7442	480

1.5% Compound Interest Factors

	Single Payment		Uniform Series				Arithmetic Gradient		
	Compound Amount Factor	Present Worth Factor	Sinking Fund Factor	Capital Recovery Factor	Compound Amount Factor	Present Worth Factor	Uniform Series	Present Worth	
N	*F/P*	*P/F*	*A/F*	*A/P*	*F/A*	*P/A*	*A/G*	*P/G*	*N*
1	1.0150	0.9852	1.0000	1.0150	1.0000	0.9852	0.0000	0.0000	1
2	1.0302	0.9707	0.4963	0.5113	2.0150	1.9559	0.4963	0.9707	2
3	1.0457	0.9563	0.3284	0.3434	3.0452	2.9122	0.9901	2.8833	3
4	1.0614	0.9422	0.2444	0.2594	4.0909	3.8544	1.4814	5.7098	4
5	1.0773	0.9283	0.1941	0.2091	5.1523	4.7826	1.9702	9.4229	5
6	1.0934	0.9145	0.1605	0.1755	6.2296	5.6972	2.4566	13.9956	6
7	1.1098	0.9010	0.1366	0.1516	7.3230	6.5982	2.9405	19.4018	7
8	1.1265	0.8877	0.1186	0.1336	8.4328	7.4859	3.4219	25.6157	8
9	1.1434	0.8746	0.1046	0.1196	9.5593	8.3605	3.9008	32.6125	9
10	1.1605	0.8617	0.0934	0.1084	10.7027	9.2222	4.3772	40.3675	10
11	1.1779	0.8489	0.0843	0.0993	11.8633	10.0711	4.8512	48.8568	11
12	1.1956	0.8364	0.0767	0.0917	13.0412	10.9075	5.3227	58.0571	12
13	1.2136	0.8240	0.0702	0.0852	14.2368	11.7315	5.7917	67.9454	13
14	1.2318	0.8118	0.0647	0.0797	15.4504	12.5434	6.2582	78.4994	14
15	1.2502	0.7999	0.0599	0.0749	16.6821	13.3432	6.7223	89.6974	15
16	1.2690	0.7880	0.0558	0.0708	17.9324	14.1313	7.1839	101.5178	16
17	1.2880	0.7764	0.0521	0.0671	19.2014	14.9076	7.6431	113.9400	17
18	1.3073	0.7649	0.0488	0.0638	20.4894	15.6726	8.0997	126.9435	18
19	1.3270	0.7536	0.0459	0.0609	21.7967	16.4262	8.5539	140.5084	19
20	1.3469	0.7425	0.0432	0.0582	23.1237	17.1686	9.0057	154.6154	20
21	1.3671	0.7315	0.0409	0.0559	24.4705	17.9001	9.4550	169.2453	21
22	1.3876	0.7207	0.0387	0.0537	25.8376	18.6208	9.9018	184.3798	22
23	1.4084	0.7100	0.0367	0.0517	27.2251	19.3309	10.3462	200.0006	23
24	1.4295	0.6995	0.0349	0.0499	28.6335	20.0304	10.7881	216.0901	24
25	1.4509	0.6892	0.0333	0.0483	30.0630	20.7196	11.2276	232.6310	25
26	1.4727	0.6790	0.0317	0.0467	31.5140	21.3986	11.6646	249.6065	26
27	1.4948	0.6690	0.0303	0.0453	32.9867	22.0676	12.0992	267.0002	27
28	1.5172	0.6591	0.0290	0.0440	34.4815	22.7267	12.5313	284.7958	28
29	1.5400	0.6494	0.0278	0.0428	35.9987	23.3761	12.9610	302.9779	29
30	1.5631	0.6398	0.0266	0.0416	37.5387	24.0158	13.3883	321.5310	30
31	1.5865	0.6303	0.0256	0.0406	39.1018	24.6461	13.8131	340.4402	31
32	1.6103	0.6210	0.0246	0.0396	40.6883	25.2671	14.2355	359.6910	32
33	1.6345	0.6118	0.0236	0.0386	42.2986	25.8790	14.6555	379.2691	33
34	1.6590	0.6028	0.0228	0.0378	43.9331	26.4817	15.0731	399.1607	34
35	1.6839	0.5939	0.0219	0.0369	45.5921	27.0756	15.4882	419.3521	35
36	1.7091	0.5851	0.0212	0.0362	47.2760	27.6607	15.9009	439.8303	36
40	1.8140	0.5513	0.0184	0.0334	54.2679	29.9158	17.5277	524.3568	40
48	2.0435	0.4894	0.0144	0.0294	69.5652	34.0426	20.6667	703.5462	48
50	2.1052	0.4750	0.0136	0.0286	73.6828	34.9997	21.4277	749.9636	50
52	2.1689	0.4611	0.0128	0.0278	77.9249	35.9287	22.1794	796.8774	52
60	2.4432	0.4093	0.0104	0.0254	96.2147	39.3803	25.0930	988.1674	60
70	2.8355	0.3527	0.0082	0.0232	122.3638	43.1549	28.5290	1231.1658	70
72	2.9212	0.3423	0.0078	0.0228	128.0772	43.8447	29.1893	1279.7938	72
80	3.2907	0.3039	0.0065	0.0215	152.7109	46.4073	31.7423	1473.0741	80
84	3.4926	0.2863	0.0060	0.0210	166.1726	47.5786	32.9668	1568.5140	84
90	3.8189	0.2619	0.0053	0.0203	187.9299	49.2099	34.7399	1709.5439	90
96	4.1758	0.2395	0.0047	0.0197	211.7202	50.7017	36.4381	1847.4725	96
100	4.4320	0.2256	0.0044	0.0194	228.8030	51.6247	37.5295	1937.4506	100
104	4.7040	0.2126	0.0040	0.0190	246.9341	52.4944	38.5890	2025.7052	104
120	5.9693	0.1675	0.0030	0.0180	331.2882	55.4985	42.5185	2359.7114	120
240	35.6328	0.0281	0.0004	0.0154	2308.8544	64.7957	59.7368	3870.6912	240
360	212.7038	0.0047	0.0001	0.0151	14113.5854	66.3532	64.9662	4310.7165	360
480	1269.6975	0.0008	0.0000	0.0150	84579.8363	66.6142	66.2883	4415.7412	480

1.75% Compound Interest Factors

	Single Payment		Uniform Series				Arithmetic Gradient		
	Compound Amount Factor	Present Worth Factor	Sinking Fund Factor	Capital Recovery Factor	Compound Amount Factor	Present Worth Factor	Uniform Series	Present Worth	
N	*F/P*	*P/F*	*A/F*	*A/P*	*F/A*	*P/A*	*A/G*	*P/G*	*N*
1	1.0175	0.9828	1.0000	1.0175	1.0000	0.9828	0.0000	0.0000	1
2	1.0353	0.9659	0.4957	0.5132	2.0175	1.9487	0.4957	0.9659	2
3	1.0534	0.9493	0.3276	0.3451	3.0528	2.8980	0.9884	2.8645	3
4	1.0719	0.9330	0.2435	0.2610	4.1062	3.8309	1.4783	5.6633	4
5	1.0906	0.9169	0.1931	0.2106	5.1781	4.7479	1.9653	9.3310	5
6	1.1097	0.9011	0.1595	0.1770	6.2687	5.6490	2.4494	13.8367	6
7	1.1291	0.8856	0.1355	0.1530	7.3784	6.5346	2.9306	19.1506	7
8	1.1489	0.8704	0.1175	0.1350	8.5075	7.4051	3.4089	25.2435	8
9	1.1690	0.8554	0.1036	0.1211	9.6564	8.2605	3.8844	32.0870	9
10	1.1894	0.8407	0.0924	0.1099	10.8254	9.1012	4.3569	39.6535	10
11	1.2103	0.8263	0.0832	0.1007	12.0148	9.9275	4.8266	47.9162	11
12	1.2314	0.8121	0.0756	0.0931	13.2251	10.7395	5.2934	56.8489	12
13	1.2530	0.7981	0.0692	0.0867	14.4565	11.5376	5.7573	66.4260	13
14	1.2749	0.7844	0.0637	0.0812	15.7095	12.3220	6.2184	76.6227	14
15	1.2972	0.7709	0.0589	0.0764	16.9844	13.0929	6.6765	87.4149	15
16	1.3199	0.7576	0.0547	0.0722	18.2817	13.8505	7.1318	98.7792	16
17	1.3430	0.7446	0.0510	0.0685	19.6016	14.5951	7.5842	110.6926	17
18	1.3665	0.7318	0.0477	0.0652	20.9446	15.3269	8.0338	123.1328	18
19	1.3904	0.7192	0.0448	0.0623	22.3112	16.0461	8.4805	136.0783	19
20	1.4148	0.7068	0.0422	0.0597	23.7016	16.7529	8.9243	149.5080	20
21	1.4395	0.6947	0.0398	0.0573	25.1164	17.4475	9.3653	163.4013	21
22	1.4647	0.6827	0.0377	0.0552	26.5559	18.1303	9.8034	177.7385	22
23	1.4904	0.6710	0.0357	0.0532	28.0207	18.8012	10.2387	192.5000	23
24	1.5164	0.6594	0.0339	0.0514	29.5110	19.4607	10.6711	207.6671	24
25	1.5430	0.6481	0.0322	0.0497	31.0275	20.1088	11.1007	223.2214	25
26	1.5700	0.6369	0.0307	0.0482	32.5704	20.7457	11.5274	239.1451	26
27	1.5975	0.6260	0.0293	0.0468	34.1404	21.3717	11.9513	255.4210	27
28	1.6254	0.6152	0.0280	0.0455	35.7379	21.9870	12.3724	272.0321	28
29	1.6539	0.6046	0.0268	0.0443	37.3633	22.5916	12.7907	288.9623	29
30	1.6828	0.5942	0.0256	0.0431	39.0172	23.1858	13.2061	306.1954	30
31	1.7122	0.5840	0.0246	0.0421	40.7000	23.7699	13.6188	323.7163	31
32	1.7422	0.5740	0.0236	0.0411	42.4122	24.3439	14.0286	341.5097	32
33	1.7727	0.5641	0.0226	0.0401	44.1544	24.9080	14.4356	359.5613	33
34	1.8037	0.5544	0.0218	0.0393	45.9271	25.4624	14.8398	377.8567	34
35	1.8353	0.5449	0.0210	0.0385	47.7308	26.0073	15.2412	396.3824	35
36	1.8674	0.5355	0.0202	0.0377	49.5661	26.5428	15.6399	415.1250	36
40	2.0016	0.4996	0.0175	0.0350	57.2341	28.5942	17.2066	492.0109	40
48	2.2996	0.4349	0.0135	0.0310	74.2628	32.2938	20.2084	652.6054	48
50	2.3808	0.4200	0.0127	0.0302	78.9022	33.1412	20.9317	693.7010	50
52	2.4648	0.4057	0.0119	0.0294	83.7055	33.9597	21.6442	735.0322	52
60	2.8318	0.3531	0.0096	0.0271	104.6752	36.9640	24.3885	901.4954	60
70	3.3683	0.2969	0.0074	0.0249	135.3308	40.1779	27.5856	1108.3333	70
72	3.4872	0.2868	0.0070	0.0245	142.1263	40.7564	28.1948	1149.1181	72
80	4.0064	0.2496	0.0058	0.0233	171.7938	42.8799	30.5329	1309.2482	80
84	4.2943	0.2329	0.0053	0.0228	188.2450	43.8361	31.6442	1387.1584	84
90	4.7654	0.2098	0.0046	0.0221	215.1646	45.1516	33.2409	1500.8798	90
96	5.2882	0.1891	0.0041	0.0216	245.0374	46.3370	34.7556	1610.4716	96
100	5.6682	0.1764	0.0037	0.0212	266.7518	47.0615	35.7211	1681.0886	100
104	6.0755	0.1646	0.0034	0.0209	290.0265	47.7373	36.6521	1749.6749	104
120	8.0192	0.1247	0.0025	0.0200	401.0962	50.0171	40.0469	2003.0269	120
240	64.3073	0.0156	0.0003	0.0178	3617.5602	56.2543	53.3518	3001.2678	240
360	515.6921	0.0019	0.0000	0.0175	29410.9747	57.0320	56.4434	3219.0833	360
480	4135.4292	0.0002	0.0000	0.0175	236253.0975	57.1290	57.0268	3257.8839	480

2% Compound Interest Factors

	Single Payment		Uniform Series				Arithmetic Gradient		
	Compound Amount Factor	Present Worth Factor	Sinking Fund Factor	Capital Recovery Factor	Compound Amount Factor	Present Worth Factor	Uniform Series	Present Worth	
N	*F/P*	*P/F*	*A/F*	*A/P*	*F/A*	*P/A*	*A/G*	*P/G*	*N*
1	1.0200	0.9804	1.0000	1.0200	1.0000	0.9804	0.0000	0.0000	1
2	1.0404	0.9612	0.4950	0.5150	2.0200	1.9416	0.4950	0.9612	2
3	1.0612	0.9423	0.3268	0.3468	3.0604	2.8839	0.9868	2.8458	3
4	1.0824	0.9238	0.2426	0.2626	4.1216	3.8077	1.4752	5.6173	4
5	1.1041	0.9057	0.1922	0.2122	5.2040	4.7135	1.9604	9.2403	5
6	1.1262	0.8880	0.1585	0.1785	6.3081	5.6014	2.4423	13.6801	6
7	1.1487	0.8706	0.1345	0.1545	7.4343	6.4720	2.9208	18.9035	7
8	1.1717	0.8535	0.1165	0.1365	8.5830	7.3255	3.3961	24.8779	8
9	1.1951	0.8368	0.1025	0.1225	9.7546	8.1622	3.8681	31.5720	9
10	1.2190	0.8203	0.0913	0.1113	10.9497	8.9826	4.3367	38.9551	10
11	1.2434	0.8043	0.0822	0.1022	12.1687	9.7868	4.8021	46.9977	11
12	1.2682	0.7885	0.0746	0.0946	13.4121	10.5753	5.2642	55.6712	12
13	1.2936	0.7730	0.0681	0.0881	14.6803	11.3484	5.7231	64.9475	13
14	1.3195	0.7579	0.0626	0.0826	15.9739	12.1062	6.1786	74.7999	14
15	1.3459	0.7430	0.0578	0.0778	17.2934	12.8493	6.6309	85.2021	15
16	1.3728	0.7284	0.0537	0.0737	18.6393	13.5777	7.0799	96.1288	16
17	1.4002	0.7142	0.0500	0.0700	20.0121	14.2919	7.5256	107.5554	17
18	1.4282	0.7002	0.0467	0.0667	21.4123	14.9920	7.9681	119.4581	18
19	1.4568	0.6864	0.0438	0.0638	22.8406	15.6785	8.4073	131.8139	19
20	1.4859	0.6730	0.0412	0.0612	24.2974	16.3514	8.8433	144.6003	20
21	1.5157	0.6598	0.0388	0.0588	25.7833	17.0112	9.2760	157.7959	21
22	1.5460	0.6468	0.0366	0.0566	27.2990	17.6580	9.7055	171.3795	22
23	1.5769	0.6342	0.0347	0.0547	28.8450	18.2922	10.1317	185.3309	23
24	1.6084	0.6217	0.0329	0.0529	30.4219	18.9139	10.5547	199.6305	24
25	1.6406	0.6095	0.0312	0.0512	32.0303	19.5235	10.9745	214.2592	25
26	1.6734	0.5976	0.0297	0.0497	33.6709	20.1210	11.3910	229.1987	26
27	1.7069	0.5859	0.0283	0.0483	35.3443	20.7069	11.8043	244.4311	27
28	1.7410	0.5744	0.0270	0.0470	37.0512	21.2813	12.2145	259.9392	28
29	1.7758	0.5631	0.0258	0.0458	38.7922	21.8444	12.6214	275.7064	29
30	1.8114	0.5521	0.0246	0.0446	40.5681	22.3965	13.0251	291.7164	30
31	1.8476	0.5412	0.0236	0.0436	42.3794	22.9377	13.4257	307.9538	31
32	1.8845	0.5306	0.0226	0.0426	44.2270	23.4683	13.8230	324.4035	32
33	1.9222	0.5202	0.0217	0.0417	46.1116	23.9886	14.2172	341.0508	33
34	1.9607	0.5100	0.0208	0.0408	48.0338	24.4986	14.6083	357.8817	34
35	1.9999	0.5000	0.0200	0.0400	49.9945	24.9986	14.9961	374.8826	35
36	2.0399	0.4902	0.0192	0.0392	51.9944	25.4888	15.3809	392.0405	36
40	2.2080	0.4529	0.0166	0.0366	60.4020	27.3555	16.8885	461.9931	40
48	2.5871	0.3865	0.0126	0.0326	79.3535	30.6731	19.7556	605.9657	48
50	2.6916	0.3715	0.0118	0.0318	84.5794	31.4236	20.4420	642.3606	50
52	2.8003	0.3571	0.0111	0.0311	90.0164	32.1449	21.1164	678.7849	52
60	3.2810	0.3048	0.0088	0.0288	114.0515	34.7609	23.6961	823.6975	60
70	3.9996	0.2500	0.0067	0.0267	149.9779	37.4986	26.6632	999.8343	70
72	4.1611	0.2403	0.0063	0.0263	158.0570	37.9841	27.2234	1034.0557	72
80	4.8754	0.2051	0.0052	0.0252	193.7720	39.7445	29.3572	1166.7868	80
84	5.2773	0.1895	0.0047	0.0247	213.8666	40.5255	30.3616	1230.4191	84
90	5.9431	0.1683	0.0040	0.0240	247.1567	41.5869	31.7929	1322.1701	90
96	6.6929	0.1494	0.0035	0.0235	284.6467	42.5294	33.1370	1409.2973	96
100	7.2446	0.1380	0.0032	0.0232	312.2323	43.0984	33.9863	1464.7527	100
104	7.8418	0.1275	0.0029	0.0229	342.0919	43.6239	34.7994	1518.0873	104
120	10.7652	0.0929	0.0020	0.0220	488.2582	45.3554	37.7114	1710.4160	120
240	115.8887	0.0086	0.0002	0.0202	5744.4368	49.5686	47.9110	2374.8800	240
360	1247.5611	0.0008	0.0000	0.0200	62328.0564	49.9599	49.7112	2483.5679	360
480	13430.1989	0.0001	0.0000	0.0200	671459.9468	49.9963	49.9643	2498.0268	480

2.5% Compound Interest Factors

	Single Payment		Uniform Series				Arithmetic Gradient		
	Compound Amount Factor	Present Worth Factor	Sinking Fund Factor	Capital Recovery Factor	Compound Amount Factor	Present Worth Factor	Uniform Series	Present Worth	
N	*F/P*	*P/F*	*A/F*	*A/P*	*F/A*	*P/A*	*A/G*	*P/G*	*N*
1	1.0250	0.9756	1.0000	1.0250	1.0000	0.9756	0.0000	0.0000	1
2	1.0506	0.9518	0.4938	0.5188	2.0250	1.9274	0.4938	0.9518	2
3	1.0769	0.9286	0.3251	0.3501	3.0756	2.8560	0.9835	2.8090	3
4	1.1038	0.9060	0.2408	0.2658	4.1525	3.7620	1.4691	5.5269	4
5	1.1314	0.8839	0.1902	0.2152	5.2563	4.6458	1.9506	9.0623	5
6	1.1597	0.8623	0.1565	0.1815	6.3877	5.5081	2.4280	13.3738	6
7	1.1887	0.8413	0.1325	0.1575	7.5474	6.3494	2.9013	18.4214	7
8	1.2184	0.8207	0.1145	0.1395	8.7361	7.1701	3.3704	24.1666	8
9	1.2489	0.8007	0.1005	0.1255	9.9545	7.9709	3.8355	30.5724	9
10	1.2801	0.7812	0.0893	0.1143	11.2034	8.7521	4.2965	37.6032	10
11	1.3121	0.7621	0.0801	0.1051	12.4835	9.5142	4.7534	45.2246	11
12	1.3449	0.7436	0.0725	0.0975	13.7956	10.2578	5.2062	53.4038	12
13	1.3785	0.7254	0.0660	0.0910	15.1404	10.9832	5.6549	62.1088	13
14	1.4130	0.7077	0.0605	0.0855	16.5190	11.6909	6.0995	71.3093	14
15	1.4483	0.6905	0.0558	0.0808	17.9319	12.3814	6.5401	80.9758	15
16	1.4845	0.6736	0.0516	0.0766	19.3802	13.0550	6.9766	91.0801	16
17	1.5216	0.6572	0.0479	0.0729	20.8647	13.7122	7.4091	101.5953	17
18	1.5597	0.6412	0.0447	0.0697	22.3863	14.3534	7.8375	112.4951	18
19	1.5987	0.6255	0.0418	0.0668	23.9460	14.9789	8.2619	123.7546	19
20	1.6386	0.6103	0.0391	0.0641	25.5447	15.5892	8.6823	135.3497	20
21	1.6796	0.5954	0.0368	0.0618	27.1833	16.1845	9.0986	147.2575	21
22	1.7216	0.5809	0.0346	0.0596	28.8629	16.7654	9.5110	159.4556	22
23	1.7646	0.5667	0.0327	0.0577	30.5844	17.3321	9.9193	171.9230	23
24	1.8087	0.5529	0.0309	0.0559	32.3490	17.8850	10.3237	184.6391	24
25	1.8539	0.5394	0.0293	0.0543	34.1578	18.4244	10.7241	197.5845	25
26	1.9003	0.5262	0.0278	0.0528	36.0117	18.9506	11.1205	210.7403	26
27	1.9478	0.5134	0.0264	0.0514	37.9120	19.4640	11.5130	224.0887	27
28	1.9965	0.5009	0.0251	0.0501	39.8598	19.9649	11.9015	237.6124	28
29	2.0464	0.4887	0.0239	0.0489	41.8563	20.4535	12.2861	251.2949	29
30	2.0976	0.4767	0.0228	0.0478	43.9027	20.9303	12.6668	265.1205	30
31	2.1500	0.4651	0.0217	0.0467	46.0003	21.3954	13.0436	279.0739	31
32	2.2038	0.4538	0.0208	0.0458	48.1503	21.8492	13.4166	293.1408	32
33	2.2589	0.4427	0.0199	0.0449	50.3540	22.2919	13.7856	307.3073	33
34	2.3153	0.4319	0.0190	0.0440	52.6129	22.7238	14.1508	321.5602	34
35	2.3732	0.4214	0.0182	0.0432	54.9282	23.1452	14.5122	335.8868	35
40	2.6851	0.3724	0.0148	0.0398	67.4026	25.1028	16.2620	408.2220	40
45	3.0379	0.3292	0.0123	0.0373	81.5161	26.8330	17.9185	480.8070	45
50	3.4371	0.2909	0.0103	0.0353	97.4843	28.3623	19.4839	552.6081	50
55	3.8888	0.2572	0.0087	0.0337	115.5509	29.7140	20.9608	622.8280	55
60	4.3998	0.2273	0.0074	0.0324	135.9916	30.9087	22.3518	690.8656	60
65	4.9780	0.2009	0.0063	0.0313	159.1183	31.9646	23.6600	756.2806	65
70	5.6321	0.1776	0.0054	0.0304	185.2841	32.8979	24.8881	818.7643	70
75	6.3722	0.1569	0.0047	0.0297	214.8883	33.7227	26.0393	878.1152	75
80	7.2096	0.1387	0.0040	0.0290	248.3827	34.4518	27.1167	934.2181	80
85	8.1570	0.1226	0.0035	0.0285	286.2786	35.0962	28.1235	987.0269	85
90	9.2289	0.1084	0.0030	0.0280	329.1543	35.6658	29.0629	1036.5499	90
95	10.4416	0.0958	0.0026	0.0276	377.6642	36.1692	29.9382	1082.8381	95
100	11.8137	0.0846	0.0023	0.0273	432.5487	36.6141	30.7525	1125.9747	100

3% Compound Interest Factors

	Single Payment		Uniform Series				Arithmetic Gradient		
	Compound Amount Factor	Present Worth Factor	Sinking Fund Factor	Capital Recovery Factor	Compound Amount Factor	Present Worth Factor	Uniform Series	Present Worth	
N	*F/P*	*P/F*	*A/F*	*A/P*	*F/A*	*P/A*	*A/G*	*P/G*	*N*
1	1.0300	0.9709	1.0000	1.0300	1.0000	0.9709	0.0000	0.0000	1
2	1.0609	0.9426	0.4926	0.5226	2.0300	1.9135	0.4926	0.9426	2
3	1.0927	0.9151	0.3235	0.3535	3.0909	2.8286	0.9803	2.7729	3
4	1.1255	0.8885	0.2390	0.2690	4.1836	3.7171	1.4631	5.4383	4
5	1.1593	0.8626	0.1884	0.2184	5.3091	4.5797	1.9409	8.8888	5
6	1.1941	0.8375	0.1546	0.1846	6.4684	5.4172	2.4138	13.0762	6
7	1.2299	0.8131	0.1305	0.1605	7.6625	6.2303	2.8819	17.9547	7
8	1.2668	0.7894	0.1125	0.1425	8.8923	7.0197	3.3450	23.4806	8
9	1.3048	0.7664	0.0984	0.1284	10.1591	7.7861	3.8032	29.6119	9
10	1.3439	0.7441	0.0872	0.1172	11.4639	8.5302	4.2565	36.3088	10
11	1.3842	0.7224	0.0781	0.1081	12.8078	9.2526	4.7049	43.5330	11
12	1.4258	0.7014	0.0705	0.1005	14.1920	9.9540	5.1485	51.2482	12
13	1.4685	0.6810	0.0640	0.0940	15.6178	10.6350	5.5872	59.4196	13
14	1.5126	0.6611	0.0585	0.0885	17.0863	11.2961	6.0210	68.0141	14
15	1.5580	0.6419	0.0538	0.0838	18.5989	11.9379	6.4500	77.0002	15
16	1.6047	0.6232	0.0496	0.0796	20.1569	12.5611	6.8742	86.3477	16
17	1.6528	0.6050	0.0460	0.0760	21.7616	13.1661	7.2936	96.0280	17
18	1.7024	0.5874	0.0427	0.0727	23.4144	13.7535	7.7081	106.0137	18
19	1.7535	0.5703	0.0398	0.0698	25.1169	14.3238	8.1179	116.2788	19
20	1.8061	0.5537	0.0372	0.0672	26.8704	14.8775	8.5229	126.7987	20
21	1.8603	0.5375	0.0349	0.0649	28.6765	15.4150	8.9231	137.5496	21
22	1.9161	0.5219	0.0327	0.0627	30.5368	15.9369	9.3186	148.5094	22
23	1.9736	0.5067	0.0308	0.0608	32.4529	16.4436	9.7093	159.6566	23
24	2.0328	0.4919	0.0290	0.0590	34.4265	16.9355	10.0954	170.9711	24
25	2.0938	0.4776	0.0274	0.0574	36.4593	17.4131	10.4768	182.4336	25
26	2.1566	0.4637	0.0259	0.0559	38.5530	17.8768	10.8535	194.0260	26
27	2.2213	0.4502	0.0246	0.0546	40.7096	18.3270	11.2255	205.7309	27
28	2.2879	0.4371	0.0233	0.0533	42.9309	18.7641	11.5930	217.5320	28
29	2.3566	0.4243	0.0221	0.0521	45.2189	19.1885	11.9558	229.4137	29
30	2.4273	0.4120	0.0210	0.0510	47.5754	19.6004	12.3141	241.3613	30
31	2.5001	0.4000	0.0200	0.0500	50.0027	20.0004	12.6678	253.3609	31
32	2.5751	0.3883	0.0190	0.0490	52.5028	20.3888	13.0169	265.3993	32
33	2.6523	0.3770	0.0182	0.0482	55.0778	20.7658	13.3616	277.4642	33
34	2.7319	0.3660	0.0173	0.0473	57.7302	21.1318	13.7018	289.5437	34
35	2.8139	0.3554	0.0165	0.0465	60.4621	21.4872	14.0375	301.6267	35
40	3.2620	0.3066	0.0133	0.0433	75.4013	23.1148	15.6502	361.7499	40
45	3.7816	0.2644	0.0108	0.0408	92.7199	24.5187	17.1556	420.6325	45
50	4.3839	0.2281	0.0089	0.0389	112.7969	25.7298	18.5575	477.4803	50
55	5.0821	0.1968	0.0073	0.0373	136.0716	26.7744	19.8600	531.7411	55
60	5.8916	0.1697	0.0061	0.0361	163.0534	27.6756	21.0674	583.0526	60
65	6.8300	0.1464	0.0051	0.0351	194.3328	28.4529	22.1841	631.2010	65
70	7.9178	0.1263	0.0043	0.0343	230.5941	29.1234	23.2145	676.0869	70
75	9.1789	0.1089	0.0037	0.0337	272.6309	29.7018	24.1634	717.6978	75
80	10.6409	0.0940	0.0031	0.0331	321.3630	30.2008	25.0353	756.0865	80
85	12.3357	0.0811	0.0026	0.0326	377.8570	30.6312	25.8349	791.3529	85
90	14.3005	0.0699	0.0023	0.0323	443.3489	31.0024	26.5667	823.6302	90
95	16.5782	0.0603	0.0019	0.0319	519.2720	31.3227	27.2351	853.0742	95
100	19.2186	0.0520	0.0016	0.0316	607.2877	31.5989	27.8444	879.8540	100

3.5% Compound Interest Factors

	Single Payment		Uniform Series				Arithmetic Gradient		
	Compound Amount Factor	Present Worth Factor	Sinking Fund Factor	Capital Recovery Factor	Compound Amount Factor	Present Worth Factor	Uniform Series	Present Worth	
N	*F/P*	*P/F*	*A/F*	*A/P*	*F/A*	*P/A*	*A/G*	*P/G*	*N*
1	1.0350	0.9662	1.0000	1.0350	1.0000	0.9662	0.0000	0.0000	1
2	1.0712	0.9335	0.4914	0.5264	2.0350	1.8997	0.4914	0.9335	2
3	1.1087	0.9019	0.3219	0.3569	3.1062	2.8016	0.9771	2.7374	3
4	1.1475	0.8714	0.2373	0.2723	4.2149	3.6731	1.4570	5.3517	4
5	1.1877	0.8420	0.1865	0.2215	5.3625	4.5151	1.9312	8.7196	5
6	1.2293	0.8135	0.1527	0.1877	6.5502	5.3286	2.3997	12.7871	6
7	1.2723	0.7860	0.1285	0.1635	7.7794	6.1145	2.8625	17.5031	7
8	1.3168	0.7594	0.1105	0.1455	9.0517	6.8740	3.3196	22.8189	8
9	1.3629	0.7337	0.0964	0.1314	10.3685	7.6077	3.7710	28.6888	9
10	1.4106	0.7089	0.0852	0.1202	11.7314	8.3166	4.2168	35.0691	10
11	1.4600	0.6849	0.0761	0.1111	13.1420	9.0016	4.6568	41.9185	11
12	1.5111	0.6618	0.0685	0.1035	14.6020	9.6633	5.0912	49.1981	12
13	1.5640	0.6394	0.0621	0.0971	16.1130	10.3027	5.5200	56.8710	13
14	1.6187	0.6178	0.0566	0.0916	17.6770	10.9205	5.9431	64.9021	14
15	1.6753	0.5969	0.0518	0.0868	19.2957	11.5174	6.3607	73.2586	15
16	1.7340	0.5767	0.0477	0.0827	20.9710	12.0941	6.7726	81.9092	16
17	1.7947	0.5572	0.0440	0.0790	22.7050	12.6513	7.1791	90.8245	17
18	1.8575	0.5384	0.0408	0.0758	24.4997	13.1897	7.5799	99.9766	18
19	1.9225	0.5202	0.0379	0.0729	26.3572	13.7098	7.9753	109.3394	19
20	1.9898	0.5026	0.0354	0.0704	28.2797	14.2124	8.3651	118.8882	20
21	2.0594	0.4856	0.0330	0.0680	30.2695	14.6980	8.7495	128.5996	21
22	2.1315	0.4692	0.0309	0.0659	32.3289	15.1671	9.1284	138.4517	22
23	2.2061	0.4533	0.0290	0.0640	34.4604	15.6204	9.5019	148.4240	23
24	2.2833	0.4380	0.0273	0.0623	36.6665	16.0584	9.8701	158.4970	24
25	2.3632	0.4231	0.0257	0.0607	38.9499	16.4815	10.2328	168.6526	25
26	2.4460	0.4088	0.0242	0.0592	41.3131	16.8904	10.5903	178.8735	26
27	2.5316	0.3950	0.0229	0.0579	43.7591	17.2854	10.9424	189.1438	27
28	2.6202	0.3817	0.0216	0.0566	46.2906	17.6670	11.2893	199.4485	28
29	2.7119	0.3687	0.0204	0.0554	48.9108	18.0358	11.6310	209.7734	29
30	2.8068	0.3563	0.0194	0.0544	51.6227	18.3920	11.9674	220.1055	30
31	2.9050	0.3442	0.0184	0.0534	54.4295	18.7363	12.2987	230.4324	31
32	3.0067	0.3326	0.0174	0.0524	57.3345	19.0689	12.6249	240.7427	32
33	3.1119	0.3213	0.0166	0.0516	60.3412	19.3902	12.9460	251.0257	33
34	3.2209	0.3105	0.0158	0.0508	63.4532	19.7007	13.2620	261.2714	34
35	3.3336	0.3000	0.0150	0.0500	66.6740	20.0007	13.5731	271.4706	35
40	3.9593	0.2526	0.0118	0.0468	84.5503	21.3551	15.0545	321.4907	40
45	4.7024	0.2127	0.0095	0.0445	105.7817	22.4955	16.4170	369.3081	45
50	5.5849	0.1791	0.0076	0.0426	130.9979	23.4556	17.6661	414.3700	50
55	6.6331	0.1508	0.0062	0.0412	160.9469	24.2641	18.8078	456.3530	55
60	7.8781	0.1269	0.0051	0.0401	196.5169	24.9447	19.8481	495.1050	60
65	9.3567	0.1069	0.0042	0.0392	238.7629	25.5178	20.7932	530.5987	65
70	11.1128	0.0900	0.0035	0.0385	288.9379	26.0004	21.6495	562.8962	70
75	13.1986	0.0758	0.0029	0.0379	348.5300	26.4067	22.4232	592.1213	75
80	15.6757	0.0638	0.0024	0.0374	419.3068	26.7488	23.1203	618.4385	80
85	18.6179	0.0537	0.0020	0.0370	503.3674	27.0368	23.7468	642.0370	85
90	22.1122	0.0452	0.0017	0.0367	603.2050	27.2793	24.3085	663.1189	90
95	26.2623	0.0381	0.0014	0.0364	721.7808	27.4835	24.8109	681.8902	95
100	31.1914	0.0321	0.0012	0.0362	862.6117	27.6554	25.2592	698.5547	100

4% Compound Interest Factors

	Single Payment		Uniform Series				Arithmetic Gradient		
	Compound Amount Factor	Present Worth Factor	Sinking Fund Factor	Capital Recovery Factor	Compound Amount Factor	Present Worth Factor	Uniform Series	Present Worth	
N	*F/P*	*P/F*	*A/F*	*A/P*	*F/A*	*P/A*	*A/G*	*P/G*	*N*
1	1.0400	0.9615	1.0000	1.0400	1.0000	0.9615	0.0000	0.0000	1
2	1.0816	0.9246	0.4902	0.5302	2.0400	1.8861	0.4902	0.9246	2
3	1.1249	0.8890	0.3203	0.3603	3.1216	2.7751	0.9739	2.7025	3
4	1.1699	0.8548	0.2355	0.2755	4.2465	3.6299	1.4510	5.2670	4
5	1.2167	0.8219	0.1846	0.2246	5.4163	4.4518	1.9216	8.5547	5
6	1.2653	0.7903	0.1508	0.1908	6.6330	5.2421	2.3857	12.5062	6
7	1.3159	0.7599	0.1266	0.1666	7.8983	6.0021	2.8433	17.0657	7
8	1.3686	0.7307	0.1085	0.1485	9.2142	6.7327	3.2944	22.1806	8
9	1.4233	0.7026	0.0945	0.1345	10.5828	7.4353	3.7391	27.8013	9
10	1.4802	0.6756	0.0833	0.1233	12.0061	8.1109	4.1773	33.8814	10
11	1.5395	0.6496	0.0741	0.1141	13.4864	8.7605	4.6090	40.3772	11
12	1.6010	0.6246	0.0666	0.1066	15.0258	9.3851	5.0343	47.2477	12
13	1.6651	0.6006	0.0601	0.1001	16.6268	9.9856	5.4533	54.4546	13
14	1.7317	0.5775	0.0547	0.0947	18.2919	10.5631	5.8659	61.9618	14
15	1.8009	0.5553	0.0499	0.0899	20.0236	11.1184	6.2721	69.7355	15
16	1.8730	0.5339	0.0458	0.0858	21.8245	11.6523	6.6720	77.7441	16
17	1.9479	0.5134	0.0422	0.0822	23.6975	12.1657	7.0656	85.9581	17
18	2.0258	0.4936	0.0390	0.0790	25.6454	12.6593	7.4530	94.3498	18
19	2.1068	0.4746	0.0361	0.0761	27.6712	13.1339	7.8342	102.8933	19
20	2.1911	0.4564	0.0336	0.0736	29.7781	13.5903	8.2091	111.5647	20
21	2.2788	0.4388	0.0313	0.0713	31.9692	14.0292	8.5779	120.3414	21
22	2.3699	0.4220	0.0292	0.0692	34.2480	14.4511	8.9407	129.2024	22
23	2.4647	0.4057	0.0273	0.0673	36.6179	14.8568	9.2973	138.1284	23
24	2.5633	0.3901	0.0256	0.0656	39.0826	15.2470	9.6479	147.1012	24
25	2.6658	0.3751	0.0240	0.0640	41.6459	15.6221	9.9925	156.1040	25
26	2.7725	0.3607	0.0226	0.0626	44.3117	15.9828	10.3312	165.1212	26
27	2.8834	0.3468	0.0212	0.0612	47.0842	16.3296	10.6640	174.1385	27
28	2.9987	0.3335	0.0200	0.0600	49.9676	16.6631	10.9909	183.1424	28
29	3.1187	0.3207	0.0189	0.0589	52.9663	16.9837	11.3120	192.1206	29
30	3.2434	0.3083	0.0178	0.0578	56.0849	17.2920	11.6274	201.0618	30
31	3.3731	0.2965	0.0169	0.0569	59.3283	17.5885	11.9371	209.9556	31
32	3.5081	0.2851	0.0159	0.0559	62.7015	17.8736	12.2411	218.7924	32
33	3.6484	0.2741	0.0151	0.0551	66.2095	18.1476	12.5396	227.5634	33
34	3.7943	0.2636	0.0143	0.0543	69.8579	18.4112	12.8324	236.2607	34
35	3.9461	0.2534	0.0136	0.0536	73.6522	18.6646	13.1198	244.8768	35
40	4.8010	0.2083	0.0105	0.0505	95.0255	19.7928	14.4765	286.5303	40
45	5.8412	0.1712	0.0083	0.0483	121.0294	20.7200	15.7047	325.4028	45
50	7.1067	0.1407	0.0066	0.0466	152.6671	21.4822	16.8122	361.1638	50
55	8.6464	0.1157	0.0052	0.0452	191.1592	22.1086	17.8070	393.6890	55
60	10.5196	0.0951	0.0042	0.0442	237.9907	22.6235	18.6972	422.9966	60
65	12.7987	0.0781	0.0034	0.0434	294.9684	23.0467	19.4909	449.2014	65
70	15.5716	0.0642	0.0027	0.0427	364.2905	23.3945	20.1961	472.4789	70
75	18.9453	0.0528	0.0022	0.0422	448.6314	23.6804	20.8206	493.0408	75
80	23.0498	0.0434	0.0018	0.0418	551.2450	23.9154	21.3718	511.1161	80
85	28.0436	0.0357	0.0015	0.0415	676.0901	24.1085	21.8569	526.9384	85
90	34.1193	0.0293	0.0012	0.0412	827.9833	24.2673	22.2826	540.7369	90
95	41.5114	0.0241	0.0010	0.0410	1012.7846	24.3978	22.6550	552.7307	95
100	50.5049	0.0198	0.0008	0.0408	1237.6237	24.5050	22.9800	563.1249	100

4.5% Compound Interest Factors

	Single Payment		Uniform Series				Arithmetic Gradient		
N	Compound Amount Factor F/P	Present Worth Factor P/F	Sinking Fund Factor A/F	Capital Recovery Factor A/P	Compound Amount Factor F/A	Present Worth Factor P/A	Uniform Series A/G	Present Worth P/G	N
1	1.0450	0.9569	1.0000	1.0450	1.0000	0.9569	0.0000	0.0000	1
2	1.0920	0.9157	0.4890	0.5340	2.0450	1.8727	0.4890	0.9157	2
3	1.1412	0.8763	0.3188	0.3638	3.1370	2.7490	0.9707	2.6683	3
4	1.1925	0.8386	0.2337	0.2787	4.2782	3.5875	1.4450	5.1840	4
5	1.2462	0.8025	0.1828	0.2278	5.4707	4.3900	1.9120	8.3938	5
6	1.3023	0.7679	0.1489	0.1939	6.7169	5.1579	2.3718	12.2333	6
7	1.3609	0.7348	0.1247	0.1697	8.0192	5.8927	2.8242	16.6423	7
8	1.4221	0.7032	0.1066	0.1516	9.3800	6.5959	3.2694	21.5646	8
9	1.4861	0.6729	0.0926	0.1376	10.8021	7.2688	3.7073	26.9478	9
10	1.5530	0.6439	0.0814	0.1264	12.2882	7.9127	4.1380	32.7431	10
11	1.6229	0.6162	0.0722	0.1172	13.8412	8.5289	4.5616	38.9051	11
12	1.6959	0.5897	0.0647	0.1097	15.4640	9.1186	4.9779	45.3914	12
13	1.7722	0.5643	0.0583	0.1033	17.1599	9.6829	5.3871	52.1627	13
14	1.8519	0.5400	0.0528	0.0978	18.9321	10.2228	5.7892	59.1823	14
15	1.9353	0.5167	0.0481	0.0931	20.7841	10.7395	6.1843	66.4164	15
16	2.0224	0.4945	0.0440	0.0890	22.7193	11.2340	6.5723	73.8335	16
17	2.1134	0.4732	0.0404	0.0854	24.7417	11.7072	6.9534	81.4043	17
18	2.2085	0.4528	0.0372	0.0822	26.8551	12.1600	7.3275	89.1019	18
19	2.3079	0.4333	0.0344	0.0794	29.0636	12.5933	7.6947	96.9013	19
20	2.4117	0.4146	0.0319	0.0769	31.3714	13.0079	8.0550	104.7795	20
21	2.5202	0.3968	0.0296	0.0746	33.7831	13.4047	8.4086	112.7153	21
22	2.6337	0.3797	0.0275	0.0725	36.3034	13.7844	8.7555	120.6890	22
23	2.7522	0.3634	0.0257	0.0707	38.9370	14.1478	9.0956	128.6827	23
24	2.8760	0.3477	0.0240	0.0690	41.6892	14.4955	9.4291	136.6799	24
25	3.0054	0.3327	0.0224	0.0674	44.5652	14.8282	9.7561	144.6654	25
26	3.1407	0.3184	0.0210	0.0660	47.5706	15.1466	10.0765	152.6255	26
27	3.2820	0.3047	0.0197	0.0647	50.7113	15.4513	10.3905	160.5475	27
28	3.4297	0.2916	0.0185	0.0635	53.9933	15.7429	10.6982	168.4199	28
29	3.5840	0.2790	0.0174	0.0624	57.4230	16.0219	10.9995	176.2323	29
30	3.7453	0.2670	0.0164	0.0614	61.0071	16.2889	11.2945	183.9753	30
31	3.9139	0.2555	0.0154	0.0604	64.7524	16.5444	11.5834	191.6404	31
32	4.0900	0.2445	0.0146	0.0596	68.6662	16.7889	11.8662	199.2199	32
33	4.2740	0.2340	0.0137	0.0587	72.7562	17.0229	12.1429	206.7069	33
34	4.4664	0.2239	0.0130	0.0580	77.0303	17.2468	12.4137	214.0955	34
35	4.6673	0.2143	0.0123	0.0573	81.4966	17.4610	12.6785	221.3802	35
40	5.8164	0.1719	0.0093	0.0543	107.0303	18.4016	13.9172	256.0986	40
45	7.2482	0.1380	0.0072	0.0522	138.8500	19.1563	15.0202	287.7322	45
50	9.0326	0.1107	0.0056	0.0506	178.5030	19.7620	15.9976	316.1450	50
55	11.2563	0.0888	0.0044	0.0494	227.9180	20.2480	16.8597	341.3749	55
60	14.0274	0.0713	0.0035	0.0485	289.4980	20.6380	17.6165	363.5707	60
65	17.4807	0.0572	0.0027	0.0477	366.2378	20.9510	18.2782	382.9465	65
70	21.7841	0.0459	0.0022	0.0472	461.8697	21.2021	18.8543	399.7503	70
75	27.1470	0.0368	0.0017	0.0467	581.0444	21.4036	19.3538	414.2422	75
80	33.8301	0.0296	0.0014	0.0464	729.5577	21.5653	19.7854	426.6797	80
85	42.1585	0.0237	0.0011	0.0461	914.6323	21.6951	20.1570	437.3091	85
90	52.5371	0.0190	0.0009	0.0459	1145.2690	21.7992	20.4759	446.3592	90
95	65.4708	0.0153	0.0007	0.0457	1432.6843	21.8828	20.7487	454.0394	95
100	81.5885	0.0123	0.0006	0.0456	1790.8560	21.9499	20.9814	460.5376	100

5% Compound Interest Factors

	Single Payment		Uniform Series				Arithmetic Gradient		
	Compound Amount Factor	Present Worth Factor	Sinking Fund Factor	Capital Recovery Factor	Compound Amount Factor	Present Worth Factor	Uniform Series	Present Worth	
N	*F/P*	*P/F*	*A/F*	*A/P*	*F/A*	*P/A*	*A/G*	*P/G*	*N*
1	1.0500	0.9524	1.0000	1.0500	1.0000	0.9524	0.0000	0.0000	1
2	1.1025	0.9070	0.4878	0.5378	2.0500	1.8594	0.4878	0.9070	2
3	1.1576	0.8638	0.3172	0.3672	3.1525	2.7232	0.9675	2.6347	3
4	1.2155	0.8227	0.2320	0.2820	4.3101	3.5460	1.4391	5.1028	4
5	1.2763	0.7835	0.1810	0.2310	5.5256	4.3295	1.9025	8.2369	5
6	1.3401	0.7462	0.1470	0.1970	6.8019	5.0757	2.3579	11.9680	6
7	1.4071	0.7107	0.1228	0.1728	8.1420	5.7864	2.8052	16.2321	7
8	1.4775	0.6768	0.1047	0.1547	9.5491	6.4632	3.2445	20.9700	8
9	1.5513	0.6446	0.0907	0.1407	11.0266	7.1078	3.6758	26.1268	9
10	1.6289	0.6139	0.0795	0.1295	12.5779	7.7217	4.0991	31.6520	10
11	1.7103	0.5847	0.0704	0.1204	14.2068	8.3064	4.5144	37.4988	11
12	1.7959	0.5568	0.0628	0.1128	15.9171	8.8633	4.9219	43.6241	12
13	1.8856	0.5303	0.0565	0.1065	17.7130	9.3936	5.3215	49.9879	13
14	1.9799	0.5051	0.0510	0.1010	19.5986	9.8986	5.7133	56.5538	14
15	2.0789	0.4810	0.0463	0.0963	21.5786	10.3797	6.0973	63.2880	15
16	2.1829	0.4581	0.0423	0.0923	23.6575	10.8378	6.4736	70.1597	16
17	2.2920	0.4363	0.0387	0.0887	25.8404	11.2741	6.8423	77.1405	17
18	2.4066	0.4155	0.0355	0.0855	28.1324	11.6896	7.2034	84.2043	18
19	2.5270	0.3957	0.0327	0.0827	30.5390	12.0853	7.5569	91.3275	19
20	2.6533	0.3769	0.0302	0.0802	33.0660	12.4622	7.9030	98.4884	20
21	2.7860	0.3589	0.0280	0.0780	35.7193	12.8212	8.2416	105.6673	21
22	2.9253	0.3418	0.0260	0.0760	38.5052	13.1630	8.5730	112.8461	22
23	3.0715	0.3256	0.0241	0.0741	41.4305	13.4886	8.8971	120.0087	23
24	3.2251	0.3101	0.0225	0.0725	44.5020	13.7986	9.2140	127.1402	24
25	3.3864	0.2953	0.0210	0.0710	47.7271	14.0939	9.5238	134.2275	25
26	3.5557	0.2812	0.0196	0.0696	51.1135	14.3752	9.8266	141.2585	26
27	3.7335	0.2678	0.0183	0.0683	54.6691	14.6430	10.1224	148.2226	27
28	3.9201	0.2551	0.0171	0.0671	58.4026	14.8981	10.4114	155.1101	28
29	4.1161	0.2429	0.0160	0.0660	62.3227	15.1411	10.6936	161.9126	29
30	4.3219	0.2314	0.0151	0.0651	66.4388	15.3725	10.9691	168.6226	30
31	4.5380	0.2204	0.0141	0.0641	70.7608	15.5928	11.2381	175.2333	31
32	4.7649	0.2099	0.0133	0.0633	75.2988	15.8027	11.5005	181.7392	32
33	5.0032	0.1999	0.0125	0.0625	80.0638	16.0025	11.7566	188.1351	33
34	5.2533	0.1904	0.0118	0.0618	85.0670	16.1929	12.0063	194.4168	34
35	5.5160	0.1813	0.0111	0.0611	90.3203	16.3742	12.2498	200.5807	35
40	7.0400	0.1420	0.0083	0.0583	120.7998	17.1591	13.3775	229.5452	40
45	8.9850	0.1113	0.0063	0.0563	159.7002	17.7741	14.3644	255.3145	45
50	11.4674	0.0872	0.0048	0.0548	209.3480	18.2559	15.2233	277.9148	50
55	14.6356	0.0683	0.0037	0.0537	272.7126	18.6335	15.9664	297.5104	55
60	18.6792	0.0535	0.0028	0.0528	353.5837	18.9293	16.6062	314.3432	60
65	23.8399	0.0419	0.0022	0.0522	456.7980	19.1611	17.1541	328.6910	65
70	30.4264	0.0329	0.0017	0.0517	588.5285	19.3427	17.6212	340.8409	70
75	38.8327	0.0258	0.0013	0.0513	756.6537	19.4850	18.0176	351.0721	75
80	49.5614	0.0202	0.0010	0.0510	971.2288	19.5965	18.3526	359.6460	80
85	63.2544	0.0158	0.0008	0.0508	1245.0871	19.6838	18.6346	366.8007	85
90	80.7304	0.0124	0.0006	0.0506	1594.6073	19.7523	18.8712	372.7488	90
95	103.0347	0.0097	0.0005	0.0505	2040.6935	19.8059	19.0689	377.6774	95
100	131.5013	0.0076	0.0004	0.0504	2610.0252	19.8479	19.2337	381.7492	100

6% Compound Interest Factors

	Single Payment		Uniform Series				Arithmetic Gradient		
N	Compound Amount Factor *F/P*	Present Worth Factor *P/F*	Sinking Fund Factor *A/F*	Capital Recovery Factor *A/P*	Compound Amount Factor *F/A*	Present Worth Factor *P/A*	Uniform Series *A/G*	Present Worth *P/G*	*N*
1	1.0600	0.9434	1.0000	1.0600	1.0000	0.9434	0.0000	0.0000	1
2	1.1236	0.8900	0.4854	0.5454	2.0600	1.8334	0.4854	0.8900	2
3	1.1910	0.8396	0.3141	0.3741	3.1836	2.6730	0.9612	2.5692	3
4	1.2625	0.7921	0.2286	0.2886	4.3746	3.4651	1.4272	4.9455	4
5	1.3382	0.7473	0.1774	0.2374	5.6371	4.2124	1.8836	7.9345	5
6	1.4185	0.7050	0.1434	0.2034	6.9753	4.9173	2.3304	11.4594	6
7	1.5036	0.6651	0.1191	0.1791	8.3938	5.5824	2.7676	15.4497	7
8	1.5938	0.6274	0.1010	0.1610	9.8975	6.2098	3.1952	19.8416	8
9	1.6895	0.5919	0.0870	0.1470	11.4913	6.8017	3.6133	24.5768	9
10	1.7908	0.5584	0.0759	0.1359	13.1808	7.3601	4.0220	29.6023	10
11	1.8983	0.5268	0.0668	0.1268	14.9716	7.8869	4.4213	34.8702	11
12	2.0122	0.4970	0.0593	0.1193	16.8699	8.3838	4.8113	40.3369	12
13	2.1329	0.4688	0.0530	0.1130	18.8821	8.8527	5.1920	45.9629	13
14	2.2609	0.4423	0.0476	0.1076	21.0151	9.2950	5.5635	51.7128	14
15	2.3966	0.4173	0.0430	0.1030	23.2760	9.7122	5.9260	57.5546	15
16	2.5404	0.3936	0.0390	0.0990	25.6725	10.1059	6.2794	63.4592	16
17	2.6928	0.3714	0.0354	0.0954	28.2129	10.4773	6.6240	69.4011	17
18	2.8543	0.3503	0.0324	0.0924	30.9057	10.8276	6.9597	75.3569	18
19	3.0256	0.3305	0.0296	0.0896	33.7600	11.1581	7.2867	81.3062	19
20	3.2071	0.3118	0.0272	0.0872	36.7856	11.4699	7.6051	87.2304	20
21	3.3996	0.2942	0.0250	0.0850	39.9927	11.7641	7.9151	93.1136	21
22	3.6035	0.2775	0.0230	0.0830	43.3923	12.0416	8.2166	98.9412	22
23	3.8197	0.2618	0.0213	0.0813	46.9958	12.3034	8.5099	104.7007	23
24	4.0489	0.2470	0.0197	0.0797	50.8156	12.5504	8.7951	110.3812	24
25	4.2919	0.2330	0.0182	0.0782	54.8645	12.7834	9.0722	115.9732	25
26	4.5494	0.2198	0.0169	0.0769	59.1564	13.0032	9.3414	121.4684	26
27	4.8223	0.2074	0.0157	0.0757	63.7058	13.2105	9.6029	126.8600	27
28	5.1117	0.1956	0.0146	0.0746	68.5281	13.4062	9.8568	132.1420	28
29	5.4184	0.1846	0.0136	0.0736	73.6398	13.5907	10.1032	137.3096	29
30	5.7435	0.1741	0.0126	0.0726	79.0582	13.7648	10.3422	142.3588	30
31	6.0881	0.1643	0.0118	0.0718	84.8017	13.9291	10.5740	147.2864	31
32	6.4534	0.1550	0.0110	0.0710	90.8898	14.0840	10.7988	152.0901	32
33	6.8406	0.1462	0.0103	0.0703	97.3432	14.2302	11.0166	156.7681	33
34	7.2510	0.1379	0.0096	0.0696	104.1838	14.3681	11.2276	161.3192	34
35	7.6861	0.1301	0.0090	0.0690	111.4348	14.4982	11.4319	165.7427	35
40	10.2857	0.0972	0.0065	0.0665	154.7620	15.0463	12.3590	185.9568	40
45	13.7646	0.0727	0.0047	0.0647	212.7435	15.4558	13.1413	203.1096	45
50	18.4202	0.0543	0.0034	0.0634	290.3359	15.7619	13.7964	217.4574	50
55	24.6503	0.0406	0.0025	0.0625	394.1720	15.9905	14.3411	229.3222	55
60	32.9877	0.0303	0.0019	0.0619	533.1282	16.1614	14.7909	239.0428	60
65	44.1450	0.0227	0.0014	0.0614	719.0829	16.2891	15.1601	246.9450	65
70	59.0759	0.0169	0.0010	0.0610	967.9322	16.3845	15.4613	253.3271	70
75	79.0569	0.0126	0.0008	0.0608	1300.9487	16.4558	15.7058	258.4527	75
80	105.7960	0.0095	0.0006	0.0606	1746.5999	16.5091	15.9033	262.5493	80
85	141.5789	0.0071	0.0004	0.0604	2342.9817	16.5489	16.0620	265.8096	85
90	189.4645	0.0053	0.0003	0.0603	3141.0752	16.5787	16.1891	268.3946	90
95	253.5463	0.0039	0.0002	0.0602	4209.1042	16.6009	16.2905	270.4375	95
100	339.3021	0.0029	0.0002	0.0602	5638.3681	16.6175	16.3711	272.0471	100

7% Compound Interest Factors

	Single Payment		Uniform Series				Arithmetic Gradient		
	Compound Amount Factor	Present Worth Factor	Sinking Fund Factor	Capital Recovery Factor	Compound Amount Factor	Present Worth Factor	Uniform Series	Present Worth	
N	F/P	P/F	A/F	A/P	F/A	P/A	A/G	P/G	N
1	1.0700	0.9346	1.0000	1.0700	1.0000	0.9346	0.0000	0.0000	1
2	1.1449	0.8734	0.4831	0.5531	2.0700	1.8080	0.4831	0.8734	2
3	1.2250	0.8163	0.3111	0.3811	3.2149	2.6243	0.9549	2.5060	3
4	1.3108	0.7629	0.2252	0.2952	4.4399	3.3872	1.4155	4.7947	4
5	1.4026	0.7130	0.1739	0.2439	5.7507	4.1002	1.8650	7.6467	5
6	1.5007	0.6663	0.1398	0.2098	7.1533	4.7665	2.3032	10.9784	6
7	1.6058	0.6227	0.1156	0.1856	8.6540	5.3893	2.7304	14.7149	7
8	1.7182	0.5820	0.0975	0.1675	10.2598	5.9713	3.1465	18.7889	8
9	1.8385	0.5439	0.0835	0.1535	11.9780	6.5152	3.5517	23.1404	9
10	1.9672	0.5083	0.0724	0.1424	13.8164	7.0236	3.9461	27.7156	10
11	2.1049	0.4751	0.0634	0.1334	15.7836	7.4987	4.3296	32.4665	11
12	2.2522	0.4440	0.0559	0.1259	17.8885	7.9427	4.7025	37.3506	12
13	2.4098	0.4150	0.0497	0.1197	20.1406	8.3577	5.0648	42.3302	13
14	2.5785	0.3878	0.0443	0.1143	22.5505	8.7455	5.4167	47.3718	14
15	2.7590	0.3624	0.0398	0.1098	25.1290	9.1079	5.7583	52.4461	15
16	2.9522	0.3387	0.0359	0.1059	27.8881	9.4466	6.0897	57.5271	16
17	3.1588	0.3166	0.0324	0.1024	30.8402	9.7632	6.4110	62.5923	17
18	3.3799	0.2959	0.0294	0.0994	33.9990	10.0591	6.7225	67.6219	18
19	3.6165	0.2765	0.0268	0.0968	37.3790	10.3356	7.0242	72.5991	19
20	3.8697	0.2584	0.0244	0.0944	40.9955	10.5940	7.3163	77.5091	20
21	4.1406	0.2415	0.0223	0.0923	44.8652	10.8355	7.5990	82.3393	21
22	4.4304	0.2257	0.0204	0.0904	49.0057	11.0612	7.8725	87.0793	22
23	4.7405	0.2109	0.0187	0.0887	53.4361	11.2722	8.1369	91.7201	23
24	5.0724	0.1971	0.0172	0.0872	58.1767	11.4693	8.3923	96.2545	24
25	5.4274	0.1842	0.0158	0.0858	63.2490	11.6536	8.6391	100.6765	25
26	5.8074	0.1722	0.0146	0.0846	68.6765	11.8258	8.8773	104.9814	26
27	6.2139	0.1609	0.0134	0.0834	74.4838	11.9867	9.1072	109.1656	27
28	6.6488	0.1504	0.0124	0.0824	80.6977	12.1371	9.3289	113.2264	28
29	7.1143	0.1406	0.0114	0.0814	87.3465	12.2777	9.5427	117.1622	29
30	7.6123	0.1314	0.0106	0.0806	94.4608	12.4090	9.7487	120.9718	30
31	8.1451	0.1228	0.0098	0.0798	102.0730	12.5318	9.9471	124.6550	31
32	8.7153	0.1147	0.0091	0.0791	110.2182	12.6466	10.1381	128.2120	32
33	9.3253	0.1072	0.0084	0.0784	118.9334	12.7538	10.3219	131.6435	33
34	9.9781	0.1002	0.0078	0.0778	128.2588	12.8540	10.4987	134.9507	34
35	10.6766	0.0937	0.0072	0.0772	138.2369	12.9477	10.6687	138.1353	35
40	14.9745	0.0668	0.0050	0.0750	199.6351	13.3317	11.4233	152.2928	40
45	21.0025	0.0476	0.0035	0.0735	285.7493	13.6055	12.0360	163.7559	45
50	29.4570	0.0339	0.0025	0.0725	406.5289	13.8007	12.5287	172.9051	50
55	41.3150	0.0242	0.0017	0.0717	575.9286	13.9399	12.9215	180.1243	55
60	57.9464	0.0173	0.0012	0.0712	813.5204	14.0392	13.2321	185.7677	60
65	81.2729	0.0123	0.0009	0.0709	1146.7552	14.1099	13.4760	190.1452	65
70	113.9894	0.0088	0.0006	0.0706	1614.1342	14.1604	13.6662	193.5185	70
75	159.8760	0.0063	0.0004	0.0704	2269.6574	14.1964	13.8136	196.1035	75
80	224.2344	0.0045	0.0003	0.0703	3189.0627	14.2220	13.9273	198.0748	80
85	314.5003	0.0032	0.0002	0.0702	4478.5761	14.2403	14.0146	199.5717	85
90	441.1030	0.0023	0.0002	0.0702	6287.1854	14.2533	14.0812	200.7042	90
95	618.6697	0.0016	0.0001	0.0701	8823.8535	14.2626	14.1319	201.5581	95
100	867.7163	0.0012	0.0001	0.0701	12381.6618	14.2693	14.1703	202.2001	100

8% Compound Interest Factors

	Single Payment		Uniform Series				Arithmetic Gradient		
N	Compound Amount Factor F/P	Present Worth Factor P/F	Sinking Fund Factor A/F	Capital Recovery Factor A/P	Compound Amount Factor F/A	Present Worth Factor P/A	Uniform Series A/G	Present Worth P/G	N
1	1.0800	0.9259	1.0000	1.0800	1.0000	0.9259	0.0000	0.0000	1
2	1.1664	0.8573	0.4808	0.5608	2.0800	1.7833	0.4808	0.8573	2
3	1.2597	0.7938	0.3080	0.3880	3.2464	2.5771	0.9487	2.4450	3
4	1.3605	0.7350	0.2219	0.3019	4.5061	3.3121	1.4040	4.6501	4
5	1.4693	0.6806	0.1705	0.2505	5.8666	3.9927	1.8465	7.3724	5
6	1.5869	0.6302	0.1363	0.2163	7.3359	4.6229	2.2763	10.5233	6
7	1.7138	0.5835	0.1121	0.1921	8.9228	5.2064	2.6937	14.0242	7
8	1.8509	0.5403	0.0940	0.1740	10.6366	5.7466	3.0985	17.8061	8
9	1.9990	0.5002	0.0801	0.1601	12.4876	6.2469	3.4910	21.8081	9
10	2.1589	0.4632	0.0690	0.1490	14.4866	6.7101	3.8713	25.9768	10
11	2.3316	0.4289	0.0601	0.1401	16.6455	7.1390	4.2395	30.2657	11
12	2.5182	0.3971	0.0527	0.1327	18.9771	7.5361	4.5957	34.6339	12
13	2.7196	0.3677	0.0465	0.1265	21.4953	7.9038	4.9402	39.0463	13
14	2.9372	0.3405	0.0413	0.1213	24.2149	8.2442	5.2731	43.4723	14
15	3.1722	0.3152	0.0368	0.1168	27.1521	8.5595	5.5945	47.8857	15
16	3.4259	0.2919	0.0330	0.1130	30.3243	8.8514	5.9046	52.2640	16
17	3.7000	0.2703	0.0296	0.1096	33.7502	9.1216	6.2037	56.5883	17
18	3.9960	0.2502	0.0267	0.1067	37.4502	9.3719	6.4920	60.8426	18
19	4.3157	0.2317	0.0241	0.1041	41.4463	9.6036	6.7697	65.0134	19
20	4.6610	0.2145	0.0219	0.1019	45.7620	9.8181	7.0369	69.0898	20
21	5.0338	0.1987	0.0198	0.0998	50.4229	10.0168	7.2940	73.0629	21
22	5.4365	0.1839	0.0180	0.0980	55.4568	10.2007	7.5412	76.9257	22
23	5.8715	0.1703	0.0164	0.0964	60.8933	10.3711	7.7786	80.6726	23
24	6.3412	0.1577	0.0150	0.0950	66.7648	10.5288	8.0066	84.2997	24
25	6.8485	0.1460	0.0137	0.0937	73.1059	10.6748	8.2254	87.8041	25
26	7.3964	0.1352	0.0125	0.0925	79.9544	10.8100	8.4352	91.1842	26
27	7.9881	0.1252	0.0114	0.0914	87.3508	10.9352	8.6363	94.4390	27
28	8.6271	0.1159	0.0105	0.0905	95.3388	11.0511	8.8289	97.5687	28
29	9.3173	0.1073	0.0096	0.0896	103.9659	11.1584	9.0133	100.5738	29
30	10.0627	0.0994	0.0088	0.0888	113.2832	11.2578	9.1897	103.4558	30
31	10.8677	0.0920	0.0081	0.0881	123.3459	11.3498	9.3584	106.2163	31
32	11.7371	0.0852	0.0075	0.0875	134.2135	11.4350	9.5197	108.8575	32
33	12.6760	0.0789	0.0069	0.0869	145.9506	11.5139	9.6737	111.3819	33
34	13.6901	0.0730	0.0063	0.0863	158.6267	11.5869	9.8208	113.7924	34
35	14.7853	0.0676	0.0058	0.0858	172.3168	11.6546	9.9611	116.0920	35
40	21.7245	0.0460	0.0039	0.0839	259.0565	11.9246	10.5699	126.0422	40
45	31.9204	0.0313	0.0026	0.0826	386.5056	12.1084	11.0447	133.7331	45
50	46.9016	0.0213	0.0017	0.0817	573.7702	12.2335	11.4107	139.5928	50
55	68.9139	0.0145	0.0012	0.0812	848.9232	12.3186	11.6902	144.0065	55
60	101.2571	0.0099	0.0008	0.0808	1253.2133	12.3766	11.9015	147.3000	60
65	148.7798	0.0067	0.0005	0.0805	1847.2481	12.4160	12.0602	149.7387	65
70	218.6064	0.0046	0.0004	0.0804	2720.0801	12.4428	12.1783	151.5326	70
75	321.2045	0.0031	0.0002	0.0802	4002.5566	12.4611	12.2658	152.8448	75
80	471.9548	0.0021	0.0002	0.0802	5886.9354	12.4735	12.3301	153.8001	80
85	693.4565	0.0014	0.0001	0.0801	8655.7061	12.4820	12.3772	154.4925	85
90	1018.9151	0.0010	0.0001	0.0801	12723.9386	12.4877	12.4116	154.9925	90
95	1497.1205	0.0007	0.0001	0.0801	18701.5069	12.4917	12.4365	155.3524	95
100	2199.7613	0.0005	0.0000	0.0800	27484.5157	12.4943	12.4545	155.6107	100

9% Compound Interest Factors

	Single Payment		Uniform Series				Arithmetic Gradient		
	Compound Amount Factor	Present Worth Factor	Sinking Fund Factor	Capital Recovery Factor	Compound Amount Factor	Present Worth Factor	Uniform Series	Present Worth	
N	*F/P*	*P/F*	*A/F*	*A/P*	*F/A*	*P/A*	*A/G*	*P/G*	*N*
1	1.0900	0.9174	1.0000	1.0900	1.0000	0.9174	0.0000	0.0000	1
2	1.1881	0.8417	0.4785	0.5685	2.0900	1.7591	0.4785	0.8417	2
3	1.2950	0.7722	0.3051	0.3951	3.2781	2.5313	0.9426	2.3860	3
4	1.4116	0.7084	0.2187	0.3087	4.5731	3.2397	1.3925	4.5113	4
5	1.5386	0.6499	0.1671	0.2571	5.9847	3.8897	1.8282	7.1110	5
6	1.6771	0.5963	0.1329	0.2229	7.5233	4.4859	2.2498	10.0924	6
7	1.8280	0.5470	0.1087	0.1987	9.2004	5.0330	2.6574	13.3746	7
8	1.9926	0.5019	0.0907	0.1807	11.0285	5.5348	3.0512	16.8877	8
9	2.1719	0.4604	0.0768	0.1668	13.0210	5.9952	3.4312	20.5711	9
10	2.3674	0.4224	0.0658	0.1558	15.1929	6.4177	3.7978	24.3728	10
11	2.5804	0.3875	0.0569	0.1469	17.5603	6.8052	4.1510	28.2481	11
12	2.8127	0.3555	0.0497	0.1397	20.1407	7.1607	4.4910	32.1590	12
13	3.0658	0.3262	0.0436	0.1336	22.9534	7.4869	4.8182	36.0731	13
14	3.3417	0.2992	0.0384	0.1284	26.0192	7.7862	5.1326	39.9633	14
15	3.6425	0.2745	0.0341	0.1241	29.3609	8.0607	5.4346	43.8069	15
16	3.9703	0.2519	0.0303	0.1203	33.0034	8.3126	5.7245	47.5849	16
17	4.3276	0.2311	0.0270	0.1170	36.9737	8.5436	6.0024	51.2821	17
18	4.7171	0.2120	0.0242	0.1142	41.3013	8.7556	6.2687	54.8860	18
19	5.1417	0.1945	0.0217	0.1117	46.0185	8.9501	6.5236	58.3868	19
20	5.6044	0.1784	0.0195	0.1095	51.1601	9.1285	6.7674	61.7770	20
21	6.1088	0.1637	0.0176	0.1076	56.7645	9.2922	7.0006	65.0509	21
22	6.6586	0.1502	0.0159	0.1059	62.8733	9.4424	7.2232	68.2048	22
23	7.2579	0.1378	0.0144	0.1044	69.5319	9.5802	7.4357	71.2359	23
24	7.9111	0.1264	0.0130	0.1030	76.7898	9.7066	7.6384	74.1433	24
25	8.6231	0.1160	0.0118	0.1018	84.7009	9.8226	7.8316	76.9265	25
26	9.3992	0.1064	0.0107	0.1007	93.3240	9.9290	8.0156	79.5863	26
27	10.2451	0.0976	0.0097	0.0997	102.7231	10.0266	8.1906	82.1241	27
28	11.1671	0.0895	0.0089	0.0989	112.9682	10.1161	8.3571	84.5419	28
29	12.1722	0.0822	0.0081	0.0981	124.1354	10.1983	8.5154	86.8422	29
30	13.2677	0.0754	0.0073	0.0973	136.3075	10.2737	8.6657	89.0280	30
31	14.4618	0.0691	0.0067	0.0967	149.5752	10.3428	8.8083	91.1024	31
32	15.7633	0.0634	0.0061	0.0961	164.0370	10.4062	8.9436	93.0690	32
33	17.1820	0.0582	0.0056	0.0956	179.8003	10.4644	9.0718	94.9314	33
34	18.7284	0.0534	0.0051	0.0951	196.9823	10.5178	9.1933	96.6935	34
35	20.4140	0.0490	0.0046	0.0946	215.7108	10.5668	9.3083	98.3590	35
40	31.4094	0.0318	0.0030	0.0930	337.8824	10.7574	9.7957	105.3762	40
45	48.3273	0.0207	0.0019	0.0919	525.8587	10.8812	10.1603	110.5561	45
50	74.3575	0.0134	0.0012	0.0912	815.0836	10.9617	10.4295	114.3251	50
55	114.4083	0.0087	0.0008	0.0908	1260.0918	11.0140	10.6261	117.0362	55
60	176.0313	0.0057	0.0005	0.0905	1944.7921	11.0480	10.7683	118.9683	60
65	270.8460	0.0037	0.0003	0.0903	2998.2885	11.0701	10.8702	120.3344	65
70	416.7301	0.0024	0.0002	0.0902	4619.2232	11.0844	10.9427	121.2942	70
75	641.1909	0.0016	0.0001	0.0901	7113.2321	11.0938	10.9940	121.9646	75
80	986.5517	0.0010	0.0001	0.0901	10950.5741	11.0998	11.0299	122.4306	80
85	1517.9320	0.0007	0.0001	0.0901	16854.8003	11.1038	11.0551	122.7533	85
90	2335.5266	0.0004	0.0000	0.0900	25939.1842	11.1064	11.0726	122.9758	90
95	3593.4971	0.0003	0.0000	0.0900	39916.6350	11.1080	11.0847	123.1287	95
100	5529.0408	0.0002	0.0000	0.0900	61422.6755	11.1091	11.0930	123.2335	100

10% Compound Interest Factors

	Single Payment		Uniform Series				Arithmetic Gradient		
	Compound Amount Factor	Present Worth Factor	Sinking Fund Factor	Capital Recovery Factor	Compound Amount Factor	Present Worth Factor	Uniform Series	Present Worth	
N	*F/P*	*P/F*	*A/F*	*A/P*	*F/A*	*P/A*	*A/G*	*P/G*	*N*
1	1.1000	0.9091	1.0000	1.1000	1.0000	0.9091	0.0000	0.0000	1
2	1.2100	0.8264	0.4762	0.5762	2.1000	1.7355	0.4762	0.8264	2
3	1.3310	0.7513	0.3021	0.4021	3.3100	2.4869	0.9366	2.3291	3
4	1.4641	0.6830	0.2155	0.3155	4.6410	3.1699	1.3812	4.3781	4
5	1.6105	0.6209	0.1638	0.2638	6.1051	3.7908	1.8101	6.8618	5
6	1.7716	0.5645	0.1296	0.2296	7.7156	4.3553	2.2236	9.6842	6
7	1.9487	0.5132	0.1054	0.2054	9.4872	4.8684	2.6216	12.7631	7
8	2.1436	0.4665	0.0874	0.1874	11.4359	5.3349	3.0045	16.0287	8
9	2.3579	0.4241	0.0736	0.1736	13.5795	5.7590	3.3724	19.4215	9
10	2.5937	0.3855	0.0627	0.1627	15.9374	6.1446	3.7255	22.8913	10
11	2.8531	0.3505	0.0540	0.1540	18.5312	6.4951	4.0641	26.3963	11
12	3.1384	0.3186	0.0468	0.1468	21.3843	6.8137	4.3884	29.9012	12
13	3.4523	0.2897	0.0408	0.1408	24.5227	7.1034	4.6988	33.3772	13
14	3.7975	0.2633	0.0357	0.1357	27.9750	7.3667	4.9955	36.8005	14
15	4.1772	0.2394	0.0315	0.1315	31.7725	7.6061	5.2789	40.1520	15
16	4.5950	0.2176	0.0278	0.1278	35.9497	7.8237	5.5493	43.4164	16
17	5.0545	0.1978	0.0247	0.1247	40.5447	8.0216	5.8071	46.5819	17
18	5.5599	0.1799	0.0219	0.1219	45.5992	8.2014	6.0526	49.6395	18
19	6.1159	0.1635	0.0195	0.1195	51.1591	8.3649	6.2861	52.5827	19
20	6.7275	0.1486	0.0175	0.1175	57.2750	8.5136	6.5081	55.4069	20
21	7.4002	0.1351	0.0156	0.1156	64.0025	8.6487	6.7189	58.1095	21
22	8.1403	0.1228	0.0140	0.1140	71.4027	8.7715	6.9189	60.6893	22
23	8.9543	0.1117	0.0126	0.1126	79.5430	8.8832	7.1085	63.1462	23
24	9.8497	0.1015	0.0113	0.1113	88.4973	8.9847	7.2881	65.4813	24
25	10.8347	0.0923	0.0102	0.1102	98.3471	9.0770	7.4580	67.6964	25
26	11.9182	0.0839	0.0092	0.1092	109.1818	9.1609	7.6186	69.7940	26
27	13.1100	0.0763	0.0083	0.1083	121.0999	9.2372	7.7704	71.7773	27
28	14.4210	0.0693	0.0075	0.1075	134.2099	9.3066	7.9137	73.6495	28
29	15.8631	0.0630	0.0067	0.1067	148.6309	9.3696	8.0489	75.4146	29
30	17.4494	0.0573	0.0061	0.1061	164.4940	9.4269	8.1762	77.0766	30
31	19.1943	0.0521	0.0055	0.1055	181.9434	9.4790	8.2962	78.6395	31
32	21.1138	0.0474	0.0050	0.1050	201.1378	9.5264	8.4091	80.1078	32
33	23.2252	0.0431	0.0045	0.1045	222.2515	9.5694	8.5152	81.4856	33
34	25.5477	0.0391	0.0041	0.1041	245.4767	9.6086	8.6149	82.7773	34
35	28.1024	0.0356	0.0037	0.1037	271.0244	9.6442	8.7086	83.9872	35
40	45.2593	0.0221	0.0023	0.1023	442.5926	9.7791	9.0962	88.9525	40
45	72.8905	0.0137	0.0014	0.1014	718.9048	9.8628	9.3740	92.4544	45
50	117.3909	0.0085	0.0009	0.1009	1163.9085	9.9148	9.5704	94.8889	50
55	189.0591	0.0053	0.0005	0.1005	1880.5914	9.9471	9.7075	96.5619	55
60	304.4816	0.0033	0.0003	0.1003	3034.8164	9.9672	9.8023	97.7010	60
65	490.3707	0.0020	0.0002	0.1002	4893.7073	9.9796	9.8672	98.4705	65
70	789.7470	0.0013	0.0001	0.1001	7887.4696	9.9873	9.9113	98.9870	70
75	1271.8954	0.0008	0.0001	0.1001	12708.9537	9.9921	9.9410	99.3317	75
80	2048.4002	0.0005	0.0000	0.1000	20474.0021	9.9951	9.9609	99.5606	80
85	3298.9690	0.0003	0.0000	0.1000	32979.6903	9.9970	9.9742	99.7120	85
90	5313.0226	0.0002	0.0000	0.1000	53120.2261	9.9981	9.9831	99.8118	90
95	8556.6760	0.0001	0.0000	0.1000	85556.7605	9.9988	9.9889	99.8773	95
100	13780.6123	0.0001	0.0000	0.1000	137796.1234	9.9993	9.9927	99.9202	100

11% Compound Interest Factors

	Single Payment		Uniform Series				Arithmetic Gradient		
	Compound Amount Factor	Present Worth Factor	Sinking Fund Factor	Capital Recovery Factor	Compound Amount Factor	Present Worth Factor	Uniform Series	Present Worth	
N	*F/P*	*P/F*	*A/F*	*A/P*	*F/A*	*P/A*	*A/G*	*P/G*	*N*
1	1.1100	0.9009	1.0000	1.1100	1.0000	0.9009	0.0000	0.0000	1
2	1.2321	0.8116	0.4739	0.5839	2.1100	1.7125	0.4739	0.8116	2
3	1.3676	0.7312	0.2992	0.4092	3.3421	2.4437	0.9306	2.2740	3
4	1.5181	0.6587	0.2123	0.3223	4.7097	3.1024	1.3700	4.2502	4
5	1.6851	0.5935	0.1606	0.2706	6.2278	3.6959	1.7923	6.6240	5
6	1.8704	0.5346	0.1264	0.2364	7.9129	4.2305	2.1976	9.2972	6
7	2.0762	0.4817	0.1022	0.2122	9.7833	4.7122	2.5863	12.1872	7
8	2.3045	0.4339	0.0843	0.1943	11.8594	5.1461	2.9585	15.2246	8
9	2.5580	0.3909	0.0706	0.1806	14.1640	5.5370	3.3144	18.3520	9
10	2.8394	0.3522	0.0598	0.1698	16.7220	5.8892	3.6544	21.5217	10
11	3.1518	0.3173	0.0511	0.1611	19.5614	6.2065	3.9788	24.6945	11
12	3.4985	0.2858	0.0440	0.1540	22.7132	6.4924	4.2879	27.8388	12
13	3.8833	0.2575	0.0382	0.1482	26.2116	6.7499	4.5822	30.9290	13
14	4.3104	0.2320	0.0332	0.1432	30.0949	6.9819	4.8619	33.9449	14
15	4.7846	0.2090	0.0291	0.1391	34.4054	7.1909	5.1275	36.8709	15
16	5.3109	0.1883	0.0255	0.1355	39.1899	7.3792	5.3794	39.6953	16
17	5.8951	0.1696	0.0225	0.1325	44.5008	7.5488	5.6180	42.4095	17
18	6.5436	0.1528	0.0198	0.1298	50.3959	7.7016	5.8439	45.0074	18
19	7.2633	0.1377	0.0176	0.1276	56.9395	7.8393	6.0574	47.4856	19
20	8.0623	0.1240	0.0156	0.1256	64.2028	7.9633	6.2590	49.8423	20
21	8.9492	0.1117	0.0138	0.1238	72.2651	8.0751	6.4491	52.0771	21
22	9.9336	0.1007	0.0123	0.1223	81.2143	8.1757	6.6283	54.1912	22
23	11.0263	0.0907	0.0110	0.1210	91.1479	8.2664	6.7969	56.1864	23
24	12.2392	0.0817	0.0098	0.1198	102.1742	8.3481	6.9555	58.0656	24
25	13.5855	0.0736	0.0087	0.1187	114.4133	8.4217	7.1045	59.8322	25
26	15.0799	0.0663	0.0078	0.1178	127.9988	8.4881	7.2443	61.4900	26
27	16.7386	0.0597	0.0070	0.1170	143.0786	8.5478	7.3754	63.0433	27
28	18.5799	0.0538	0.0063	0.1163	159.8173	8.6016	7.4982	64.4965	28
29	20.6237	0.0485	0.0056	0.1156	178.3972	8.6501	7.6131	65.8542	29
30	22.8923	0.0437	0.0050	0.1150	199.0209	8.6938	7.7206	67.1210	30
31	25.4104	0.0394	0.0045	0.1145	221.9132	8.7331	7.8210	68.3016	31
32	28.2056	0.0355	0.0040	0.1140	247.3236	8.7686	7.9147	69.4007	32
33	31.3082	0.0319	0.0036	0.1136	275.5292	8.8005	8.0021	70.4228	33
34	34.7521	0.0288	0.0033	0.1133	306.8374	8.8293	8.0836	71.3724	34
35	38.5749	0.0259	0.0029	0.1129	341.5896	8.8552	8.1594	72.2538	35
40	65.0009	0.0154	0.0017	0.1117	581.8261	8.9511	8.4659	75.7789	40
45	109.5302	0.0091	0.0010	0.1110	986.6386	9.0079	8.6763	78.1551	45
50	184.5648	0.0054	0.0006	0.1106	1668.7712	9.0417	8.8185	79.7341	50
55	311.0025	0.0032	0.0004	0.1104	2818.2042	9.0617	8.9135	80.7712	55
60	524.0572	0.0019	0.0002	0.1102	4755.0658	9.0736	8.9762	81.4461	60
65	883.0669	0.0011	0.0001	0.1101	8018.7903	9.0806	9.0172	81.8819	65
70	1488.0191	0.0007	0.0001	0.1101	13518.3557	9.0848	9.0438	82.1614	70
75	2507.3988	0.0004	0.0000	0.1100	22785.4434	9.0873	9.0610	82.3397	75
80	4225.1128	0.0002	0.0000	0.1100	38401.0250	9.0888	9.0720	82.4529	80
85	7119.5607	0.0001	0.0000	0.1100	64714.1881	9.0896	9.0790	82.5245	85
90	11996.8738	0.0001	0.0000	0.1100	109053.3983	9.0902	9.0834	82.5695	90
95	20215.4301	0.0000	0.0000	0.1100	183767.5459	9.0905	9.0862	82.5978	95
100	34064.1753	0.0000	0.0000	0.1100	309665.2297	9.0906	9.0880	82.6155	100

12% Compound Interest Factors

	Single Payment		Uniform Series				Arithmetic Gradient		
N	Compound Amount Factor *F/P*	Present Worth Factor *P/F*	Sinking Fund Factor *A/F*	Capital Recovery Factor *A/P*	Compound Amount Factor *F/A*	Present Worth Factor *P/A*	Uniform Series *A/G*	Present Worth *P/G*	*N*
1	1.1200	0.8929	1.0000	1.1200	1.0000	0.8929	0.0000	0.0000	1
2	1.2544	0.7972	0.4717	0.5917	2.1200	1.6901	0.4717	0.7972	2
3	1.4049	0.7118	0.2963	0.4163	3.3744	2.4018	0.9246	2.2208	3
4	1.5735	0.6355	0.2092	0.3292	4.7793	3.0373	1.3589	4.1273	4
5	1.7623	0.5674	0.1574	0.2774	6.3528	3.6048	1.7746	6.3970	5
6	1.9738	0.5066	0.1232	0.2432	8.1152	4.1114	2.1720	8.9302	6
7	2.2107	0.4523	0.0991	0.2191	10.0890	4.5638	2.5515	11.6443	7
8	2.4760	0.4039	0.0813	0.2013	12.2997	4.9676	2.9131	14.4714	8
9	2.7731	0.3606	0.0677	0.1877	14.7757	5.3282	3.2574	17.3563	9
10	3.1058	0.3220	0.0570	0.1770	17.5487	5.6502	3.5847	20.2541	10
11	3.4785	0.2875	0.0484	0.1684	20.6546	5.9377	3.8953	23.1288	11
12	3.8960	0.2567	0.0414	0.1614	24.1331	6.1944	4.1897	25.9523	12
13	4.3635	0.2292	0.0357	0.1557	28.0291	6.4235	4.4683	28.7024	13
14	4.8871	0.2046	0.0309	0.1509	32.3926	6.6282	4.7317	31.3624	14
15	5.4736	0.1827	0.0268	0.1468	37.2797	6.8109	4.9803	33.9202	15
16	6.1304	0.1631	0.0234	0.1434	42.7533	6.9740	5.2147	36.3670	16
17	6.8660	0.1456	0.0205	0.1405	48.8837	7.1196	5.4353	38.6973	17
18	7.6900	0.1300	0.0179	0.1379	55.7497	7.2497	5.6427	40.9080	18
19	8.6128	0.1161	0.0158	0.1358	63.4397	7.3658	5.8375	42.9979	19
20	9.6463	0.1037	0.0139	0.1339	72.0524	7.4694	6.0202	44.9676	20
21	10.8038	0.0926	0.0122	0.1322	81.6987	7.5620	6.1913	46.8188	21
22	12.1003	0.0826	0.0108	0.1308	92.5026	7.6446	6.3514	48.5543	22
23	13.5523	0.0738	0.0096	0.1296	104.6029	7.7184	6.5010	50.1776	23
24	15.1786	0.0659	0.0085	0.1285	118.1552	7.7843	6.6406	51.6929	24
25	17.0001	0.0588	0.0075	0.1275	133.3339	7.8431	6.7708	53.1046	25
26	19.0401	0.0525	0.0067	0.1267	150.3339	7.8957	6.8921	54.4177	26
27	21.3249	0.0469	0.0059	0.1259	169.3740	7.9426	7.0049	55.6369	27
28	23.8839	0.0419	0.0052	0.1252	190.6989	7.9844	7.1098	56.7674	28
29	26.7499	0.0374	0.0047	0.1247	214.5828	8.0218	7.2071	57.8141	29
30	29.9599	0.0334	0.0041	0.1241	241.3327	8.0552	7.2974	58.7821	30
31	33.5551	0.0298	0.0037	0.1237	271.2926	8.0850	7.3811	59.6761	31
32	37.5817	0.0266	0.0033	0.1233	304.8477	8.1116	7.4586	60.5010	32
33	42.0915	0.0238	0.0029	0.1229	342.4294	8.1354	7.5302	61.2612	33
34	47.1425	0.0212	0.0026	0.1226	384.5210	8.1566	7.5965	61.9612	34
35	52.7996	0.0189	0.0023	0.1223	431.6635	8.1755	7.6577	62.6052	35
40	93.0510	0.0107	0.0013	0.1213	767.0914	8.2438	7.8988	65.1159	40
45	163.9876	0.0061	0.0007	0.1207	1358.2300	8.2825	8.0572	66.7342	45
50	289.0022	0.0035	0.0004	0.1204	2400.0182	8.3045	8.1597	67.7624	50
55	509.3206	0.0020	0.0002	0.1202	4236.0050	8.3170	8.2251	68.4082	55
60	897.5969	0.0011	0.0001	0.1201	7471.6411	8.3240	8.2664	68.8100	60
65	1581.8725	0.0006	0.0001	0.1201	13173.9374	8.3281	8.2922	69.0581	65
70	2787.7998	0.0004	0.0000	0.1200	23223.3319	8.3303	8.3082	69.2103	70
75	4913.0558	0.0002	0.0000	0.1200	40933.7987	8.3316	8.3181	69.3031	75
80	8658.4831	0.0001	0.0000	0.1200	72145.6925	8.3324	8.3241	69.3594	80
85	15259.2057	0.0001	0.0000	0.1200	127151.7140	8.3328	8.3278	69.3935	85
90	26891.9342	0.0000	0.0000	0.1200	224091.1185	8.3330	8.3300	69.4140	90
95	47392.7766	0.0000	0.0000	0.1200	394931.4719	8.3332	8.3313	69.4263	95
100	83522.2657	0.0000	0.0000	0.1200	696010.5477	8.3332	8.3321	69.4336	100

13% Compound Interest Factors

	Single Payment		Uniform Series				Arithmetic Gradient		
	Compound Amount Factor	Present Worth Factor	Sinking Fund Factor	Capital Recovery Factor	Compound Amount Factor	Present Worth Factor	Uniform Series	Present Worth	
N	*F/P*	*P/F*	*A/F*	*A/P*	*F/A*	*P/A*	*A/G*	*P/G*	*N*
1	1.1300	0.8850	1.0000	1.1300	1.0000	0.8850	0.0000	0.0000	1
2	1.2769	0.7831	0.4695	0.5995	2.1300	1.6681	0.4695	0.7831	2
3	1.4429	0.6931	0.2935	0.4235	3.4069	2.3612	0.9187	2.1692	3
4	1.6305	0.6133	0.2062	0.3362	4.8498	2.9745	1.3479	4.0092	4
5	1.8424	0.5428	0.1543	0.2843	6.4803	3.5172	1.7571	6.1802	5
6	2.0820	0.4803	0.1202	0.2502	8.3227	3.9975	2.1468	8.5818	6
7	2.3526	0.4251	0.0961	0.2261	10.4047	4.4226	2.5171	11.1322	7
8	2.6584	0.3762	0.0784	0.2084	12.7573	4.7988	2.8685	13.7653	8
9	3.0040	0.3329	0.0649	0.1949	15.4157	5.1317	3.2014	16.4284	9
10	3.3946	0.2946	0.0543	0.1843	18.4197	5.4262	3.5162	19.0797	10
11	3.8359	0.2607	0.0458	0.1758	21.8143	5.6869	3.8134	21.6867	11
12	4.3345	0.2307	0.0390	0.1690	25.6502	5.9176	4.0936	24.2244	12
13	4.8980	0.2042	0.0334	0.1634	29.9847	6.1218	4.3573	26.6744	13
14	5.5348	0.1807	0.0287	0.1587	34.8827	6.3025	4.6050	29.0232	14
15	6.2543	0.1599	0.0247	0.1547	40.4175	6.4624	4.8375	31.2617	15
16	7.0673	0.1415	0.0214	0.1514	46.6717	6.6039	5.0552	33.3841	16
17	7.9861	0.1252	0.0186	0.1486	53.7391	6.7291	5.2589	35.3876	17
18	9.0243	0.1108	0.0162	0.1462	61.7251	6.8399	5.4491	37.2714	18
19	10.1974	0.0981	0.0141	0.1441	70.7494	6.9380	5.6265	39.0366	19
20	11.5231	0.0868	0.0124	0.1424	80.9468	7.0248	5.7917	40.6854	20
21	13.0211	0.0768	0.0108	0.1408	92.4699	7.1016	5.9454	42.2214	21
22	14.7138	0.0680	0.0095	0.1395	105.4910	7.1695	6.0881	43.6486	22
23	16.6266	0.0601	0.0083	0.1383	120.2048	7.2297	6.2205	44.9718	23
24	18.7881	0.0532	0.0073	0.1373	136.8315	7.2829	6.3431	46.1960	24
25	21.2305	0.0471	0.0064	0.1364	155.6196	7.3300	6.4566	47.3264	25
26	23.9905	0.0417	0.0057	0.1357	176.8501	7.3717	6.5614	48.3685	26
27	27.1093	0.0369	0.0050	0.1350	200.8406	7.4086	6.6582	49.3276	27
28	30.6335	0.0326	0.0044	0.1344	227.9499	7.4412	6.7474	50.2090	28
29	34.6158	0.0289	0.0039	0.1339	258.5834	7.4701	6.8296	51.0179	29
30	39.1159	0.0256	0.0034	0.1334	293.1992	7.4957	6.9052	51.7592	30
31	44.2010	0.0226	0.0030	0.1330	332.3151	7.5183	6.9747	52.4380	31
32	49.9471	0.0200	0.0027	0.1327	376.5161	7.5383	7.0385	53.0586	32
33	56.4402	0.0177	0.0023	0.1323	426.4632	7.5560	7.0971	53.6256	33
34	63.7774	0.0157	0.0021	0.1321	482.9034	7.5717	7.1507	54.1430	34
35	72.0685	0.0139	0.0018	0.1318	546.6808	7.5856	7.1998	54.6148	35
40	132.7816	0.0075	0.0010	0.1310	1013.7042	7.6344	7.3888	56.4087	40
45	244.6414	0.0041	0.0005	0.1305	1874.1646	7.6609	7.5076	57.5148	45
50	450.7359	0.0022	0.0003	0.1303	3459.5071	7.6752	7.5811	58.1870	50
55	830.4517	0.0012	0.0002	0.1302	6380.3979	7.6830	7.6260	58.5909	55
60	1530.0535	0.0007	0.0001	0.1301	11761.9498	7.6873	7.6531	58.8313	60
65	2819.0243	0.0004	0.0000	0.1300	21677.1103	7.6896	7.6692	58.9732	65
70	5193.8696	0.0002	0.0000	0.1300	39945.1510	7.6908	7.6788	59.0565	70
75	9569.3681	0.0001	0.0000	0.1300	73602.8316	7.6915	7.6845	59.1051	75
80	17630.9405	0.0001	0.0000	0.1300	135614.9266	7.6919	7.6878	59.1333	80
85	32483.8649	0.0000	0.0000	0.1300	249868.1918	7.6921	7.6897	59.1496	85

14% Compound Interest Factors

	Single Payment		Uniform Series				Arithmetic Gradient		
N	Compound Amount Factor F/P	Present Worth Factor P/F	Sinking Fund Factor A/F	Capital Recovery Factor A/P	Compound Amount Factor F/A	Present Worth Factor P/A	Uniform Series A/G	Present Worth P/G	N
1	1.1400	0.8772	1.0000	1.1400	1.0000	0.8772	0.0000	0.0000	1
2	1.2996	0.7695	0.4673	0.6073	2.1400	1.6467	0.4673	0.7695	2
3	1.4815	0.6750	0.2907	0.4307	3.4396	2.3216	0.9129	2.1194	3
4	1.6890	0.5921	0.2032	0.3432	4.9211	2.9137	1.3370	3.8957	4
5	1.9254	0.5194	0.1513	0.2913	6.6101	3.4331	1.7399	5.9731	5
6	2.1950	0.4556	0.1172	0.2572	8.5355	3.8887	2.1218	8.2511	6
7	2.5023	0.3996	0.0932	0.2332	10.7305	4.2883	2.4832	10.6489	7
8	2.8526	0.3506	0.0756	0.2156	13.2328	4.6389	2.8246	13.1028	8
9	3.2519	0.3075	0.0622	0.2022	16.0853	4.9464	3.1463	15.5629	9
10	3.7072	0.2697	0.0517	0.1917	19.3373	5.2161	3.4490	17.9906	10
11	4.2262	0.2366	0.0434	0.1834	23.0445	5.4527	3.7333	20.3567	11
12	4.8179	0.2076	0.0367	0.1767	27.2707	5.6603	3.9998	22.6399	12
13	5.4924	0.1821	0.0312	0.1712	32.0887	5.8424	4.2491	24.8247	13
14	6.2613	0.1597	0.0266	0.1666	37.5811	6.0021	4.4819	26.9009	14
15	7.1379	0.1401	0.0228	0.1628	43.8424	6.1422	4.6990	28.8623	15
16	8.1372	0.1229	0.0196	0.1596	50.9804	6.2651	4.9011	30.7057	16
17	9.2765	0.1078	0.0169	0.1569	59.1176	6.3729	5.0888	32.4305	17
18	10.5752	0.0946	0.0146	0.1546	68.3941	6.4674	5.2630	34.0380	18
19	12.0557	0.0829	0.0127	0.1527	78.9692	6.5504	5.4243	35.5311	19
20	13.7435	0.0728	0.0110	0.1510	91.0249	6.6231	5.5734	36.9135	20
21	15.6676	0.0638	0.0095	0.1495	104.7684	6.6870	5.7111	38.1901	21
22	17.8610	0.0560	0.0083	0.1483	120.4360	6.7429	5.8381	39.3658	22
23	20.3616	0.0491	0.0072	0.1472	138.2970	6.7921	5.9549	40.4463	23
24	23.2122	0.0431	0.0063	0.1463	158.6586	6.8351	6.0624	41.4371	24
25	26.4619	0.0378	0.0055	0.1455	181.8708	6.8729	6.1610	42.3441	25
26	30.1666	0.0331	0.0048	0.1448	208.3327	6.9061	6.2514	43.1728	26
27	34.3899	0.0291	0.0042	0.1442	238.4993	6.9352	6.3342	43.9289	27
28	39.2045	0.0255	0.0037	0.1437	272.8892	6.9607	6.4100	44.6176	28
29	44.6931	0.0224	0.0032	0.1432	312.0937	6.9830	6.4791	45.2441	29
30	50.9502	0.0196	0.0028	0.1428	356.7868	7.0027	6.5423	45.8132	30
31	58.0832	0.0172	0.0025	0.1425	407.7370	7.0199	6.5998	46.3297	31
32	66.2148	0.0151	0.0021	0.1421	465.8202	7.0350	6.6522	46.7979	32
33	75.4849	0.0132	0.0019	0.1419	532.0350	7.0482	6.6998	47.2218	33
34	86.0528	0.0116	0.0016	0.1416	607.5199	7.0599	6.7431	47.6053	34
35	98.1002	0.0102	0.0014	0.1414	693.5727	7.0700	6.7824	47.9519	35
40	188.8835	0.0053	0.0007	0.1407	1342.0251	7.1050	6.9300	49.2376	40
45	363.6791	0.0027	0.0004	0.1404	2590.5648	7.1232	7.0188	49.9963	45
50	700.2330	0.0014	0.0002	0.1402	4994.5213	7.1327	7.0714	50.4375	50
55	1348.2388	0.0007	0.0001	0.1401	9623.1343	7.1376	7.1020	50.6912	55
60	2595.9187	0.0004	0.0001	0.1401	18535.1333	7.1401	7.1197	50.8357	60
65	4998.2196	0.0002	0.0000	0.1400	35694.4260	7.1414	7.1298	50.9173	65
70	9623.6450	0.0001	0.0000	0.1400	68733.1785	7.1421	7.1356	50.9632	70
75	18529.5064	0.0001	0.0000	0.1400	132346.4742	7.1425	7.1388	50.9887	75
80	35676.9818	0.0000	0.0000	0.1400	254828.4415	7.1427	7.1406	51.0030	80
85	68692.9810	0.0000	0.0000	0.1400	490657.0073	7.1428	7.1416	51.0108	85

15% Compound Interest Factors

	Single Payment		Uniform Series				Arithmetic Gradient		
	Compound Amount Factor	Present Worth Factor	Sinking Fund Factor	Capital Recovery Factor	Compound Amount Factor	Present Worth Factor	Uniform Series	Present Worth	
N	*F/P*	*P/F*	*A/F*	*A/P*	*F/A*	*P/A*	*A/G*	*P/G*	*N*
1	1.1500	0.8696	1.0000	1.1500	1.0000	0.8696	0.0000	0.0000	1
2	1.3225	0.7561	0.4651	0.6151	2.1500	1.6257	0.4651	0.7561	2
3	1.5209	0.6575	0.2880	0.4380	3.4725	2.2832	0.9071	2.0712	3
4	1.7490	0.5718	0.2003	0.3503	4.9934	2.8550	1.3263	3.7864	4
5	2.0114	0.4972	0.1483	0.2983	6.7424	3.3522	1.7228	5.7751	5
6	2.3131	0.4323	0.1142	0.2642	8.7537	3.7845	2.0972	7.9368	6
7	2.6600	0.3759	0.0904	0.2404	11.0668	4.1604	2.4498	10.1924	7
8	3.0590	0.3269	0.0729	0.2229	13.7268	4.4873	2.7813	12.4807	8
9	3.5179	0.2843	0.0596	0.2096	16.7858	4.7716	3.0922	14.7548	9
10	4.0456	0.2472	0.0493	0.1993	20.3037	5.0188	3.3832	16.9795	10
11	4.6524	0.2149	0.0411	0.1911	24.3493	5.2337	3.6549	19.1289	11
12	5.3503	0.1869	0.0345	0.1845	29.0017	5.4206	3.9082	21.1849	12
13	6.1528	0.1625	0.0291	0.1791	34.3519	5.5831	4.1438	23.1352	13
14	7.0757	0.1413	0.0247	0.1747	40.5047	5.7245	4.3624	24.9725	14
15	8.1371	0.1229	0.0210	0.1710	47.5804	5.8474	4.5650	26.6930	15
16	9.3576	0.1069	0.0179	0.1679	55.7175	5.9542	4.7522	28.2960	16
17	10.7613	0.0929	0.0154	0.1654	65.0751	6.0472	4.9251	29.7828	17
18	12.3755	0.0808	0.0132	0.1632	75.8364	6.1280	5.0843	31.1565	18
19	14.2318	0.0703	0.0113	0.1613	88.2118	6.1982	5.2307	32.4213	19
20	16.3665	0.0611	0.0098	0.1598	102.4436	6.2593	5.3651	33.5822	20
21	18.8215	0.0531	0.0084	0.1584	118.8101	6.3125	5.4883	34.6448	21
22	21.6447	0.0462	0.0073	0.1573	137.6316	6.3587	5.6010	35.6150	22
23	24.8915	0.0402	0.0063	0.1563	159.2764	6.3988	5.7040	36.4988	23
24	28.6252	0.0349	0.0054	0.1554	184.1678	6.4338	5.7979	37.3023	24
25	32.9190	0.0304	0.0047	0.1547	212.7930	6.4641	5.8834	38.0314	25
26	37.8568	0.0264	0.0041	0.1541	245.7120	6.4906	5.9612	38.6918	26
27	43.5353	0.0230	0.0035	0.1535	283.5688	6.5135	6.0319	39.2890	27
28	50.0656	0.0200	0.0031	0.1531	327.1041	6.5335	6.0960	39.8283	28
29	57.5755	0.0174	0.0027	0.1527	377.1697	6.5509	6.1541	40.3146	29
30	66.2118	0.0151	0.0023	0.1523	434.7451	6.5660	6.2066	40.7526	30
31	76.1435	0.0131	0.0020	0.1520	500.9569	6.5791	6.2541	41.1466	31
32	87.5651	0.0114	0.0017	0.1517	577.1005	6.5905	6.2970	41.5006	32
33	100.6998	0.0099	0.0015	0.1515	664.6655	6.6005	6.3357	41.8184	33
34	115.8048	0.0086	0.0013	0.1513	765.3654	6.6091	6.3705	42.1033	34
35	133.1755	0.0075	0.0011	0.1511	881.1702	6.6166	6.4019	42.3586	35
40	267.8635	0.0037	0.0006	0.1506	1779.0903	6.6418	6.5168	43.2830	40
45	538.7693	0.0019	0.0003	0.1503	3585.1285	6.6543	6.5830	43.8051	45
50	1083.6574	0.0009	0.0001	0.1501	7217.7163	6.6605	6.6205	44.0958	50
55	2179.6222	0.0005	0.0001	0.1501	14524.1479	6.6636	6.6414	44.2558	55
60	4383.9987	0.0002	0.0000	0.1500	29219.9916	6.6651	6.6530	44.3431	60
65	8817.7874	0.0001	0.0000	0.1500	58778.5826	6.6659	6.6593	44.3903	65
70	17735.7200	0.0001	0.0000	0.1500	118231.4669	6.6663	6.6627	44.4156	70
75	35672.8680	0.0000	0.0000	0.1500	237812.4532	6.6665	6.6646	44.4292	75
80	71750.8794	0.0000	0.0000	0.1500	478332.5293	6.6666	6.6656	44.4364	80
85	144316.6470	0.0000	0.0000	0.1500	962104.3133	6.6666	6.6661	44.4402	85

16% Compound Interest Factors

	Single Payment		Uniform Series				Arithmetic Gradient		
N	Compound Amount Factor F/P	Present Worth Factor P/F	Sinking Fund Factor A/F	Capital Recovery Factor A/P	Compound Amount Factor F/A	Present Worth Factor P/A	Uniform Series A/G	Present Worth P/G	N
1	1.16000	0.86207	1.00000	1.16000	1.00000	0.86207	0.00000	0.00000	1
2	1.34560	0.74316	0.46296	0.62296	2.16000	1.60523	0.46296	0.74316	2
3	1.56090	0.64066	0.28526	0.44526	3.50560	2.24589	0.90141	2.02448	3
4	1.81064	0.55229	0.19738	0.35738	5.06650	2.79818	1.31562	3.68135	4
5	2.10034	0.47611	0.14541	0.30541	6.87714	3.27429	1.70596	5.58580	5
6	2.43640	0.41044	0.11139	0.27139	8.97748	3.68474	2.07288	7.63801	6
7	2.82622	0.35383	0.08761	0.24761	11.41387	4.03857	2.41695	9.76099	7
8	3.27841	0.30503	0.07022	0.23022	14.24009	4.34359	2.73879	11.89617	8
9	3.80296	0.26295	0.05708	0.21708	17.51851	4.60654	3.03911	13.99979	9
10	4.41144	0.22668	0.04690	0.20690	21.32147	4.83323	3.31868	16.03995	10
11	5.11726	0.19542	0.03886	0.19886	25.73290	5.02864	3.57832	17.99412	11
12	5.93603	0.16846	0.03241	0.19241	30.85017	5.19711	3.81890	19.84721	12
13	6.88579	0.14523	0.02718	0.18718	36.78620	5.34233	4.04129	21.58993	13
14	7.98752	0.12520	0.02290	0.18290	43.67199	5.46753	4.24643	23.21747	14
15	9.26552	0.10793	0.01936	0.17936	51.65951	5.57546	4.43523	24.72844	15
16	10.74800	0.09304	0.01641	0.17641	60.92503	5.66850	4.60864	26.12405	16
17	12.46768	0.08021	0.01395	0.17395	71.67303	5.74870	4.76757	27.40737	17
18	14.46251	0.06914	0.01188	0.17188	84.14072	5.81785	4.91295	28.58282	18
19	16.77652	0.05961	0.01014	0.17014	98.60323	5.87746	5.04568	29.65575	19
20	19.46076	0.05139	0.00867	0.16867	115.37975	5.92884	5.16662	30.63207	20
21	22.57448	0.04430	0.00742	0.16742	134.84051	5.97314	5.27663	31.51803	21
22	26.18640	0.03819	0.00635	0.16635	157.41499	6.01133	5.37651	32.31997	22
23	30.37622	0.03292	0.00545	0.16545	183.60138	6.04425	5.46705	33.04422	23
24	35.23642	0.02838	0.00467	0.16467	213.97761	6.07263	5.54899	33.69696	24
25	40.87424	0.02447	0.00401	0.16401	249.21402	6.09709	5.62303	34.28412	25
26	47.41412	0.02109	0.00345	0.16345	290.08827	6.11818	5.68983	34.81139	26
27	55.00038	0.01818	0.00296	0.16296	337.50239	6.13636	5.75000	35.28412	27
28	63.80044	0.01567	0.00255	0.16255	392.50277	6.15204	5.80414	35.70731	28
29	74.00851	0.01351	0.00219	0.16219	456.30322	6.16555	5.85279	36.08565	29
30	85.84988	0.01165	0.00189	0.16189	530.31173	6.17720	5.89643	36.42345	30
31	99.58586	0.01004	0.00162	0.16162	616.16161	6.18724	5.93555	36.72469	31
32	115.51959	0.00866	0.00140	0.16140	715.74746	6.19590	5.97057	36.99305	32
33	134.00273	0.00746	0.00120	0.16120	831.26706	6.20336	6.00188	37.23185	33
34	155.44317	0.00643	0.00104	0.16104	965.26979	6.20979	6.02985	37.44414	34
35	180.31407	0.00555	0.00089	0.16089	1120.71295	6.21534	6.05481	37.63270	35
40	378.72116	0.00264	0.00042	0.16042	2360.75724	6.23350	6.14410	38.29924	40
45	795.44383	0.00126	0.00020	0.16020	4965.27391	6.24214	6.19336	38.65982	45
50	1670.70380	0.00060	0.00010	0.16010	10435.64877	6.24626	6.22005	38.85207	50
55	3509.04880	0.00028	0.00005	0.16005	21925.30498	6.24822	6.23432	38.95341	55
60	7370.20137	0.00014	0.00002	0.16002	46057.50853	6.24915	6.24186	39.00632	60
65	15479.94095	0.00006	0.00001	0.16001	96743.38095	6.24960	6.24580	39.03373	65
70	32513.16484	0.00003	0.00000	0.16000	203201.03025	6.24981	6.24785	39.04784	70

18% Compound Interest Factors

	Single Payment		Uniform Series				Arithmetic Gradient		
	Compound Amount Factor	Present Worth Factor	Sinking Fund Factor	Capital Recovery Factor	Compound Amount Factor	Present Worth Factor	Uniform Series	Present Worth	
N	*F/P*	*P/F*	*A/F*	*A/P*	*F/A*	*P/A*	*A/G*	*P/G*	*N*
1	1.1800	0.8475	1.0000	1.1800	1.0000	0.8475	0.0000	0.0000	1
2	1.3924	0.7182	0.4587	0.6387	2.1800	1.5656	0.4587	0.7182	2
3	1.6430	0.6086	0.2799	0.4599	3.5724	2.1743	0.8902	1.9354	3
4	1.9388	0.5158	0.1917	0.3717	5.2154	2.6901	1.2947	3.4828	4
5	2.2878	0.4371	0.1398	0.3198	7.1542	3.1272	1.6728	5.2312	5
6	2.6996	0.3704	0.1059	0.2859	9.4420	3.4976	2.0252	7.0834	6
7	3.1855	0.3139	0.0824	0.2624	12.1415	3.8115	2.3526	8.9670	7
8	3.7589	0.2660	0.0652	0.2452	15.3270	4.0776	2.6558	10.8292	8
9	4.4355	0.2255	0.0524	0.2324	19.0859	4.3030	2.9358	12.6329	9
10	5.2338	0.1911	0.0425	0.2225	23.5213	4.4941	3.1936	14.3525	10
11	6.1759	0.1619	0.0348	0.2148	28.7551	4.6560	3.4303	15.9716	11
12	7.2876	0.1372	0.0286	0.2086	34.9311	4.7932	3.6470	17.4811	12
13	8.5994	0.1163	0.0237	0.2037	42.2187	4.9095	3.8449	18.8765	13
14	10.1472	0.0985	0.0197	0.1997	50.8180	5.0081	4.0250	20.1576	14
15	11.9737	0.0835	0.0164	0.1964	60.9653	5.0916	4.1887	21.3269	15
16	14.1290	0.0708	0.0137	0.1937	72.9390	5.1624	4.3369	22.3885	16
17	16.6722	0.0600	0.0115	0.1915	87.0680	5.2223	4.4708	23.3482	17
18	19.6733	0.0508	0.0096	0.1896	103.7403	5.2732	4.5916	24.2123	18
19	23.2144	0.0431	0.0081	0.1881	123.4135	5.3162	4.7003	24.9877	19
20	27.3930	0.0365	0.0068	0.1868	146.6280	5.3527	4.7978	25.6813	20
21	32.3238	0.0309	0.0057	0.1857	174.0210	5.3837	4.8851	26.3000	21
22	38.1421	0.0262	0.0048	0.1848	206.3448	5.4099	4.9632	26.8506	22
23	45.0076	0.0222	0.0041	0.1841	244.4868	5.4321	5.0329	27.3394	23
24	53.1090	0.0188	0.0035	0.1835	289.4945	5.4509	5.0950	27.7725	24
25	62.6686	0.0160	0.0029	0.1829	342.6035	5.4669	5.1502	28.1555	25
26	73.9490	0.0135	0.0025	0.1825	405.2721	5.4804	5.1991	28.4935	26
27	87.2598	0.0115	0.0021	0.1821	479.2211	5.4919	5.2425	28.7915	27
28	102.9666	0.0097	0.0018	0.1818	566.4809	5.5016	5.2810	29.0537	28
29	121.5005	0.0082	0.0015	0.1815	669.4475	5.5098	5.3149	29.2842	29
30	143.3706	0.0070	0.0013	0.1813	790.9480	5.5168	5.3448	29.4864	30
31	169.1774	0.0059	0.0011	0.1811	934.3186	5.5227	5.3712	29.6638	31
32	199.6293	0.0050	0.0009	0.1809	1103.4960	5.5277	5.3945	29.8191	32
33	235.5625	0.0042	0.0008	0.1808	1303.1253	5.5320	5.4149	29.9549	33
34	277.9638	0.0036	0.0006	0.1806	1538.6878	5.5356	5.4328	30.0736	34
35	327.9973	0.0030	0.0006	0.1806	1816.6516	5.5386	5.4485	30.1773	35
40	750.3783	0.0013	0.0002	0.1802	4163.2130	5.5482	5.5022	30.5269	40
45	1716.6839	0.0006	0.0001	0.1801	9531.5771	5.5523	5.5293	30.7006	45
50	3927.3569	0.0003	0.0000	0.1800	21813.0937	5.5541	5.5428	30.7856	50
55	8984.8411	0.0001	0.0000	0.1800	49910.2284	5.5549	5.5494	30.8268	55
60	20555.1400	0.0000	0.0000	0.1800	114189.6665	5.5553	5.5526	30.8465	60
65	47025.1809	0.0000	0.0000	0.1800	261245.4494	5.5554	5.5542	30.8559	65
70	107582.2224	0.0000	0.0000	0.1800	597673.4576	5.5555	5.5549	30.8603	70

20% Compound Interest Factors

	Single Payment		Uniform Series				Arithmetic Gradient		
N	Compound Amount Factor *F/P*	Present Worth Factor *P/F*	Sinking Fund Factor *A/F*	Capital Recovery Factor *A/P*	Compound Amount Factor *F/A*	Present Worth Factor *P/A*	Uniform Series *A/G*	Present Worth *P/G*	*N*
1	1.2000	0.8333	1.0000	1.2000	1.0000	0.8333	0.0000	0.0000	1
2	1.4400	0.6944	0.4545	0.6545	2.2000	1.5278	0.4545	0.6944	2
3	1.7280	0.5787	0.2747	0.4747	3.6400	2.1065	0.8791	1.8519	3
4	2.0736	0.4823	0.1863	0.3863	5.3680	2.5887	1.2742	3.2986	4
5	2.4883	0.4019	0.1344	0.3344	7.4416	2.9906	1.6405	4.9061	5
6	2.9860	0.3349	0.1007	0.3007	9.9299	3.3255	1.9788	6.5806	6
7	3.5832	0.2791	0.0774	0.2774	12.9159	3.6046	2.2902	8.2551	7
8	4.2998	0.2326	0.0606	0.2606	16.4991	3.8372	2.5756	9.8831	8
9	5.1598	0.1938	0.0481	0.2481	20.7989	4.0310	2.8364	11.4335	9
10	6.1917	0.1615	0.0385	0.2385	25.9587	4.1925	3.0739	12.8871	10
11	7.4301	0.1346	0.0311	0.2311	32.1504	4.3271	3.2893	14.2330	11
12	8.9161	0.1122	0.0253	0.2253	39.5805	4.4392	3.4841	15.4667	12
13	10.6993	0.0935	0.0206	0.2206	48.4966	4.5327	3.6597	16.5883	13
14	12.8392	0.0779	0.0169	0.2169	59.1959	4.6106	3.8175	17.6008	14
15	15.4070	0.0649	0.0139	0.2139	72.0351	4.6755	3.9588	18.5095	15
16	18.4884	0.0541	0.0114	0.2114	87.4421	4.7296	4.0851	19.3208	16
17	22.1861	0.0451	0.0094	0.2094	105.9306	4.7746	4.1976	20.0419	17
18	26.6233	0.0376	0.0078	0.2078	128.1167	4.8122	4.2975	20.6805	18
19	31.9480	0.0313	0.0065	0.2065	154.7400	4.8435	4.3861	21.2439	19
20	38.3376	0.0261	0.0054	0.2054	186.6880	4.8696	4.4643	21.7395	20
21	46.0051	0.0217	0.0044	0.2044	225.0256	4.8913	4.5334	22.1742	21
22	55.2061	0.0181	0.0037	0.2037	271.0307	4.9094	4.5941	22.5546	22
23	66.2474	0.0151	0.0031	0.2031	326.2369	4.9245	4.6475	22.8867	23
24	79.4968	0.0126	0.0025	0.2025	392.4842	4.9371	4.6943	23.1760	24
25	95.3962	0.0105	0.0021	0.2021	471.9811	4.9476	4.7352	23.4276	25
26	114.4755	0.0087	0.0018	0.2018	567.3773	4.9563	4.7709	23.6460	26
27	137.3706	0.0073	0.0015	0.2015	681.8528	4.9636	4.8020	23.8353	27
28	164.8447	0.0061	0.0012	0.2012	819.2233	4.9697	4.8291	23.9991	28
29	197.8136	0.0051	0.0010	0.2010	984.0680	4.9747	4.8527	24.1406	29
30	237.3763	0.0042	0.0008	0.2008	1181.8816	4.9789	4.8731	24.2628	30
31	284.8516	0.0035	0.0007	0.2007	1419.2579	4.9824	4.8908	24.3681	31
32	341.8219	0.0029	0.0006	0.2006	1704.1095	4.9854	4.9061	24.4588	32
33	410.1863	0.0024	0.0005	0.2005	2045.9314	4.9878	4.9194	24.5368	33
34	492.2235	0.0020	0.0004	0.2004	2456.1176	4.9898	4.9308	24.6038	34
35	590.6682	0.0017	0.0003	0.2003	2948.3411	4.9915	4.9406	24.6614	35
40	1469.7716	0.0007	0.0001	0.2001	7343.8578	4.9966	4.9728	24.8469	40
45	3657.2620	0.0003	0.0001	0.2001	18281.3099	4.9986	4.9877	24.9316	45
50	9100.4382	0.0001	0.0000	0.2000	45497.1908	4.9995	4.9945	24.9698	50
55	22644.8023	0.0000	0.0000	0.2000	113219.0113	4.9998	4.9976	24.9868	55
60	56347.5144	0.0000	0.0000	0.2000	281732.5718	4.9999	4.9989	24.9942	60

25% Compound Interest Factors

	Single Payment		Uniform Series				Arithmetic Gradient		
	Compound Amount Factor	Present Worth Factor	Sinking Fund Factor	Capital Recovery Factor	Compound Amount Factor	Present Worth Factor	Uniform Series	Present Worth	
N	*F/P*	*P/F*	*A/F*	*A/P*	*F/A*	*P/A*	*A/G*	*P/G*	*N*
1	1.2500	0.8000	1.0000	1.2500	1.0000	0.8000	0.0000	0.0000	1
2	1.5625	0.6400	0.4444	0.6944	2.2500	1.4400	0.4444	0.6400	2
3	1.9531	0.5120	0.2623	0.5123	3.8125	1.9520	0.8525	1.6640	3
4	2.4414	0.4096	0.1734	0.4234	5.7656	2.3616	1.2249	2.8928	4
5	3.0518	0.3277	0.1218	0.3718	8.2070	2.6893	1.5631	4.2035	5
6	3.8147	0.2621	0.0888	0.3388	11.2588	2.9514	1.8683	5.5142	6
7	4.7684	0.2097	0.0663	0.3163	15.0735	3.1611	2.1424	6.7725	7
8	5.9605	0.1678	0.0504	0.3004	19.8419	3.3289	2.3872	7.9469	8
9	7.4506	0.1342	0.0388	0.2888	25.8023	3.4631	2.6048	9.0207	9
10	9.3132	0.1074	0.0301	0.2801	33.2529	3.5705	2.7971	9.9870	10
11	11.6415	0.0859	0.0235	0.2735	42.5661	3.6564	2.9663	10.8460	11
12	14.5519	0.0687	0.0184	0.2684	54.2077	3.7251	3.1145	11.6020	12
13	18.1899	0.0550	0.0145	0.2645	68.7596	3.7801	3.2437	12.2617	13
14	22.7374	0.0440	0.0115	0.2615	86.9495	3.8241	3.3559	12.8334	14
15	28.4217	0.0352	0.0091	0.2591	109.6868	3.8593	3.4530	13.3260	15
16	35.5271	0.0100	0.0072	0.2572	138.1085	3.8874	3.5366	13.7482	16
17	44.4089	0.0225	0.0058	0.2558	173.6357	3.9099	3.6084	14.1085	17
18	55.5112	0.0180	0.0046	0.2546	218.0446	3.9279	3.6698	14.4147	18
19	69.3889	0.0144	0.0037	0.2537	273.5558	3.9424	3.7222	14.6741	19
20	86.7362	0.0115	0.0029	0.2529	342.9447	3.9539	3.7667	14.8932	20
21	108.4202	0.0092	0.0023	0.2523	429.6809	3.9631	3.8045	15.0777	21
22	135.5253	0.0074	0.0019	0.2519	538.1011	3.9705	3.8365	15.2326	22
23	169.4066	0.0059	0.0015	0.2515	673.6264	3.9764	3.8634	15.3625	23
24	211.7582	0.0047	0.0012	0.2512	843.0329	3.9811	3.8861	15.4711	24
25	264.6978	0.0038	0.0009	0.2509	1054.7912	3.9849	3.9052	15.5618	25
26	330.8722	0.0030	0.0008	0.2508	1319.4890	3.9879	3.9212	15.6373	26
27	413.5903	0.0024	0.0006	0.2506	1650.3612	3.9903	3.9346	15.7002	27
28	516.9879	0.0019	0.0005	0.2505	2063.9515	3.9923	3.9457	15.7524	28
29	646.2349	0.0015	0.0004	0.2504	2580.9394	3.9938	3.9551	15.7957	29
30	807.7936	0.0012	0.0003	0.2503	3227.1743	3.9950	3.9628	15.8316	30
31	1009.7420	0.0010	0.0002	0.2502	4034.9678	3.9960	3.9693	15.8614	31
32	1262.1774	0.0008	0.0002	0.2502	5044.7098	3.9968	3.9746	15.8859	32
33	1577.7218	0.0006	0.0002	0.2502	6306.8872	3.9975	3.9791	15.9062	33
34	1972.1523	0.0005	0.0001	0.2501	7884.6091	3.9980	3.9828	15.9229	34
35	2465.1903	0.0004	0.0001	0.2501	9856.7613	3.9984	3.9858	15.9367	35
40	7523.1638	0.0001	0.0000	0.2500	30088.6554	3.9995	3.9947	15.9766	40
45	22958.8740	0.0000	0.0000	0.2500	91831.4962	3.9998	3.9980	15.9915	45
50	70064.9232	0.0000	0.0000	0.2500	280255.6929	3.9999	3.9993	15.9969	50
55	213821.1768	0.0000	0.0000	0.2500	855280.7072	4.0000	3.9997	15.9989	55

30% Compound Interest Factors

	Single Payment		Uniform Series				Arithmetic Gradient		
N	Compound Amount Factor F/P	Present Worth Factor P/F	Sinking Fund Factor A/F	Capital Recovery Factor A/P	Compound Amount Factor F/A	Present Worth Factor P/A	Uniform Series A/G	Present Worth P/G	N
1	1.3000	0.7692	1.0000	1.3000	1.0000	0.7692	0.0000	0.0000	1
2	1.6900	0.5917	0.4348	0.7348	2.3000	1.3609	0.4348	0.5917	2
3	2.1970	0.4552	0.2506	0.5506	3.9900	1.8161	0.8271	1.5020	3
4	2.8561	0.3501	0.1616	0.4616	6.1870	2.1662	1.1783	2.5524	4
5	3.7129	0.2693	0.1106	0.4106	9.0431	2.4356	1.4903	3.6297	5
6	4.8268	0.2072	0.0784	0.3784	12.7560	2.6427	1.7654	4.6656	6
7	6.2749	0.1594	0.0569	0.3569	17.5828	2.8021	2.0063	5.6218	7
8	8.1573	0.1226	0.0419	0.3419	23.8577	2.9247	2.2156	6.4800	8
9	10.6045	0.0943	0.0312	0.3312	32.0150	3.0190	2.3963	7.2343	9
10	13.7858	0.0725	0.0235	0.3235	42.6195	3.0915	2.5512	7.8872	10
11	17.9216	0.0558	0.0177	0.3177	56.4053	3.1473	2.6833	8.4452	11
12	23.2981	0.0429	0.0135	0.3135	74.3270	3.1903	2.7952	8.9173	12
13	30.2875	0.0330	0.0102	0.3102	97.6250	3.2233	2.8895	9.3135	13
14	39.3738	0.0254	0.0078	0.3078	127.9125	3.2487	2.9685	9.6437	14
15	51.1859	0.0195	0.0060	0.3060	167.2863	3.2682	3.0344	9.9172	15
16	66.5417	0.0150	0.0046	0.3046	218.4722	3.2832	3.0892	10.1426	16
17	86.5042	0.0116	0.0035	0.3035	285.0139	3.2948	3.1345	10.3276	17
18	112.4554	0.0089	0.0027	0.3027	371.5180	3.3037	3.1718	10.4788	18
19	146.1920	0.0068	0.0021	0.3021	483.9734	3.3105	3.2025	10.6019	19
20	190.0496	0.0053	0.0016	0.3016	630.1655	3.3158	3.2275	10.7019	20
21	247.0645	0.0040	0.0012	0.3012	820.2151	3.3198	3.2480	10.7828	21
22	321.1839	0.0031	0.0009	0.3009	1067.2796	3.3230	3.2646	10.8482	22
23	417.5391	0.0024	0.0007	0.3007	1388.4635	3.3254	3.2781	10.9009	23
24	542.8008	0.0018	0.0006	0.3006	1806.0026	3.3272	3.2890	10.9433	24
25	705.6410	0.0014	0.0004	0.3004	2348.8033	3.3286	3.2979	10.9773	25
26	917.3333	0.0011	0.0003	0.3003	3054.4443	3.3297	3.3050	11.0045	26
27	1192.5333	0.0008	0.0003	0.3003	3971.7776	3.3305	3.3107	11.0263	27
28	1550.2933	0.0006	0.0002	0.3002	5164.3109	3.3312	3.3153	11.0437	28
29	2015.3813	0.0005	0.0001	0.3001	6714.6042	3.3317	3.3189	11.0576	29
30	2619.9956	0.0004	0.0001	0.3001	8729.9855	3.3321	3.3219	11.0687	30
31	3405.9943	0.0003	0.0001	0.3001	11349.9811	3.3324	3.3242	11.0775	31
32	4427.7926	0.0002	0.0001	0.3001	14755.9755	3.3326	3.3261	11.0845	32
33	5756.1304	0.0002	0.0001	0.3001	19183.7681	3.3328	3.3276	11.0901	33
34	7482.9696	0.0001	0.0000	0.3000	24939.8985	3.3329	3.3288	11.0945	34
35	9727.8604	0.0001	0.0000	0.3000	32422.8681	3.3330	3.3297	11.0980	35
40	36118.8648	0.0000	0.0000	0.3000	120392.8827	3.3332	3.3322	11.1071	40
45	134106.8167	0.0000	0.0000	0.3000	447019.3890	3.3333	3.3330	11.1099	45

35% Compound Interest Factors

	Single Payment		Uniform Series				Arithmetic Gradient		
	Compound Amount Factor	Present Worth Factor	Sinking Fund Factor	Capital Recovery Factor	Compound Amount Factor	Present Worth Factor	Uniform Series	Present Worth	
N	*F/P*	*P/F*	*A/F*	*A/P*	*F/A*	*P/A*	*A/G*	*P/G*	*N*
1	1.3500	0.7407	1.0000	1.3500	1.0000	0.7407	0.0000	0.0000	1
2	1.8225	0.5487	0.4255	0.7755	2.3500	1.2894	0.4255	0.5487	2
3	2.4604	0.4064	0.2397	0.5897	4.1725	1.6959	0.8029	1.3616	3
4	3.3215	0.3011	0.1508	0.5008	6.6329	1.9969	1.1341	2.2648	4
5	4.4840	0.2230	0.1005	0.4505	9.9544	2.2200	1.4220	3.1568	5
6	6.0534	0.1652	0.0693	0.4193	14.4384	2.3852	1.6698	3.9828	6
7	8.1722	0.1224	0.0488	0.3988	20.4919	2.5075	1.8811	4.7170	7
8	11.0324	0.0906	0.0349	0.3849	28.6640	2.5982	2.0597	5.3515	8
9	14.8937	0.0671	0.0252	0.3752	39.6964	2.6653	2.2094	5.8886	9
10	20.1066	0.0497	0.0183	0.3683	54.5902	2.7150	2.3338	6.3363	10
11	27.1439	0.0368	0.0134	0.3634	74.6967	2.7519	2.4364	6.7047	11
12	36.6442	0.0273	0.0098	0.3598	101.8406	2.7792	2.5205	7.0049	12
13	49.4697	0.0202	0.0072	0.3572	138.4848	2.7994	2.5889	7.2474	13
14	66.7841	0.0150	0.0053	0.3553	187.9544	2.8144	2.6443	7.4421	14
15	90.1585	0.0111	0.0039	0.3539	254.7385	2.8255	2.6889	7.5974	15
16	121.7139	0.0082	0.0029	0.3529	344.8970	2.8337	2.7246	7.7206	16
17	164.3138	0.0061	0.0021	0.3521	466.6109	2.8398	2.7530	7.8180	17
18	221.8236	0.0045	0.0016	0.3516	630.9247	2.8443	2.7756	7.8946	18
19	299.4619	0.0033	0.0012	0.3512	852.7483	2.8476	2.7935	7.9547	19
20	404.2736	0.0025	0.0009	0.3509	1152.2103	2.8501	2.8075	8.0017	20
21	545.7693	0.0018	0.0006	0.3506	1556.4838	2.8519	2.8186	8.0384	21
22	736.7886	0.0014	0.0005	0.3505	2102.2532	2.8533	2.8272	8.0669	22
23	994.6646	0.0010	0.0004	0.3504	2839.0418	2.8543	2.8340	8.0890	23
24	1342.7973	0.0007	0.0003	0.3503	3833.7064	2.8550	2.8393	8.1061	24
25	1812.7763	0.0006	0.0002	0.3502	5176.5037	2.8556	2.8433	8.1194	25
26	2447.2480	0.0004	0.0001	0.3501	6989.2800	2.8560	2.8465	8.1296	26
27	3303.7848	0.0003	0.0001	0.3501	9436.5280	2.8563	2.8490	8.1374	27
28	4460.1095	0.0002	0.0001	0.3501	12740.3128	2.8565	2.8509	8.1435	28
29	6021.1478	0.0002	0.0001	0.3501	17200.4222	2.8567	2.8523	8.1481	29
30	8128.5495	0.0001	0.0000	0.3500	23221.5700	2.8568	2.8535	8.1517	30
31	10973.5418	0.0001	0.0000	0.3500	31350.1195	2.8569	2.8543	8.1545	31
32	14814.2815	0.0001	0.0000	0.3500	42323.6613	2.8569	2.8550	8.1565	32
33	19999.2800	0.0001	0.0000	0.3500	57137.9428	2.8570	2.8555	8.1581	33
34	26999.0280	0.0000	0.0000	0.3500	77137.2228	2.8570	2.8559	8.1594	34
35	36448.6878	0.0000	0.0000	0.3500	104136.2508	2.8571	2.8562	8.1603	35

40% Compound Interest Factors

	Single Payment		Uniform Series				Arithmetic Gradient		
	Compound Amount Factor	Present Worth Factor	Sinking Fund Factor	Capital Recovery Factor	Compound Amount Factor	Present Worth Factor	Uniform Series	Present Worth	
N	*F/P*	*P/F*	*A/F*	*A/P*	*F/A*	*P/A*	*A/G*	*P/G*	*N*
1	1.4000	0.7143	1.0000	1.4000	1.0000	0.7143	0.0000	0.0000	1
2	1.9600	0.5102	0.4167	0.8167	2.4000	1.2245	0.4167	0.5102	2
3	2.7440	0.3644	0.2294	0.6294	4.3600	1.5889	0.7798	1.2391	3
4	3.8416	0.2603	0.1408	0.5408	7.1040	1.8492	1.0923	2.0200	4
5	5.3782	0.1859	0.0914	0.4914	10.9456	2.0352	1.3580	2.7637	5
6	7.5295	0.1328	0.0613	0.4613	16.3238	2.1680	1.5811	3.4278	6
7	10.5414	0.0949	0.0419	0.4419	23.8534	2.2628	1.7664	3.9970	7
8	14.7579	0.0678	0.0291	0.4291	34.3947	2.3306	1.9185	4.4713	8
9	20.6610	0.0484	0.0203	0.4203	49.1526	2.3790	2.0422	4.8585	9
10	28.9255	0.0346	0.0143	0.4143	69.8137	2.4136	2.1419	5.1696	10
11	40.4957	0.0247	0.0101	0.4101	98.7391	2.4383	2.2215	5.4166	11
12	56.6939	0.0176	0.0072	0.4072	139.2348	2.4559	2.2845	5.6106	12
13	79.3715	0.0126	0.0051	0.4051	195.9287	2.4685	2.3341	5.7618	13
14	111.1201	0.0090	0.0036	0.4036	275.3002	2.4775	2.3729	5.8788	14
15	155.5681	0.0064	0.0026	0.4026	386.4202	2.4839	2.4030	5.9688	15
16	217.7953	0.0046	0.0018	0.4018	541.9883	2.4885	2.4262	6.0376	16
17	304.9135	0.0033	0.0013	0.4013	759.7837	2.4918	2.4441	6.0901	17
18	426.8789	0.0023	0.0009	0.4009	1064.6971	2.4941	2.4577	6.1299	18
19	597.6304	0.0017	0.0007	0.4007	1491.5760	2.4958	2.4682	6.1601	19
20	836.6826	0.0012	0.0005	0.4005	2089.2064	2.4970	2.4761	6.1828	20
21	1171.3556	0.0009	0.0003	0.4003	2925.8889	2.4979	2.4821	6.1998	21
22	1639.8978	0.0006	0.0002	0.4002	4097.2445	2.4985	2.4866	6.2127	22
23	2295.8569	0.0004	0.0002	0.4002	5737.1423	2.4989	2.4900	6.2222	23
24	3214.1997	0.0003	0.0001	0.4001	8032.9993	2.4992	2.4925	6.2294	24
25	4499.8796	0.0002	0.0001	0.4001	11247.1990	2.4994	2.4944	6.2347	25
26	6299.8314	0.0002	0.0001	0.4001	15747.0785	2.4996	2.4959	6.2387	26
27	8819.7640	0.0001	0.0000	0.4000	22046.9099	2.4997	2.4969	6.2416	27
28	12347.6696	0.0001	0.0000	0.4000	30866.6739	2.4998	2.4977	6.2438	28
29	17286.7374	0.0001	0.0000	0.4000	43214.3435	2.4999	2.4983	6.2454	29
30	24201.4324	0.0000	0.0000	0.4000	60501.0809	2.4999	2.4988	6.2466	30
31	33882.0053	0.0000	0.0000	0.4000	84702.5132	2.4999	2.4991	6.2475	31
32	47434.8074	0.0000	0.0000	0.4000	118584.5185	2.4999	2.4993	6.2482	32
33	66408.7304	0.0000	0.0000	0.4000	166019.3260	2.5000	2.4995	6.2487	33
34	92972.2225	0.0000	0.0000	0.4000	232428.0563	2.5000	2.4996	6.2490	34
35	130161.1116	0.0000	0.0000	0.4000	325400.2789	2.5000	2.4997	6.2493	35

45% Compound Interest Factors

	Single Payment		Uniform Series				Arithmetic Gradient		
	Compound Amount Factor	Present Worth Factor	Sinking Fund Factor	Capital Recovery Factor	Compound Amount Factor	Present Worth Factor	Uniform Series	Present Worth	
N	*F/P*	*P/F*	*A/F*	*A/P*	*F/A*	*P/A*	*A/G*	*P/G*	*N*
1	1.4500	0.6897	1.0000	1.4500	1.0000	0.6897	0.0000	0.0000	1
2	2.1025	0.4756	0.4082	0.8582	2.4500	1.1653	0.4082	0.4756	2
3	3.0486	0.3280	0.2197	0.6697	4.5525	1.4933	0.7578	1.1317	3
4	4.4205	0.2262	0.1316	0.5816	7.6011	1.7195	1.0528	1.8103	4
5	6.4097	0.1560	0.0832	0.5332	12.0216	1.8755	1.2980	2.4344	5
6	9.2941	0.1076	0.0543	0.5043	18.4314	1.9831	1.4988	2.9723	6
7	13.4765	0.0742	0.0361	0.4861	27.7255	2.0573	1.6612	3.4176	7
8	19.5409	0.0512	0.0243	0.4743	41.2019	2.1085	1.7907	3.7758	8
9	28.3343	0.0353	0.0165	0.4665	60.7428	2.1438	1.8930	4.0581	9
10	41.0847	0.0243	0.0112	0.4612	89.0771	2.1681	1.9728	4.2772	10
11	59.5728	0.0168	0.0077	0.4577	130.1618	2.1849	2.0344	4.4450	11
12	86.3806	0.0116	0.0053	0.4553	189.7346	2.1965	2.0817	4.5724	12
13	125.2518	0.0080	0.0036	0.4536	276.1151	2.2045	2.1176	4.6682	13
14	181.6151	0.0055	0.0025	0.4525	401.3670	2.2100	2.1447	4.7398	14
15	263.3419	0.0038	0.0017	0.4517	582.9821	2.2138	2.1650	4.7929	15
16	381.8458	0.0026	0.0012	0.4512	846.3240	2.2164	2.1802	4.8322	16
17	553.6764	0.0018	0.0008	0.4508	1228.1699	2.2182	2.1915	4.8611	17
18	802.8308	0.0012	0.0006	0.4506	1781.8463	2.2195	2.1998	4.8823	18
19	1164.1047	0.0009	0.0004	0.4504	2584.6771	2.2203	2.2059	4.8978	19
20	1687.9518	0.0006	0.0003	0.4503	3748.7818	2.2209	2.2104	4.9090	20
21	2447.5301	0.0004	0.0002	0.4502	5436.7336	2.2213	2.2136	4.9172	21
22	3548.9187	0.0003	0.0001	0.4501	7884.2638	2.2216	2.2160	4.9231	22
23	5145.9321	0.0002	0.0001	0.4501	11433.1824	2.2218	2.2178	4.9274	23
24	7461.6015	0.0001	0.0001	0.4501	16579.1145	2.2219	2.2190	4.9305	24
25	10819.3222	0.0001	0.0000	0.4500	24040.7161	2.2220	2.2199	4.9327	25
26	15688.0172	0.0001	0.0000	0.4500	34860.0383	2.2221	2.2206	4.9343	26
27	22747.6250	0.0000	0.0000	0.4500	50548.0556	2.2221	2.2210	4.9354	27
28	32984.0563	0.0000	0.0000	0.4500	73295.6806	2.2222	2.2214	4.9362	28
29	47826.8816	0.0000	0.0000	0.4500	106279.7368	2.2222	2.2216	4.9368	29
30	69348.9783	0.0000	0.0000	0.4500	154106.6184	2.2222	2.2218	4.9372	30
31	100556.0185	0.0000	0.0000	0.4500	223455.5967	2.2222	2.2219	4.9375	31
32	145806.2269	0.0000	0.0000	0.4500	324011.6152	2.2222	2.2220	4.9378	32
33	211419.0289	0.0000	0.0000	0.4500	469817.8421	2.2222	2.2221	4.9379	33
34	306557.5920	0.0000	0.0000	0.4500	681236.8710	2.2222	2.2221	4.9380	34
35	444508.5083	0.0000	0.0000	0.4500	987794.4630	2.2222	2.2221	4.9381	35

50% Compound Interest Factors

	Single Payment		Uniform Series				Arithmetic Gradient		
	Compound Amount Factor	Present Worth Factor	Sinking Fund Factor	Capital Recovery Factor	Compound Amount Factor	Present Worth Factor	Uniform Series	Present Worth	
N	*F/P*	*P/F*	*A/F*	*A/P*	*F/A*	*P/A*	*A/G*	*P/G*	*N*
1	1.5000	0.6667	1.0000	1.5000	1.0000	0.6667	0.0000	0.0000	1
2	2.2500	0.4444	0.4000	0.9000	2.5000	1.1111	0.4000	0.4444	2
3	3.3750	0.2963	0.2105	0.7105	4.7500	1.4074	0.7368	1.0370	3
4	5.0625	0.1975	0.1231	0.6231	8.1250	1.6049	1.0154	1.6296	4
5	7.5938	0.1317	0.0758	0.5758	13.1875	1.7366	1.2417	2.1564	5
6	11.3906	0.0878	0.0481	0.5481	20.7813	1.8244	1.4226	2.5953	6
7	17.0859	0.0585	0.0311	0.5311	32.1719	1.8829	1.5648	2.9465	7
8	25.6289	0.0390	0.0203	0.5203	49.2578	1.9220	1.6752	3.2196	8
9	38.4434	0.0260	0.0134	0.5134	74.8867	1.9480	1.7596	3.4277	9
10	57.6650	0.0173	0.0088	0.5088	113.3301	1.9653	1.8235	3.5838	10
11	86.4976	0.0116	0.0058	0.5058	170.9951	1.9769	1.8713	3.6994	11
12	129.7463	0.0077	0.0039	0.5039	257.4927	1.9846	1.9068	3.7842	12
13	194.6195	0.0051	0.0026	0.5026	387.2390	1.9897	1.9329	3.8459	13
14	291.9293	0.0034	0.0017	0.5017	581.8585	1.9931	1.9519	3.8904	14
15	437.8939	0.0023	0.0011	0.5011	873.7878	1.9954	1.9657	3.9224	15
16	656.8408	0.0015	0.0008	0.5008	1311.6817	1.9970	1.9756	3.9452	16
17	985.2613	0.0010	0.0005	0.5005	1968.5225	1.9980	1.9827	3.9614	17
18	1477.8919	0.0007	0.0003	0.5003	2953.7838	1.9986	1.9878	3.9729	18
19	2216.8378	0.0005	0.0002	0.5002	4431.6756	1.9991	1.9914	3.9811	19
20	3325.2567	0.0003	0.0002	0.5002	6648.5135	1.9994	1.9940	3.9868	20
21	4987.8851	0.0002	0.0001	0.5001	9973.7702	1.9996	1.9958	3.9908	21
22	7481.8276	0.0001	0.0001	0.5001	14961.6553	1.9997	1.9971	3.9936	22
23	11222.7415	0.0001	0.0000	0.5000	22443.4829	1.9998	1.9980	3.9955	23
24	16834.1122	0.0001	0.0000	0.5000	33666.2244	1.9999	1.9986	3.9969	24
25	25251.1683	0.0000	0.0000	0.5000	50500.3366	1.9999	1.9990	3.9979	25
26	37876.7524	0.0000	0.0000	0.5000	75751.5049	1.9999	1.9993	3.9985	26
27	56815.1287	0.0000	0.0000	0.5000	113628.2573	2.0000	1.9995	3.9990	27
28	85222.6930	0.0000	0.0000	0.5000	170443.3860	2.0000	1.9997	3.9993	28
29	127834.0395	0.0000	0.0000	0.5000	255666.0790	2.0000	1.9998	3.9995	29
30	191751.0592	0.0000	0.0000	0.5000	383500.1185	2.0000	1.9998	3.9997	30
31	287626.5888	0.0000	0.0000	0.5000	575251.1777	2.0000	1.9999	3.9998	31
32	431439.8833	0.0000	0.0000	0.5000	862877.7665	2.0000	1.9999	3.9998	32

60% Compound Interest Factors

	Single Payment		Uniform Series				Arithmetic Gradient		
	Compound Amount Factor	Present Worth Factor	Sinking Fund Factor	Capital Recovery Factor	Compound Amount Factor	Present Worth Factor	Uniform Series	Present Worth	
N	*F/P*	*P/F*	*A/F*	*A/P*	*F/A*	*P/A*	*A/G*	*P/G*	*N*
1	1.6000	0.6250	1.0000	1.6000	1.0000	0.6250	0.0000	0.0000	1
2	2.5600	0.3906	0.3846	0.9846	2.6000	1.0156	0.3846	0.3906	2
3	4.0960	0.2441	0.1938	0.7938	5.1600	1.2598	0.6977	0.8789	3
4	6.5536	0.1526	0.1080	0.7080	9.2560	1.4124	0.9464	1.3367	4
5	10.4858	0.0954	0.0633	0.6633	15.8096	1.5077	1.1396	1.7181	5
6	16.7772	0.0596	0.0380	0.6380	26.2954	1.5673	1.2864	2.0162	6
7	26.8435	0.0373	0.0232	0.6232	43.0726	1.6046	1.3958	2.2397	7
8	42.9497	0.0233	0.0143	0.6143	69.9161	1.6279	1.4760	2.4027	8
9	68.7195	0.0146	0.0089	0.6089	112.8658	1.6424	1.5338	2.5191	9
10	109.9512	0.0091	0.0055	0.6055	181.5853	1.6515	1.5749	2.6009	10
11	175.9219	0.0057	0.0034	0.6034	291.5364	1.6572	1.6038	2.6578	11
12	281.4750	0.0036	0.0021	0.6021	467.4583	1.6607	1.6239	2.6969	12
13	450.3600	0.0022	0.0013	0.6013	748.9333	1.6630	1.6377	2.7235	13
14	720.5759	0.0014	0.0008	0.6008	1199.2932	1.6644	1.6472	2.7415	14
15	1152.9215	0.0009	0.0005	0.6005	1919.8692	1.6652	1.6536	2.7537	15
16	1844.6744	0.0005	0.0003	0.6003	3072.7907	1.6658	1.6580	2.7618	16
17	2951.4791	0.0003	0.0002	0.6002	4917.4651	1.6661	1.6609	2.7672	17
18	4722.3665	0.0002	0.0001	0.6001	7868.9441	1.6663	1.6629	2.7708	18
19	7555.7864	0.0001	0.0001	0.6001	12591.3106	1.6664	1.6642	2.7732	19
20	12089.2582	0.0001	0.0000	0.6000	20147.0970	1.6665	1.6650	2.7748	20
21	19342.8131	0.0001	0.0000	0.6000	32236.3552	1.6666	1.6656	2.7758	21
22	30948.5010	0.0000	0.0000	0.6000	51579.1683	1.6666	1.6660	2.7765	22
23	49517.6016	0.0000	0.0000	0.6000	82527.6693	1.6666	1.6662	2.7769	23
24	79228.1625	0.0000	0.0000	0.6000	132045.2709	1.6666	1.6664	2.7772	24
25	126765.0600	0.0000	0.0000	0.6000	211273.4334	1.6667	1.6665	2.7774	25
26	202824.0960	0.0000	0.0000	0.6000	338038.4934	1.6667	1.6665	2.7776	26
27	324518.5537	0.0000	0.0000	0.6000	540862.5894	1.6667	1.6666	2.7776	27
28	519229.6859	0.0000	0.0000	0.6000	865381.1431	1.6667	1.6666	2.7777	28

APPENDIX D

SELECTED BIBLIOGRAPHY

The American Society for Engineering Education, and Institute of Industrial Engineers. *The Engineering Economist.* Norcross, GA: Institute of Industrial Engineers. Published quarterly by the Engineering Economy Division of the American Society for Engineering Education and the Institute of Industrial Engineers.

American Telephone and Telegraph, Engineering Department. *Engineering Economy.* 3d ed. New York: American Telephone and Telegraph, 1980.

Au, Tung, and Thomas P. Au. *Engineering Economics for Capital Investment Analysis.* 2d ed. Englewood Cliffs, NJ: Prentice-Hall, 1991.

Barish, Norman N., and Seymour Kaplan. *Economic Analysis for Engineering and Managerial Decision Making.* 2d ed. New York: McGraw-Hill, 1978.

Baumol, W. J. *Economic Theory and Operations Analysis.* 4th ed. Englewood Cliffs, NJ: Prentice-Hall, 1977.

Bierman, H., and S. Smidt. *The Capital Budgeting Decision.* 7th ed. New York: Macmillan, 1988.

Blank, Leland T., and Anthony J. Tarquin. *Engineering Economy: A Behavioral Approach.* 3d ed. New York: McGraw-Hill, 1989.

Bussey, Lynn E. *The Economic Analysis of Industrial Projects.* Englewood Cliffs, NJ: Prentice-Hall, 1978.

Canada, John R., and John A. White, Jr. *Capital Investment Decision Analysis for Management and Engineering.* Englewood Cliffs, NJ: Prentice-Hall, 1980.

Dasgupta, Agit K., and D. U. Pearce. *Cost-Benefit Analysis: Theory and Practice.* New York: Harper & Row, 1972.

DeGarmo, E. Paul, John R. Canada, and William G. Sullivan. *Engineering Economy.* 9th ed. New York: Macmillan, 1992.

Emerson, Robert, and William Taylor. *An Introduction to Engineering Economy.* 2d ed. Bozeman, MT: Cardinal, 1979.

Eschenbach, Ted. *Cases in Engineering Economy.* New York: Wiley, 1989.

Fabrycky, Wolter, and Gerald J. Thuesen. *Economic Decision Analysis.* 7th ed. Englewood Cliffs, NJ: Prentice-Hall, 1988.

Fleischer, G. A. *Capital Allocation Theory.* New York: Appleton Century Crofts, 1969.

———. *Engineering Economy: Capital Allocation Theory.* Monterey, CA: Brooks/Cole, 1984.

———. *Introduction to Engineering Economy.* Boston: PWS Publishing, 1994.

Frost, Michael J. *How to Use Benefit-Cost Analysis in Project Appraisal.* 2d ed. New York: Macmillan, 1979.

Grant, Eugene L., W. Grant Ireson, and Richard S. Leavenworth. *Principles of Engineering Economy.* 8th ed. New York: Wiley, 1990.

Hillier, F. S., and G. J. Lieberman. *Introduction to Operations Research.* 3d ed. San Francisco: Holden-Day, 1980.

Institute of Industrial Engineers. *Industrial Engineering Terminology.* Rev. ed. Norcross, GA: Industrial Engineering and Management Press, 1990.

James, L. D., and R. R. Lee. *Economics of Water Resources Planning.* New York: McGraw-Hill, 1970.

Kasner, Erick. *Essentials of Engineering Economics.* New York: McGraw-Hill, 1979.

Kleinfeld, Ira H. *Engineering and Managerial Economics.* New York: Holt, Rinehart and Winston, 1986.

Lasser, J. K. *Your Income Tax.* New York: Simon & Schuster. (See latest edition.)

Lindley, D. V. *Making Decisions.* 2d ed. New York: Wiley, 1985.

Mao, J. C. T. *Quantitative Analysis of Financial Decisions.* New York: Macmillan, 1969.

Mishan, E. J. *Economics for Social Decisions.* New York: Praeger, 1973.

Morris, William T. *Engineering Economic Analysis.* Reston, VA: Reston, 1976.

Newnan, Donald G. *Engineering Economic Analysis.* 4th ed. San Jose, CA: Engineering Press, 1991.

Oakford, R. V. *Capital Budgeting: A Quantitative Evaluation of Investment Alternatives.* New York: Wiley, 1970.

Oglesby, C. H., and R. G. Hicks. *Highway Engineering.* 4th ed. New York: Wiley, 1982.

Park, Chan S. *Contemporary Engineering Economics*, Reading, MA: Addison-Wesley, 1993.

——— and Gunther P. Sharp-Bette. *Advanced Engineering Economics*. New York: Wiley, 1990.

Park, William R., and D. E. Jackson. *Cost Engineering Analysis.* 2d ed. New York: Wiley, 1984.

Reisman, Arnold. *Managerial and Engineering Economics.* Boston: Allyn and Bacon, 1971.

Riggs, James L., and Thomas M. West. *Engineering Economics.* 3d ed. New York: McGraw-Hill, 1986.

Rose, L. M. *Engineering Investment Decisions.* Amsterdam, Netherlands: Elsevier Scientific, 1976.

Sassone, Peter G., and William A. Schaffer. *Cost-Benefit Analysis: A Handbook.* New York: Academic, 1978.

Smith, Gerald W. *Engineering Economy: The Analysis of Capital Expenditures.* 4th ed. Ames, IA: Iowa State University Press, 1987.

Steiner, H. M. *Conflict in Urban Transportation.* Lexington, MA: D.C. Heath, 1978.

———. *Public and Private Investments: Socioeconomic Analysis*. New York: Wiley, 1980 (also Glen Echo, MD: Books Associates, 1985).

———. *Basic Engineering Economy*. Rev. ed. Glen Echo, MD: Books Associates, 1989.

Stevens, G. T. *Economic and Financial Analysis of Capital Investments*, New York: Wiley, 1979.

Taylor, George A. *Managerial and Engineering Economy.* 3d ed. New York: Van Nostrand, 1980.

Thuesen, G. J., and W. J. Fabrycky. *Engineering Economy.* 7th ed. Englewood Cliffs, NJ: Prentice-Hall, 1989.

Weingartner, H. Martin. *Mathematical Programming and the Analysis of Capital Budgeting Problem.* Englewood Cliffs, NJ: Prentice-Hall, 1963 (also Chicago: Markham Publishing, 1967).

Wellington, A. M. *The Economic Theory of Railway Location.* 2d ed. New York: Wiley, 1887.

White, John A., Marvin H. Agee, and Kenneth E. Case. *Principles of Engineering Economic Analysis.* 3d ed. New York: Wiley, 1989.

INDEX